AAS
Agents and Actions Supplements
Vol. 38/III

Series Editors
K. Brune, Erlangen
M. J. Parnham, Bonn

Birkhäuser Verlag
Basel · Boston · Berlin

Recent Progress on Kinins

Pharmacological and Clinical Aspects
of the Kallikrein-Kinin System

Part II

Proceedings of the International Conference "Kinin 91 Munich",
held in Munich, September 8–14, 1991

Edited by

G. Bönner
H. Fritz
B. Schoelkens
G. Dietze
K. Luppertz
(Technical Editor)

Birkhäuser Verlag
Basel · Boston · Berlin

Volume Editors' Addresses:

Editors:

PD Dr. Gerd Bönner
Krankenhaus Köln Merheim
Medizinische Klinik
Ostmerheimer Strasse 200
D–5000 Köln 91

Prof. Dr. Hans Fritz
Abteilung für Klinische Chemie
und Klinische Biochemie in der
Chirurgischen Klinik Innenstadt
der Universität München
Nussbaumstrasse 20
D–8000 München 2

Prof. Dr. Bernward Schoelkens
HOECHST AG
Abteilung Pharmakologie H 821
Postfach 80 03 20
D–6230 Frankfurt/Main 80

Prof. Dr. Günther Dietze
Max Grundig Klinik
Schwarzwald Hochstrasse
D–7580 Bühl 13

Technical Editor:

Dr. Karin Luppertz
Abteilung für Klinische Chemie
und Klinische Biochemie in der
Chirurgischen Klinik Innenstadt
der Universität München
Nussbaumstrasse 20
D–8000 München 2

A CIP catalogue record for this book is available from the Library of Congress,
Washington D.C., USA

Deutsche Bibliothek Cataloging-in-Publication Data

Recent progress on kinins. – Basel ; Boston ; Berlin : Birkhäuser.
  (Agents and actions : Supplements ; Vol. 38)
  ISBN 3-7643-2816-9 (Basel ...)
  ISBN 0-8176-2816-9 (Boston)
NE: International Conference Kinin <1991, München>; Agents and actions / Supplements
3. Pharmacological and clinical aspects of the kallikrein kinin system.
  Pt. 2 (1992)

**Pharmacological and clinical aspects of the kallikrein kinin
system** : proceedings of the International Conference "Kinin 91
Munich", held in Munich, September 8–14, 1991 / ed. by
G. Bönner ... – Basel ; Boston ; Berlin : Birkhäuser.
  (Agents and actions : Supplements ; Vol. 38)
NE: Bönner, Gerd [Hrsg.]
Pt. 2 (1992)
  (Recent progress on kinins ; 3)
  ISBN 3-7643-2819-3 (Basel ...)
  ISBN 0-8176-2819-3 (Boston)

ISBN 3-7643-2816-9 (Vol. 38 (Set))   ISBN 0-8176-2816-9 (Vol. 38 (Set))
ISBN 3-7643-2817-7 (Vol. 38/I)   ISBN 0-8176-2817-7 (Vol. 38/I)
ISBN 3-7643-2818-5 (Vol. 38/II)   ISBN 0-8176-2818-5 (Vol. 38/II)
ISBN 3-7643-2819-3 (Vol. 38/III)   ISBN 0-8176-2819-3 (Vol. 38/III)

CONTENTS

Circulation

Vessels

Brain and Peripheral Nervous System

Heart

Kidney

Effects of ACE-Inhibitors

Hypertension

VIII

Inflammation

Respiratory Tract Diseases

Circulation

Vessels

THE KALLIKREIN - KININ SYSTEM IN BLOOD VESSELS

H. L. Nolly, M. C. Lama, O. A. Carretero and A. G. Scicli

Hypertension and Vascular Res. Div., Henry Ford Hospital, Detroit, MI, USA
and Argentine Council of Res. (CONICET), School of Medicine, Mendoza, R.A.

SUMMARY: We have previously reported that vascular tissue contains kallikrein
and kallikrein mRNA. We can now show that kallikrein is present throughout
the vascular tree and is released from arterial and venous rings incubated
"in vitro". Using the isolated perfused rat hindquarters as a model, we
found that kallikrein appeared in the perfusate in concentrations that
increased linearly with time. Treatment with puromycin inhibited kallikrein
release by 87% (p < 0.01), these data suggest that kallikrein is synthesized
and released by the vascular wall. Local generation of kinins (autocrine /
paracrine system) may contribute to the regulation of vascular homeostasis.

INTRODUCTION

It has been postulated that endocrine as well as paracrine factors may contribute to control of vascular homeostasis (1). We have previously shown

that vascular tissue contains a kininogenase with the characteristics of

glandular kallikrein (2), and that the mRNA coding for glandular kalli-

krein is present in vascular tissue (3). These data suggest the existence

of a vascular kallikrein-kinin system which may contribute to local regulation

of vascular homeostasis. However, to act as part of a paracrine system,

kallikrein should interact with kininogen, which is present in high concen-

tration in extracellular fluids and also releases kinins. The possibility of

such interaction would be supported if it could be shown that: 1) large as

well as small arteries and veins contain kallikrein and 2) whether vascular
tissue release kallikrein. In addition, although the presence of mRNA for
kallikrein suggests synthesis, it is not clear whether kallikrein released
from vascular tissue is indeed locally synthesized.

We studied whether kallikrein is present in large and small arteries
and veins, and whether it is released from vascular rings obtained from
different territories as well as from the isolated perfused rat hindquarters.
In addition, we determined whether puromycin, an inhibitor of protein synthesis, influences the rate of kallikrein release from the vascular wall.

MATERIALS AND METHODS

<u>Preparation</u> <u>of</u> <u>Tissue</u> <u>Homogenates</u>: Male Wistar rats weighing 250-300 g
were anesthetized with ether and decapitated. The thoracic and abdominal aorta,
tail artery, thoracic segment of the inferior vena cava, and tail vein were
removed and rinsed several times with ice-cold 0.01M Tris-HCl buffer (pH 7.4),
0,25M sucrose and 3 mg/ml EDTA. After being cleaned of connective tissue and
fat, the vessels were weighed, minced and homogenized with 0.1M Tris-HCl
buffer (pH 7.4). The homogenate was centrifuged at 1000 G for 10 minutes,
the supernatant separated out and the pellet washed and centrifuged again at
2000 G for 20 minutes. Both supernatants were pooled (final concentration,
100 mg wet tissue per ml) and kept at -20°C until needed for measurement of
active and total kininogenase activity.

<u>Release</u> <u>of</u> <u>Kallikrein</u> <u>from</u> <u>Vascular</u> <u>Rings</u>: Male Wistar rats weighing
250-300 g were anesthetized with ether and decapitated. Arteries (thoracic
aorta and tail artery) and veins (thoracic segment of the inferior vena cava
and tail vein) were removed, pooled separately and rinsed several times with
ice-cold 0.01M Tris-HCl buffer (pH 7.4) and 0.25M sucrose. The vessels were
cleaned of connective tissue and fat and cut into 2-3 mm rings. The arterial
and venous rings were divided into two groups. One group was not incubated,
but immediately homogenized; the other (100 mg wet tissue) was incubated in
oxygenated Krebs-Ringer solution at 37°C and gently shaken. The buffer was
changed every hour during the 3-hr incubation period. We chose that time
because in pilot experiments the oxygen consumption in a Warburg apparatus
and the kallikrein released were linear up to that point.

<u>Release</u> <u>of</u> <u>Kallikrein</u> <u>from</u> <u>the</u> <u>Isolated</u> <u>Perfused</u> <u>Rat</u> <u>Hindquarters</u>: Rats
were anesthetized with sodium pentobarbital (Nembutal, Abbot) and given 1000 U

heparin i.v. After 3 min, the animal was decapitated to facilitate blood drainage. An abdominal incision was made and the aorta and vena cava carefully dissected from the renal vessels to the bifurcation. The rectum was cut between double ligatures and the distal sigmoid displaced to the upper abdomen. All major and minor tributaries of the descending aorta and abdominal vena cava were ligated except for the femoral and tail arteries. A 20 gauge catheter was placed in the aorta just below the renal arteries. The vena cava was also catheterized, and warmed (37°C) Krebs-Henseleit solution containing 3.5% Ficoll 70 and gassed with O_2/CO_2 (95%/5%) was perfused through the aorta. The buffer was passed through a 0.45-μm filter prior to use. In the perfusion system, the medium was oxygenated by passing it through the blood compartment of a C-DAK hollow-fiber artificial kidney (Cordis-Dow, Miami, FL) and gassed with the O_2/CO_2 mixture. The isolated hindquarters were perfused in a single-pass (non-recirculating) system using a peristaltic pump. Perfusion pressure was recorded constantly. Details of the procedure were similar to those reported previously (4). During the initial perfusion period, the sample was thoroughly rinsed with the buffer for 33 min until the effluent was consistently clear. The flow rate was adjusted to obtain the desired perfusion pressure, normally 60-80 mm Hg. In the puromycin-treated group, the animals were injected with 10 mg puromycin i.p. 3 hr before the experiment.

   Kininogenase Activity: Active and total kininogenase activity were measured by incubating 400 μl of the homogenate supernatant (80 mg wet tissue), 100 ul of the medium bathing the slices, or 1000 μl of the hindquarters perfusate (previously concentrated 5 times) for 5 hr at 37°C with 200 μl partially purified dog kininogen (2000 ng kinin-releasing capability) in the presence of 1000 μl fresh 0.1 Tris-HCl buffer (pH 8.5) containing EDTA (15 mg/ml), 1-10-phenanthroline (1mg/ml), 8-OH-quinoline (1 mg/ml) and soybean trypsin inhibitor (SBTI) (100 μg/ml). Vascular kininogenase is inhibited by aprotinin and phenylmethylsulfonyl fluoride (PMSF) but is resistant to SBTI. SBTI was included in the incubation buffer to inhibit plasma kallikrein and trypsin-like enzymes that could contaminate the homogenates. Kinins generated during incubation were measured by radioimmunoassay (RIA); the lower limit of sensitivity of the kinin RIA is 1 pg/sample. Total kininogenase activity was measured by incubating 500 μl of the homogenate with 20 μg trypsin for 30 min at 37°C. The reaction was stopped by adding 100 μg SBTI, after which the homogenate was incubated with kininogen as described above for active kininogenase. Vascular kininogenase activity was expressed as the amount of kinins generated per mg protein (or wet crude weight) per min of incubation with kininogen. Under these conditions, the kininogen itself released small amounts of kinins (kininogen blank). In every assay, duplicate tubes containing the reagents but no homogenates were run to assay for kininogen blank. Vascular kininogenase activity was calculated as kininogenase activity in the measured by RIA (2).

   Results are expressed either per hour or as cumulative release during the total incubation time The rate of release was obtained by comparing

slopes (activity/incubation time) to asses whether changes in vascular
kininogenase were specific or whether a similar pattern could be observed
with a generalized cellular enzyme, lactic dehydrogenase (LDH). LDH activity
was measured as a marker of cellular injury following standard protocols.
Proteins were determined by Bradford's method (2), using albumin as a
standard.

<u>Statistics</u>: Two-sided two-sample t tests were used to assess kininogenase
activity in the different protocols. Bonferroni's adjustment for multiple
comparisons was employed. Unless otherwise noted, all results are expressed
as mean ± SEM.

RESULTS

Kininogenase was detected in both arteries and veins. Activity was higher in
the tail arteries than in the abdominal or thoracic aorta (324 ± 33 vs 108 ±
19 and 68 ± 13, respectively; p < 0.001). The tail veins had six times more
active kininogenase than the vena cava (800 ± 500 vs 118 ± 16; p < 0.001).
Total kininogenase activity (active and inactive) was distributed similary
in various arteries and veins (Figure 1). Inactive vascular kininogenase
accounted for almost 70 ± 2% of total activity in the arteries and 68 ± 4% in
the veins. Total kininogenase was highest in the tail veins (2102 ± 142) and
lowest in the thoracic aorta (332 ± 33).

Figure 2 shows that arterial rings released kallikrein into the medium
bathing the slices. Active and total kallikrein were 90 ± 13 and 170 ± 14
pg Bk/mg wet tissue (respectively) after 1 hr of incubation, 201 ± 25 and
366 ± 24 after 2 hr, and 311 ± 41 and 537 ± 40 after 3 hr. Similar results
were found in rings from veins incubated "in vitro" (Figure 3).

Figure 4 shows active and total kallikrein released into the perfusate
of the isolated perfused rat hindquarters. Cumulative release up to 5 hours
of perfusion is represented. The rate of release was quite constant when
considered as a function of time. Pretreatment with the protein synthesis
inhibitor puromycin (10 mg i.p.) reduced total kallikrein in the perfusate
from 105 ± 19 to 8.5 ± 3.6 (p < 0.005).

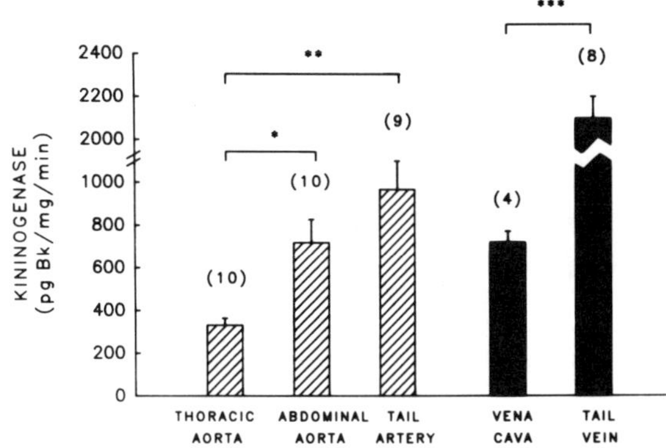

FIGURE 1. Total kallikrein content in large and small arteries and veins

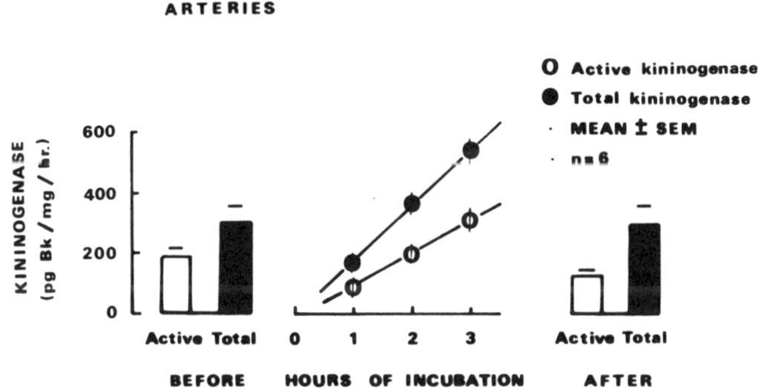

FIGURE 2. Release of kallikrein into the incubation buffer by isolated rat arterial rings. Kallikrein content in the rings before incubation and after 3 hr incubation. Difference were not significant.

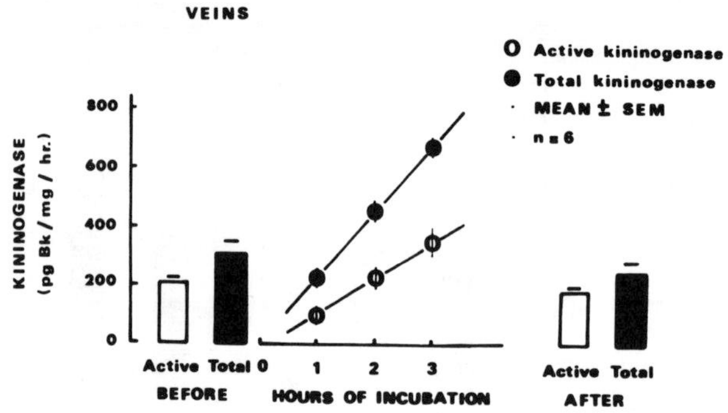

FIGURE 3. Release of kallikrein into the incubation buffer by isolated rat venous rings. Kallikrein content in the rings before incubation and after 3 hr incubation.

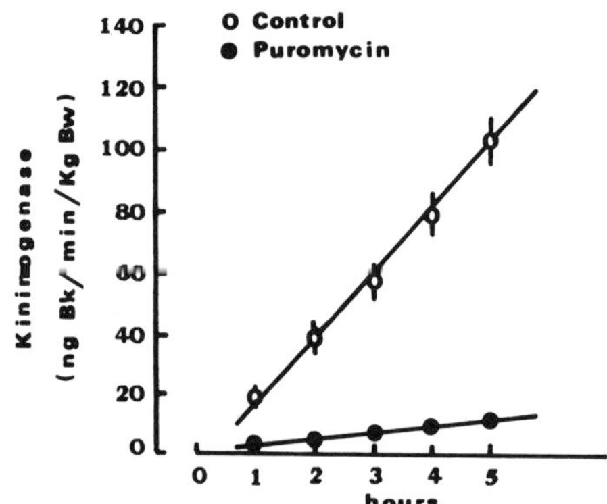

FIGURE 4. Release of total (Active Plus Trypsin-Activatable) kallikrein into the perfusate of isolated perfused rat hindquarter. Pretreatment with puromycin inhibited release of kallikrein into the perfusate (p < 0.001).

DISCUSSION

We have shown that kallikrein is present in the aorta, a large vessel, and in the small arteries of the tail. Kallikrein was also found in the tail veins and vena cava, with the highest concentration in the tail vessels. These results suggest that kallikrein, a potent kinin-generating enzyme, is distributed throughout the vascular tree and that kallikrein concentrations are higher in small vessels.

When vascular tissue is incubated "in vitro", kallikrein appears in the incubation medium. The amount of enzyme present in the medium after 3 hr incubation was higher than both the kallikrein activity contained in non-incubated vessels and that in control vessels incubated for the same lenght of time. Lactic dehydrogenase activity (LDH), a measure of cellular injury, indicate that release was not secondary to cell death. These results suggest that vascular tissue releases kallikrein into the extracellular space and that "de novo" synthesis replenishes the pools of intracellular kallikrein. The data obtained with the isolated perfused rat hindquarters are consistent with this interpretation. Since kallikrein appeared in the perfusate continuously for at least 5 hr since the concentration increased linearly with time, release appeared to proceed at a constant rate. Pretreatment with puromycin, a protein synthesis inhibitor, reduced kallikrein released into the perfusate by almost 90%. Thus release of kallikrein does not occur if protein synthesis is blocked, again indicating that "de novo" synthesis is needed for vascular kallikrein to be released. We do not know whether kallikrein is released in a constitutive or regulated form, although the constant rate of release appears to suggest a non-regulated mechanism.

We have previously reported that mRNA coding for kallikrein is present in vascular tissue and in vascular smooth muscle cells (VSMC) from the rat aorta (3). Deendothelization does not appear to cause major changes in vascular kallikrein content (5), suggesting that vascular kallikrein originates in the VSMC while we cannot ignore contribution of the endothelial cells to vascular kallikrein content, nor as a source of the released kallikrein, the present results suggest that vascular tissue synthesizes and releases kallikrein and confirm the existence of a vascular kallikrein-kinin system.

In normal rats, systemic administration of a kinin receptor antagonist has no effect or else results in only small changes in blood pressure (6). This suggests that under basal conditions the kallikrein-kinin system may not play a major role in the regulation of vascular tone. However, it has recently been found that some of the acute and chronic effects of angiotensin-converting enzyme inhibitors (ACEi) are blunted by kinin receptor antagonists, although, ACEi do not induce drastic changes in plasma kinin concentrations (6). Because of the existence of a vascular kallikrein-kinin system, it is possible that changes in the vascular concentrations of kinins may explain the kinin-mediated effects of ACEi. Thus the vascular kallikrein-kinin system may contribute to the regulation of circulatory homeostasis when kinin metabolism is affected by kininase inhibitors.

In summary, we have demonstrated that kallikrein is present in and released from vascular tissue. Released kallikrein appears to be the product of "de novo" synthesis. Kinins generated within the arterial or venous tissue wall by vascular kallikrein may participate in circulatory homeostasis and mediate part of the effects of kininase inhibitors.

REFERENCES

1. Carretero OA, Scicli AG. Kinins as regulators of blood flow and blood pressure. In: Laragh JH, Brenner BM (eds). Hypertension: Pathophysiology, Diagnosis and Management. New York, Raven Press. 1990; 805-817.

2. Nolly H, Scicli AG, Scicli G, Carretero OA. Characterization of a kininogenase from rat vascular tissue resembling tissue kallikrein. Circ Res. 1985; 54: 816-821

3. Saed GM, Carretero OA, Mac Donald RJ and Scicli AG. Kallikrein messenger
   RNA in rat arteries and veins. Circulation Research 1990; 67:510-516.

4. Murray RD, Itoh S, Inagami T, Misono K, Seto S, Scicli AG, Carretero OA.
   Effects of synthetic atrial natriuretic factor in the isolated perfused
   rat kidney. Am J Physiol 1985; 249: F603 - F 609.

5. Nolly H, Carretero OA, Scicli G, Madeddu P, Scicli AG. A kallikrein-like
   enzyme in blood vessels of one-kidney, one-clip hypertensive rats. Hypertension 1990; 16: 436-440.

6. Carretero OA and Scicli AG. Local hormonal factors (intracrine, autocrine
   and paracrine) in hypertension. Hypertension 1991; 18 (suppl I): I58 - I69.

AAS 38/III
Recent Progress on Kinins
© 1992 Birkhäuser Verlag Basel

VASCULAR WALL KININOGEN CONCENTRATION IS INVERSELY RELATED TO INTRINSIC KININOGENASE IN SPONTANEOUSLY HYPERTENSIVE RATS

Narendra B. Oza and H. Danana Goud

Renal Section, Evans Memorial Department of Clinical Research and Department of Medicine, University Hospital, Boston University School of Medicine, Boston, MA, 02118, U.S.A.

SUMMARY: We determined kininogen concentration and kininogenase activity in the blood-free, aorta homogenates of 15 week old spontaneously hypertensive rats (SHR) and in their age-matched wistar kyoto (WKY) controls. Active kininogenase in the SHR was only 18% of the WKY controls. Thus, the kininogen concentration was inversely related to kininogenase activity suggesting a local interaction of these components and possible generation of kinins within the vascular wall.

INTRODUCTION

Recent studies have suggested that tissue kallikreins can process many different natural substrates from which the generation of kinin from kininogen appears to be a major *in vivo* function (1). We have recently reported the occurence of kininogen, kininogenase and bradykininase in the smooth muscle cells of the vascular wall (2). Subsequently, the expression of rK1 (principal kininogenase) specific gene has been reported in the rat vascular wall (3). Since components of the entire kallikrein-kinin system are present in the vascular wall, it should be interesting to determine the level of these components in situations wherein the system has a probable functional role. In this study we have determined the concentration of kininogen and active kininogenase in adult spontaneously hypertensive rats (SHR) and in the age-matched wistar cyoto (WKY) control rats. We have found that kininogen concentration is inversely related to kininogenase in these groups of rats suggesting local consumption of kininogen for the release of kinins.

MATERIALS AND METHODS

Preparation of rat aorta homogenate. 15 week old SHR or WKY rats were anesthetized with penobarbital and injected with 0.1 ml heparin (5000 u/ml) into the vena cava. The heart was cut and used to lift aorta. The aorta was cut and placed in a petri dish containing saline. The tissue was rinsed with several changes of saline until free of any visible blood. The aorta was stripped of adventitia, blotted dry on a filter and cut into 1 cm ring. Aortic rings were weighed, placed in a centrifuge tube and 0.1 M sodium phosphate buffer, pH 7.4, containing 0.5 % Triton was added at 80 mg tissue/ml buffer. The tissue was frozen and thawed four times and then homogenized intermittently for 5 min on ice. The homogenate was centrifuged at 30,000 rpm for 30 min. The supernatant was dialyzed against the phosphate buffer, pH 7.4, for 18 hr at 4°C. The dialyzed sample was used for the analysis of kallikrein and kininogen.

Determination of kininogen. Aorta homogenates were incubated with TPCK-trypsin (tosyl phenylalanyl chlorometryl ketone treated trypsin). The reactions were terminated by placing the tubes in a boiling water bath and the generated kinins were isolated and estimated y the kinin-RIA. The kininogen concentration was calculated from the kinin equivalence as reported previously (4).

Kininogenase assay. Kininogenase activity was determined using our microkininogenase assay (5). Aorta homogenates were incubated with partially purified dog kininogen for 5 or 24 hours (active enzyme). The liberated kinins were purified by ethanol extraction and analyzed by a kinin RIA.

Kinin-RIA. The kinin-RIA is performed using ^{125}I-Tyr^8-bradykinin as a radiolabelled antigen and highly sensitive bradykinin antiserum (courtesy of Dr. K. Shimamoto) as detailed in a previous publication (6). The least detectable dose of kinin in this RIA is 1 pg/tube.

Protein determination. Protein microassay (Bio-Rad Chemical Division, Richmond, CA) was used according to the directions of the manufacturer to estimate protein concentration of aorta homogenates.

RESULTS

Kininogen concentration and kininogenase activity are shown in the table.

Table.   Kininogen and Kininogase activity, mean ± SEM

	WKY	SHR
Numbers of rats	9	8
Blood pressure, mmHg	122 ± 2	179 ± 5**
Kininogen	182 ± 33	313 ± 32*
Kininogenase@	105 ± 43	19 ± 11*

@ picogram kinins/mgm protein/hour
* P < 0.05, ** P < 0.01

DISCUSSION

The aortic tissue of hypertensive rats show a remarkable decrease in kininogenase activity with a simultaneous and significant increase in the concentration of kininogen. Artefact arising from plasma components is unlikely because 1) the aortic tissue was extensively washed prior to homogenization and 2) we have found that the vascular smooth muscle cells continue the biosynthesis of kininogen and kallikrein even after 15-20 passages of the primary culture (2). These results suggest that the SHR may have intrinsic deficiency in kininogenase of the aortic wall and thereby contribute to the development of hypertension. Because the SHR rats are deficient in kininogenase, the kininogen is under utilized and therefore the concentration is higher. Although there are other interpretations of our data, we hypothesize that the abnormal kininogenase activity and kininogen concentration of the vascular tissue may have implications in the maintenance of vascular tone and in the development of high blood pressure in genetically hypertensive rats.

ACKNOWLEDGEMENTS: We thank Dr. Norman G. Levinsky, M.D. for stimulating discussions and suggestions during these studies. We thank Ms. Susan Freeley for secretarial assistance.

REFERENCES

1. MacDonald, RJ, Margolius, HS, Erdös, EG. Molecular biology of tissue kallikrein. Biochem J 1988; 253:313-321.

2. Oza, NB, Schwartz, JH, Goud, HD, Levinsky, NG. Rat aortic smooth muscle cells in culture express kallikrein, kininogen and bradykininase activity. J Clin Invest 1990; 85:597-600.

3. Saed, GM, Carretero, OA, Scicli, G, Madeddu, P, Scicli, AG. Kallikrein messenger RNA in rat arteries and veins. Circ Res 1990; 67:510-516.

4. Weinberg, MS, Oza, NB, Levinsky, NG. Components of the kallikrein-kinin system in rat urine. Biochem Pharmacol 1984; 33:1779-1782.

5. Oza, NB, Murphy, CM, Kaufman, JS, Beasley, D, Levinsky, NG. A microkininogenase assay for studies of kallikreins in renal micropuncture/microperfusion. In: Kinins V. Abe, K et al eds. Plenum Publishing Corp., New York, 1989; 549-554.

6. Lieberthal, W, Oza, NB, Bernard, DB, Levinsky, NG. The effect of cations on the activity of human urinary kallikrein. J Biol Chem 1982; 257:10827-10830.

AAS 38/III
Recent Progress on Kinins
© 1992 Birkhäuser Verlag Basel

BRADYKININ SUPPRESSES ENDOTHELIN-INDUCED CONTRACTION OF CORONARY, RENAL AND FEMORAL ARTERIES THROUGH ITS B_2-RECEPTOR ON THE ENDOTHELIUM

Hironori Ohde, Shigeto Morimoto*, Keiko Ohnishi, Etsuyo Yamaoka,
Keisuke Fukuo*, Osamu Yasuda* and Toshio Ogihara*

Department of Drug Research, Fujimoto Pharmaceutical Co. Nishi-otsuka 1-3-40,
Matsubara City, Osaka 580, and *Department of Geriatric Medicine, Osaka
University Medical School, Fukushima-ku, Osaka 553, Japan

SUMMARY: Effect of bradykinin (BK) on endothelin-1 (ET-1)-induced vasoconstriction and its mechanism were investigated. The development of isometric force of arterial rings of canine coronary, renal and femoral arteries was recorded using a organ bath containing Krebs-Henseleit buffer aerated with 95% O_2 and 5% CO_2. ET-1 at more than 10^{-9} M dose-dependently induced vascular contraction similarly among the three arteries. BK at more than 10^{-8} M dose-dependently suppressed the ET-1-induced vasoconstriction only in the presence of endothelium, and the effect of BK was largest in the coronary arteries. The BK-induced suppression was not affected by addition of des-Arg^9-[Leu^8]-BK, an antagonist for B_1-receptor, but did be completely reversed by addition of B_2-receptor antagonist (10^{-6} M) [D-Arg^0,Hyp^3,$Thi^{5,8}$,D-Phe^7]-BK. The BK's suppression of the ET-1-induced vasoconstriction was partly reversed by additions of each 10^{-5} of N^g-nitro-L-arginine, a substrate inhibitor of nitric oxide, methylene blue, an inhibitor of soluble guanylate cyclase, or indomethacin, an inhibitor of cyclooxygenase. The reversing effects of methylene blue and indomethacin were additive. BK suppresses the ET-1-induced vasoconstriction through B_2-receptor on the endothelium. Both endothelial nitric oxide and prostaglandin(s) are participated in the BK's effect.

INTRODUCTION

Endothelin-1 (ET-1) is a potent vasoconstrictive peptide recently identified from the conditioned medium of cultured aortic endothelial cells (1), and has been shown to exert extremely potent contraction of the arteries including the coronary artery (2). On the other hand, bradykinin (BK), a nonapeptide produced from kininogen by the action of kallikrein, has been known to show a variety of physiological and pharmacological effects on vascular smooth muscle; the physiological properties of BK is its ability to induce relaxation of arteries (3), although it also induces contraction of arteries (4) and vein (5) _in vitro_. The presence of intact endothelium has been reported to be required for the relaxing action of BK, while not for its contracting effect (6-8). These various effects of BK on arteries have been partly explained by the presence of two different types of receptors, designated B_1 and B_2 (4). Endothelial cells were also reported to possess both a low-affinity B_1-receptor and a high-affinity B_2-receptor (9). In the present study, we investigated the effect of BK on ET-1-induced contraction in canine ringed coronary artery, and found that BK is a potent dilator of ET-1-induced contraction, through its B_2 receptor on the endothelium.

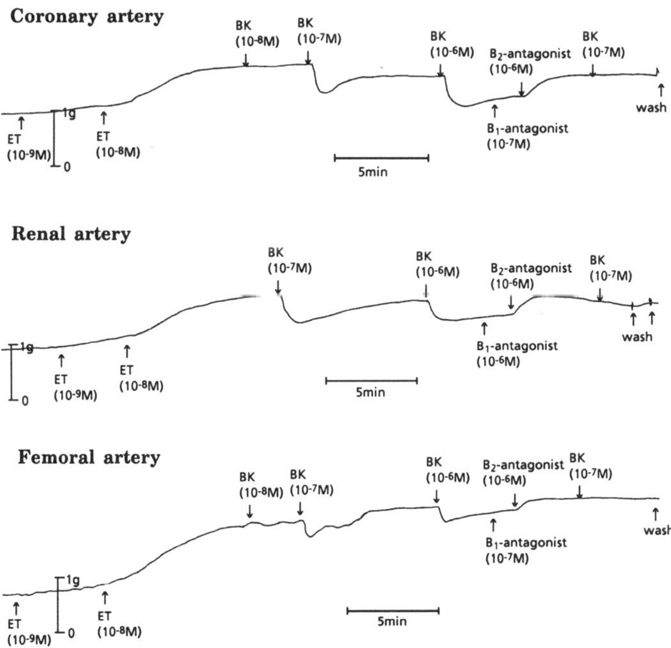

Figure 1. Effects of BK-receptor-antagonists on the BK-induced reversal of the ET-1-induced contraction in isolated preparations of canine coronary artery (top), renal artery (middle) and femoral artery (bottom).

MATERIALS AND METHODS

Materials. ET-1, BK, a B_1-receptor antagonist des-Arg^9-[Leu^8]-BK (10), a B_2-receptor antagonist [D-Arg^0,Hyp^3,$Thi^{5,8}$,D-Phe^7]-BK (3), and N^g-nitro-L-arginine [Arg(NO_2)] (11) were purchased from the Peptide Institute Inc. (Osaka, Japan). Methylene blue and indomethacin were purchased from Sigma Chemical Co. (St. Louis, MO). Other materials used were commercial products of the highest grade available.

Contraction Study. The coronary, renal and femoral arteries were excised from a dog weighing about 10 kg anesthetized by pentobarbital (30 mg/kg, i.v.), freed of connective tissue and cut into rings of about 3 mm length. The coronary artery rings were fixed vertically between two clips in a 10-ml organ bath containing Krebs-Henseleit buffer of the following composition: 130 mM NaCl, 4.7 mM KCl, 1.15 mM KH_2PO_4, 1.6 mM $CaCl_2$, 14.9 mM $NaHCO_3$ and 5.5 mM glucose. The strips were maintained at 37°C, aerated with 95% O_2-5% CO_2 and connected to an isometric force transducer (TB-612T, Nihon Koden, Tokyo,

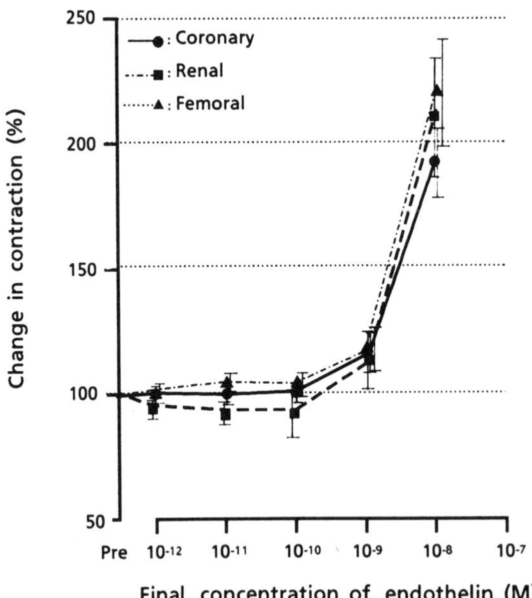

Figure 2. Comparison of dose-dependent contractile effects of ET-1 among canine coronary (circle), renal (square), and femoral (triangle) arteries. Results are given as means ± S.E.M.. Each point represents the mean of 5 to 9 experiments.

Japan) and Medicalcorder (D-6000, Nihon Koden). Endothelium was removed by gentle rubbing of the strips with a cotton swab. Lack of functional endothelium was demonstrated by absence of acetylcholine-induced relaxation and by positive staining for Evans blue dyeing.

RESULTS

ET-1-induced vasoconstriction. Figure 1 and 2 compares the dose-dependent effects of ET-1 on the contraction of the three types of canine arteries. ET-1 at concentrations of greater than 10^{-9} M dose-dependently induced slowly developing, and persistent contraction of all the three arterial rings, with similar half maximal effective concentration of 5 x 10^{-9} M. The ET-1-induced contraction was not affected by removing the endothelium from the three types of canine arteries (data not shown).

BK's effects on the ET-1-induced vasoconstriction in the three arterial rings. Addition of BK at concentrations of 10^{-8} and 10^{-7} M significantly

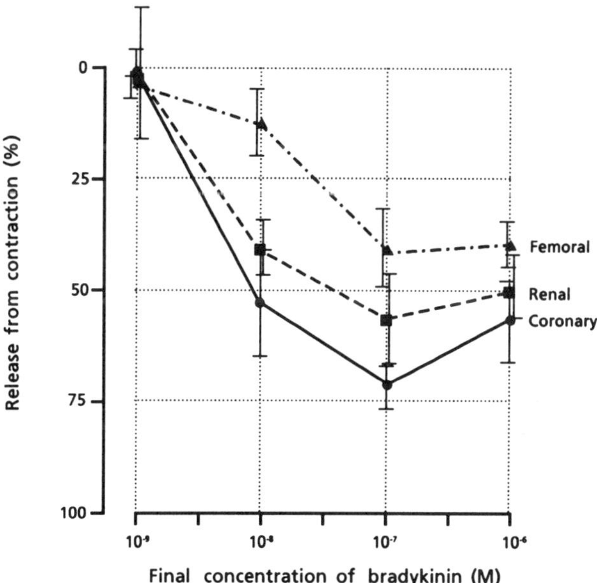

Figure 3. Comparison of dose-dependent attenuating effects of BK on ET-1 (10^{-8} M)-induced contraction among canine coronary, renal and femoral arteries. Keys as in Figure 2.

relaxed the tonic contraction of the arterial rings induced by 10^{-8} M of ET-1. The B_2 receptor antagonist significantly attenuated the inhibitory effects of BK on the contraction similarly in all the three arterial preparations (Figure 1). As shown in Figure 3, BK at concentrations of more than 10^{-8} M dose-dependently suppressed vasoconstriction with 10^{-7} M effecting the maximal suppressive effects, with the degree of attenuation occurring in the following descending order; the coronary artery was most affected, and the femoral artery, the last (Figures 3, 4). The inhibition on the ET-1-induced contraction caused by BK was not observed in the ringed coronary artery without the endothelium (data not shown). Using the artery with intact endothelium, the B_1-receptor antagonist (des-Arg^9-[Leu^8]-BK) did not show any significant effect on the ET-1-induced contraction or the inhibition on the contraction by BK (Figure 1, 4). However, the B_2-receptor antagonist ([D-Arg^0,Hyp^3,Thi^{5,8},D-Phe^7]-BK) significantly reduced the inhibitory effect of BK on the ET-1-induced contraction (Figure 1, 4). Moreover, pretreatment of the arterial rings with the B_2-receptor antagonist significantly reverse the attenuating effect of BK (Figure 4).

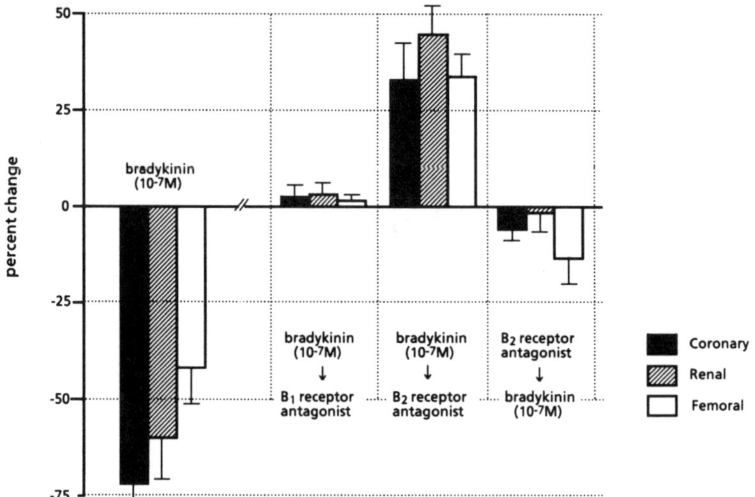

Figure 4. Comparison of the BK-induced attenuation and its reversal by BK antagonists among coronary (closed columns), renal (hatched columns) and femoral (open columns) arteries. Each column and bar represents mean $\pm$ S.E.M. of 3 to 7 experiments. Attenuation was given as a percentage against the ET-1 (10^{-8} M)-provoked contraction.

Mechanisms of the BK's reversing effect on the ET-1-induced vasoconstriction. Next we investigated the possible mechanism(s) of the attenuating effect of BK on the ET-1 (10^{-8} M)-induced vasoconstriction using renal arterial rings. Figure 5 shows the effect of Arg(NO_2), a substrate inhibitor for the production of nitric oxide (11), of methylene blue, an inhibitor of soluble guanylate cyclase (12), and of indomethacin, an inhibitor of cyclooxygenase, on the attenuating effect of BK. Compared to the control experiment, pretreatment of the arterial preparation with 10^{-5} M of Arg(NO_2) slightly attenuating effect of 10^{-7} M of BK on the contraction. Similarly, pretreatment of the preparation with 10^{-5} M of methylene blue significantly diminished the BK-induced attenuation of the contraction. Indomethacin at a concentration of 10^{-5} M also slightly reversed the attenuation induced by BK. Coaddition of 10^{-5} M of methylene blue and 10^{-5} M of indomethacin almost completely counteracted the relaxing effect of BK (Figure 5).

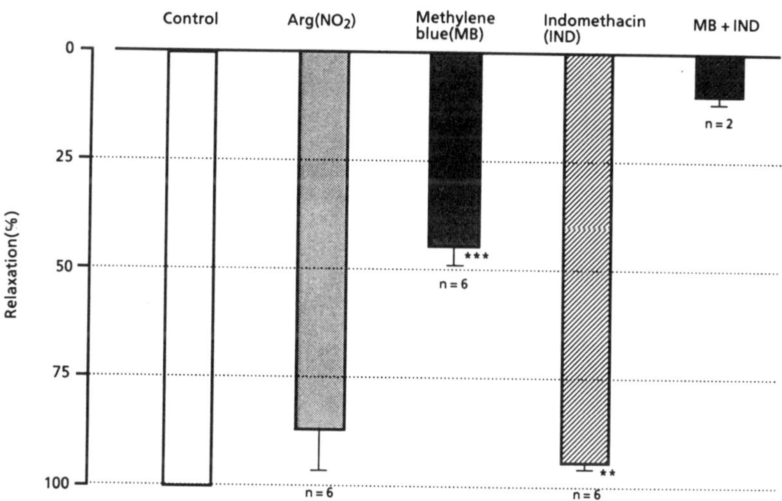

Figure 5. Reversing effects of 10^{-5} M of Arg(NO_2), 10^{-5} M of methylene blue, and 10^{-5} M of indomethacin on the BK-induced attenuation of 10^{-8} M of ET-1-provoked contraction of renal artery rings. Results are given as means ± S.E.M.. Significant differences from each control experiments (n=6): **p<0.05, ***p<0.001.

DISCUSSION

In this study, BK at concentrations of greater than 10^{-8} M significantly inhibited the ET-1-induced contraction only in the presence of intact endothelium; the requirement of intact endothelium in the present study for the relaxing action of BK was compatible with the previous reports (6–8). Moreover, a B_2-receptor antagonist ($[D-Arg^0,Hyp^3,Thi^{5,8},D-Phe^7]$-BK) inhibited the effect of BK on the ET-1-induced contraction, although an antagonist for the B_1-receptor (des-Arg^9-$[Leu^8]$-BK) did not show these effects of BK. These results indicates that inhibition of ET-1-induced contraction of the three arteries by BK is mediated by BK B_2-receptor on the endothelial cells.

Previous studies revealed that BK applied to intact cells or tissues appeared to evoke several kinds of intracellular signal transduction including: activation of phospholipase C (13) followed by increase in breakdown of polyphosphoinositides into phosphatidyl inositol (14,15) and increase in the intracellular calcium concentration (9,14,15), accumulation of cyclic AMP and cyclic GMP (16,17), activation of phospholipase A_2 (18) and release of prostaglandins (PGs) (19) including PGI_2 (20), and production of the endothelium-derived relaxing factor (9,10,20) which is known to be nitric oxide (21) stimulating soluble guanylate cyclase (22). Among these intracellular events, the elevation of intracellular calcium (9,14), the release of nitric oxide (9,20) and production of PGI_2 (20) in response to BK in endothelial cells was reported to be mediated predominantly through B_2 receptor activation. In the present study, BK's attenuation of the ET-1-induced contraction was slightly reversed by $Arg(NO_2)$, an inhibitor of nitric oxide production (11), methylene blue, an inhibitor of soluble guanylate cyclase (12), and indomethacin, an inhibitor of cyclooxygenase. Moreover, coaddition of methylene blue and indomethacin almost completely counteracted the relaxing effect of BK. These evidence suggest that both the nitric oxide/cyclic GMP system and prostaglandin metabolism participate in the relaxing effect of BK on the ET-1-induced arterial contraction.

The releasing effect of BK on ET-1-induced contraction of the three arteries _in vitro_ in this study also suggest that BK might be an endogenous factor for prevention of vasospasms of many arteries _in vivo_. The effect of BK and ET-1 on vasospasms _in vivo_ remains to be elucidated.

REFERENCES

1. Yanagisawa M, Kurihara H, Kimura S, Tomobe Y, Kobayashi M, Mitsui M, Yazaki Y and Mazaki T (1988) A novel potent vasoconstrictor peptide produced by vascular endothelial cells. Nature 332, 411–415.
2. Yamamoto Y, Tomoike H, Egashira K and Nakamura M (1987) Attenuation of endothelium-related relaxation and enhanced responsiveness of vascular smooth muscle to histamine in spastic coronary arterial segments from miniature pigs. Circ. Res. 61, 772–778.
3. Regoli D and Barabe J (1980) Pharmacology of bradykinin and related kinins. Pharmacol. Rev. 32, 1–46.
4. Regoli D, Barabe J and Park WK (1977) Receptors for bradykinin in rabbit aortae. Can. J. Physiol. Pharmacol. 55, 855–867.
5. Gaudreau P, Barabe J, St-Pierre S and Regoli D (1981) Pharmacological studies of kinins in venous smooth muscles. Can. J. Physiol. Pharmacol. 59, 371–379.
6. Regoli D, Mizrahi J, D'Orleans-Juste P and Caranikas S (1982) Effects of kinins on isolated blood vessels. Role of endothelium. Can. J. Physiol. Pharmacol. 60, 1580–1583.
7. Cherry PD, Furchgott RF, Zawadzki JV and Jothianandan D (1982) Role of endothelial cells in relaxation of isolated arteries by bradykinin. Proc. Natl. Acad. Sci. USA. 79, 2106–2110.
8. Toda N, Bian K, Akiba T and Okamura T (1987) Heterogeneity in mechanisms of bradykinin action in canine isolated blood vessels. Eur. J. Pharmacol. 135, 321–329.
9. Sung C, Arleth AJ, Shikano K and Berkoowitz BA (1988) Characterization and function of bradykinin receptors in vascular endothelial cells. J. Pharmacol. Exp. Ther. 247, 8–13.
10. Furchgott RF (1983) Role of endothelium in responses of vascular smooth muscle. Circ. Res. 53, 557–573.
11. Gross SS, Stuehr DJ, Aisaka K, Jaffe EA, Levi R and Griffith OW (1990) Macrophage and endothelial cell nitric oxide synthesis. Biochem. Biophys. Res. Commune. 170, 96–103.
12. Martin W, Villani GM, Jothianandan D and Furchgott RF (1985) Blockade of rabbit aorta by certain ferrous hemoproteins. J. Pharmacol. Exp. Ther. 233, 679–685.
13. Bell RJ, Baenzinger NJ and Majerus PW (1980) Bradykinin-stimulated release of arachidonate from phosphatidyl inositol in mouse fibrosarcoma cells. Prostaglandins 20, 269–274.
14. Derian CK and Moskowitz M (1986) Polyphosphoinositide hydrolysis in endothelial cells and carotid artery segments. Bradykinin-2 receptor stimulation is calcium-independent. J. Biol. Chem. 261, 3831–3837.
15. Martin TW and Michaelis C (1988) Bradykinin stimulates phosphodiesteratic cleavage of phosphatidylcholine in cultured endothelial cells. Biochem. Biophys. Res. Commun. 157, 1271–1279.
16. Fahey JV, Coisek CP Jr and Newcombe DS (1977) Human synovial fibroblasts: the relationships between cyclic AMP, bradykinin, and prostaglandins. Agents Actions 7, 255–264.
17. Stoner J, Manganiello VC and Vaughan M (1973) Effects of bradykinin and indomethacin on cyclic GMP and cyclic AMP in lung slices. Proc. Natl. Acad. Sci. USA. 70, 3830–3833.
18. Newcombe DS, Fahey JV and Ishikawa Y (1977) Hydrocortisone inhibition of the bradykinin activation of human synovial fibroblasts. Prostaglandins. 13, 235–244.

19. Hong SC and Levine L (1976) Stimulation of prostaglandin synthesis by bradykinin and thrombin and their mechanisms of action on MC5-5 fibroblasts. J. Biol. Chem. <u>251</u>, 5814-5816.
20. D'Orleans P, deNucci G and Vane JR (1989) Kinins act on B1 or B2 receptors to release conjointly endothelium-derived relaxing factor and prostacyclin from bovine aortic endothelial cells. Br. J. Pharmacol. <u>96</u>, 920-926.
21. Palmer RMJ, Ferrige AG and Moncada S (1987) Nitric oxide release accounts for the biological activity of endothelium-derived relaxing factor. Nature <u>327</u>, 524-526.
22. Furchgott RF and Zawadzkiki JV (1980) The obligatory role of endothelial cells in the relaxation of arterial smooth muscle by acetylcholine. Nature <u>288</u>, 373-376.

KININOGENASE ACTIVITY OF rSMT3,
A RAT SUBMANDIBULAR GLAND TONIN

Geralda W. Araujo, Ana M. P. Silva, João B. Pesquero,
Charles J. Lindsey, Wilson T. Beraldo and Jorge L. Pesquero

Department of Physiology and Biophysics, Universidade Federal de Minas Gerais, Belo
Horizonte; Department of Physiology, Universidade Federal do Rio Grande do Norte and
Department of Biophysics, Escola Paulista de Medicina, São Paulo, Brazil.

SUMMARY: rSMT3, a tonin like angiotensin II generating enzyme present in rSMG presents
a potent oxytocic effect on the isolated rat uterus, which was blocked by a B_2 bradykinin receptor
antagonist. Under optimal conditions of pH, rSMT3 liberates kinin at rate 19-fold greater than
angiotensin II.

INTRODUCTION

Tonins and kallikreins are serine proteinases that generate vasoactive peptides involved in the
regulation of blood flow in several tissues. Tonin, present in different rat tissues, cleaves the
vasoconstrictor peptide angiotensin II directly from angiotensinogen (1), whereas kallikrein
cleaves the vasodilatador peptides bradykinin and lysyl-bradykinin from the precursor kininogen
(2). However, the capabilities of kallikrein to liberate angiotensin II from angiotensinogen and
tonin to liberate bradykinin from kininogen has been demonstrated (3-4). The number of members
of the tonin family has evolved in the last years with the identification of potencially novel tonin
genes (5-8) and purification and characterization of novel enzymes with

---

ABBREVIATIONS: BSA, bovine serum albumin; HPLC, high performance liquid
chromatography; RIA, radioimmunoassay; rSMG, rat submandibular gland; TFA, trifluoroacetic
acid.

tonin activity from different rat tissues (9-10). In addition two enzymes with tonin-like activity designated rSMT3 and rSMT4 were purified from rSMG. The observed physicochemical properties for rSMT4 are identical to that of tonin described by Boucher et al (11) whereas rSMT3 is a novel serine proteinase that may have a function similar to that of tonin.

In 1984, Ikeda and Arakawa (3) demonstrated that tonin is capable of forming a substance with oxytocic activity identified as bradykinin. Therefore, we decided to investigate whether rSMT3 and rSMT4 display an oxytocic activity. Our results (12), show that rSMT3 has a potent oxytocic action when tested on isolated rat uterus whereas rSMT4 was devoid of oxytocic activity. In addition, in experiments utilizing a parallel uterus system we observed that the oxytocic effect showed by rSMT3 is not due to a direct action upon the uterine horn. When incubated with rat uterus homogenate, rSMT3 liberates a substance with oxytocic activity whose retention time in HPLC coincided with that of standard bradykinin. In order to characterize the kininogenase activity of rSMT3, we investigated the capability of the enzyme to act upon kininogen and the products identified by biological assay and HPLC.

MATERIALS AND METHODS

The peptides were synthesized in the Department of Biophysics, Escola Paulista de Medicina, São Paulo, Brasil. Electrophoretically homogeneous rSMT3 and rat plasma angiotensinogen was obtained as previously (9). Antisera to angiotensin II was prepared by immunizing in rabbits with (4-8)-angiotensinamide coupled to BSA. Coupling and RIA was carried out as earlier described (13). A standard curve was obtained by plotting the B/B_0 against logarithm of angiotensin II concentration (14). Kininogen was purified acording to Diniz and Carvalho (15).

rSMT3 (1.2 µg) was incubated with angiotensinogen or kininogen (300 µg) in a final volume of 250 µL of one of the following buffer solutions: 0.2 M sodium acetate for pH 5.0 and 5.5, 0.2 M sodium phosphate for pH 6.0, 6.5 and 7.0 and 0.2 M Tris-HCl for pH 7.5, 8.0 and 8.5. All buffer solutions contained 5 mM EDTA, 8 mM dipyridil, 0.6 M sodium tetrathionate and 1.5 mM 8-hydroxyquinoline. Reactions were carried out at 37°C for 18 h angiotensinogen and 4 h for kininogen and terminated by adding 10 µL of 6 M HCl and the generated products determined directly by RIA or alternatively the reaction was terminated by adding 100 µL of 20% TFA and

incubates applied on a reversed phase column in a HPLC system and fractions analyzed by RIA
or biological assays.

Biological assays

The isolated rat uterus was prepared as previously described (16). Isotonic contractions
were recorded with a frontal writing lever and the oxytocic effect of samples were compared to
those of synthetic standards of angiotensin II or bradykinin. For arterial blood pressure assays
four-month old male normotensive Wistar rats and with a mean arterial pressure of 116 ± 12 mm
Hg, were prepared with permanent catheters in the carotid artery for drug administration in the
femoral artery for blood pressure recording (17) two days before the experiment. Incubates (0.05
mL) or saline containing standard concentrations of bradykinin were administrated intraarterially
with 15 min intervals. The amount of bradykinin generated was quantified to interpolation of
standard dose response curves obtained in the same animals.

High performance liquid chromatography

The incubates with angiotensinogen and kininogen, after addition of 20% TFA were filtered
and applied on a MinoRPC S 5/20 (Pharmacia Biosystem BA Broma, Sweden) column in the
HPLC system. Samples were eluted with a 30 to 60% acetonitrile gradient in TFA 1%.

RESULTS

Electrophoretically pure rSMT3 (1.2 µg) was incubated with 300 µg of rat plasma
angiotensinogen or dog plasma kininogen. Samples were applied to HPLC and oxytocic activity
of the different fractions were analysed. Figure 1A shows the retention times for lysyl-
bradykinin, bradykinin and angiotensin II standards. The retention times for lysyl-bradykinin,
bradykinin and angiotensin II were 9.28, 12.02 and 15.12 minutes, respectively. Oxytocic activity
of the angiotensinogen incubate was only observed in the fractions corresponding to the elution
time of angiotensin II (figure 1B), whereas with the kininogen incubate, uterin contracting activity
was found in the fractions corresponded to the bradykinin elution time (figure 1C). The
quantitation of angiotensin II in the active fractions was by RIA.

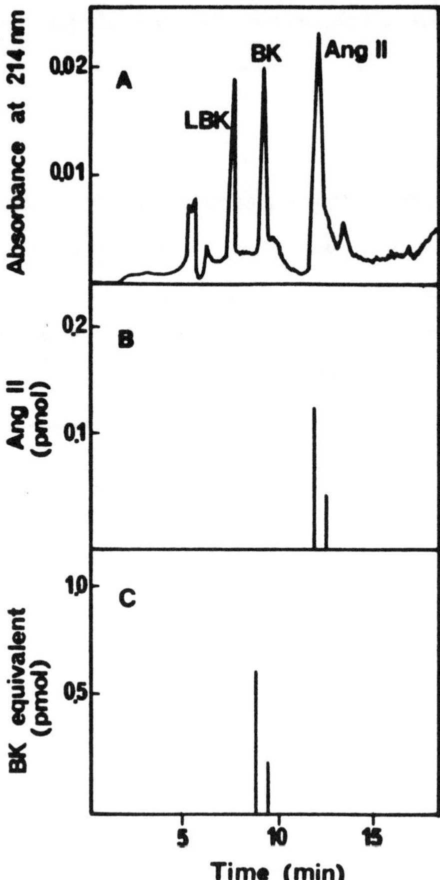

Figure 1 - Elution profiles obtained by HPLC of the products liberated by rSMT3 from angiotensinogen and kininogen. A) synthetic standards bradykinin (BK), lysyl-bradykinin (LBK) and angiotensin II (Ang II); B) angiotensin II immunoreactivity of the fractions from angiotensinogen incubate, determined by RIA and C) bradykinin equivalents in the fractions from kininogen incubate, determined by bioassay.

The oxytocic activity of the kininogen incubates with rSMT3 was blocked by previous addition to the organ bath of D-Arg^0,[Hyp^3,Thi^{5,8},D-Phe^7]-bradykinin (5 μM), a B$_2$receptor antagonist. Saralasin (0.2 μM) inhibited the uterin contractions caused by the incubate of angiotensinogen with rSMT3. Furthermore saralasin had no effect on the contractions induced by

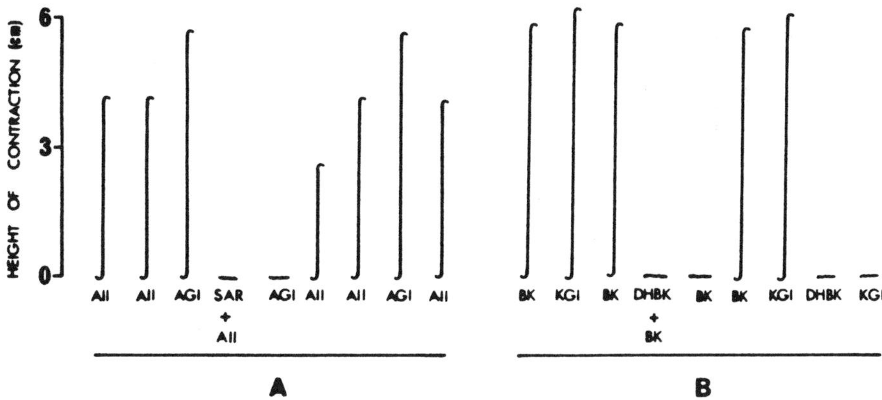

Figure 2 - Effect of (A) the 0.2 µmolar of angiotensin II antagonist, saralasin, (SAR) on the contraction induced by 76 nM of angiotensin II (AII) or rSMT3 incubated with angiotensinogen (AGI) and (B) effect of the 5.0 µM bradykinin antagonist, D-Arg^0,[Hyp^3,Thi^{5,8},D-Phe^7]-bradykinin, (DHBK) on the contraction induced by 37 nM of bradykinin or by rSMT3 incubated with kininogen (KGI).

kininogen incubates and the bradykinin antagonist did not affect the contractions caused by angiotensinogen incubates with rSMT3 applied to the rat uterus (figure 2A,B).

The generation of kinins by rSMT3 from kininogen was also assayed by measuring the hypotensive effect of incubates in the unanesthetized rats pretreated with converting enzyme inhibitor (enalaprilat 0.5 mg^{-1}.Kg^{-1}, ia), (figure 3). Using this method, the generation of bradykinin was estimate to be 19 pmol.min^{-1}.mg^{-1} of protein. When angiotensinogen was the substrate, approximately 1 pmol.min^{-1}.mg^{-1} of protein of angiotensin II was generated by rSMT3 as determined by RIA. The optima pH for kinin liberating activity was 8.0 while angiotensin II was optimally generated from angiotensinogen at pH 6.0.

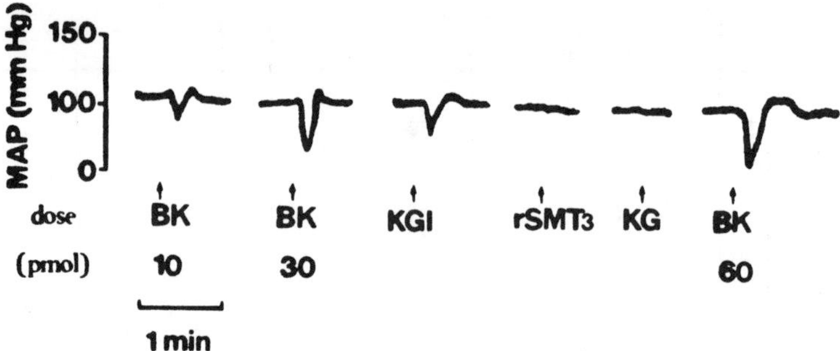

Figure 3 - Effects upon the mean arterial pressure (MAP) of unanesthetized follw in intraarterially administration of different doses of bradykinin (BK) compared to hypotensive action of rSMT3 or kininogen alone (KGI).

DISCUSSION

In a recent paper, we showed the presence in the rat submadibular gland of two serine proteinases with tonin-like activity, designated rSMT3 and rSMT4 (9). The physicochemical properties observed for rSMT4 are identical to those of tonin described by Boucher at al. (11) and therefore may be related with the S2 mRNA described by Ashley and MacDonald (5-6). rSMT3 presented different physicochemical properties to rSMT4 and therefore represents a novel tonin. We have also observed that in addition to angiotensin II liberating properties from angiotensinogen, rSMT3 is able to contract the isolated rat uterus and this effect was shown to be due to release of a kinin like substance from a substrate in the rat uterus (12). The contraction of the rat uterus elicited by rSMT3 was inhibited by bradykinin receptor antagonist but no by saralasin, and since the presence of the kininogen has been demonstrated in the rat uterus (18), its possible that rSMT3 acts upon kininogen present in rat uterus to liberate a bradykinin-like substance. The present study also showed that rSMT3 is able to cleave the kinin like substance from kininogen. As demonstrated by biological assay and HPLC the kinin liberated by rSMT3 was identical with bradykinin. Although rSMT3 is capable of cleaving kininogen it is not a true kallikrein since its showed negligible activity towards the synthetic kallikrein substrate, Ac.Phe.Arg.Nan (9). Another difference between kallikreins and tonins is that kallikrein does not act upon tonin substrate such as angiotensinogen (1-14) fragment and angiotensin I (9). The previously reported kinin liberating activity of tonin (3) could be explained by a possible contamination of tonin with rSMT3. Recently another serine proteinase was purified from rat submandibular gland. This enzyme demonstrated a potent vasoconstrictor action related to the liberation of angiotensin II (10). Despite some differences, many physicochemical properties of rSMT3 and the latter enzyme are similar and possibly are the same enzyme. Both enzymes are probably encoded for by the gene product S3 mRNA described by Ashley and MacDonald (6).

Wines et al. (8), have pointed out that short gene conversions between serine proteinase members may generate novel assortments of variants in amino acid coding regions that may affect substrate specificity contributing to the diversity of enzyme activity that could be important to a concerted evolution. rSMT3 is an enzyme that has capability of forming a kinin-tensin system and may represent a evolution in a energetic concept to the control of numerous physiological process involving angiotensin and bradykinin.

ACKNOWLEDGEMENTS: The authors are greatly indebted to Ronaldo L. Nunes and Mercia P. Lima for able technical assistence. This work was supported by grants UFMG Research Council (PRPq/UFMG) and the Minas Gerais State Research Foudation (FAPEMIG).

REFERENCES

1.   Grisé C, Boucher R, Thibault G, Genest J. Formation of angiotensin II by tonin from partially purified human angiotensinogen. Can J Biochem 1981; 59:250-5.

2.   Fiedler F. Enzymology of glandular kallikreins. Handb Exp Pharmacol 1979; 25:103-61.

3.   Ikeda M, Arakawa K. Kinonogenase activity of tonin. Hypertension 1984; 6:222-8.

4.   Maruta H, Arakawa K. Confirmation of direct angiotensin formation by kallikrein. Biochem J 1983; 213:193-200.

5.   Ashley PL, MacDonald RJ. Rat submandibular gland kallikrein-related mRNAs: nucleotide sequence of four distintic types including tonin. Biochemistry 1985; 24:4512-20.

6.   Ashley PL, MacDonald RJ. Tissue-specific expression of kallikrein-related genes in the rat. Biochemistry 1985; 24: 4   4520-7.

7.   Clements JA. The glandular kallikrein family of enzymes: Tissue-specific expression and hormonal regulation. Endocr Rev 1989; 10:393-419.

8.   Wines DR, Brady JM, Southard EM, MacDonald RJ. Evolutiion of the rat kallikrein gene family: gene convertion leads to functional diversity. J Mol Evol 1991; 32:476-92.

9.   Araujo GW, Pesquero JB, Lindsey CJ, Paiva ACM, Pesquero JL. Identification and characterization of serine proteinases with tonin-like activity in the rat submandibular and prostate glands. Bioch Biophys Acta 1991; 167-71.

10.   Yamaguchi T, Carretero OA, Scicli AG. A novel serine protease with vasoconstritor activity coded by the kallikrein gene S3. J Biol Chem 1991; 266:5011-17.

11.   Boucher R, Asselin J, Genest J. A new enzyme leading to the direct formation of angiotensin II. Circ Res 1974; 34 (Suppl. I):203-9.

12.   Pesquero JL, Araujo GW, Lima MP, Beraldo WT. The tonin-kinin system. Agents and Actions (in press).

13.   Vieira JGH, Noguti KO, Russo EMK, Maciel RMB. Radioimmunoassay for the measurement of plasma renin activity: technical aspects. Rev Bras Pat Clin 1981; 17:195-200.

14.   Rodbard D. Bridson W, Rayford PI. Rapid calculation of radioimmunoassay results. J Lab Clin Med 1969; 74:770-7.

15.   Diniz CR, Carvalho IF. A micro-method for determination of bradykininogen under several conditions. Ann NY Acad Sci 1963; 104:77-89.

16.   Feitosa MH, Pesquero JL, Ferreira MAD, Oliveira GM, Rogana E, Beraldo WT. Tonin and kallikrein-kinin system. Adv Exp Med Biol 1989; 247A:573-80.

17.   Lindsey CJ, De Paula UM, Paiva ACM. Protracted effect of converting enzyme inhibitor on the rat's reponse to intraarterial bradykinin. Hypertension 5, 1983: (Suppl V1). V134-V137.

18.   Beraldo WT, Lauar NS, Siqueira G, Heneine IF, Catanzaro OL. Peculiarities of the oxytocic action of rat urinary kallikrein. In: Chemistry and Biology of the kallikrein-kinin System in Health and Disease. Pisano JJ, Austen KF, editors. Bethesda: Fogarty International Center Proceedings, US Printing Office, 1976; 27:103-5.

Circulation

Brain and Peripheral Nervous System

TISSUE KALLIKREIN-KININ SYSTEM IN THE BRAIN AND THE REGULATION OF ARTERIAL PRESSURE

D.R. Fior, D.T.O. Martins and C.J. Lindsey

Department of Biophysics, Escola Paulista de Medicina, 04034 São Paulo, SP, Brazil

SUMMARY: The intracerebral administration of kininase inhibitors produced an increase in arterial pressure of the SHR but not in normotensive animals. The SHR were several fold more sensitive to the pressor response elicited by the intracerebral injection of bradykinin, however SHR with low blood pressure showed decreased sensitivity to the bradykinin pressor effect. The results suggest that an endogenous kinin system plays a role in the central regulation of blood pressure of the SHR.

INTRODUCTION

The existence of a functional kallikrein-kinin system in the central nervous system has been a matter of speculation since the identification of different components of the system in brain tissue. Kallikrein (1), kininogen (2), kininase II (3), and kinin receptors (4,5) are present in central tissue and bradykinin (6) or bradykinin-like immunoreactivity (7) were found in the hypothalamus or in the cerebrospinal fluid (8). Although the presence of enzymes involved in the synthesis and degradation of kinins as well as the existence of kinin receptors point towards a functional kininergic system in the brain very little is known on the possible physiological functions of such a system. When injected into the central nervous system bradykinin produces a variety of behavioral and physiological effects such as sedation, excitation, electroencephalographic alterations (9), antinociceptive and antidiuretic actions (10,11), and cardiovascular alterations. The best studied central action of bradykinin is the pressor effect which occurs in response to intracerebroventricular (icv) administration of kinins (12,13). This effect is obtained in rats, rabbits, dogs and cats and is probably common to mammals. This effect can also be obtained by administering bradykinin to the cerebral circulation of different species (14) and when this occurs the pressor effect is generally preceded by a hypotensive effect. The pressor component of the response is certainly of central nature since it is opposite to the systemic action of bradykinin which causes vasorelaxation and hypotension. The central mediation of the bradykinin pressor action was effectively demonstrated in the cross circulation preparation (15). The neuronal structures of the kinin pressor effect has long been a subject of controversy. The pressor response has been reported to occur in response to the stimulation of the septal area (16) or the lateral hypothalamus (17) or in the ventral part of the third ventricle. Other findings suggest that bradykinin may act by stimulating rhomboencephalic structures (18). The injection of the peptide in the fourth cerebral ventricle of rats showed that bradykinin is about forty times more potent than when injected in the lateral ventricle, has a larger maximal effect and a shorter latency for the manifestation of the pressor response (19 and unpublished observations), suggesting that the pressor effect is mediated by periventricular structures. Microinjections of bradykinin in to different structures of the *medulla oblongata* showed that the pressor effect is mediated by injections in the nucleus tractus solitarius (NTS) and the spinal trigeminal tract (STT). In these structures pressor responses were obtained by the injections of 5 pmol of bradykinin which contrasts with the high doses (nmols) necessary to obtain pressor responses following injection of bradykinin in the forebrain structures. The pressor action of bradykinin is apparently a selective effect of bradykinin since other vasoactive agents such as noradrenaline, serotonin, acetylcholine, prostaglandins E_2 and $F_{2\alpha}$ and atrial natriuretic factor did not alter blood pressure following injection in the fourth cerebral ventricle. Angiotensin II and Lysyl-vasopressin caused pressor responses following injections to the fourth cerebral ventricle, however the doses necessary to elicit such responses were in the nmol range (ED_{50} = 1.1 and 1.4 nmol respectively), and approximately 140 larger than the doses necessary to produce pressor effects following iv injections (ED_{50} = 8 and 9 pmol respectively). The kinin receptors in the *medulla* are of the B_2 subtype (4) which interact competitively with the kinin antagonist DArg^0 - Hyp^3 -Thi^{5,8} -DPhe^7 - bradykinin with an *in vivo* pA_2 of 9.9 . The existence of sensitive kinin receptors in the medulla suggests that bradykinin may play a role in the central regulation of blood pressure. An indication that bradykinin may participate in the central control of blood pressure comes from experiments in the spontaneously hypertensive rat (SHR) which show that these animals, with respect to normotensive Wistar rats, have an increased sensitivity to the central pressor action of bradykinin. The increased sensitivity to bradykinin was shown to result in part from a decrease of the central kininase activity in these animals (5). It has been shown that in the brain, eighty percent of bradykinin degradation is due to the activity of an endopeptidase (EC 3.4.24.15) which is inhibited by CPP-Ala-Ala-Phe-pAB (20). In order to investigate the participation of the kinins in the central blood pressure regulation the effect of kininase inhibitors enalaprilat and CPP-ALa-Ala-Phe-pAB was examined in normotensive and hypertensive rats. In another experiment the pressor effect of bradykinin was examined in strains of rats with different blood pressure levels.

METHODS

Four month old female normotensive Wistar rats weighing approximately 200g were anaesthetized with a mixture of pentobarbital and chloral hydrate. Permanent cannulae were placed in the posterior portion of the fourth ventricle (11.7 mm posterior to bregma and 7.6 mm deep from skull surface). The cannulae were made from gauge 7 needles and were placed 2 mm above the injection site. Gingival needles [30G] were used for injections and when placed in the cannula the end of the needle protruded 2 mm from the cannula. The cannulae were anchored to the skull by jeweller's screws embedded in dental acrylic cement. Following the intracerebroventricular surgery, a polyethylene catheter (PE-10 connected to PE-50) filled with heparinized saline was placed in the abdominal aorta through the femoral artery. The other end of the catheter was slipped beneath the skin and exteriorized on the back of the animal. After surgery the animals were individually housed in plastic cages (30x20x10 cm) that also served as recording chambers. Two days after implantation the effect of centrally administered bradykinin on the MAP was recorded in the unanesthetized and unrestrained animals with a Narco P-1000 B pressure transducer and a DMP-4B Narco Physiograph (Narco Biosystems, Houston, TX, USA). Intracerebroventricular dose-responsecurves were obtained by injecting one µl of saline containing different concentrations of bradykinin. Each animal received no more than three injections and doses were administered in random order. Delivery of the peptide volume took slightly less than 1s. Latency for the manifestation of the pressor response to peptides was defined as the time elapsed between the moment of the injection and the onset of the pressor effect of the ED_{50} dose of the peptide. Cannulae placement in ventricles was verified in frozen brain slices following injection of the dye (1μl of 1% Evans Blue). The criteria for discarding an animal were the lack of diffusion of the dye into the expected ventricular space and or the observation of injection into the tissue adjacent to the ventricles. Regression lines obtained for the linear parts of the dose-response curves were compared for critical differences following covariance analyses and tests of regression, linearity and parallelism (21). The effective dose values and respective confidence intervals were estimated by a linear calibration method (21). For differences between independent means, student's t test was used, preceded by analyses of variance in the case of multiple comparisons. Paired *t* test was used to compare related means.

RESULTS

The administration of endopeptidase inhibitor CPP-Ala-Ala-Phe-pAB (2.4 μmol) and enalaprilat (0.6 μmol) to the fourth cerebral ventricle of unanesthetized SHR produced a mean increase in the arterial pressure of 12±9 mm Hg (n=12). The mean latency for the onset of the pressor effect of the kininase inhibitors was 42±15 seconds and the mean duration of the pressor response was 4±2 minutes. Figure 1 shows examples of the effect caused by the kininase inhibitors compared to the effect caused by the injection of 15 pmols of bradykinin in the fourth cerebral ventricle of unanaesthetized SHR. The pressor response to bradykinin has a shorter onset than that observed for the enzyme inhibitors (Figure 1). The enzyme inhibitors did not produce an increase in the arterial pressure of the normotensive rats. The pressor effect of bradykinin was examined in the NWR, the Wistar Kyoto (WKY) the SHR and in SHR, selected from our breeding colony, which did not develop high blood pressure (SHR*). The NWR had a mean arterial pressure (MAP) of 112±15, the WKY of 105±8, the SHR of 160±15 and the SHR* of 128±16. Dose response curves for the pressor effect of bradykinin injected in the fourth cerebral of the four groups of animals are shown in figure 2. The mean maximal effect of approximately 30 mmHg was obtained for the NWR, SHR and the SHR*. The WKY animals had a maximal effect of 23±11 mmHg. The ED_{50} for the pressor effect was 3 pmol in the SHR, 22 pmol in the

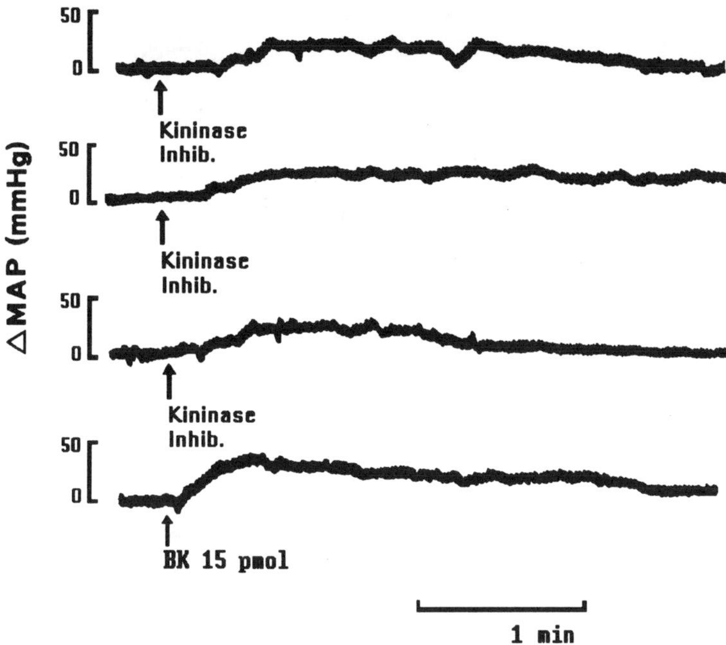

Figure 1.   Effects caused by the kininase inhibitors compared to the effect caused by the injection of 15 pmol of bradykinin (BK 15 pmol) in the fourth cerebral ventricle of unanaestetized SHR

NWR, 17 pmol in the WKY and 23 pmol in the SHR[*]. There was no difference in the latency for the manifestation of the pressor effect between the different groups.

DISCUSSION

The endopeptidase inhibitor CPP-Ala-Ala-pAB and enalaprilat injected in the fourth cerebral ventricle of unanaesthetized SHR produced a mean increase in arterial pressure approximately 40 seconds after administration. The mean duration of the response was of approximately 4 minutes. The pressor effect caused by the inhibitors was found to be comparable to the pressor response caused by 10 pmol of bradykinin injected in the fourth ventricle. The administration of

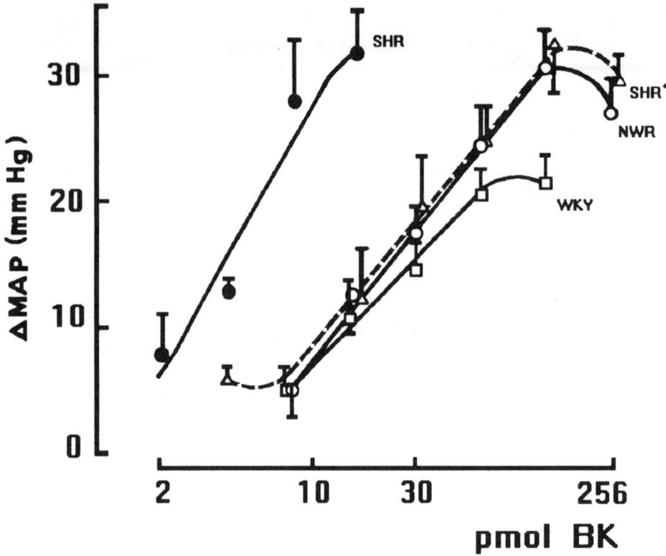

Figure 2. Log dose response curves for the pressor effect of bradykinin injected in the fourth cerebral ventricle of normotensive Wistar (NWR), Wistar Kyoto (WKY) and spontaneously hypertensive (SHR) rats. The SHR* are spontaneously hypertensive rats which did not develop high blood pressure.

enalaprilat alone did not cause alterations in blood pressure. Eighty percent of bradykinin metabolization in the brain is due to an endopeptidase activity of EC 3.4.24.15 which is inhibited by CPP-Ala-Ala-Phe-pAB (18). EC 3.4.24.15 is an endopeptidase which can degrade a variety of neuropeptides including neurotensin, vasopressin and angiotensin II. Therefore it is not possibly to ensure whether or not the effect observed is due to the inhibition of bradykinin degradation. However the simultaneous administration of the angiotensin converting enzyme inhibitor enalaprilat would exclude the participation of angiotensin II in the response to the enzyme inhibitors. Another aspect to be considered is that the pressor response to bradykinin is fairly specific. A vasoactive agents such as substance P, atrial natriuretic factor, noradrenaline, serotonin, prostaglandins E_2 or $F_{2\alpha}$, acetylcholine do not alter blood pressure following injections to the fourth ventricle (unpublished observations). Angiotensin and vasopressin only cause pressor responses following injection into the fourth ventricle of the SHR in doses which exceeded in fifty times the dose of bradykinin necessary to produce a similar effect (unpublished observation). Thus it is probable that the increase in the arterial pressure resulting from the administration of kininase inhibitors results from a decrease in bradykinin breakdown in the brain. The SHR are 7 times more sensitive to the pressor effect of bradykinin than the NWR or WKY. However, rats from the spontaneously hypertensive strain which did not develop high blood pressure showed similar sensitivity to the bradykinin pressor effect. These results suggest that the increased sensitivity of the SHR may be related to the expression of high blood pressure. The present data suggests that there is an endogenous kinin system in the brain related to the control of blood pressure and that tissue kallikrein-kinin system may be related to the expression of high blood pressure in the SHR.

REFERENCES

1. CHAO, J.; CHAO, L.; SWAIN, C.C.; TSAI, J. e MARGOLIUS, H. - Tissue kallkrein in rat brain and pituitary: Regional distribution and estrogen induction in the anterior pituitary. Endocrinology, 1987; 120: 475-482.

2. SCICLI, A.G.; FORBES, G.; NOLLY, H.; DUJOVNI, M. e CARRETERO, O.A. Kallikrein-kinins in the central nervous system. Clin. and Exp. - Theory and Practice 1984; A6 (10 & 11): 1731-1738.

3. MENDELSOHN, F.A.D.; SIEW, Y.C. e DUNBAR, M. - In vitro autoradiographic localization of angiotensin-converting enzyme in rat brain using 125I-labelled MK 351A. J. Hypertens.1984; 2 (Suppl III): 41-44.

4. LINDSEY, C.J.; NAKAIE, C.R. AND MARTINS, D.T.O. Central nervous system kinin receptors and the hypertensive response mediated by bradykinin. Br. J. Pharmacol. 1989; 97: 763-768.

5. MARTINS, D.T.O.; FIOR, D.R.; NAKAIE, C. R. AND LINDSEY, C.J. - Kinin receptors of the central nervous system of spontaneously hypertensive rats related to the pressor response to bradykinin. Br. J. Pharmacol. 1991; 103: 1851-1856.

6. PERRY, D. C. AND SNYDER, S.H. Identification of bradykinin in mammalian brain. J. Neurochem, 1984; 43: 1072-1080.

7. CORREA, F.M.A.; INNIS, R.B.; UHL, G.R. e SNYDER, S.H. - Bradykinin-like immunoreactive neuronal systems localized histochemically in rat brain. Proc. Natl. Acad. Sci. 1979; 76: 1489-1493.

8. HERMANN, K.; SCHACHETELIN, G. e MARIN-GREZ, M. - Kinin in cerebrospinal fluid: reduced concentration in spontaneously hypertensive rats. Experientia 1986; 42: 1238-1239.

9. KARIYA, K. e YAMAUCHI, A. - Effects of intraventricular injection of bradykinin on the EEG and the blood pressure in concious rats. Neuropharmacology 1981; 20: 1221-1224.

10. RIBEIRO, S.A. e ROCHA e SILVA, M. - Antinociceptive action of bradykinin and related kinins of longer molecular weight by the intraventricular route. Br. J. Pharmacol. 1973; 47: 517-528.

11. HOFFMAN, W.E. e SCHIMID, P.G. - Separation of pressor and antidiuretic effects of intraventricular bradykinin. Neuropharmacology 1978; 17: 999-1002.

12. CORREA, F.M.A. e GRAEFF, F.G. - Central mechanisms of the hypertensive action of intraventricular bradykinin in the unanesthetized rat. Neuropharmacology, 1967; 13: 65-75.

13. PEARSON, L.; LAMBERT, G.A. e LANG, W.J. - Centrally mediated cardiovascular and EEG responses to bradykinin and eledoisin. Eur. J. Pharmacol. 1969; 8: 153-158.

14. PEARSON, L. E LANG, W.J. - A comparison in concious and anaesthetized dogs of the effect on blood pressure of bradykinin, kallidin, eledoisin and kallikrein. Eur. J. Pharmacol. 1967; 2: 83-87.

15. BENETATO, G.; HAULICA, I.; MUSCALU, I.; BUIBUIANU, C. e GALESANU, S. - On the central nervous action of bradykinin. Rev. Rouman. Physiol. 1964; 1: 313.

16. CORREA, F.M.A. e GRAEFF, F.G. - On the mechanism of the hypertensive action of intraseptal bradykinin in the rat. Neuropharmacology 1976; 15: 713-717.

17. DIZ, D.I. e JACOBOWITZ, D.M. - Cardiovascular effects of discrete intrahypothalamic and preoptic injections of bradykinin. Brain Res. Bull. 1984; 12: 409-417.

18. WILKINSON, D. L. e SCROOP, G.C. - Role of prostaglandins and the area postrema in the central pressor action of bradykinin Eur. J. Pharmacol. 1985; 113: 287-290.

19. LINDSEY, C.J.; FUJITA, K. e MARTINS, D.T.O. - The central pressor effect of bradykinin in normotensive and hypertensive rats. Hypertension 1988; 2: 126-129.

20. McDERMOTT, J.R.; GIBSON, A.M. e TURNER, J.D. Involvement of endopeptidase 24.15 in the inactivation of bradykinin by rat brain slices. Bichem. Biophys. Res. Comm. 1987; 146: 154-158.

21. SNEDECOR, G.W. AND COCHRAN, W.G. Statistical Methods. Iowa City, USA. The Iowa State University Press, 1980 pp. 149-393.

BRAIN KALLIKREIN-KININ SYSTEM IN ARTERIAL PRESSURE REGULATION

Philip J. Privitera

Department of Cell and Molecular Pharmacology, Medical University of South Carolina,
171 Ashley Ave., Charleston, S.C., 29425, U.S.A.

INTRODUCTION

During the past decade, each of the components necessary for the formation and degradation of kinins has been reported to be present in brain tissue and/or cerebrospinal fluid (CSF) (1,2,3,4). A functional role for this brain peptide system, however, remains to be defined. There is increasing evidence to suggest that brain kinins may be involved in central cardiovascular regulatory mechanisms. For example, a number of laboratories have demonstrated that intracerebroventricular (ICV) administration of bradykinin into the brain causes increases in blood pressure and heart rate (5,6,7). In addition, activation of the endogenous brain kallikrein-kinin system by the central administration of melittin, an activator of membrane-bound kallikrein, elevates blood pressure and heart rate in association with an increase in CSF kinin level (4). Moreover, kinin receptor blockade attenuates the melittin-induced hypertensive response (8). Similarly, the ICV administration of porcine pancreatic kallikrein produces an increase in B.P. and an elevation in brain kinin level (9). Inhibition of kininase II activity in the brain by ICV captopril also induces a hypertensive response (10). Collectively, these findings suggest a linkage between the brain tissue kallikrein-kinin system and the central nervous system regulation of blood pressure.

ROSTRAL VENTROLATERAL MEDULLA AS A SITE OF ACTION FOR KININS

Although the brain kallikrein-kinin system may be involved in arterial pressure control mechanisms, we still do not have a complete picture regarding the specific brain areas or nuclei involved in mediating the central cardiovascular actions of BK. Sites in the lateral septal region of the forebrain, hypothalamus and the fourth ventricular region have previously been identified where BK can act to increase mean arterial pressure and heart rate (5,11,12). In our studies we focused on the area of the rostral ventrolateral medulla (RVLM) containing the C1 group of adrenaline containing cell bodies as a possible site of kinin action because of its established importance in the central cardiovascular regulatory mechanisms(13). This vasomotor area receives input from the NTS, a major site of termination of baroreceptor and cardiopulmonary afferent neurons (14). In addition, neurons originating in the RVLM project directly to cell bodies of preganglionic sympathetic neurons in the intermediolateral cell column of the spinal cord and provide excitatory input to the sympathetic nervous system (15).

In rats anesthetized with Inactin (120 mg/kg), microinjections (100 nl) of drugs were made stereotaxically into the RVLM of 16-18 week old male Wistar-Kyoto (WKY) and spontaneously hypertensive rats (SHR) as described in previous studies (16). Mean arterial pressure was measured with a Statham transducer and heart rate was determined with a tachograph (Grass Instruments Co.) triggered by a lead II electrocardiogram.

Unilateral microinjection of 500 pmoles of BK into the RVLM produced increases in mean arterial pressure and heart rate in both normotensive WKY (n=6) and in SHR (n=9) (Table 1).

Table 1. Cardiovascular effects of bradykinin (500 pmoles) microinjected into the rostral ventrolateral medulla of Wistar-Kyoto (WKY) and spontaneously hypertensive rats (SHR)

	WKY			SHR		
Measurements	Basal	Response	Change	Basal	Response	Change
Mean Art. Press. mm Hg.	117±6	124±5	+7±2	172±3*	191±2	+20±3*
Heart Rate bpm	301±12	307±11	+6±1	311±9	318±10	+7±3

*$p < 0.05$ compared to WKY

The hypertensive response to BK in SHR was significantly greater ($p < 0.05$) than in WKY rats whereas the heart rate response to BK, which was relatively small, was similar in both groups.

We also examined the effects of the selective kinin B_2 receptor antagonist DArg^0-Hyp^3-Thi^{5,8}-DPhe^7-BK, which was kindly supplied by Drs. Stewart and Vavrek. When 4 nmoles of kinin antagonist was injected into the RVLM prior to BK administration, the pressor responses to BK, 500 pmoles, were inhibited by approximately 60% in both the WKY and SHR. It is also of interest to note that microinjection of 4 nmoles of the kinin receptor antagonist alone produced a lowering of mean arterial pressure in both groups of rats, though the hypotensive effect observed was significantly greater in the SHR. RVLM injection of DArg^0-Hyp^3-Thi^{5,8}-DPhe^7-BK decreased mean arterial pressure by 39±5% in the SHR but only lowered arterial pressure by 11±4% in the WKY rat.

Our observation of an enhanced pressor response to BK in the SHR is consistent with findings of other investigators that pressor effect of ICV or 4th ventricular BK administration is augmented in the SHR (6,17,18). The mechanism for this enhanced responsiveness of the SHR to BK is unknown, but may be related to changes in receptor density or affinity, or to differences in the activity of kininases in the region (17). The finding that DArg^0-Hyp^3-Thi^{5,8}-DPhe^7-BK reduced the hypertensive responses to BK suggests that this response is mediated by the kinin B$_2$ receptor subtype (19,20). The reduction in arterial pressure produced by the kinin receptor antagonist in both goups of rats suggests that endogenous kinins in the RVLM might have a role in the central control of blood pressure.

AUTORADIOGRAPHIC LOCALIZATION OF BRADYKININ BINDING SITES IN BRAIN

The presence of kinin receptors in the brain is suggested by a number of pharmacologic studies in which the central cardiovascular actions of bradykinin can be blocked by specific kinin receptor antagonists (19,20). There is, however, virtually no data regarding the distribution or location of BK receptors in the brain. Specific BK binding sites have been found in primary brain cultures from neonatal rats (21) and in adult guinea pig brain homogenates (22) using radioligand binding assays. We conducted studies to determine the distribution of BK receptors in the guinea pig brain using quantitative receptor autoradiography (23). Our goal was to determine whether high affinity specific BK binding sites are found in brain areas associated with cardiovascular control mechanisms.

Autoradiographic experiments were performed as previously described (23). Briefly, whole brains from Dunkin-Hartley guinea pigs (250-350 g) were removed and serial 10 μm coronal sections of brain tissue were cut at -15°C in a cryostat, mounted onto gelatinized microscope slides and stored at -20°C. On the day of the experiment, slide-mounted tissue sections were warmed to room temperature and preincubated for 30 min at 37°C in 25 mM TES buffer (pH 6.8) in order to degrade endogenous kinins present in the sections. Slides sections were then cooled to 4°C and 2.0 ml of 25 mM TES incubation buffer, pH 6.8, containing [^{125}I]BK (10-1000 pM) was applied to each slide. The incubation buffer contained 1 mM 1,10 phenanthroline, 140 μg/ml bacitracin, 0.1 mM captopril, 1 mM dithiothreitol (DTT), 300 mM sucrose and 0.2% BSA. Adjacent sections placed on separate slides were exposed to the same concentrations of radioligand in the presence of 1 μM bradykinin to define non-specific binding. Incubation was carried out for 120 min at 4°C.

Slide sections were then rinsed 3 x 2 min with 25 mM TES buffer (pH 6.8) containing 300 mM sucrose at 4°C and dipped into ice-cold deionized water to remove buffer salts. Slide-mounted tissue sections, together with a set of [^{125}I]-brain paste standards containing known amounts of radioactivity, were apposed to a sheet of Hyperfilm (Amersham) and exposed for 5-7 days. Development of the autoradiograms and subsequent quantitative analyses using a Quantimet 920 computer assisted image analysis system have previously been described (23).

Autoradiographic studies showed that specific binding of [^{125}I]BK was localized to the caudal portion of the medulla oblongata (Fig. 1). High density specific [^{125}I]BK binding sites were observed in the nucleus of the solitary tract (nTS), the dorsal motor nucleus of the vagus nerve (X), the area prostrema (AP) and the caudal subnucleus of the spinal trigeminal nucleus (Sp5). Specific high density [^{125}I]BK binding sites were not detected in any other brain regions.

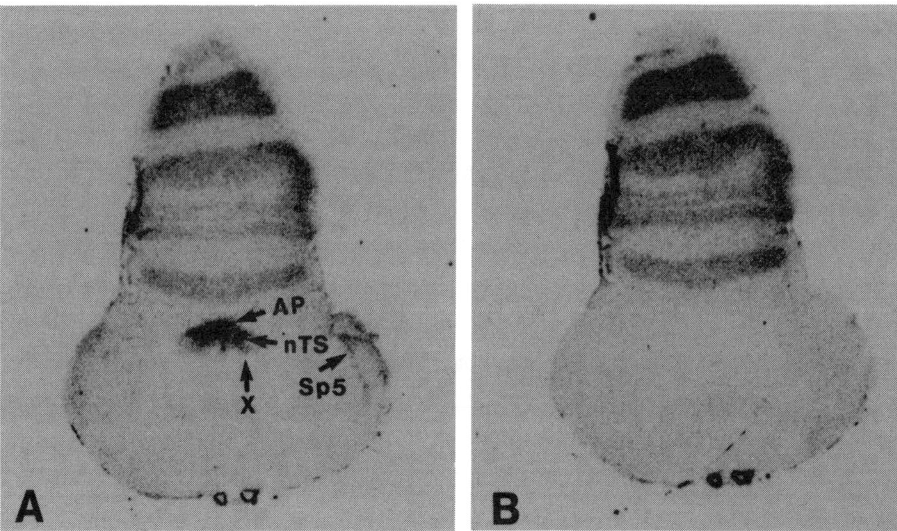

Figure 1. Photomicrograph of the distribution of [^{125}I]bradykinin binding sites in the guinea pig medulla oblongata. Panel A shows total binding. Panel B shows nonspecific binding (binding remaining in the presence of 1 μM Bk). Note the high level of binding in nucleus of the solitary tract (nTS), the dorsal motor nucleus of the vagus (X), the area prostrema (AP), and the nucleus commissuralis (n.c.). The caudal subnucleus of the spinal trigeminal nucleus (Sp5) also shows a moderate density of specific labelling.

Saturation binding experiments performed on consecutive brain sections using increasing concentrations of [125I]BK (0.01-1.0 nM) indicated the presence of a single high affinity binding site with an equilibrium dissociation constant (k_D) value of 73.5 ± 9.9 pM and a maximal number of binding sites (B_{max}) of 27.8 ± 1.9 attomoles per mm^2. Competition studies with unlabelled BK and related analogues demonstrated concentration-dependent inhibition of specific [125I]BK binding. The rank order of potency was BK> Met-Lys-BK> Lys-BK> D-Arg^0-Hyp^3-Thi^{5,8}-D-Phe^7-BK with the k_i values for these compounds being in the very low nanomolar range. The bradykinin B_1 receptor agonist Des-Arg^9-BK and antagonist Des-Arg^9-[Leu^8]-BK, however, did not significantly inhibit [125I]BK binding up to concentrations of 1 μM.

These autoradiographic studies demonstrated for the first time the location of specific [125I]BK binding sites in areas of the brain associated with central cardiovascular control mechanisms. For example, the nTS represents an essential link in the reflex and tonic regulation of blood pressure (13) while the X contains the cell bodies of parasympathetic preganglionic neurons and, as such, is involved in the central regulation of parasympathetic tone (24). The AP is contiguous to the nTS and appears to be involved in regulation of central sympathetic outflow (25). The presence of specific BK binding sites in these brainstem nuclei support a role for the brain kallikrein-kinin system in the central control of blood pressure. It is interesting to note that specific BK binding sites were not found in regions such as the lateral septal area (26), anterior third ventricular region (12) and the RVLM where bradykinin has been demonstrated to produce changes in cardiovascular function in the rat. The rank order of potency for kinin analogues in inhibiting [125I]BK binding suggest that specific BK binding sites localized in the guinea pig medulla are of the B_2 receptor type.

Our finding of specific high affinity [125I]BK binding sites in the caudal subnucleus of Sp5 is noteworthy since this region is continuous with the substantia gelatinosa of the dorsal horn of the spinal cord and neurons in this region of Sp5 receive sensory information from nociceptors and thermoreceptors (27). Steranka et al. (28) have previously localized [3H]BK receptors in the spinal cord of guinea pigs to the substantia gelatinosa of the dorsal horn, the dorsal root ganglia and the trigeminal ganglia. Their study supports a role for BK in pain mediation since the [3H]BK receptors were localized to areas in the spinal cord involved in nociception.

CONCLUSION

Our studies indicate that the RVLM is a site for the central cardiovascular actions of kinins. Moreover, the data suggest that the central cardiovascular actions of kinins are mediated by kinin receptors of the B_2 subtype. In addition, our autoradiographic studies provide evidence for the presence of kinin B_2 receptors in areas of the brainstem that are associated with central cardiovascular control mechanisms. Collectively, the data support a role for the brain kallikrein-kinin system in the central regulation of cardiovascular function.

ACKNOWLEDGEMENTS

This work was supported in part by NIH Grant Hl 41921, and by the Medical University of South Carolina Biomedical Research Support Grant.

REFERENCES

1. Chao J, Woodley C, Chao L, Margolius HS. Identification of tissue kallikrein in brain and in cell-free translation product encoded by brain mRNA. J Biol Chem 1983; 258:15173-15178.

2. Perry DC, Snyder SH. Identification of bradykinin in mammalian brain. J Neurochem 1984; 43:1072-1080.

3. Scicli AG, Forbes G, Nolly H, Dujovny M, Carretero OA. Kallikreins-kinins in the central nervous system. Clin Exper Hyperten 1984; 10 & 11:1731-1738.

4. Thomas GR, Thibodeaux H, Margolius HS, Privitera PJ. Cerebrospinal fluid kinins and cardiovascular function: Effects of cerebroventricular melittin. Hypertension 1984; 6 (Supp II) 46-50.

5. Corrêa FMA, Graeff FG. Central mechanisms of the hypertensive action of intraventricular bradykinin in the unanesthetized rat. Neuropharmacology 1974; 13:65-75.

6. Lindsey CJ, Fujita K, Martins DTO. The central pressor effect of bradykinin in normotensive and hypertensive rats. Hypertension 1988; 11 (Suppl I) :I-126-I-129

7. Thomas GR, Hiley CR. Cardiovascular effects of intracerebroventricular bradykinin and melittin in the rat. J Pharm Pharmacol 1988; 40:721-723.

8. Yang X-P, Carretero OA, Jacobsen G, Scicli AG. Role of endogenous brain kinins in the cardiovascular response to intracerebroventricular melittin. Hypertension 1989; 14: 629-635.

9. Kariya K, Yamauchi A. Relationship between hypertensive response and brain kinin level in the rat injected intraventricularly with glandular kallikrein. Japan J Pharmacol 1987; 43: 129-132.

10. Madeddu P, Glorioso N, Soro A, Tonolo G, Manunta P, Troffa C, Demontis MP, Varoni MV, Anania V. Brain kinins are responsible for the pressor effect of intracerebroventricular captopril in spontaneously hypertensive rats. Hypertension 1990;15: 407-412.

11. Diz DI, Jacobowitz DM. Cardiovascular effects of discrete intrahypothalamic and preoptic injections of bradykinin. Brain Res Bull 1984; 12: 409-417.

12. Lewis RE, Phillips MI. Localization of the central pressor action of bradykinin to the cerebral third ventricle. Am J Physiol 1984; 247: R63-R68.

13. Reis DJ, Granata AR, Joh TH, Ross CA, Ruggiero DA, Park DH. Brain stem catecholamine mechanisms in tonic and reflex control of blood pressure, Hypertension 1984; (Suppl II) 6: II-7-II-14.

14. Ross CA, Ruggiero DA, Reis DJ. Projections from the nucleus tractus solitarii to the rostral ventrolateral medulla. J Comp Neurol 1985; 242: 511-534.

15. Ross CA, Ruggiero DA, Joh TH, Park DH, Reis DJ. Rostral ventrolateral medulla: Selective projections to the thoracic autonomic cell column from the region containing C1 adrenaline neurons. J Comp Neurol 1984; 228: 168-185.

16. Privitera PJ, Granata AR, Underwood M, Reis DJ, Gaffney TE. C1 Area of the rostral ventrolateral medulla as a site for the central hypotensive action of propranolol. J Pharmacol Exp Therap 1988; 246: 529-533.

17. Martins DTO, Fior DR, Nakaie CR, Lindsey. CJ. Kinin receptors of the central nervous system of spontaneously hypertensive rats related to the pressor response to bradykinin. Brit J Pharmacol 1991; 103: 1851-1856.

18. Unger T, Rockhold RW, Yukimura T, Rettig R, Rascher W, Ganten D. Role of kinins and substance P in the central blood pressure regulation of normotensive and spontaneously hypertensive rats. In: Central Nervous System Mechanisms in Hypertension. Buckley JP, Ferrario CM, eds. New York: Raven Press, 1981: 115-127.

19. Yang X-P, Carretero OA, Akahoshi M, Scicli AG. Effects of blood pressure of intracerebroventricular administration of a kinin antagonist. In: Advances in Experimental Medicine and Biology. Kinins V, Vol. 247A, Abe K, Moriya H, Fujii S, eds. New York: Plenum Press, 1989: 439-445.

20. Lindsey CJ, Nakaie CR, Martins, DTO. Central nervous system kinin receptors and the hypertensive response mediated by bradykinin. Brit J Pharmacol 1989; 97: 763-768.

21. Lewis RE, Childers SR, Phillips MI. ^{125}I-Tyr-bradykinin binding in primary rat brain cultures. Brain Res 1985; 346: 263-272.

22. Fujiwara Y, Mantione CR, Vavrek RJ, Stewart JM, Yamamura HI. Characterization of $[^3H]$ bradykinin binding sites in guinea pig central nervous system: Possible existence of B_2 subtypes. Life Sci 1989; 44: 1645-1653.

23. Privitera PJ, Daum PR, Hill DR, Hiley CR. Autoradiographic visualization and characteristics of $[^{125}I]$bradykinin binding sites in guinea pig brain. Brain Res 1992; in press

24 Bystrzycka EK, Nail BS. Brain stem nuclei associated with respiratory, cardiovascular and other anatomic functions. In: The Rat Nervous System. Vol 2, Hindbrain and Spinal cord. Paxinos G.editor. New York: Academic Press, 1985: 95-110.

25. Barnes KL, Ferrario CM, Conomy JP. Comparison of the hemodynamic changes produced by electrical stimulation of the area postrema and nucleus tractus solitarii in the dog. Circ Res 1979; 45: 136-143.

26. Corrêa FMA, Graeff FG. Central site of the hypertensive action of bradykinin. J Pharmacol Exp Ther 1975; 192: 670-676.

27. Tracey DJ, Somatosensory system. In: The Rat Nervous System. Vol 2, Hindbrain and Spinal Cord. Paxinos, G, editor, New York: Academic Press, 1985: 121-153.

28. Steranka LR, Manning DC, Dehaas CJ, Ferkany JW, Borosky SA, Conner JR, Vavrek RJ, Stewart JM, Snyder SH. Bradykinin as a pain mediator: receptors are localized to sensory neurons and antagonists have analgesic actions. Proc Natl Acad Sci USA 1988; 85: 3245-3249.

INTERACTIONS OF T-KININ (ILE-SER-BRADYKININ) WITH NEUROGENIC, AUTACOIDAL AND EFFECTOR SYSTEMS IN AFFECTING CARDIOVASCULAR FUNCTION

A.R. Volpe[1], M. Carmignani[2,3], G. Porcelli[1]

Inst. of Chemistry[1] (CNR Receptor Chemistry Center) and Pharmacology[2], Catholic University School of Medicine, I-00168 Rome, and Dept. of Cell Biology and Physiology[3], University of L'Aquila, I-67010 Coppito (L'Aquila), Italy

INTRODUCTION

T-kinin (Ile-Ser-Bradykinin;TK), a trypsin-releasable pharmacologically active peptide, is formed in physiological conditions by action of T-kininogenases, but not of kallikreins, on T-kininogen, a protein making up over 70% of the kinin-carrying kininogens in rat blood. TK and T-kininogen, usually not found in human plasma, increase greatly in blood and exudates during acute inflammatory and neoplastic diseases[1,2]. As regard to pharmacological properties, TK from rat plasma was reported to be less active than bradykinin (BK) in contracting both rat uterus and guinea-pig ileum and more potent than BK (but less potent than Met-Lys-BK) in depressing blood pressure (BP) in the rat. The enhanced potency of TK, as compared to that of BK, in BP depression was referred to a higher resistance to converting enzyme degradation during pulmonary passage[1,3]. However, although kinins have a broad spectrum of biological activities (including effects on the cardiovascular -CV- system, capillary permeability, ion transport, prostaglandin synthesis and smooth muscle), a pharmacological profile of TK as factor influencing systemic haemodynamics remains to be defined[4].

The aim of this study was to further assess *in vivo* the CV effects of TK in relation to those of BK and to characterize these effects as a consequence of possible actions of TK on some neurogenic, autacoidal and effector systems.

MATERIALS AND METHODS

Aortic BP (systolic, diastolic and mean), maximum rate of rise of the left ventricular isovolumetric pressure (dP/dt, an index of cardiac inotropism) and heart rate (HR) were monitored polygraphically under thiopental anesthesia in male Wistar rats (average weight:262±3 g; n=96; mean±S.E.M.), as previously described[5].

In a first series of experiments, TK (0.6-77.6 nM/kg, ratio=2.0) and BK (0.7-89.6 nM/kg,ratio=2.0) were administered by i.v. injection in order to assess the dose-response relationship with reference to their effects on BP, dP/dt and HR. BK and TK were also tested, at the same doses, following bilateral vagotomy (carried out at the neck below the nodose ganglion) or treatment with the ganglioplegic drug hexamethonium (HEX; 10 μM/kg, by i.v. route) or during i.v. infusion of the myolitic drug papaverine (PAP; 2.7 μM/kg/min). Moreover, the above doses of BK and TK were given by i.v. route during i.v. infusion of TK (10 nM/kg/min) or BK(20 nM/kg/min), respectively. The infusion doses of TK and BK and the doses of HEX and PAP were able to induce minor changes of BP (5-10 mmHg), dP/dt and HR.

In a second series of experiments, carried out to assess some pharmacodynamic mechanisms determining CV effects of TK (also in comparison with BK), the two peptides were administered by i.v. route (at the doses used for assessing dose-response relationship) after treatment with the anti-inflammatory drugs hydrocortisone (HYD; 200 μM/kg, by i.v. route) or indomethacin (IND; 22 μM/kg, by i.v. route), the anti-hypertensive drug reserpine (RES; 4 μM/kg/day for three days, by i.p. route) and the anti-calmodulin drug naphthalenesulfonamide (W7; 4 μM/Kg,by i.p. route). TK and BK were also tested during i.v. infusion of ouabain (OUA; 170 nM/kg/min) or verapamil (VER; 100 nM/kg/min; a drug blocking the receptor-operated calcium channels). Moreover, dibutyryl-cyclic-AMP (dcAMP; 2.5-20 μM/kg; an analogue of cAMP able to cross plasma membrane) and PAP (1.3-5.2 μM/kg; a drug inhibiting the phosphodiesterase activity) were tested by i.v. route before and during infusion of TK (10 nM/kg/min). Finally, several drugs (nM/kg) were injected by i.v. route before and during i.v. infusion of TK (1.2-12 nM/Kg/min): acetylcholine (ACH;14), angiotensin I (A1;0.5) and II (A2;0.5), histamine (HIS;34), isoprenaline (ISO;1), phenylephrine (PE; O.5), serotonin (SER;2.7) and tyramine (TYR;144).

All doses were expressed in terms of free bases and peak CV responses were considered. Each consecutive test was not performed until all CV parameters had returned to the basal values (i.e. preceding the first administration) and had stabilized[5].

Data were expressed as means±S.E.M. and compared by analysis of variance. Only a p value less than 0.05 was considered to be significant.

RESULTS

TK and BK were equipotent in reducing systolic and diastolic BP and in increasing dP/dt and HR (Table 1).

DRUG (nM/kg)	BLOOD PRESSURE (-ΔmmHg)		dP/dt (+ΔmmHg/secx10^{-1})	HEART RATE (+Δbeats/min)
	SYSTOLIC	DIASTOLIC		
T-KININ				
9.7	25±2	36±3	187±25	16±6
38.8	37±3	49±5	269±20	20±5
77.6	41±4	60±5	274±24	27±5
9.7 (I)	10±1*	15±2*	150±27	19±3
77.6 (I)	48±5	7±3	267±31	32±6
BRADYKININ				
2.8	11±2	16±3	175±10	10±4
11.2	25±2	38±4	258±26	14±7
44.4	36±5	47±5	306±32	22±5
2.8 (I)	5±1*	7±0.3*	71± 3*	11±3
44.4 (I)	18±2*	25±2*	203± 7*	29±6

Table 1. Cardiovascular responses of rats to i.v. T-kinin alone or during i.v. infusion (I) of bradykinin (20nM/kg/min) and to i.v. bradykinin alone or during i.v. infusion (I) of T-kinin (10 nM/kg/min).

*Values are means±S.E.M. (n=12 in each group) *Signicantly different from the control (i.e. T-kinin or bradykinin alone; p<0.05)*

The effects of TK and BK on BP and cardiac inotropism were potentiated by vagotomy (TK>BK), while those of TK were not changed and those of BK were reduced by HEX (Table 2). BK did not oppose CV responses to TK, with the exception of the hypotensive responses induced by lower doses (0.6-9.7 nM/kg). On the contrary, TK antagonised the effects of BK on BP and dP/dt in a dose-related manner (Table 1). PAP greatly reduced both hypotensive and positive inotropic effects of TK without relation to the dose and opposed those induced by lower doses of BK (0.7-11.2 μM/kg) (Table 2).

Table 2.
Effects of vagotomy (VAG), i.v. hexamethonium (HEX;10μM/kg), papaverine (PAP; 2.7 μM/kg/min), hydrocortisone (HYD;200 μM/kg) and verapamil (VER; 100 μM/kg/min) and i.p. indomethacin (IND; 22 μM/kg) on the cardiovascular responses of rats to i.v. T-kinin (TK) and bradykinin (BK).

Values are means$\pm$S.E.M. (n=6-8 in each group)
**Significantly different from the control (i.e. TK and BK alone; p<0.05)*

Treatment	DRUG (nM/kg)		BLOOD PRESSURE (-ΔmmHg)		dP/dt (+ΔmmHG/sec10^{-1})
	TK	BK	SYSTOLIC	DIASTOLIC	
	0.6		12±1	16±2	79± 6
	9.7		32±3	45±4	195±20
	38.8		37±4	51±3	234±16
		5.6	21±2	30±3	112±12
		11.2	25±3	37±3	181± 8
		44.4	38±4	64±4	216±24
VAG	0.6		23±2*	30±2*	147±11*
VAG		5.6	26±3	42±2*	166± 8*
HEX	38.8		32±4	53±4	198±27
HEX		44.4	21±1*	47±3*	98±11*
PAP	38.8		11±2*	12±1*	107±10*
PAP		11.2	19±4	24±2*	185±14
HYD	38.8		22±3*	35±3*	379±31*
HYD		11.2	16±6	32±4	234±10*
VER	9.7		8±1*	14±1*	176±10
VER		11.2	8±2*	9±1*	170± 3
IND	9.7		10±2*	12±2*	492±37*

HYD and IND (IND>HYD) strongly antagonised, with next appearance of a hypertensive response to both TK and BK, the TK-induced arterial hypotension and increased the effects of the two peptides on dP/dt and HR (Table 2). The hypotensive, positive inotropic and tachycardic responses to dcAMP and PAP were markedly opposed (dcAMP>PAP) by TK in a dose-indipendent manner (Table 3).

DRUG (nM/kg)	TK infusion	BLOOD PRESSURE (-ΔmmHg)		dP/dt (+ΔmmHg/secx10^{-1})
		SYSTOLIC	DIASTOLIC	
Acetylcholine (14)	---	-44±2	-52±4	-109±12
	A	-25±1*	-29±2*	- 47± 4*
Histamine (34)	---	-23±1	-28±3	+289±27
	A	-10±4*	-11±1*	+240±25
Isoprenaline (1)	---	-25±2	-36±1	+642±47
	A	- 7±0.4*	-14±2*	+365±17*
Tyramine (144)	---	+52±5	+17±4	+278±22
	A	+20±3*	+20±5	+133±18*
Dibutyryl-cAMP (10000)	---	-30±2*	-44±4	+449±38
	B	- 4±0.4*	-13±2*	+ 51±51*
Papaverine (5200)	---	-17±3	-47±3	+214±13
	B	- 7±1*	-12±1*	+ 84± 6*

Table 3. Cardiovascular responses of rats to several drugs, injected by i.v. route, before and during i.v.infusion of T-kinin (TK) at the doses of 1.2 nM/kg/min (A) or 10 nM/kg/min (B).

Values are means$\pm$S.E.M. (n=16)
**Significantly different from the control (i.e. in the absence of TK infusion ;p<0.05).*

On the other hand, VER counteracted, in a similar manner and depending on the dose, BP responses to TK and BK, while OUA, RES and W7 did not change CV responses to these peptides (Table 2). When considering the interactions between TK and various physiological agonists and the effects of this peptide on the TYR-sensible release of noradrenaline from the postganglionic sympathetic endings, TK reduced CV responses to ACH and HIS and, mostly, to ISO and TYR and unaffected those to PE, A1 and A2 (Table3).

DISCUSSION

TK was shown to be, in the rat, a physiological factor regulating CV homeostasis and as potent as BK in determining systemic vasodilation and increase of both cardiac inotropism and HR. Therefore, in humans, TK may assure an increase of cardiac output during inflammatory and neoplastic diseases[2,6].

The CV effects of TK were found to depend, as those of BK, on the vagal nerve activity but not, on the contrary of the CV responses to BK, on the sympathetic one, as it was suggested by the results obtained in the rats treated with HEX. However, considering that TK opposes CV responses to TYR and that RES does not affect those to TK, one may assume that TK increases the TYR-sensible release of noradrenaline from the peripheral sympathetic endings indipendently on both nerve activity and nerve content of this amine.

The observed CV responses to TK during infusion of BK show that TK does not interact significantly with the BK_2 cardiac receptors and poorly affects vascular reactivity to the activation of BK_1 receptors. On the other hand, since TK opposes CV responses to BK, it may be thought that mechanisms other than receptor interactions are involved in such effect of TK[4]. In this respect, on the basis of the CV responses to TK and BK during PAP infusion and to dcAMP and PAP during TK infusion, it is likely that CV effects of TK depend, more than those of BK, upon an increase of the cAMP levels in both vascular and cardiac myocells. Moreover, as suggested by the results obtained in the rats treated with HYD and IND, TK (but not BK) induced vasodilation is related to activity of the prostaglandin system, which also plays a role in the stimulatory effects of the TK and BK on the heart. The observed effects of VER,OUA and W7 indicate that both TK and BK oppose Ca^{2+} entry through the receptor-operated channels of plasma membrane in vascular myocells and do not interact, in vascular and cardiac myocells, with Na, K-ATPase and calmodulin. The TK-induced changes of Ca^{2+} availability for concractile processes in these cells (following interactions with the cAMP and prostaglandin systems and at the calcium channels) may explain the observed reduction of CV responses to ACH and HIS and, mostly, to ISO[5].

This study shows that TK affects CV function by involving specific neurogenic and autacoidal mechanisms and by changing Ca^{2+} homeostasis in vascular myocells and myocardium.Only in part these mechanisms are activated by BK.

ACKNOWLEDGMENTS:This work was supported by grants of Italian MURST to Marco Carmignani (40% and 60%, 1989-1990 and 1990-1991) and of CNR (Progetto Finalizzato Chimica Fine II).

REFERENCES

1. H. Okamoto, L.M. Greenbaum. Biochem. Pharmacol. 1983; 32:2637
2. H. Okamoto, L.M. Greenbaum. Adv. Exp. Med. Biol. 1986; 198:69

3. J. Roblero, J.W. Ryan, J.M. Stewart. Res. Comm. Chem. Path. Pharmac. 1973; 6:207
4. A.R. Volpe, M. Carmignani. Pharmacol. Res. 1990; 22 (Suppl. 2):503
5. M. Carmignani, V.N. Finelli, P. Boscolo. Toxicol. Appl. Pharmacol. 1983; 69:442
6. M. Schachter. Pharmacol. Rev. 1980; 31:1.

AAS 38/III
Recent Progress on Kinins

MODULATION OF PRESYNAPTIC SYMPATHETIC ACTIVITY BY KININS AND RELATED COMPOUNDS: INFLUENCE OF CONVERTING ENZYME INHIBITION

P. Dominiak, M. Simon, A. Blöchl and P. Brenner

Institute of Pharmacology, Medical University of Lübeck, Ratzeburger Allee 160, W-2400 Lübeck

SUMMARY: Since converting enzyme and kininase II are identical enzymes and probably influences both, the biosynthesis of Ang II and the metabolism of bradykinin we investigated the effects of bradykinin, desArg-bradykinin and some bradykinin antagonists (desArg[9]-Leu[8]-bradykinin, HOE S 890307) on the sympathetic outflow of pithed SHR or Brown-Norway-Rats before and after acute or chronic inhibition of the converting enzyme by ramipril. bradykinin increased dose dependently the noradrenaline and adrenaline release in particular when the converting enzyme was inhibited. DesArg-bradykinin caused a dose-dependent increase in adrenaline release only after converting enzyme inhibition. The bradykinin-antagonists led to an increase in adrenaline release during ramipril administration. The weak but significant stimulation of adrenaline release by the bradykinin antagonists after converting enzyme inhibition might be due to unspecific actions on the adrenal medulla possibly induced by histamine release from mast cells.

INTRODUCTION

Angiotensin II (Ang II) increases the release of noradrenaline from the presynaptic sympathetic nerves and of adrenaline from the adrenal medulla (1-3). Consequently, the blockade of the Ang II biosynthesis by inhibition of the converting enzyme should lead to a decrease in catecholamine release as discussed by (3). However, the investigations and results concerning sympathetic activity during converting enzyme inhibition seem to be somewhat controversely: a decrease, no significant changes and an increase in noradrenaline and adrenaline release after acute or chronic inhibition of the converting enzyme have been reported (4-6). These different results could be due to the different methods and animals used in those investigations. However, there are some reasons why unchanged or increased catecholamine release is more likely. Very recently we demonstrated that acute or chronic converting enzyme inhibition by ramipril increased the uptake blocking effects of desipramine on the sympathetic varicosities leading to an enhanced sympathetic outflow (7). On the other hand, bradykinin and Ang I may accumulate during converting enzyme (kininase II) inhibition, both peptides are also known to be capable of changing the catecholamine release (8-10). They could therefore compensate for Ang II on noradrenaline and adrenaline release. In experiments in which we induced the sympathetic outflow by electrical stimulation during infusions of Ang I, we were able to demonstrate that Ang I did not further enhance the catecholamine release after chronic inhibition of converting enzyme (11). Concerning the changes in catecholamine release, Ang I could not therefore compensate for the reduction in Ang II on the sympathetic nerves. Therefore the aim of our present study was to test the hypothesis that bradykinin releases catecholamines in particular when converting enzyme is inhibited. Since the activity of kininase I may increase during the inhibition of kininase II it was also of interest to investigate desArg-bradykinin, the kininase I- mediated metabolite of bradykinin and the respective bradykinin antagonists for their effects on sympathetic activity.

MATERIALS AND METHODS

Animals: Male spontaneously hypertensive rats (SHR, strain: SHR/NCrl BR, SAVO Kißlegg) aged about 12 weeks and weighing about 200 g and male Brown-Norway-Rats (BNR, kininogen-defect strain:BN/Mai pft f, catholic University of Leuven and control strain: BN/HAN, Central-Institute for laboratory animal breeding Hannover) weighing about 250 g were used for the experiments. The animals were housed in plastic cages (Makrolon^R) and given water and standard diet (Altromin^R) ad libitum.

Drugs: Ramipril was administered either intravenously (0.1 mg/kg) or orally (1 mg/kg/d for 14 days) by gavage to achieve acute or chronic inhibition of the converting enzyme. Controls received the same volume of water. To block the neuromuscular junction in pithed animals, d-tubocurarin (3 mg/kg) was injected i.v. 30 min prior to the stimulation experiments. During the stimulation experiments the following test substances were administered to the animals: bradykinin (120 - 1200 ng/kg/min), desArg-bradykinin (0.9 - 27 μg/kg/min), the bradykinin-1-antagonist (B$_1$) desArg(9)-Leu(8)-bradykinin (5 - 150 μg/kg/min) and the B$_2$-antagonist HOE S 890307 (0.1 - 10 μg/kg/min).

Experimental procedure: Rats were pithed under ether anesthesia using a steel rod (12), which was coated with enamel except for the length of the thoracolumbar spinal cord (Th4 - Th12 segment). The steel rod served as a positive electrode. A negative electrode was placed into the skin of the back. Both vagus nerves were cut at the neck. Sympathetic outflow was induced by preganglionic electrical stimulation (50 mA, 0.5 Hz, 1 ms, 3 min) and assessed by measuring stimulation dependent circulating noradrenaline and adrenaline (13) at the end of each stimulation period in blood samples obtained from a PE-50 catheter which was inserted into the left carotid artery. Drugs were administered through a vena femoralis catheter (PE-10) using an Infors^R infusion pump.

Determination of circulating catecholamines: Blood samples taken from the left carotid artery were mixed with heparin (5.000 I.U.) and centrifugated at 4°C. Reduced glutathione (0.2m) and EDTA (0.25m) were added to the yielded plasma and the samples were kept frozen (-80°C) until assayed (14). After thawing, a glutathione solution (300 mm EDTA and 50 mm reduced glutathione [pH 7.0]) was added to the plasma and adsorbed to alumina. Desorption was performed by perchloric acid (200 mm), and after centrifugation the supernatant was injected into the chromatographic system consisting of a reversed phase Macherey and Nagel column (Nucleosil^R; flow rate: 1ml/min) and an electrochemical detector (Waters M 460^R).

RESULTS

Bradykinin: In control Brown-Norway-Rats (cBNR) bradykinin induced a dose dependent increase in noradrenaline and adrenaline release. Before bradykinin infusion the stimulation-dependent circulating noradrenaline was about 115 pg/ml and reached about 370 pg/ml when the highest dose (3.6 μg/kg/min) of bradykinin was administered. The respective values for adrenaline were 105 pg/ml and 380 pg/ml. Bradykinin also increased noradrenaline and adrenaline dose dependently in defect BNR (dBNR), but to a lesser extent. Stimulation-dependent circulating noradrenaline before bradykinin was about 85 pg/ml and significantly lower (p < 0.01) when compared to cBNR (see above) and increased up to 200 pg/ml during 3.6 μg/ml/min bradykinin infusion. In comparison to cBNR all circulating noradrenaline concentrations of dBNR were found to be significantly lower (p < 0.01). The respective stimulation- dependent adrenaline concentrations were 100 pg/ml before and 265 pg/ml during the highest bradykinin dose (values are not depicted).

Figure 1 shows the influence of chronic converting enzyme inhibition on catecholamine release in dBNR. It is evident from the diagram that the converting enzyme inhibitor Ramipril increases clear cut the bradykinin induced adrenaline release by about 800%.

In figure 2 the influence of bradykinin and ramipril on noradrenaline release of SHR is depicted. All administered bradykinin doses lead to a significant increase in noradrenaline release by about 80%. However, after converting enzyme inhibition the noradrenaline increase was dose dependently to bradykinin. Stimulation dependent circulating adrenaline reached 5.000 pg/ml during the highest bradykinin dose (1.200 ng/kg/min) when Ramipril was administered chronically (Fig. 3).

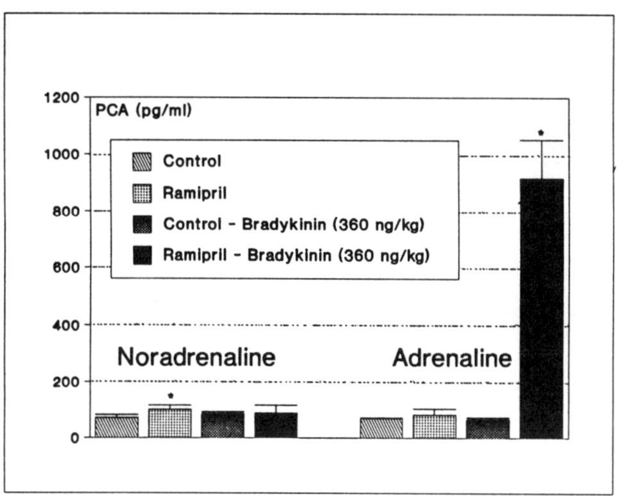

Figure 1. Stimulation-dependent plasma catecholamine concentrations (PCA) of dBNR. The converting enzyme inhibitor ramipril was administered orally by gavage for 14 days (1 mg/kg/day). Bradykinin was infused intravenously at a dose of 360 ng/kg/min during electrical stimulation of the spinal cord. Stimulation parameters: 50 mA, 0.5 Hz, 1 ms, 3 min. Significance values: * = p< 0.05.

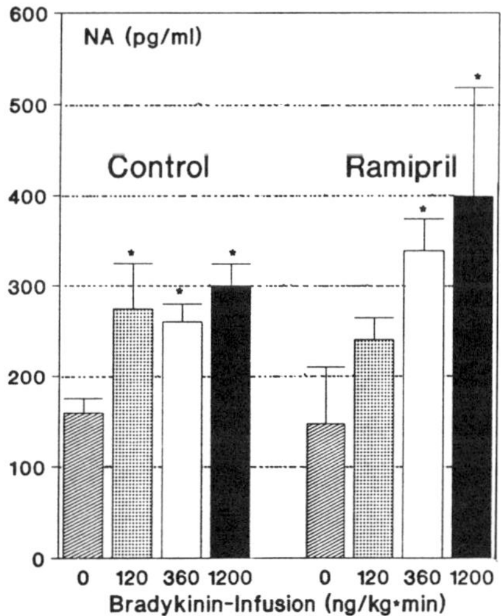

Figure 2. Stimulation-dependent plasma noradrenaline concentrations of SHR. Ramipril was administered orally by gavage for 14 days (1 mg/kg/day). Bradykinin was (i.v.) infused during electrical stimulation of the spinal cord. Stimulation parameters : 50 mA, 0.5 Hz, 1 ms, 3 min. Significance values: * = p< 0.05.

DesArg-bradykinin: Before acute inhibition of the converting enzyme the stimulation dependent noradrenaline and adrenaline concentrations in plasma of SHR reached 90 and 50 pg/ml respectively. DesArg-bradykinin did not change noradrenaline or adrenaline release into the plasma. However, plasma adrenaline concentrations after acute Ramipril administration (0.1 mg/kg i.v.) remained unchanged during desArg-bradykinin but on a 30% higher level (not depicted).

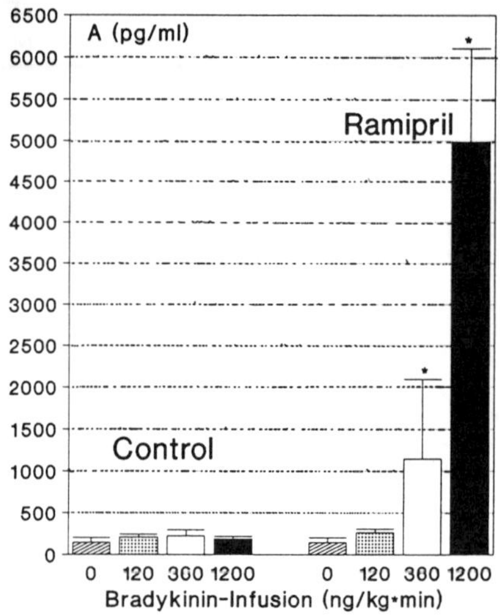

Figure 3. Stimulation-dependent plasma adrenaline concentrations of SHR. Ramipril was administered orally by gavage for 14 days (1 mg/kg/day). Bradykinin was (i.v.) infused during electrical stimulation of the spinal cord. Stimulation parameters: 50 mA, 0.5 Hz, 1 ms, 3 min. Significance values: * = $p < 0.05$.

Figure 4 shows the effects of desArg-bradykinin on adrenaline release of SHR after chronic inhibition of the converting enzyme. Whereas adrenaline increased in. a dose dependent pattern, stimulation-dependent noradrenaline concentration did not change (not depicted).

DesArg(9)-Leu(8)-bradykinin (B_1-antagonist): Before acute administration of Ramipril the B_1-antagonist did not change catecholamine concentrations in plasma of SHR. After acute converting enzyme inhibition, however, stimulation dependent adrenaline but not noradrenaline increased slightly but significantly in a dose dependent pattern (not depicted). The effect of the B_1-antagonist on adrenaline release was enhanced after chronic Ramipril treatment when compared to the acute experiment (figure 5).

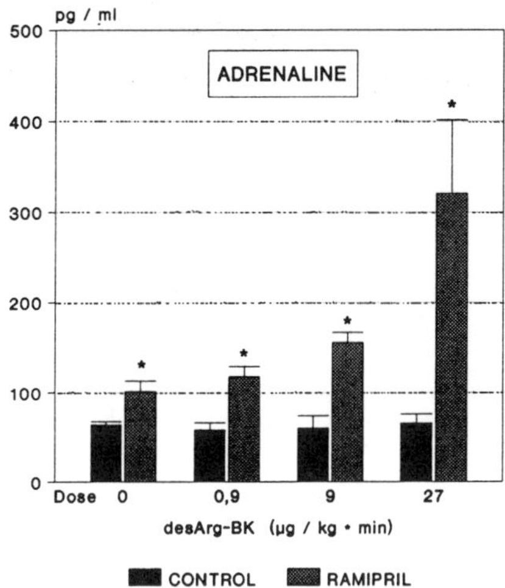

Figure 4. Stimulation-dependent plasma adrenaline concentrations of SHR. Ramipril was administered orally by gavage for 14 days (1 mg/kg/day). desArg-bradykinin (desArg-BK) was (i.v.) infused during electrical stimulation of the spinal cord. Stimulation parameters: 50 mA, 0.5 Hz, 1 ms, 3 min. Significance values: * = $p < 0.05$.

HOE S 890307 (B_2-antagonist): As was observed with the B_1-antagonist, HOE S 890307 did not change noradrenaline or adrenaline release before the converting enzyme was inhibited. Subsequent to acute Ramipril administration, stimulation- dependent noradrenaline was slightly but significantly increased after all doses of HOE injected, whereas the adrenaline release shifted significantly to a dose dependent pattern from 85 pg/ml upto 225 pg/ml (not depicted).

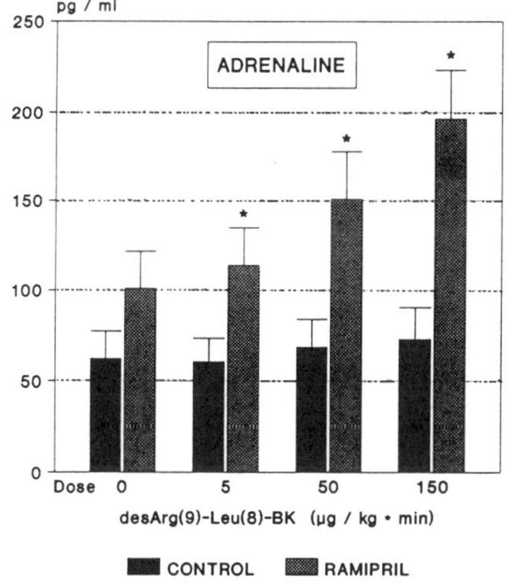

Figure 5. Stimulation dependent plasma adrenaline concentrations of SHR. Ramipril was administered orally by gavage for 14 days (1 mg/kg/day). The B_1-receptor antagonist desArg(9)-Leu(8)-bradykinin was (i.v.) infused during electrical stimulation of the spinal cord. Stimulation parameters: 50 mA, 0.5 Hz, 1 ms, 3 min. Significance values: * = $p < 0.05$.

Chronic inhibition of the converting enzyme changed adrenaline release in a behaviour similar to the acute experiment as shown in figure 6. However, noradrenaline remained unchanged after chronic ramipril treatment (not shown).

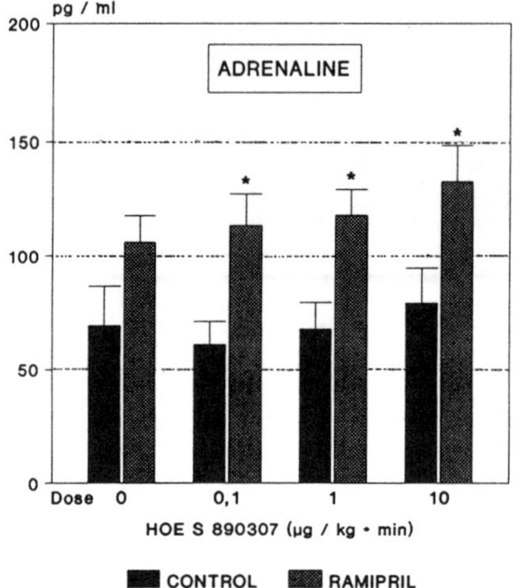

Figure 6. Stimulation dependent plasma adrenaline concentrations of SHR. Ramipril was administered orally by gavage for 14 days (1 mg/kg/day). The B_2-receptor antagonist HOE S 890307 was (i.v.) infused during electrical stimulation of the spinal cord. Stimulation parameters: 50 mA, 0.5 Hz, 1 ms, 3 min. Significance values: * = $p<$ 0.05, ** = $p<$ 0.01

DISCUSSION

The results about the catecholamine releasing effects of bradykinin, reported in the literature, are somewhat contradictory. It was demonstrated by (15,16) that bradykinin is able to release adrenaline from the adrenal medulla. Furthermore, (17) confirmed the results of an increase in adrenaline stimulation and reported about a bradykinin-dependent noradrenaline release which still existed after adrenalectomy. In contrast, (10) detected a decrease in ^{3}H-noradrenaline release when bradykinin was administered. These differences may reflect the different techniques of catecholamine measurements used: determination of endogenous catecholamines versus tritiated noradrenaline. Our results obtained in Brown-Norway rats and spontaneously hypertensive rats proved the former that bradykinin stimulates the release of noradrenaline and adrenaline. In addition, we demonstrated for the first time that the catecholamine release was dose dependent to bradykinin. The clear-cut increase in adrenaline release after inhibition of the converting enzyme by ramipril (figures 1, 2 and 3) and additionally infused bradykinin is further evidence for the stimulatory effects of bradykinin on the sympathetic system since the converting enzyme is identical to kininase II (18) and bradykinin is reported to accumulate during converting enzyme inhibition (19).

On the other hand, bradykinin can also be degraded by the enzyme kininase I, especially when the kininase II is inhibited possibly resulting in an increase in the peptide desArg-bradykinin. This is why we investigated the actions of desArg-bradykinin on the sympathetic system. As was demonstrated by figure 4, desArg-bradykinin stimulated adrenaline release dose dependently only after inhibition of the converting enzyme. Assuming that the concentration of desArg-bradykinin increases during converting enzyme inhibition, its effects at least on adrenaline release, are additive to those of bradykinin, and vice versa.

The question arises which bradykinin receptor type mediates the catecholamine releasing effects. It is known that bradykinin is a potent B_2-receptor agonist with weak effects for the B_1-receptor, and that desArg-bradykinin possesses high affinity for the B_1-receptor but only a small affinity for the B_2-receptor (20). Since, in the present study, bradykinin singularly affected noradrenaline release and stimulated adrenaline release much more when compared to desArg-bradykinin, particularly after inhibition of the converting enzyme, the effects of both peptides on the sympathetic system seem to be mediated by the B_2-receptor.

As was demonstrated for desArg-bradykinin, the B_1-antagonist desArg(9)-Leu(8)-bradykinin and the B_2-antagonist HOE S 890307 stimulated adrenaline release in a dose dependent pattern only after converting enzyme inhibition. These results are difficult to rationalize. However, a number of studies have shown that bradykinin antagonists may act via non-receptor-mediated mechanisms. (17) reported about an increased blood pressure in rats by stimulation of catecholamine release from the adrenal medulla. Furthermore, a histamine release from rat peritoneal mast cells (21) and from human skin mast cells (22) after various kinin analogues was demonstrated. Histamine, on the other hand, is capable of releasing adrenaline from the adrenal medulla.

In conclusion, while bradykinin and, to a less degree, desArg-bradykinin, induce catecholamine release from sympathetic varicosities and the adrenal medulla probably B_2-receptor mediated, bradykinin antagonists cause adrenaline release possibly by degranulation of histamine-containing mast cells.

Since bradykinin antagonists affect adrenaline release itself, experiments undertaken with the latter to antagonize the catecholamine releasing actions of bradykinin and desArg-bradykinin are difficult to interpret. However, our present results support our hypothesis that bradykinin and/or desArg-bradykinin may contribute to the unchanged sympathetic outflow during converting enzyme inhibition compensating for the lack of effect of Ang II.

P. Dominiak et al.

REFERENCES

1. Peach MJ. Renin-angiotensin system: Biochemistry and mechanism of action. Physiol Rev 1977; 57: 313-370.

2. Zimmermann BG. Actions of angiotensin on adrenergic nerve endings. Fed Proc 1978; 37: 199-202.

3. De Jonge A, Knape JThA, van Meel JCA, Kalkman HO, Wilfert, B,Thoolen, MJMC, van Brummelen, P, Timmermans, PBMWM, van Zwieten, PA. Effect of captopril on sympathetic neurotransmission in pithed normotensive rats. Eur J Pharmacol 1983; 88: 231-240.

4. Majewski H, Hedler L, Schurr C, Starke K. Modulation of noradrenaline release in the pithed rabbit: a role for angiotensin II. J Cardiovasc Pharmacol 1984; 6: 888-896.

5. Dominiak P, Elfrath A, Türck D. Effects of chronic treatment with ramipril, a new ACE blocking agent, on presynaptic sympathetic nervous system of SHR. Clin Exp Hypertens A-Theor 1987; 9: 369-373.

6. Kuo Y-JJ, Keeton TK. Captopril increases norepinephrine spillover rate in conscious spontaneously hypertensive rats. J Pharmacol Exp Ther 1991; 258: 223-231.

7. Dominiak P, Blöchl A. Does converting enzyme inhibition change the neuronal and extraneuronal uptake of catecholamines? Basic Res Cardiol 1991; 86: (Suppl. 3) 149-156.

8. Nussberger J, Brunner DB, Waeber B, Brunner HR. Specific measurement of angiotensin metabolites and in vitro generated angiotensin II in plasma. Hypertension 1986; 8: 476-482.

9. Iimura O, Shimamoto K, Tanaka S, Hosoda S, Nishitani T, Ando T, Masuda A. The mechanism of the hypotensive effect of captopril (converting enzyme inhibitor) with special reference to the kallikrein-kinin and renin-angiotensin system. Jpn J Med 1986; 25: 34-39.

10. Starke K, Peskar BA, Schumacher KA, Taube HD. Bradykinin and postganglionic sympathetic transmission. Naunyn-Schmiedeberg's Arch Pharmacol 1977; 299: 23-32.

11. Dominiak P, Brenner P, Blöchl A, Simon M. Wirkungen von Angiotensin I, Angiotensin II und Bradykinin auf die Freisetzung endogener Katecholamine unter chronischer Hemmung des Conversionsenzyms. FOCUS MUL 1991; 8: 210-218.

12. Gillespie JS, Muir TC. A method of stimulating the complete sympathetic outflow from the spinal cord to blood vessels in the pithed rat. Br J Pharmac Chemother 1967; 30: 78-87.

13. Yamaguchi I, Kopin IJ. Plasma catecholamine and blood pressure responses to sympathetic stimulation in pithed rats. Am J Physiol 1979; 237: H305-H310.

14. Hjemdahl P. Inter-laboratory comparison of plasma catecholamine determinations using several different assays. Acta Physiol Scand 1984; 527: 43-54.

15. Robinson RL. Stimulation of the catecholamine output of the isolated, perfused adrenal gland of the dog by angiotensin and bradykinin. J Pharmacol Exp Ther 1967; 156: 252-257.

16. Staszewska-Barczak J, Van JR. The release of catecholamines from the adrenal medulla by peptides. Br J Pharmacol Chemother 1967; 30: 655-667.

17. Mulinari R, Benetos A, Gavras I, Gavras H. Vascular and sympathoadrenal responses to bradykinin and a bradykinin analogue. Hypertension 1988; 11: 754-757.

18. Erdös EG. The kinins. Biochem Pharmacol 1976; 25: 1563-1569.

19. Graf K, Bossaller C, Auch-Schwelk W, Gräfe M, Baumgarten C, Fleck E. ACE-inhibitors diminish the degradation of bradykinin and enhance endothelium-dependent relaxations in isolated coronary arteries. Pharm Pharmacol Lett 1991; 1: 71-73.

20. Bathon JM, Proud D. Bradykinin antagonists. Annu Rev Pharmacol Toxicol 1991; 31: 129-162.

21. Devillier P, Renoux M, Drapeau G, Regoli D. Histamine release from rat peritoneal mast cells by kinin antagonists. Eur J Pharmacol 1988; 149: 137-140.

22. Lawrence ID, Warner JA, Cohan VL, Lichtenstein LM, Kagey-Sobotka A et al. Induction of histamine release from human skin mast cells by bradykinin analogs. Biochem Pharmacol 1989; 38: 227-233.

Circulation
Heart

AAS 38/III
Recent Progress on Kinins
© 1992 Birkhäuser Verlag Basel

THE KALLIKREIN - KININ SYSTEM IN CARDIAC TISSUE

H. L. Nolly, G. Saed, G. Scicli, O. A. Carretero and A. G. Scicli

Hypertension and Vascular Res. Div., Henry Ford Hospital, Detroit, MI, USA
and Argentine Council of Res. (CONICET), School of Medicine, Mendoza, R. A.

SUMMARY: Kallikrein and minute amounts of kininogen have been found in rat cardiac tissue. The mRNA for kallikrein was also determined by the polymerase chain reaction using KK-specific probe. The existence of an intrinsic kallikrein-kinin system in the heart raises the possibility that the enzyme-peptide system is involved in local regulation of cardiac function and metabolism.

INTRODUCTION

Recent reports indicate that kinins may be important mediators of the cardio-protective effects of angiotensin-converting enzyme inhibitors (ACEi) (1). Since some of these data were obtained using isolated heart preparations (2), the pathway by which kinins are generated in heart tissue is not clear. Preliminary data suggest the presence of an endogenous cardiac kallikrein-kinin system (3). Such a system may provide an anatomical basis to clarify (at least in part) the mechanisms by which ACEi exert their cardiac actions.

It has recently been found that glandular kallikrein, a potent kininogenase, is present in vascular tissue (4). Thus it is possible that this enzyme is also present in the heart. We studied glandular kallikrein and its mRNA are present in rat cardiac tissue. Our findings indicate that the heart contains, releases and possibly synthesizes glandular kallikrein.

MATERIALS AND METHODS

Preparation of rat cardiac extracts: Male Wistar rats (Charles River, Wilmington, MA) weighing 250-300 g were anesthetized with sodium pentobarbital (50 mg/kg i.p.). Catheters were placed in the left ventricle and vena cava. Phosphate-buffered saline (pH 7.4) containing heparin (12.5 U/ml) was infused through the ventricle at a rate of 10 ml/min and blood collected via the vena cava. The infusion was continued until the effluent appeared to be blood-free (usually 5 min). The heart was removed, separated from the aortic and pulmonary trunks, and placed in a Petri dish containing Krebs-Henseleit buffer, then washed by renewing the buffer several times. (Most hearts continue beating during these washings). It was minced, rinsed further and frozen at -20°C until used. At this time it was frozen and thawed four times and then twice homogenized in the cold for 10 sec with a Polytron homogenizer (position 7). The homogenate (200 mg wet tissue/ml) was centrifuged at 1,000 G for 10 min to eliminate debris and the supernatant dialyzed overnight at 4°C against 0.01M Tris-HCl buffer (pH 7.4).

Incubation procedure: The homogenate supernatant (400 μl) from tissue extract (80 mg wet tissue weight) was incubated for 5 hr at 37°C with 200 μl partially purified dog kininogen (2,000 ng kinin-releasing capability) in the presence of 1,000 μl fresh 0.1M Tris-HCl buffer (pH 8.5) containing EDTA (15 mg/ml), 1-10 phenanthroline (1mg/ml), 8-OH-quinoline (1 mg/ml) and soybean trypsin inhibitor (SBTI) (100 μg/ml). SBTI was used to inhibit plasma kallikrein or trypsin-like enzymes that might contaminate the preparations. The kinins generated during the 5-hour incubation period were measured by radioimmunoassay (RIA) (5). Bradykinin recovery was 81 $\pm$ 4% (n = 6). Results are expressed as pg bradykinin per mg protein per min incubation. To determine the optimum pH, aliquots of the cardiac homogenate were incubated with kininogen and peptidase inhibitors at pH levels ranging from 5 to 9 using different buffers (0.1 M acetate, 0.1 M phosphate and 0.1 M Tris-HCl).

Trypsin activation: To determine whether inactive kallikrein was present, the homogenates were trypsinized. To establish the best radio between trypsin and tissue homogenate for activation of inactive kininogenase, a fixed amount of tissue was preincubated for 30 min at 37°C with varying concentrations of trypsin ranging from 0.05 to 1.0 μg/mg tissue in 500μl 0.1M Tris-HCl buffer (pH 8.5). A ratio equal to 0.2 μg trypsin per mg tissue was found to be optimal for our use. In short, the equivalent of 100 mg tissue homogenate in 500 μl Tris was incubated with trypsin (20μg) in 0.5 ml 0.1M Tris-HCl buffer (pH 8.5) for 30 min at 37°C, after which the reaction was stopped by adding SBTI (100 μg).

Immunological characterization: Inhibition of kininogenase activity was assessed by incubating the tissue homogenates with globulin purified from rabbit antiserum against rat urinary kallikrein or from non-immunized rabbits (6). Prior to incubation with kininogen, 400 μl of the homogenate supernatant from the cardiac tissue (equivalent to 80 mg tissue) was incubated with 200 μg of either globulin, after which kininogenase activity was measured. Immunoreactive glandular kallikrein was measured by RIA as described previously (7).

<u>Affinity chromatography of immobilized kallikrein antibodies:</u> Immunoaffinity chromatography of the rat cardiac homogenate was performed as described previously (8). The antikallikrein-sepharose gel was equilibrated with 0.1M sodium phosphate buffer (pH 7.4). The cardiac homogenate supernatant was mixed with the gel for 2 hr at room temperature and then for 24 hr at 4°C. Unbound proteins were separated out by successive washings, first with 0.1M sodium phosphate buffer alone (pH 7.4) and then with 1M NaCl added (pH 6.0). Kininogenase activity was eluted with 0.1M sodium acetate buffer (pH 3.5) containing 1M NaCl. Samples were dialyzed against distilled water, concentrated and frozen at -10°C.

<u>Gel filtration on Ultrogel AcA$_{54}$</u> : 2 ml of the supernatant homogenate (35 mg protein) was applied to an Ultrogel Aca$_{54}$ column (100 x 1 cm) equilibrated and eluted with 0.1M phosphate buffer (pH 7.4). The column was eluted at a rate of 18 ml/hr; 3-ml fractions were collected and kininogenase activity monitored. To determine molecular weight, the elution volume of the cardiac enzyme was compared with standards of known molecular weight, namely gamma globulin (MW 158,000), ovalbumin (MW 43,000), chymotrypsinogen (MW 25,000) and myoglobin (MW 17,000). Rat submandibular gland kallikrein, used as a control, was purified using a modification of a previous method (9). Purified rat kallikrein was homogeneous on 12% alkaline polyacrylamide gel electrophoresis and released 1.2 μg of kinin per 1 μg protein per min when incubated with 2,000 ng semipurified dog kininogen.

<u>Inhibition studies:</u> Cardiac homogenates (equivalent to 80 mg wet tissue in 400 μl Tris) were preincubated at 37°C for 30 min together with SBTI (100 μg/ml final concentration), phenylmethylsulfonyl fluoride (PMSF, 2mM), aprotinin (1000 KIU), and D-Phe-Phe-Arg-chloromethyl ketone (D-PPACK, 10^{-6}M). The inhibitors were dissolved in 0.1M Tris-HCl (pH 7.4), except for PMSF which was dissolved in methanol. After the preincubation period, samples were incubated with kininogen for 5 hr at 37°C to assay kininogenase activity. Proteins were determined by Bradford's method (10). As a control, we used a dilution of purified rat submandibular kallikrein which gave kininogenase activity similar to that observed with the cardiac kininogenase.

<u>Release of kallikrein from heart slices:</u> Slices 2 mm thick were obtained from the surface of ventricles taken from control rats and rats treated with puromycin (10 mg i.p., 3 hr before the experiment). After rinsing the slices several times with Krebs-Henseleit buffer, they were incubated in oxygenated Krebs-Henseleit solution at 37°C. The buffer was completely renewed every 30 min (total incubation time, 120 min). Kininogenase activity was measured in each supernatant and in the tissue.

<u>Detection of kallikrein mRNA in cardiac tissue:</u> Tissue were obtained as described above (Preparation·of rat cardiac extracts). After washing the samples several times, ventricles and atria were separated into 2 pools and immediately frozen at -70°C until needed.

Extraction of RNA: Total RNA was isolated from rat heart (ventricles and atria) by the guanidium thiocyanate extraction method (11).

Kallikrein complementary DNA probe: The recombinant plasmid pcXP39, bearing a rat pancreatic kallikrein complementary DNA (cDNA) insert, was prepared as described previously (11).

Kallikrein RNA analysis: Northern and slot blotting were performed as described previously (11).

Amplification method: cDNA was synthesized using cloned Moloney murine leukemia virus reverse transcriptase (Bethesda Research Laboratories, Gaithesburg, MD) according to the manufacturer's protocol) and as previously described (11). Southern blotting was performed as described previously (11).

Measurement of kininogen in cardiac tissue: The kininogen content in cardiac tissue and in the incubation medium was assayed by a mofification of the technique of Dinis et al (12). The samples (1000 μl) were boiled during 10 minutes, allowed to cool, and incubated with trypsin (20 μg) for 60 minutes at 37°C, pH 8.5 in the presence of 3 mM 1-10 phenanthroline. The reaction was stopped adding SBTI (100 μg) and boiling. The samples were further processed to measure the kinins generated.

Reagents: Molecular weight markers (BioRad, NY) Ultrogel (LKB, NJ), Trypsin and SBTI (Worthington Biochemical Corp., NJ), captopril (Squibb), aprotinin (Bayer), and polyacrylamide (Eastman Kodak, NY) were all analytical grade.

To assess whether changes in cardiac kininogenase were specific, or whether a similar pattern could be observed with a generalized cellular enzyme, lactic dehydrogenase activity (LDH) was measured following standard protocols. Proteins were determined by Bradford's method (10), using albumin as a standard.

Statistics: Two-sided, two sample t tests were used to assay kininogenase activity in rat heart slices. Bonferroni's adjustment for multiple comparisons was employed. Unless otherwise noted, all results are expressed as mean $\pm$ SEM.

RESULTS

The rat heart contained kininogenase activity which was resistant to inhibition by SBTI. Most of the enzyme was present in an inactive form. Only 10% was already active; the rest required activation by trypsin. Cardiac kininogenase activity was inhibited by preincubation with antiglandular kallikrein anti- bodies (Fig 1). In addition, the kininogenase activity of cardiac homogenates was strongly adsorbed to immobilized anti-kallikrein antibodies, requiring 1M Na Cl and pH 3.5 to elute (Fig 2). The molecular weight of the cardiac kininogenase was calculated as 33,000 daltons, based on its elution volume on gel filtration. The optimum pH was 8.5 ± 0.5 (data not shown). Mobility on disc polyacrylamide gel electrophoresis of both cardiac kininogenase and glandular kallikrein was similar (Fig 3).

Both active and trypsin-activatable kininogenase were released into the incubation medium by rat cardiac slices incubated "in vitro". Figure 4 shows that release of total kallikrein (active plus inactive) was continuous over 3 hr incubation (control group). In a second group of rats previously injected with puromycin (10 mg i.p.), the protein synthesis inhibitor almost completely blunted the release of cardiac kininogenase.

mRNA for kallikrein was demonstrated by the polymerase chain reaction (PCR) using a kallikrein-specific probe (Fig 5). A hybridization band of the predicted sized was obtained from ventricular but not atrial mRNA, suggesting that kallikrein is synthesized in the ventricles.

Kininogen content in cardiac tissue was 323 ± 27 pg Bk/mg (n = 7). Kininogen released into the medium after 2 hr incubation was 3 times higher than that in non-incubated slices: 989 ± 58 pg Bk/mg (n = 7) (Fig 6).

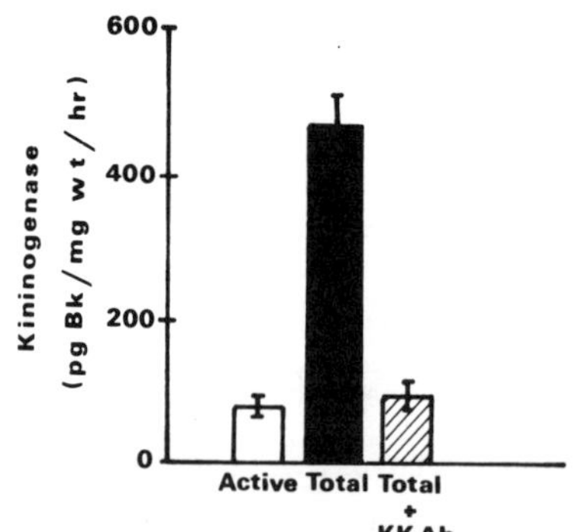

FIGURE 1. Kininogenase activity resistant to inhibition by SBTI in cardiac tissue. Total kininogenase activity was determined after activation of inactive kallikrein with trypsin.
KK Ab = Kallikrein Antibodies

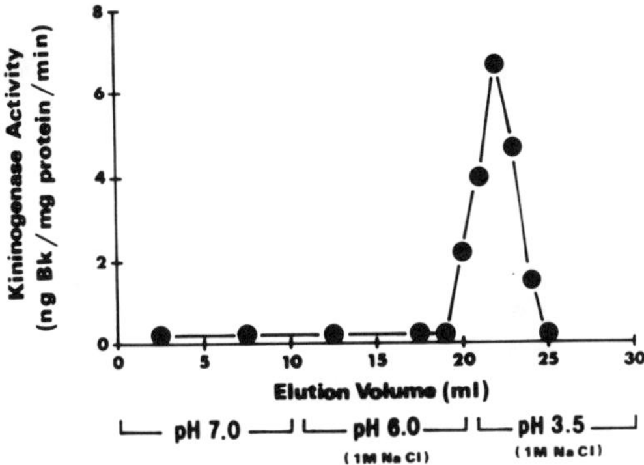

FIGURE 2. Affinity chromatography of cardiac extract on Antikallikrein-CH-Sepharose. The cardiac kininogenase was eluted with 0.1M sodium acetate buffer, 1M NaCl (pH 3.5).

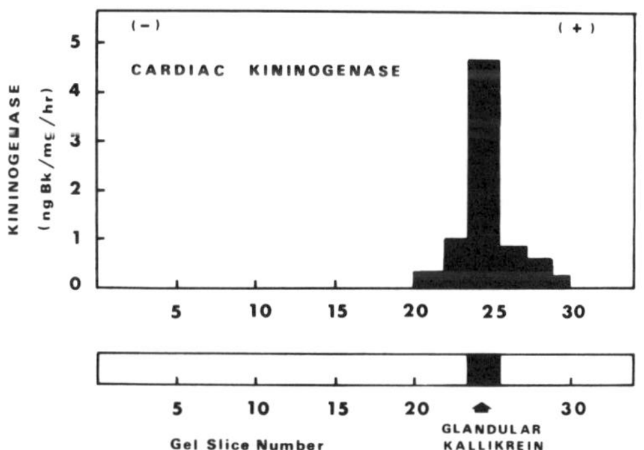

FIGURE 3. Alkaline disc gel electrophoresis of the purified cardiac kininogenase and of glandular kallikrein. Each slice (2 mm) was homogenized and centrifuged; the kininogenase activity was measured in the supernatant.

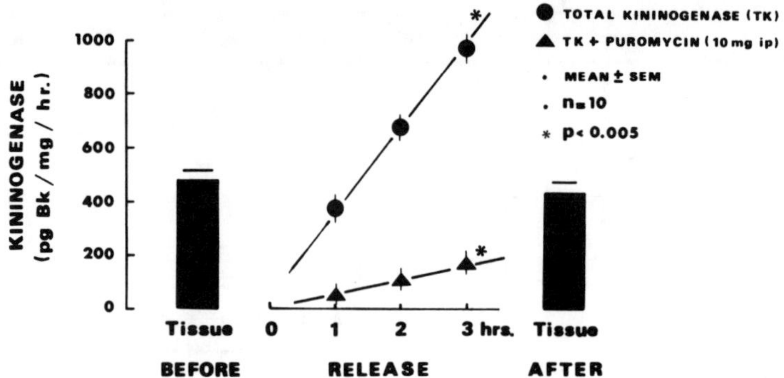

FIGURE 4. Release of kallikrein into the incubation buffer by isolated rat cardiac slices. Kallikrein content in the slices before incubation and after 3 hr incubation. Difference were not significant.

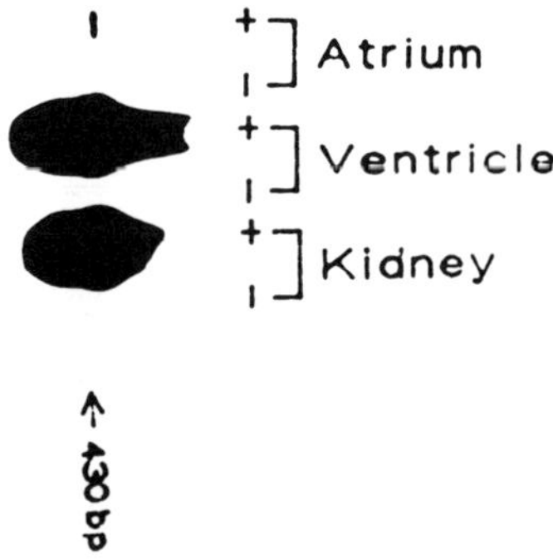

FIGURE 5. Hybridization of the kallikrein cDNA probe to a southern blot of polymerase chain reaction (PCR)-amplified products obtained using the kallikrein gene family primers from 1 μg atrium, ventricle and kidney. Size of the PCR-amplified fragments was estimated to be 430 bp.

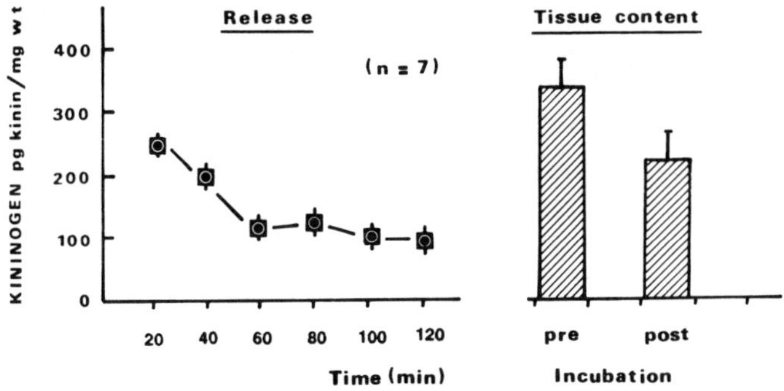

FIGURE 6: Release of kininogen into the incubation buffer by
isolated rat cardiac slices. Kininogen content in the slices
before incubation and after 2 hr incubation.

DISCUSSION

The present data demonstrate that a serine protease having the characteristics
of glandular kallikrein is present in rat cardiac tissue.   The cardiac kininogenase and purified glandular kallikrein have similar molecular weight,
electrophoretic mobility and optimum pH and identical resistance to inhibition
by SBTI. The cardiac kininogenase is strongly attached to immobilized polyclonal antibodies raised against rat urinary kallikrein (6). In addition
its enzymatic activity is blocked by the same antibodies, indicating that it is
either kallikrein or a closely related enzyme. We also found that the mRNA
which codes for glandular kallikrein is present in the ventricles. Thus the
presence of both the enzyme and its mRNA suggests that the heart synthesizes
kallikrein.

When epicardial slices were incubated "in vitro", kallikrein was released into the medium in concentrations too high to be the result of nonspecific release due to damage, as indicated also by the lack of association between kallikrein and LDH activity. This data suggest that cardiac tissue synthesizes and releases kallikrein. This is also suggested by the fact that kallikrein release is blocked by pretreatment with puromycin, a well-known protein synthesis inhibitor.

In addition to kallikrein, we found preliminary evidence that the heart contains and releases kininogen, the kinin precursor. The amount of kininogen released into the medium after 2 hr incubation was almost 3 times higher than the present in non-incubated tissue. A bimodal pattern was seen, with an early high-release phase followed by slow continuous release. Under our experimental conditions, we cannot distinguish between plasma (or lymph) contamination and tissue kininogen. Thus the initial phase may be due to washout of contaminating plasma kininogen, while the slow phase may be due to either kininogen synthesized and released by cardiac tissue or washout of kininogen trapped in a slowly exchanging compartment. The kininogen content of cardiac tissue was slightly higher before 2 hr incubation than after ward ($p < 0.05$). The difference may indicate the amount of kininogen trapped in the plasma and washed out from the tissue during the incubation period. However, it has been reported that mRNA for kininogen is present in endothelial cells, while Oza et al reported that vascular smooth muscle cells in culture release kininogen (13). These data as well as our own suggest that the heart synthesizes both kallikrein and kininogen.

The concentration of active kallikrein in tissue is very small. Most of the enzyme is present in an active form, susceptible to trypsin activation. Thus clarification of the functional significance of our findings requires further work, since a precondition for a cardiac kallikrein-kinin system is released during the "in vitro" studies. It may be that activation occurs during release or in the extracellular space. Since kinins are potents coronary vasodilators, the presence of active cardiac kallikrein as well as kininogen would strengthen the possibility that locally released kinins help regulate coronary blood flow. Kinins have been shown to reduce the incidence and duration of ventricular fibrillation following ischemic reperfusion (2). Using the same model, it has been reported that angiotensin-converting enzyme inhibitors (ACEi) also protect against ventricular fibrillation, and that this protective effect is abolished by a kinin antagonist (14). Kinin antagonists have also been reported to reverse the protective effects of ACEi on experimental cardiac infarction. An important issue is the origin of the kinins that mediate the effects of ACEi. The present results suggest they are due to the glandular kallikrein-kinin system present in the heart.

In summary, we have found evidence that the heart contains a local kallikrein-kinin pathway. This extends our previous observations (3 , 15) and confirms the report of Xiong et al (16). This cardiac kallikrein-kinin system may play an important role in pathological conditions such as ischemic heart disease and/or regulation of cardiovascular homeostasis during treatment with ACE inhibitors.

REFERENCES

1. Linz W, Schölkens BA, Han YF. Beneficial effects of the converting enzyme inhibitor ramipril, in ischemic rat hearts. J Cardiovasc Pharmacol 1986; 8 (suppl 10):S91-9.

2. Linz W, Schölkens BA. Influence of local converting enzyme inhibition on angiotensin and bradykinin effects in ischemic rat hearts. J Cardiovasc Pharmacol. 1987; 10 (suppl 7):S75-82.

3. Britos J, Nolly H. Kinin-forming enzyme of rat cardiac tissue. Hypertension 1981; 3 (suppl II):42-45.

4. Nolly H, Carretero OA, Scicli G, Madeddu P and Scicli AG. A kallikrein-like enzyme in blood vessels of one-kidney, one clip hypertensive rats. Hypertension 1990; 16:436-440.

5. Carretero OA, Oza NB, Piwonska A, Ocholik T, Scicli AG.   Measurements of urinary kallikrein activity by kinin radioimmunoassay. Biochem Pharmacol 1976; 25: 2265-2270.

6. Oza NB, Amin VM, McGregor RK, Scicli AG, Carretero OA. Isolation of rat urinary kallikrein and properties of its antibodies. Biochem Pharmacol 1976; 25: 1607-1612.

7. Rabito SF, Scicli AG, Kher V, Carretero OA. Imnumoreactive glandular kallikrein in rat plasma: A radioimmunoassay for its determination. Am J Physiol 1982; 242: H602-H610.

8. Nolly HL, Scicli AG, Scicli G, Carretero OA. Characterization of a kininogenase from rat vascular tissue resembling tissue kallikrein. Circ Res 1985; 56: 817-821.

9. Nustad K, Pierce JV. Purification of rat urinary kallikrein and their specific antibody. Biochemistry 1974; 13: 2312-2319.

10. Bradford MM. A rapid and sensitive method for the quantification of microgram quantities of protein utilizing the principle of protein-dye binding. Anal Biochem 1976; 72: 248-254.

11. Saed GM, Carretero OA, Mac Donald RJ and Scicli AG. Kallikrein messenger RNA in rat arteries and veins. Circulation Research 1990; 67: 510-516.

12. Diniz CR and Carvalho IF. Micromethod for the determination of bradykinin-ogen under several conditions. Ann N Y Acad Sci 1963; 104: 77-89.

13. Oza NB, Schwartz JH, Goud HD, Levinsky NG. Rat aortic smooth muscle cells inculture express kallikrein, kininogen and bradykininase activity. J Clin Invest 1990; 85: 597-600.

14. Linz W, Martorana PA and Schölkens BA. Local inhibition of bradykinin degradation in ischemic hearts. J Cardiov Pharmacol 1990; 15(suppl 6): S99-S109.

15. Nolly H, De Vito E, Cabrera R, Koninckx. Kinin-releasing enzyme in cardiac tissue. In Hypertension (Villareal, H., ed) 1981; pp 75-84.

16. Xiong W, Chen LM, Woodley-Miller C, Simson JAV and Chao J. Identification, purification and localization of tissue kallikrein in rat heart. Biochem J 1990; 267, 639-646.

LACK OF SPECIFIC BINDING OF BRADYKININ AND D-ARG [HYP^3, THI^5, D-TIC, OIC^8] BRADYKININ IN MEMBRANES FROM RAT HEART

Udo Albus*, Xiaoxing Gao and Lowell M. Greenbaum

Dept. Pharmacol., Medical College of Georgia, Augusta, Ga. 30912, U.S.A., and *Hoechst AG, Postfach 800 320, 6230 Frankfurt/M. 80, Germany

SUMMARY: Recent data demonstrated effects of bradykinin (BK) in the isolated perfused rat heart which could be blocked by BK antagonists. However, so far BK receptors in heart tissue have not been characterized. To search for BK receptors in rat hearts iodinated D-Arg[Hyp^3,Thi^5,D-Tic,Oic^8]BK (Hoe 140) was used as a ligand for binding due to its high affinity and to its resistance to degradation. The labeled antagonist bound well to membranes prepared from rat ileum and could be displaced by the unlabeled antagonist as well as by BK and T-kinin in a concentration dependent manner. However, there was no specific binding detectable when membranes from rat left cardiac ventricle were used. To assure integrity of at least one other receptor in these membranes, binding of ^{3}H-methylscopolamine and its displacement by the respective cold ligand was demonstrated. To rule out an occupation of the BK receptors by endogenous formed BK, the tissue was treated with an acidic buffer with high ionic strength to remove surface bound BK. However, this treatment did not unmask any specific binding for the BK antagonist. In view of the possibility that the structure of the labeled antagonist prevented its binding, experiments were carried out using tritiated BK itself. These experiments also failed to demonstrate specific binding sites which indicate that the structural aspects of Hoe 140 are probably not interfering in binding of the antagonist, especially since it binds to ileum. It is concluded, that there are too few binding sites for BK in the rat heart homogenate for ordinary binding studies. Autoradiographic studies are suggested to elucidate whether the effects of BK in rat hearts are exclusively mediated by receptors on the myocardial vascular tissue.

INTRODUCTION

Earlier findings demonstrated that in experimental studies, angiotensin-converting enzyme (ACE) inhibitors were able to limit infarct size in dogs (1), and in isolated rat hearts with postischemic insults they attenuated the incidence and duration of reperfusion arrhythmias (2).

Later it was shown that the effect of the ACE inhibitors on the heart was probably the result of reduced degradation of bradykinin (BK) rather than angiotensin I (Ang I) conversion. Infusion of BK in isolated rat heart with regional ischemia produced virtually the same effects as ACE inhibitors, whereby very low concentrations of BK produced effects without influencing coronary flow (3). BK-induced vasodilation in the rat hearts is mediated through activation of kinin B_2 receptors as reported by another group (4).

The present study was undertaken to determine the the nature of cardiac BK receptors in terms of their binding properties useing the iodinated BK B_2 receptor antagonist D-Arg[Hyp^3,Thi^5,D-Tic,Oic^8]BK (Hoe 140) to search for BK receptors in rat hearts because of its high affinity and resistance to degradation (5, 6).

METHODS

The BK receptor antagonist Hoe 140 was synthesized in the Dept. of Pharmasynthesis of the Hoechst AG, Frankfurt, Germany. Iodination of the compound by the method of Bolton and Hunter was performed in the radiochemical laboratory of the Hoechst AG.

Cardiac membranes were prepared according to Manning et al. (7) with some modifications from rat hearts of female Spraque-Dawley rats weighing about 200 g. The left ventricles were prepared and homogenized in 25 mmol/l TES buffer, pH 6.8, containing 1 mmol/l 1,10-phenanthroline using a Brinkman Polytron PT-10 (Brinkmann Instuments, Inc. Westbury NY) at the setting 11 for 3 times 10 seconds. The homogenate was centrifuged for 15 min at 1,500 x g. The pellet was then resuspended and centrifuged another and two times at 50,000 x g. The final pellet was resuspended in assay buffer (25 mmol/l TES, pH 6.8 containing 1 mmol/l 1,10-phenanthroline, 140 μg/ml bacitracin, 1 μmol/l ramipril, and 0.1 μmol/l PMSF) and stored at -70° C. Protein was determined using BCA protein assay reagent (Pierce, Rockford, Il).

Binding experiments were performed in assay buffer containing 0.1 % BSA in a volume of 0.5 ml in the case of ^{125}I-Hoe 140 with 20 pmol/l tracer, if not otherwise stated, and 1 ml in the case of ^{3}H-BK with 0.8 nmol/l tracer at 4° C for 90 min. At the end of the incubation time, ice-cold washing buffer (TES 25 mmol/l, pH 6.8) was added and the total volume was filtered through Whatman GF/B filters, pretreated with 0.1 % polyethylenimine. The filter was washed four additional times with 4 ml washing buffer. Filter bound radioactivity was determined in a Packard gamma counter or in a Beckmann liquid scintillation counter, respectively.

Binding of ^{3}H-methylscopolamine (MS) was carried out in Tris/HCl, pH 7.4, with 2 mmol/l $MgCl_2$ was used. The incubation of 10^{-9} mol/l ^{3}H-MS with the cardiac membranes took place in a volume of 1 ml at room temperature. After one hour the suspension was passed through Whatman GF/B filters and the radioactivity retained on the filters was determined. The results are presented in mean $\pm$ S.D.

RESULTS

Iodination of Hoe 140 did not destroy the binding ability of the compound since it bound to ileum membranes and could be displaced by unlabeled Hoe 140 as well as by BK and T-kinin (fig 1). However, when the same method was applied to membranes prepared from rat hearts, binding of ^{125}I-Hoe 140 did increase with the amount of added tracer, but there was no significant displacement in the presence of 1 μmol/l unlabeled Hoe 140. A significant displacement from 101 ± 10.5 fmol/mg prot. to 45 ± 2.5 fmol/mg prot. was only achieved at a concentration as high as 10 μmol/l of the unlabeled compound.

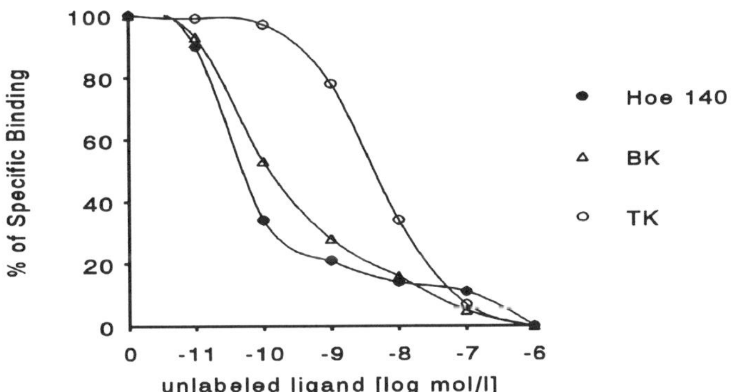

Fig. 1. Displacement of ^{125}I-Hoe 140 from ileum membranes by unlabeled Hoe 140, bradykinin and T-kinin. Unspecific binding had been estimated in the presence of 1 μmol/l unlabeled Hoe 140.

Endogenous BK, liberated during tissue preparation, could have affected the binding experiments. However, pretreatment of the membranes by a buffer containing 200 mmol/l acetic acid, pH 3, and 500 mmol/l sodium chloride, which had been shown to remove all surface bound BK (8), lowered the overall binding but did not change the pattern of displacement by unlabeled Hoe 140.

In order to rule out the possibility that the process of iodination of Hoe 140 might have affected its ability to bind to a cardiac receptor, ^{3}H-BK was used to study binding. Tritiated

BK bound well to ileum membranes in a displaceable manner (fig. 2), but when incubated with cardiac membranes only 7.3±0.3 fmol/mg protein bound, even when high concentrations

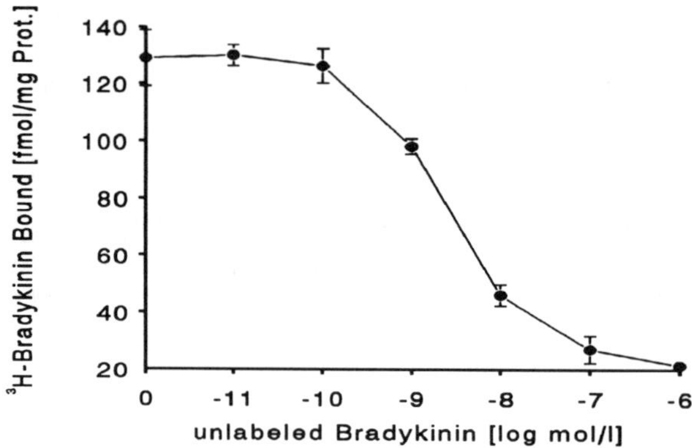

Fig. 2. Binding of ^{3}H-bradykinin to ileum membranes in the presence of increasing concentrations of unlabeled bradykinin.

of membrane protein were introduced. In the presence of 1 μmol/l unlabeled Hoe 140 the binding of ^{3}H-BK was 5.3 $\pm$ 0.7 fmol/mg protein.

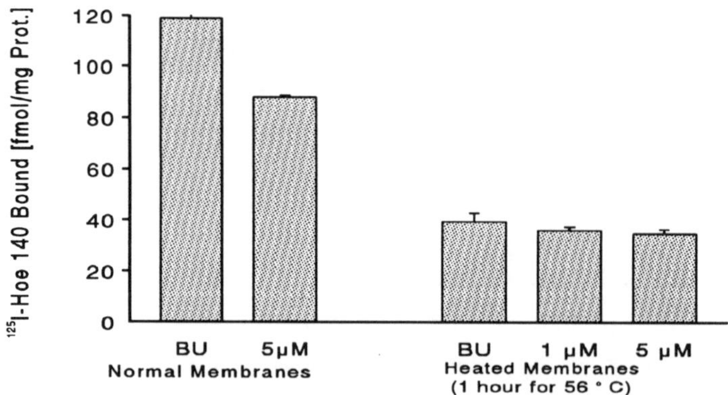

Fig. 3. Binding of ^{125}I-Hoe 140 to heat inactivated cardiac membranes in the absence (BU) or presence of 1 or 5 μmol/l unlabeled Hoe 140. Cardiac membranes were incubated for one hour on ice or at 56° C before incubation with ^{125}I-Hoe 140.

Surprisingly, total binding of ^{125}I-Hoe 140 to membranes from rat heart and rat ileum as well as ^{3}H-BK binding to ileum were about in the same range, whereas binding of ^{3}H-BK binding to rat heart membranes was much lower.

Inactivation of the membranes should not affect unspecific binding, however when cardiac membranes were exposed to 56° C for one hour, total binding was significantly reduced and could not be further reduced by 5 μmol/l unlabeled compound as it was the case in untreated membranes (fig. 3).

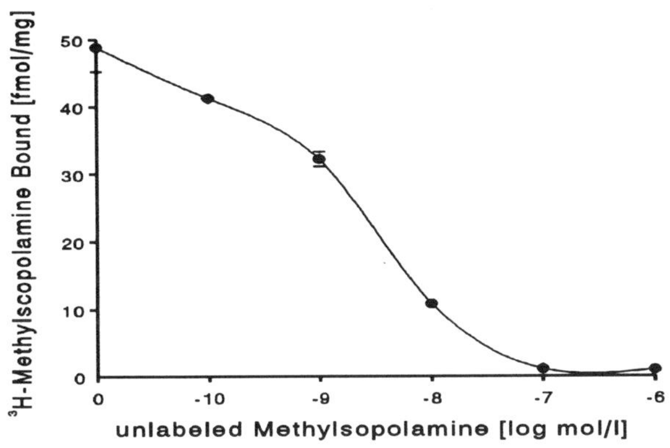

Fig. 4. Binding of ^{3}H-methylscopolamine in the presence of increasing concentrations of unlabeled methylscopolamine.

To assure that cardiac membranes had not been completely inactivated during the process of preparation, the integrity of the acetylcholine receptor was demonstrated by the binding of ^{3}H-methylscopolamine and the displacement by the unlabeled tracer (fig. 4).

DISCUSSION

This study fails to demonstrate specific binding in cardiac tissue of the rat. However, it had been previously shown that BK exerts effects in the isolated rat heart, which could be blocked by BK-receptor blockers. The reason for such kind of discrepancies are not unrevealed by the present study.

The method itself seems to be not the reason for the lack of specific binding, since specific binding of ^{125}I-Hoe 140 to ileum membranes, an organ with a high density of BK receptors, could be performed and displacement of the tracer was shown with unlabeled Hoe 140, BK and T-kinin.

It is interesting, that total binding of ^{125}I-Hoe 140 to cardiac membranes occurs in about the same quantity as to ileum membranes, however, it was displaceable from ileum with an IC_{50} of 1 nmol/l whereas in cardiac membranes there was only little displacement and only at an excess of 1 μmol/l of unlabeled Hoe 140.

This undisplaceable binding was, however, destroyed by an incubation of one hour at 56° C to a remainder of about 25 %, which is rather unusual for unspecific binding. From the remaining binding no further ligand could be displaced by an excess of unlabeled Hoe 140. It is not clear which structures are responsible for this heat sensitive binding. Binding to ACE is excluded due to the presence of the converting enzym inhibitor ramiprilat and further it should also be displaceable by unlabeled Hoe 140.

At 4° C dissociation of ^{3}H-BK from its receptor is very slow with a half live from 90 min in various mammalian tissues (9) to 180 min in rat-13-cells (8). Furthermore, BK is easily released when tisssue is damadged or comes into contact with foreign surfaces. Therefore it is assumeable that tissue preparation leads to a release of kinins which could block a large part of the receptors for some time when the tissue is kept on ice. However, when cardiac tissue in our study was incubated with acetic acid and a high salt concentration to remove all surface bound BK accordig to Roberts and Gullick (8) specific binding was not increased.

To rule out the possibility of damadged cardiac membranes due to a different occurence of proteases in cardiac tissue or by physical impact, we proved the ability of the acethylcholine receptor in the same membrane preparation by binding of ^{3}H-methylscopolamine and displacing it with unlabeled compound. This does not rule out the possibility that the BK receptor might be selectively inactivated, but it shows that the membrane is not completely destroyed.

It is conceivable that BK receptor in rat heart might be slightly different than in rat ileum. This could mean that ^{125}I-Hoe 140 binds to ileum and that unlabeled Hoe 140 still blocks the effects of BK in the isolated rat heart, while labeled Hoe 140 might have lost its ability to bind to a cardiac receptor due to the structural changes after the labeleing. This possibility should be ruled out by ^{3}H-BK, since tritium is to small to create a big change in structure. However, using ^{3}H-BK also failed to show specific binding in cardiac membranes. In fact, total binding of ^{3}H-BK was much lower per mg of protein than binding of ^{125}I-Hoe 140.

After all, the effects of BK in the isolated rat heart prove the presence of BK receptors in cardiac tissue. However, it could be speculated that these receptors occur only on endothelial cell of the cardiac vascular system and that all effects are mediated by EDRF and prostacyclin released by the endothelial cell after BK stimulation. In this case the number of receptors in cardiac tissue could be very limited and it could be that there are too few BK receptors in rat heart homogenate for oridinary binding studies.

Autoradiographic studies would be a method of choise to elucidate whether the effects of BK in rat hearts are exclusively mediated by receptors on the myocardial vascular tissue.

REFERENCES

1.   Ertl G, Kloner RA, Alexander RW, Braunwald E. Limitation of experimental infarct size by an angiotensin converting enzyme inhibitor. Circulation 1982; 65:40-48

2.   Linz W, Schölkens BA, Han YF. Beneficial effects of the converting enzyme inhibitor ramipril in ischemic rat hearts: J Cardiovasc Pharmacol 1986; 8(Suppl):S91-S99

3.   Linz W, Martorana PA, Schölkens BA. Local inhibiton of bradykinin degradation in ischemic hearts. J Cardiovasc Pharmacol 1990; 15(Suppl.6):S99-S109

4.   Baydoun AR, Woodward B. Effects of bradykinin in the rat isolated perfused heart: role of kinin receptors and endothelium-derived relaxing factor. Br J Pharmacol 1991; 103:1829-1833

5.   Hock FJ, Wirth K, Albus U, Linz W, Gerhards HJ, Wiemer G, Henke S, Breipohl G, Köng W, Knolle J, Schölkens BA. Hoe 140 a new potent and long acting bradykinin-antagonist: in vitro studies. Br J Pharmacol 1991; 102:769-773

6.   Wirth K, Hock FJ, Albus U, Linz W, Alpermann HG, Anagnostopoulos H, Henke S, Breipohl G, Köng W, Knolle J, Schölkens BA. Hoe 140 a new potent and long acting bradykinin-antagonist: in vitro studies. Br J Pharmacol 1991; 102:774-777

7.   Manning DC, Vavrek R, Stewart JM, Snyder SH. Two bradykinin sites with picomolar affinities. J Pharmacol Exp Therap 1986; 237:504-512

8.   Roberts RA, Gullick WJ. Bradykinin receptor number and sensitivity to ligand of mitogenesis is increased by expression of a mutant ras oncogene. J Cell Sci 1989; 94:527-535

9.   Innis RB, Manning DC, Stewart JM, Snyder SH. [^{3}H]Bradykinin receptor binding in mammalian tissue membranes. Proc Natl Acad Sci USA 1981; 78:2630-2634

10.   Roberts RA, Gullick WJ. Bradykinin receptors undergo ligand-induced desensitization. Biochemistry 1990; 29:1975-1979

AAS 38/III
Recent Progress on Kinins
© 1992 Birkhäuser Verlag Basel

EARLY ACE-INHIBITION IN MYOCRDIAL INFARCTION
Possible role of bradykinin

Wiek H. van Gilst and P.A. de Graeff

Department of Clinical Pharmacology, University of Groningen, Bloemsingel 1,
9713 BZ Groningen, The Netherlands

SUMMARY: Restoration of coronary blood flow in the ischemic myocardium is absolutely needed to prevent irreversible cellular damage but on the other hand may have potentially hazardous consequences. Since thrombolysis during myocardial infarction is designed to salvage a maximal number of myocardial cells threatened by ischemia, a concommitant intervention which reduces cellular damage due to reperfusion will improve the net result of such procedure.

The adjunctive use of ACE-inhibitors with thrombolytic therapy early during acute myocardial infarction offers theoretic advantages. This article summarizes the results indicating that ACE-inhibitors do play an important role in cardioprotection in the acute phase of myocardial ischemia followed by reperfusion. Probably, their effect on bradykinin breakdown is at least partly responsible for this effect.

INTRODUCTION

Restoration of coronary blood flow in the ischemic myocardium is absolutely needed to prevent irreversible cellular damage but on the other hand may have potentially hazardous consequences. This paradox is illustrated by recent progress in methods that improve myocardial oxygen supply to the acutely ischemic myocardium and thus reduce cellular damage (1,2,3) on the one hand and the early demonstration by Tennant and Wiggers that providing such supply by reperfusion of the ischemic myocardium is associated with malignant ventricular arrhythmias largely caused by additional cellular damage (4).

Though the latter finding dates from 50 years ago, elucidation of reperfusion phenomena has only during the last ten years taken on a new significance because of the above mentioned clinical developments, especially the general use of thrombolytic therapy in the setting of acute myocardial infarction (5).

Since this procedure is designed to salvage a maximal number of myocardial cells threatened by ischemia, a concommitant intervention which reduces cellular damage due to reperfusion will improve the net result of such procedure.

Ischemia preceding reperfusion is a dynamic process involving a wide range of cellular events, such as loss of ionic homeostasis, ultrastructural damage, depletion of ATP and creatine kinase, increase in cyclic AMP, increase in lysophosphoglycerides and long-chain acyl carnitines, increase in alpha adrenoreceptors, prostaglandin release, catecholamine release, loss of intracellular components, increase in myocardial lactate, etc, etc. All these events together will set the stage for the occurrence of reperfusion phenomena and, since their progression is dependent on both duration and severity of ischemia, it is also conceivable that the extent of reperfusion damage is related to these two factors. If studied in detail, the curve representing duration of ischemia versus incidence of reperfusion damage appears to be bell shaped. The decline in the occurrence of reperfusion damage is due to the onset of irreversible cellular injury during ischemia (6) and the occurrence is maximal when a maximal number of cells in the ischemic zone is in the late phase of reversible injury (7).

The cascade of events involved in the process of ischemia and subsequent reperfusion, makes it an exceedingly complex process. On the other hand this multifactorial aspect offers the opportunity for a broad range of drugs to affect the process. Nevertheless, the comparative efficacy of various drugs on reperfusion damage, especially in the clinical situation, is not well documented.

Clearly, despite the benefits of thrombolytic therapy, the use of adjuvant therapies, including betablockers, calcium antagonists, converting enzyme inhibitors, etc., must be explored. In this paper we will discuss the experimental evidence that converting enzyme inhibitors may affect reperfusion damage beneficially, special attention will be paid to the role of bradykinin.

CARDIOPROTECTION BY CONVERTING ENZYME INHIBITORS

The extent to which the renin-angiotensin system participates in the process of myocardial ischemia and reperfusion is unknown. Recent evidence indicates that the renin-angiotensin system is activated by acute coronary occlusion in vivo (8,9). Since blockade of this system lowers systemic blood pressure and improves cardiac output (8), it may play a role in the extension of ischemic damage of the myocardium (10-12). It has also been shown that converting enzyme inhibitors, at least captopril, interfere with the response to noradrenaline by an angiotensin-II-independent mechanism (13,14) and that these agents may facilitate prostacyclin synthesis (15-17). In isolated rat hearts, it appeared that captopril effectively reduced cellular damage and the incidence of ventricular fibrillation upon reperfusion (18) in a dose dependent manner (19). Furthermore, to a lesser degree, this property was shared by another angiotensin converting enzyme inhibitor, ramipril. The effects of ramipril were limited to the active enzyme inhibiting form and were not found for the inactive prodrug. The effects of captopril were associated with an abolishment of noradrenaline overflow upon reperfusion. Simultaneous administration of indomethacin attenuated the effects of captopril. Several other investigators did show cardioprotective effects of enalaprilat in vivo (20-22).

These beneficial effects of ACE-inhibition in vitro were confirmed in vivo. In the closed-chest pig model, captopril reduced purine and catecholamine overflow in the coronary sinus (23-24). The incidence and duration of reperfusion arrhythmias (mainly idioventricular rhythm) did not differ between the captopril-treated and control groups. Because mean arterial pressure and cardiac output were comparable between the treated and control groups, a direct action of captopril on the heart was suggested, probably by potentiation of (locally generated) bradykinin and its vasodilatory effect (23-24). Other investigators were able to confirm these cardioprotective effects using different animal models (20,21,54-57).

Several mechanisms may be responsible for the observed effects. As mentioned above, captopril has been shown to possess direct antiadrenergic properties (13,14). According to the findings of Sheridan and colleagues (25), blockade of alpha-adrenergic receptors, which are increased in number during ischemia (26) will lead to a reduced incidence of reperfusion phenomena. However, our results show that the reduction in reperfusion damage is associated with a reduced noradrenaline overflow. This suggest that captopril is active at the presynaptic site, which is difficult to explain in terms of a blockade of adrenergic receptors.

Interestingly, certain endogenous substances, such as bradykinin and prostaglandins, have also been identified as potentially cardioprotective. Preservation of adrenergic nerve endings during ischemia has been demonstrated in the presence of increased prostacyclin concentrations (27). Because converting enzyme inhibitors are known to prevent bradykinin breakdown by inhibition of kininase II (= angiotensin converting enzyme) and, subsequently, are able to facilitate prostaglandin synthesis, it has been suggested that these compounds exert their cardioprotective effects by increasing (local) bradykinin and prostaglandin levels during ischaemia and reperfusion.

BRADYKININ

Bradykinin is a vasoactive compound that is able to decrease vascular tone of coronary arteries and other blood vessels (28-30). Several investigators have shown that this decrease in coronary vascular tone is associated with a release of prostaglandins (28,31,32). This prostaglandin-mediated vasodilation of bradykinin can be blocked by cyclooxygenase inhibitors (31-33). It has been demonstrated that the major prostaglandin responsible for the vasodilatory effects of bradykinin is prostacyclin (34). However, other investigators have failed to show a prostacyclin-dependent vasodilatory mechanism (35). Therefore, an additional prostacyclin-independent mechanism may also be involved in this bradykinin-induced coronary vasodilation. Later, it was shown that bradykinin can stimulate the release of endothelium derived relaxing factor (EDRF), a vasodilatory compound which is probably nitric oxide (30). Under most experimental conditions, prostacyclin and EDRF are released simultaneously, and the bradykinin receptor-mediated release of these two autacoids seems to be tightly coupled (36).

Little is known about the cardioprotective properties of bradykinin. In the isolated ischaemic working rat heart, bradykinin reduced the incidence and duration of reperfusion-induced ventricular arrhythmias (37). There also is some indirect evidence of a possible cardioprotective effect of bradykinin (38,39). The underlying mechanism of this action is still unknown; potentiation of EDRF and prostacyclin production by bradykinin may be involved, but more studies are needed to confirm this hypothesis.

It is known that prostaglandins, which can affect coronary perfusion and myocardial function, are generated in the heart itself. This production has been demonstrated in cardiac muscle cells, the interstitium and coronary vessels (40,41). The physiological importance of prostaglandins and other arachidonic acid metabolites synthesized in the heart is not yet entirely clear. One should realize that these compounds generated elsewhere in the body, especially in blood platelets, leucocytes, the kidneys and the lungs, are also able to influence cardiac performance. The coronary vasodilating prostaglandins, PGI_2 (=prostacyclin) and PGE_1, show cardioprotective effects in ischaemic hearts, both in vivo and in vitro (40,42). The same results are observed after administration of iloprost, a stable PGI_2 analogue (42-45). The mechanism of this protective effect is unknown, but several factors seem to be involved (40-42). The following mechanisms have been mentioned in the literature:(1) increase in collateral blood flow to the ischaemic myocardium;(2) reduction in afterload and oxygen demand of the ischaemic heart;(3) inhibition of platelet aggregation;(4) inhibition of intracellular lysosome disruption;(5) inhibition of myocardial catecholamine release from the ischaemic myocardium;(6) preservation of free radical systems (e.g. superoxide dismutase), and finally (7) inhibition of neutrophil activitation.

PROSTAGLANDINS

Some authors reported an increased production of PGI_2 by different blood vessels after administration of captopril, but others failed to detect such a stimulatory action (46). Moreover, it is unknown whether this stimulation of PGI_2 synthesis by captopril is caused by a direct or indirect action on vascular cells. Because bradykinin is known to be a strong stimulator of PGI_2 production, this effect of captopril might be mediated by prevention of bradykinin breakdown (47-50). Also, the production of PGE_2 is stimulated by captopril and its stereoisomer S,R-captopril, both in vitro and in vivo (31,50,51). On the other hand, enalaprilat failed to stimulate PGE_2 synthesis in vitro (51). The stimulatory action of captopril and its stereoisomer appeared to be dependent on the presence of their sulfhydryl group. This difference between captopril and enalaprilat has also been shown in man (50,52). During administration of teprotide in hypertensive patients, an elevation of PGE concentration was observed without changes in angiotensin II and bradykinin levels indicating a direct action (53). In the kidney, kinin-mediated prostaglandin production was potentiated by captopril (48). Furthermore, captopril and bradykinin had more than additive effects on arachidonic acid release and PGE_2 synthesis in cultured renomedullary interstitial cells, suggesting that these compounds are synergistic. All these studies indicate that ACE-inhibitors can stimulate prostaglandin synthesis, both directly by an unknown mechanism and indirectly by preventing bradykinin breakdown.

CONCLUSION

The injury due to reperfusion is multifactorial but is currently largely ascribed to massive accumulation of calcium in ischemic myocardium and the release of oxygen free radicals. Since ischemic myocardium cannot survive without reperfusion, a strategy to maximize myocardial salvage must include interventions to decrease injury during both ischemia and reperfusion. Various pharmacologic agents have shown promise in animal models, and we are now in need of clinical trials to confirm their benefits and demonstrate the optimal thrombolytic procedure.

The adjunctive use of ACE-inhibitors with thrombolytic therapy early during acute myocardial infarction offers theoretic advantages both in the acute and in the chronic phase. In the early phase, they may blunt the catecholamine response, can elicit coronary vasodilation and stimulate beneficial autocoids such as prostacyclin and bradykinin. Furthermore, in the chronic phase ACE-inhibitors may diminish left ventricular volume, prevent ventricular dilation and improve patient survival. The CATS pilot has shown that at least part of these advantageous effects in the acute phase are demonstrable in the clinical setting of myocardial infarction followed by reperfusion (66-68).

At present, a large number of controlled clinical trials mainly focusing on the effects of ACE-inhibition in the chronic phase is underway. Only few studies concentrate on the effect of acute intervention with ACE inhibitors in ischemia-reperfusion i.e. thrombolysis in myocardial infarction.

In April 1990, a large nationwide acute intervention trial with captopril in 280 patients receiving thrombolytic therapy was started in the Netherlands, the Captopril and Thrombolysis Study (CATS). The primary hypothesis of CATS supposes a very early effect of converting enzyme inhibition on evolving myocardial damage due to ischemia and the consequences of early reperfusion. This will be evaluated by serial echocardiography, Holtermonitoring and neurohumoral measurements immediately upon thrombolysis and during the first year after myocardial infarction. The results of the CATStudy will provide further information on the role of ACE-inhition as adjunctive therapy to thrombolysis and clinical relevance of the above discussed mechanisms of action.

REFERENCES

1   Rentrop P, Blanke H, Karsch KR, Kaiser H, Kostering H, Leitz K (1981) Selective intracoronary thrombolysis in acute myocardial infarction and unstable angina pectoris. Circulation 63: 307

2   Mathey DG, Kuck KH, Tilsner V, Krebber HJ, Bleifeld W (1981) Nonsurgical coronary artery recanalization in acute transmural myocardial infarction. Circulation 63: 489

3   Markis JE, Malagold M, Parker JA, Silverman KJ, Barry WH, Als AV, Paulen S, Grossman W, Braunwald E (1981) Myocardial salvage after intracoronary thrombolysis with streptokinase in acute myocardial infarction, assessment with intracoronary thallium-201. N Eng J Med 305: 777

4   Tennant R, Wiggers CJ (1935) The effect of coronary occlusion on myocardial contraction. Am J Physiol 112: 351-361

5   Goldberg S, Greenspon AJ, Urban PL, Muza B, Berger B, Walinsky P, Maroko PR (1983) Reperfusion arrhythmia: A marker of restoration of antegrade flow during intracoronary thrombolysis for acute myocardial infarction. Am Heart J 105: 26

6   Manning AS, Hearse DJ (1984) Reperfusion-induced arrhythmias: mechanisms and prevention. J Mol Cell Cardiol 16: 497-518

7   Hearse DJ (1984) Critical distinctions in the modification of myocardial cell injury In: Opie LH (ed) Calcium Antagonists and cardiovascular disease. New York: Raven Press

8   Liang C, Gavras H, Black J, Sherman LG, Hood WB (1982) Renin- angiotensin system inhibition in acute myocardial infarction in dogs. Circulation 66: 1249-1255

9   Ertl G, Alexander RW, Kloner RA (1983) Interactions between coronary occlusion and the renin-angiotensin system in the dog. Bas Res Cardiol 78: 518

10   Hock CE, Ribeiro LGT, Lefer AM (1985) Preservation of ischemic myocardium by a new converting enzyme inhibitor, enalaprilic acid, in acute myocardial infarction. Am Heart J 109: 222-228

11   Lefer AM, Peck RC (1984) Cardioprotective effects of enalapril in acute myocardial ischemia. Pharmacology 29: 61

12   Ertl G, Kloner RA, Alexander RW, Braunwald E (1982) Limitation of experimental infarct size by an angiotensin-converting enzyme inhibitor. Circulation 65: 40

13   Clough DP, Collis MG, Conway J, Hatton R, Keddie JR (1982) Interaction of angiotensin-converting enzyme inhibitors with the function of the sympathetic nervous system. Am J Cardiol 49: 1410-1414

14   Saruta T, Suzuki H, Okundo T, Kondo K (1982) Effects of angiotensin-converting enzyme inhibitors on the vascular response to norepinephrine. Am J Cardiol 49: 1535-1536

15   Schwartz PJ, Stone HL (1980) Left stellectomy in the prevention of ventricular fibrillation caused by acute myocardial ischemia in conscious dogs with anterior infarction. Circulation 62: 1256-1265

16   Dusing R, Scherag R, Landsberg G, Glanzer K, Kramer HJ (1983) The convertingenzyme inhibitor captopril stimulates prostacyclin synthesis by isolated rat aorta. Eur J Pharmacol 91: 501-504

17   Mullane KM, Moncada S (1980) Prostacyclin mediates the potentiated hypotensive effect of bradykinin following captopril treatment. Eur J Pharmacol 66: 355-365

18   Gilst WH van, Graeff PA de, Kingma JH, Wesseling H, Langen CDJ de (1984) Captopril reduces purine loss and reperfusion arrhythmias in the rat heart after coronary artery occlusion. Eur J Pharmacol 100: 113-117

19   Graeff PA de, Gilst WH van, Kingma JH, Langen CDJ de, Wesseling H (1986) Effects of captopril in a closed-chest pig model against ischemia-reperfusion injury. N-S Archiv Pharmacol 280: 181-193

20   Lefer AM, Peck RC. Cardioprotective effects of enalapril in acute myocardial ischemia. Pharmacology 1984; 29: 61-69.

21   Hock CE, Ribeiro LGT, Lefer AM. Preservation of ischemic myocardium by a new converting enzyme inhibitor, enalaprilic acid, in acute myocardial infarction. Am Heart J 1985; 109: 222-228.

22   Li K, Chen X. Protective effect of captopril and enalapril on myocardial ischemia and reperfusion damage of rat. J Mol Cell Cardiol 1987; 19: 909-915.

23   Graeff PA de, Gilst WH van, Bel K, Langen CDJ de, Kingma JH, Wesseling H. Concentration-dependent protection by captopril against myocardial damage during ischemia and reperfusion in a closed-chest pig model. J Cardiovasc Pharmacol 1987; 9(suppl 2): S37-S42.

24   Graeff PA de, Langen CDJ de, Gilst WH van et al. Protective effects of captopril against ischemia/reperfusion-induced ventricular arrhythmias in vitro and in vivo. Am J Med 1988; 84(suppl 3A): 67-74.

25   Sheridan DJ, Penkoske PA, Sobel BE, Corr PB (1980) Alpha-adrenergic contributions to dysrhythmia during myocardial ischaemia and reperfusion in cats. J Clin Invest 65: 161-171

26   Corr PB, Hayman JA, Kramer JB, Kipnis RJ (1981) Increased alpha- adrenergic receptors in ischaemic cat myocardium - a potential mediator of electrophysiological derangements. J Clin Invest 67: 1232-1236

27   Schror K, Funke K (1985) Prostaglandins and myocardial noradrenaline overflow after sympathetic nerve stimulation during ischemia and reperfusion. J Cardiovasc Pharmacol 7 (suppl. 5): S50- S54

28   Needleman P, Marshall GR, Sobel BE. Hormone interactions in the isolated rabbit heart. Synthesis and coronary vasomotor effects of prostaglandins, angiotensin, and bradykinin. Circ Res 1975; 37: 802-8.

29   Young MA, Vatner SF. Regulation of large coronary arteries. Circ Res 1986; 59: 579-96.

30   Furchgott RF. Role of endothelium in responses of vascular smooth muscle. Circ Res 1983; 53: 557-73.

31   Moore TJ, Crantz FR, Hollenberg NK et al. Contribution of prostaglandins to the antihypertensive action of captopril in essential hypertension. Hypertension 1981; 3: 168-73.

32   Swartz SL, Williams GH. Angiotensin-converting enzyme inhibition and prostaglandins. Am J Cardiol 1982; 49: 1405-09.

33   Türker RK, Ercan ZS, Ersoy A, Zengil H. Inhibition by nicotine of the vasodilator effect of bradykinin: evidence for a prostacyclin-dependent mechanism. Arch Int Pharmacodyn Ther 1982; 257: 94-103.

34   Needleman P, Kaley G. Cardiac and coronary prostaglandin synthesis and function. N Engl J Med 1978; 298: 1122-28.

35   Schrör K, Metz U, Krebs R. The bradykinin-induced coronary vasodilation. Evidence for an additional prostacyclin-independent mechanism. Naunyn-Schmiedeberg's Arch Pharmacol 1979; 307: 213-21.

36   Nucci G de, Gryglewski RJ, Warner TD, Vane JR. Receptor-mediated release of endothelium derived relaxing factor and prostacyclin from bovine aortic endothelial cells is coupled. Proc Natl Acad Sci USA 1988; 85: 2334-38.

37   Linz W, Schölkens BA, Han YF. Beneficial effects of the converting enzyme inhibitor, ramipril, in ischemic rat hearts. J Cardiovasc Pharmacol 1986; 8(suppl 10): S91-9.

38   Schölkens BA, Linz W, König W. Effects of the angiotensin converting enzyme inhibitor, ramipril, in isolated ischaemic rat heart are abolished by a bradykinin antagonist. J Hypertension 1988; 6(suppl 4): S25-8.

39   Schölkens BA, Linz W, König W. Role of bradykinin in the cardiac action of the converting enzyme inhibitor ramipril. Eur Heart J 1988; 9(Suppl A): P398.

40   Schrör K. Actions of prostaglandins on the heart. In: Prostacyclin and its stable analogue iloprost. Gryglewski RJ, Stock G (eds). Springer-Verlag, Berlin Heidelberg 1987; 159-178.

41   Mehta J, Mehta P. Prostacyclin and thromboxane A2 production by human cardiac atrial tissues. Am Heart J 1985; 109: 1-3.

42   Simpson PJ, Lucchesi BR. Myocardial ischemia: the potential therapeutic role of prostacyclin and its analogues. In: Prostacyclin and its stable analogue iloprost. Gryglewski RJ, Stock G (eds). Springer-Verlag, Berlin Heidelberg 1987; 179-194.

43   Gilst WH van, Boonstra PW, Terpstra JA, Wildevuur ChRH, Langen CDJ de. Improved functional recovery of the isolated rat heart after 24 hours of hypothermic arrest with a stable prostacyclin analogue (ZK 36 374). J Mol Cell Cardiol 1983; 15: 789-792.

44   Gilst WH van, Boonstra PW, Terpstra JA, Wildevuur ChRH, Langen CDJ de. Improved recovery of cardiac function after 24 h of hypothermic arrest in the isolated rat heart: comparison of a prostacyclin analogue (ZK 36 374) and a calcium entry blocker (diltiazem). J Cardiovasc Pharmacol 985; 7: 520-524.

45   Langen CDJ de, Gilst WH van, Wesseling H. Sustained protection by iloprost of the porcine heart in the acute and chronic phases of myocardial infarction. J Cardiovasc Pharmacol 1985; 7: 924-928.

46   Boeynaems JM. Drugs influencing the vascular production of prostacyclin. Prost Leuk Ess Fatty Acids 1988; 34: 197-204.

47   Heavy D, Barrow SE, Hickling NE, Ritter JM. Aspirin causes short-lived inhibition of bradykinin-stimulated prostacyclin production in man. Nature 1985; 318: 186.

48   Mullane KM, Moncada S. PGI$_2$ mediates the potentiated hypotensive effect of bradykinin following captopril treatment. Eur J Pharmacol 1980; 66: 355-365.

49   Mullane KM, Moncada S, Vane JR. Does prostaglandin release contribute to the hypotension induced by inhibitors of angiotensin converting enzyme? In: Prostaglandins and the Kidney. Dunn MJ, Patrono C, Cinotti GA (eds). New York, Plenum Publishing Co, 1982; 213-233.

50   Swartz SL, Williams GH, Hollenberg NK et al. Increase in prostaglandins during converting enzyme inhibition. Clin Sci 1980; 59(suppl 6): 133s-135s.

51   Zusman RM. Effects of converting-enzyme inhibitors on the renin-angiotensin-aldosterone, bradykinin, and arachidonic-prostaglandin systems: Correlation of chemical structure and biologic activity. Am J Kidney Dis 1987; X(suppl 1): 13-23.

52   Shoback DM, Williams GH, Swartz SL et al. Time course and effect of sodium intake on vascular and hormonal responses to enalapril (MK 421) in normal subjects. J Cardiovasc Pharmacol 1983; 5: 1010-1018.

53   Vinci JM, Horowitz D, Zusman RM et al. The effect of converting enzyme inhibition with SQ 20881 on plasma and urinary kinins, protaglandin E and angiotensin II in hypertensive man. Hypertension 1979; 1: 416-426.

54   Westlin W, Mullane K. Does captopril attenuate reperfusion-induced myocardial dysfunction by scavenging free radicals? Circulation 1988; 77(suppl I): I-30-I-39.

55   Elfallah MS, Ogilvie RI. Effect of vasodilator drugs on coronary occlusion and reperfusion arrhytmias in anesthetized dogs. J Cardiovasc Pharmacol 1985; 7: 826-832.

56   Ertl G, Kloner RA, Alexander W, Braunwald E. Limitation of experimental infarct size by an angiotensin-converting inhibitor. Circulation 1982; 65: 40-48.

57   Danniell HB, Carson RR, Ballard KD, Thomas GR, Privitera PJ. Effects of captopril in limiting infarct size in conscious dogs. J Cardiovasc Pharmacol 1984; 6: 1043-1047.

58   Kingma JH, Gilst WH van, Graeff PA de et al. Captopril as a possible cardioprotective agent during thrombolytic therapy in acute myocardial infarction: Dosage and clinical and neurohumoral effects. In: New frontiers in cardiovascular therapy: Focus on angiotensin converting enzyme inhibition. Sonnenblick EH, Laragh JH, Lesch M (eds). Excerpta Medica, Amsterdam 1989; 277-285.

59   Kingma JH, Gilst WH van, Peels CH, Jaarsma W, Verheugt FW. Converting enzyme inhibition during thrombolytic therapy in acute myocardial infarction. Circulation 1990;4:III-666

68   Kingma JH, Louwerenburg JH, Gilst WH van, Six AJ, Graeff PA de, Wesseling H. Concomitant use of captopril during thrombolytic therapy in acute myocardial infarction. In: New frontiers in cardiovascular therapy: Focus on angiotensin converting enzyme inhibition. Eds: Sonnenblick EH, Laragh J, Lesch M. Exerpta Medica Princeton 1989;277-286.

AAS 38/III
Recent Progress on Kinins
© 1992 Birkhäuser Verlag Basel

ACTIVATION OF KININS ON MYOCARDIAL ISCHEMIA

K. Shimamoto, T. Miura, T. Miki and O. Iimura

The Second Department of Internal Medicine, Sapporo Medical College,
S-1 W-16, Sapporo 060, Japan

SUMMARY: In experimental ischemic dog heart with coronary-constriction, no increase of coronary blood flow, Max dP/dt or MVO_2 with no change of kinin in arterial blood were exhibited. Following sympathetic nerve stimulation, remarkable increases of kinin in coronary sinus blood were observed with a significant elevation of left ventricular end-diastolic pressure and an augmented production of lactate from the heart as well as an ischemic change of ECG-ST. Infusion of kinin into the left main coronary artery resulted in no change in the mean systemic blood pressure, coronary blood flow, coronary vascular resistance, cardiac function, myocardial metabolism or ECG-ST in the control and coronary-constricted groups. These data suggest that kinin was released significantly from the ischemic heart, however, such a level of kinin has no significant effect on coronary circulation or myocardial metabolism.

In ischemia-reperfusion rabbit hearts, no significant influence of the ACE inhibitors, captopril and ramiprilat, were observed. Species differences may be responsible for the beneficial role of ACE inhibitors in the limitation of infarct size in the dog hearts, possessing collateral flow, that are not seen in the rabbit heart with poor collateral flow.

INTRODUCTION

Recently, the cardioprotective action of angiotensin converting enzyme (ACE) inhibitors has been of interest in reperfusion injury, cardiac dysfunction and limitation of experimental myocardial infarct size(1-3). There are only a few studies concerning the possible role of bradykinin in myocardial ischemia. The role of the plasma kallikrein-kinin system in patients with acute myocardial infarction has been investigated (4-11), and decreases in plasma prekallikrein and kininogen were shown in these reports. In our studies (11) and those of Hashimoto et al (10), significant increases in plasma kinin levels in patients with acute myocardial infarction have been observed.

Thus, it has been confirmed that plasma kinin levels increase in patients with acute myocardial infarction. The increase may be caused by increased kinin synthesis, as suggested by the significant decrease in plasma kininogen. The significance or origin of the increased plasma kinin was not clarified in these clinical studies.

In this manuscript, the role of kinin synthesis in ischemic myocardium is discussed from three viewpoints , 1) kinin release from the ischemic dog heart, 2) the effects of intracoronary infused kinin on cardiac function, myocardial metabolism and coronary circulation and 3) the effects of ACE-inhibitors on infarct/risk ratio in ischemia-reperfusion rabbit hearts.

PLASMA KININ IN A DOG MYOCARDIAL ISCHEMIC MODEL

The detailed mechanisms of the kinin increase in the patients with myocardial infaction (10,11) still remain unclear; therefore, the possibility of its release from the ischemic heart was evaluated by using a dog myocardial ischemia model with coronary constriction in our department (12). The left main coronary artery and coronary sinus were cannulated with a Griggs-type autoperfusing cannula in anesthetized, open-chest dogs. Coronary constriction was carried out by graded constriction. The anterior ansa of the left stellate ganglion was stimulated with an electrode to precipitate myocardial ischemia. Blood samples were taken from the femoral artery and coronary sinus. As shown in the upper panel, the average ratio of reduced coronary blood flow after coronary constriction was 40.3% in the constricted group. After sympathetic nerve stimulation, coronary blood flow was further reduced in the constricted group. The left ventricular end-diastolic pressure was not changed by coronary constriction, but increased significantly with additional sympathetic nerve stimulation. In the dog without coronary constriction, coronary blood flow increased and the left ventricular end-diastolic pressure decreased significantly with sympathetic nerve stimulation, suggesting that sympathetic nerve stimulation augments cardiac function in the non-constricted group. These data indicate that the coronary constriction did not induce myocardial ischemia, but additional sympathetic nerve stimulation induced the ischemic cardiac dysfunction. The arterio-coronary sinus difference of lactate (ΔL) and the myocardial lactate extraction ratio (ΔL:La) are shown in the lower panel (12). After sympathetic nerve stimulation, myocardial lactate release was reduced in the constricted group, as was the myocardial lactate extraction ratio. No significant change was seen in the non-constricted group. These data also suggest that in the constricted group, coronary constriction did not induce the myocardial ischenia, but the combined sympathetic nerve stimulation resulted in the ischemic cardiac dysfunction and the increase in anaerobic mechanism caused by myocardial ischemia.

The effects of sympathetic nerve stimulation on kinin levels in the femoral artery and coronary sinus are shown in Fig 1 (12). Coronary constriction did not produce any ischemic change or change in plasma kinin in either the artery or the coronary sinus. Moreover, after sympathetic nerve stimulation, no significant changes in plasma kinin

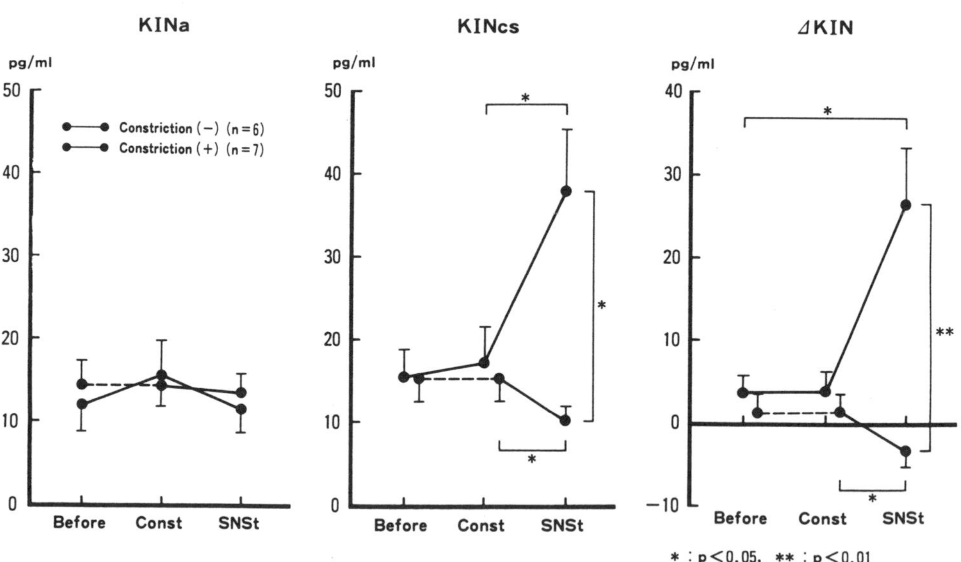

Fig.1 Effects of SNSt on kinin concentration in arterial blood (KINa) and coronary sinus blood (KINcs), and sinu-arterial difference of kinin (ΔKIN) in dogs with coronary constriction

levels in the femoral artery were found in either group. In the constricted group, plasma kinin levels in the coronary sinus increased significantly after sympathetic nerve stimulation (Fig 1). In the non-constricted group, plasma kinin levels in the coronary sinus decreased only slightly (Fig 1). Both the difference in coronary sinus-femoral artery kinin levels (ΔKIN)(Fig 1) and the ratio of ΔKIN/ femoral artery kinin (KINa) represented, for the most part, the same tendency as the coronary sinus kinin levels. These data suggest that there is a pronounced release of plasma kinin from the heart when apparent myocardial ischemia occurs. Regarding the mechanism of plasma kinin increase in this study, it still remains unclear whether an increase in production or a decrease in breakdown contributed to this process.

EFFECTS OF BRADYKININ ON CARDIAC FUNCTION, MYOCARDIAL METABOLISM AND CORONARY CIRCULATION

To clarify the pathophysiological role of increased kinin on myocardial metabolism and coronary circulation, 0.1, 1.0 and 10 ng/kg/min of bradykinin were infused for 5

Table 1. Effects of intracoronary infusion of kinin (10ng/kg/min for 5 minutes) on myocardial metabolism

		MVO$_2$	ΔL	ΔL/La	L/Pcs
		ml/min/100g	mM		
Non-Constricted n=4	Before	8.6 ± 1.5	1.14 ± 0.14	0.46 ± 0.05	14.20 ± 0.23
	KIN	8.6 ± 1.7	1.18 ± 0.16	0.49 ± 0.06	15.30 ± 1.32
Constricted n=4	Before	8.6 ± 1.2	1.36 ± 0.19	0.37 ± 0.04	13.31 ± 1.58
	Constrict.	4.7 ± 1.0*	1.23 ± 0.21	0.35 ± 0.08	13.09 ± 2.10
	KIN	4.7 ± 1.1	1.26 + 0.41	0.30 ± 0.11	12.47 ± 2.22

*: p<0.05 vs Before

min into the left main coronary artery in both non-constricted and constricted groups through the autoperfusing cannula (13). Following infusion of kinin into the left main coronary artery, neither 0.1 nor 1.0 ng/kg/min of kinin for 5 min caused any change in plasma kinin levels in the artery or coronary sinus. A dose of 10 ng/kg/min of kinin for 5 min produced a significant elevation in plasma kinin in the coronary sinus from 12.8 to 142 pg/ml without any change in plasma kinin in the artery. However, infusion of kinin (10 ng/kg/min for 5 min) into the left main coronary artery resulted in no change in the mean systemic blood pressure, coronary blood flow, coronary vascular resistance, dp/dt max, LVEDP, MVO$_2$, myocardial lactate metabolism, or ECG ST segment in either group (Table 1).

From these findings, it was demonstrated that kinin was released significantly from the ischemic heart; however, such a level of kinin (10 ng/kg/min) has no significant effect on coronary circulation, myocardial metabolism, or ECG-ST segment. Thus, we could not clarify the role of increased kinin in this dog experiment. Likewise, it is unclear whether exogenous bradykinin might have the same action as endogenous kinin, since kinin is a local hormone. Further studies including the administration of ACE inhibitors or bradykinin antagonist will be necessary to reach conclusions about the role of kinin in the heart

EFFECTS OF ACE INHIBITORS ON MYOCARDIAL INFARCT SIZE IN RABBIT HEARTS

The cardioprotective effect of ACE inhibitors against ischemia/reperfusion injury has been reported for some time. Scholkens et al.(4) recently suggested that local bradykinin degradation by ACE inhibitors may play an important role in the beneficial effect of these agents against ischemic injury. To test whether ACE inhibitors directly protect ischemic myocardium from infarction, we assessed the effect of captopril and ramiprilat on myocardial infarct size in a rabbit ischemia/reperfusion model (Fig. 2) (14). Male rabbits (Japanese White) underwent 30 min coronary occlusion and 72 hour reperfusion. The rabbits were untreated (control rabbits), or treated with 0.5 mg/kg of captopril i.v., or 0.05 mg/kg of ramiprilat i.v. at 15 min before the ischemia. The handling of the heart for analysis of infarct size and the size of ischemic area (area at risk) is illustrated in Fig. 3. In the control rabbits, there was a significant linear correlation between histological infarct size and the size of area at risk. Administration of captopril or ramiprilat, however, failed to modify that infarct size - risk area size relationship, even though elevation of plasma renin activity by the ACE inhibitors was

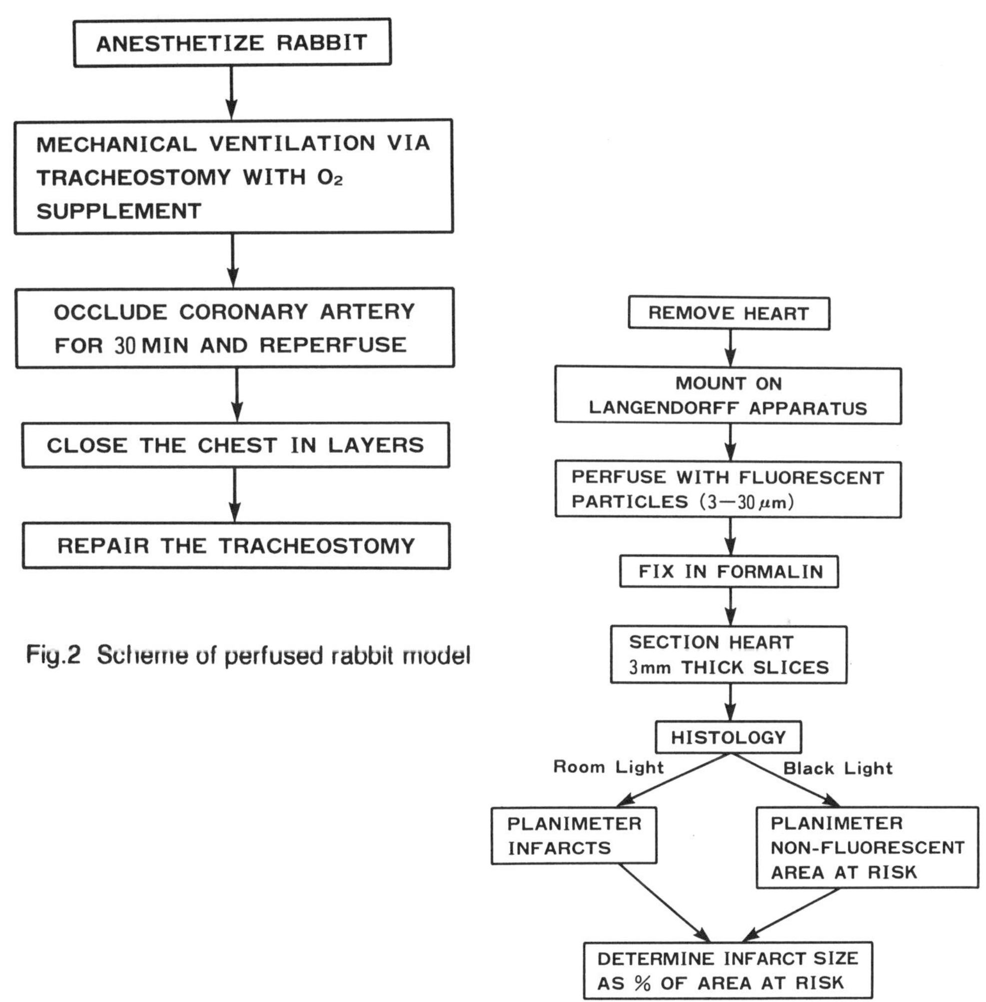

Fig.2 Scheme of perfused rabbit model

Fig.3 Scheme of tissue handling

noted in the treated animals. These findings suggest that inhibition of ACE does not achieve significant salvage of ischemic myocardium in the rabbit. This result is in contrast with a previous study by Ertl et al.(1) which reported a significant limitation of infarct size by captopril in a dog model. The reason for the conflicting results between the studies is unclear, but could be attributable to the difference in the methodology of infarct sizing (tetrazolium staining vs. histology), or species difference in the development of coronary collaterals (fair in the dog vs. poor in the rabbit), which might play a role in the protective effect of captopril in the dog heart (1).

CONCLUSION

1) The kinin level in coronary sinus blood was clearly elevated during myocardial ischemia induced by sympathetic nerve stimulation in coronary-constricted dogs.

2) Intracoronary infused bradykinin did not show any significant effect on coronary blood flow, cardiac function, or myocardial metabolism in the dog hearts.

3) No significant influence of ACE inhibitors on infarct/risk ratio was observed in the ischemia-reperfusion rabbit hearts.

Although a pronounced release of plasma kinin from the ischemic heart was observed, the pathophysiological significance of the augmented kallikrein-kinin system in ischemic heart still remains unclear.

REFERENCES

1. Ertl G, Kloner RA, Alexander RW et al., Limitation of experimental infarct size by an angiotensin-converting enzyme inhibitor. Circulation 1982; 65:40-48.

2. Linz W, Martorana PA and Scholkens BA, Local inhibition of bradykinin degradation in ischemic hearts. J Cardiovasc Pharmacol 1990; 15(Suppl 6):S99-S109.

3. Westlin W and Mullane K, Does captopril attenuate reperfusion-induced myocardial dysfunction by scavenging free radicals ? Circulation 1988; 77(Suppl 1):I-30-I-39.

4. Schölkens BA, Linz W and König W, Effects of the angiotensin converting enzyme inhibitor, ramipril, in isolated ischemic rat heart are abolished by a bradykinin antagonist, J Hypertens 1988; 6)Suppl 4): S25-S28.

5. Sicuteri F, Franci G, Del Bianco PL et al, A contribution to the interpretation of shock and pain in myocardial infarction. Mal Cardiovasc 1967; 8:343-62.

6. Dzzinskii AA, and Kuimov AD, Blood kinin system in pathogenesis and clinic of ischemic heart disease. Cor Vasa 1972; 14:9-15.

7. Gomazkov OA. Das Kallikrein-Kinin System bei Myokardischemie und Herzinfarkt-Experimentalle Untersuchungen. Med Welt 1974; 25:804-7.

8. Golikov AP, Bobkov AI, Ivleva V, Changes in sympatho-adrenal and kallikrein-kinin system in the terminal stage of extensive myocardial infarction. Cor Vasa 1977; 19:169-175.

9. Torstilo I, The plasma kinin system in acute myocardial infarction. Acta Med Scand 1978; 204(Suppl 620): 1-62.

10. Hashimoto K, Hamamoto H, Honda Y et al., Changes in components of kinin system and hemodynamics in acute myocardial infarction. Am Heart J 1978; 95:619-26.

11. Ando T, Shimamoto K, A sensitive radioimmunoassay system for blood and plasma kinin, and its clinical applications. Sapporo Med J 1983; 52:453-62 (Engl Abst).

12. Matsuki T, Shoji T, Yoshida S et al., Sympathetically induced myocardial ischemia causes the heart to release kinin. Cardiovasc Res 1987; 21:428-32.

13. Matsuki T, Shoji T, Yoshida S, Studies on experimental coronary insufficiency - dynamics of plasma kinin in myocardial ischemia and its pathophysiological roles. Sapporo Med J 1986; 55:137-51.

14. Miki T, Miura T, Urabe K, Shimamoto K et al., Does inhibition of the angiotensin converting enzyme have direct cardioprotective effect against ischemic myocardial necrosis ? Clin Exp Pharmacol Physiol: will be submitted.

AAS 38/III
Recent Progress on Kinins
© 1992 Birkhäuser Verlag Basel

THE POSSIBLE ROLE OF BRADYKININ IN THE ANTIISCHEMIC ACTIVITY OF ACE-INHIBITORS

P. A. Martorana, and B. A. Schölkens

Cassella AG - Hoechst AG, SBU Cardiovascular Agents Frankfurt/Main, Germany

SUMMARY: The ACE-inhibitor ramiprilat (40 ng/kg/min) was infused for 6 h into the left coronary artery of anesthetized dogs with ligation of the descending branch of this artery. This route of administration and the low dose were chosen to achieve local cardiac effects without affecting systemic hemodynamics. Ramiprilat significantly reduced infarct-size expressed as percentage of the area at risk. The cardioprotective effect of ramiprilat was mimicked by bradykinin and abolished by coadministration of a bradykinin antagonist. These results strongly suggest that bradykinin plays a role in the cardioprotective effect of the ACE-inhibitor ramiprilat.

INTRODUCTION

Bradykinin ist widely distributed, being synthetized in many tissues, and has a wide range of biological actions. Its wide tissue distribution, diverse actions and short half-life suggest that bradykinin works locally in a paracrine manner rather than as a classical circulating hormone.

The role of bradykinin in the myocardium has received relatively little attention. In 1970 Wilkens et al. (1) reported that the small fall in cardiac tissue pH, which follows an ischemic insult, induces an activation of the local kinin system. In 1977 Hashimoto et al. (2) found an increased concentration of bradykinin in coronary sinus blood after coronary occlusion in the

dog. Also in the dog, myocardial ischemia induced by coronary artery stenosis and sympathetic stimulation caused the heart to release plasma kinin (3). In man, kinin levels in peripheral blood were found to increase soon after myocardial infarction (4). This led Hashimoto et al. in 1978 (4) to the suggestion that kinin released in patients with infarction may have a compensatory cardioprotective effect.
Ten years later, in 1988, this concept was revived by Schölkens et al. (5) who showed in the isolated working rat heart, that bradykinin could reduce ischemia-reperfusion injury. The same group also reported that bradykinin, infused in the coronary artery of anesthetized dogs during ischemia and reperfusion, reduced lactate concentrations in coronary sinus blood and preserved tissue levels of glycogen and energy-rich phosphates in the ischemic area (6).

ACE-inhibitors inhibit kininase II, the major catabolic enzyme of bradykinin and thus, potentiate the effects of bradykinin. Consequently, it was of interest to investigate the role of bradykinin in the cardiovascular effects of ACE-inhibitors. This investigation was made possible by the development of novel, specific bradykinin antagonists. In the present study we investigated the role of local (cardiac) bradykinin in the infarct-limiting effect of the ACE-inhibitor ramiprilat, by using the novel bradykinin antagonist HOE 140 (H-D-Arg-Arg-Pro-Hyp-Gly-Thi-Ser-D-Tic-Oic-Arg-OH) (7, 8).

MATERIALS AND METHODS

Dogs were anesthetized, instrumented, ventilated and the thorax opened. The left descending coronary artery was ligated for 6 h. Five groups were formed. Group 1 received saline into the main stem of the left coronary artery starting 30 min before the occlusion and lasting for the duration of the experiment. The next group received bradykinin in a subhypotensive dose of 1 ng/kg/min, and group 3 received ramiprilat in the subhypotensive dose of 40 ng/kg/min. In group 4, ramiprilat was administered together with the bradykinin antagonist HOE 140 (0.5 ng/kg/min) and group 5 received the bradykinin antagonist alone as per above (9).

Bradykinin and ramiprilat were given following the same schedule as saline.
The bradykinin antagonist, alone or in combination, was infused from 30 min
before to 30 min after the occlusion due of its long biological half-life
(7, 8).

RESULTS

In the present study the intracoronary route of administration and the very
low doses were chosen to obtain a local (cardiac) effect with no or minimal
effects on systemic hemodynamics such as blood pressure. During the 6 h
coronary occlusion, ramiprilat infusion as well as bradykinin had no
significant effects on systemic blood pressure.

Six h after coronary occlusion the heart was removed. The left descending
coronary artery was perfused distal to the occlusion with a solution of
triphenyltetrazolium chloride and the left ostium was perfused with 0.5 %
Evans blue. This procedure allowed accurate estimation of the area at risk of
infarction and of the infarct-area. The area at risk was similar in all
groups. The size of the infarction of saline-treated dogs averaged 55 % of the
area at risk. Both ramiprilat and bradykinin significantly reduced infarct-
size (Table 1). This cardioprotective effect of the ACE-inhibitor was
abolished by the coadministration of the bradykinin antagonist.

TABLE 1. Effect of intracoronary infusion of ramiprilat (40 ng/kg/min) and
bradykinin (1 ng/kg/min) on infarct size in anesthetized dogs

Treatment (i.c.)	N	AR/(LV + S) (%)	I/AR (%)
Saline	5	38.6 ± 2.7	55.0 ± 4.0
Ramiprilat	8	38.6 ± 1.7	25.3 ± 3.2**
Bradykinin	5	41.4 ± 1.7	33.1 ± 5.6*

Mean values ± SEM are given.
i.c. = intracoronary, N = number of animals, AR = area at risk of
infarction, (LV + S) = left ventricle plus septum, I = infarct.
* P <0.02
**p <0.001

CONCLUSION

In the present study ramiprilat effectively limited infarct-size following coronary occlusion in a dose that had no significant effect on systemic hemodynamics, pointing to potential benefits of subhyopotensive doses of ACE-inhibitors in myocardial ischemia. In addition, the observation that the infarct-limiting effect of ramiprilat was reversed by a bradykinin-antagonist and that administration of bradykinin also reduced the size of the infarction provides evidence for the involvement of bradykinin in the anti-ischemic effect of the ACE inhibitor.

The mechanism of action of the antiischemic effect of bradykinin remains to be elucidated. However, it is known that in endothelial cells bradykinin stimulates the B_2 receptors. This results in an increase in intracellular Ca^{2+} concentration and in the liberation of prostacyclin (PGI_2) and endothelium-derived relaxing factor (EDRF). Both these effects are blocked by a bradykinin B_2 antagonist (10). EDRF stimulates the soluble guanylate cyclase to form cyclic GMP leading to vasodilation and inhibition of platelet aggregation (11, 12). Additionally, cyclic GMP was found to improve the energy state in the ischemic heart (13). PGI_2, perfused in isolated working hearts reduces ischemia-induced injury (14). Indomethacin reverses this effect (this laboratory, unpublished results), but only attenuates the cardioprotective effect of bradykinin (14). Finally, bradykinin has favorable cardiac metabolic effects by increasing myocardial glucose uptake (15). Thus, bradykinin may be the first step in a chain of events with potential antiischemic activity.

REFERENCES

1. Wilkens H, Back N, Steger R, Karn J. The influence of blood pH on peripheral vascular tone: possible role of proteases and vaso-active peptides. "Shock, Biochemical, Pharmacological and Clinical Aspects", eds. Bertelli A and Back N. (Plenum Press) (1970): page 201

2. Hashimoto K, Hirose M, Furukawa S, Hayakawa H, Kimura C. Changes in hemodynamics and bradykinin concentration in coronary sinus blood in experimental cocornary occlusion. Jap Heart J (1977); 18:679-689

3. Matsuki T, Shoji T, Yoshida S, Kudoh Y, Motoe M, Inoue M, Nakata T, Hosoda S, Shimamoto K, Yellon D, Imura O. Sympathetically induced myocardial ischemia causes the heart to release plasma kinin. Cardiovasc Res (1987); 21:128-132

4. Hashimoto K, Hamamoto H, Honda Y, Hirose M, Furukawa S, Kimura E. Changes
    in components of kinin system and hemodynamics in acute myocardial
    infarction. Am Heart J (1978); 95:619-626

5. Schölkens BA, Linz W, König W. Effects of the angiotensin converting
    enzyme inhibitor, ramipril in isolated rat heart are abolished by a
    bradykinin antagonist. J Hypert (1988); 6 (Suppl. 4) S 25 - S 28

6. Linz W, Martorana PA, Schölkens BA. Local inhibition of bradykinin
    degradation in ischemic hearts. J Cardiovasc Pharmacol (1978); 15
    (Suppl. 6) S 99 - S 109

7. Hock FJ, Wirth K, Albus U, Linz W, Gerhards HJ, Wiemer G, Henke St,
    Breipohl G, König W, Knolle J, Schölkens BA. HOE 140 a new potent and long
    acting bradykinin-antagonist: in vitro studies. Brit J Pharmacol (1991);
    102:769-773

8. Wirth K, Hock FJ, Albus U, Linz W, Alpermann HG, Anagnostopoulos H,
    Henke St, Breipohl G, König W, Knolle J, Schölkens BA. HOE 140 a new
    potent and long acting bradykinin-antagonist: in vivo studies. Brit J
    Pharmacol (1991); 102:774-777

9. Martorana PA, Kettenbach B, Breipohl G, Linz W, Schölkens BA. Reduction of
    infarct size by local angiotensin-converting enzyme inhibition is
    abolished by a bradykinin antagonist. Eur J Pharmacol (1990); 182:395-396.

10. Wiemer G, Becker RHA. Die Ramiprilat-stimulierte cGMP- und PGI_2-Bildung in
    Endothelzellen ist Bradikinin-vermittelt. Kardiol (1991); 80 (Suppl 3):120

11. Busse R, Lückhoff A, Bassenge E. Endothelium-derived relaxing factor
    inhibits platelet activation. Naunyn Schmiedebergs Arch Pharmacol (1987);
    336:566-571

12. Palmer RMJ, Ferrige AG Moncada S. Nitric oxide release accounts for the
    biological activity of endothelium-derived relaxing factor. Nature (1987);
    327:524-526

13. Vuorinen P, Laustiola K, Metsä-Ketelä T. The effects of cyclic AMP and
    cyclic GMP on redox state and energy state in hypoxic rat atria. Life Sci
    (1984); 35:155-161

14. Linz W, Martorana PA, and Schölkens BA. Local inhibition of bradykinin
    degradation in ischemic hearts. J Cardiovasc Pharmacol (1990); 15
    (Suppl. 6) S 99 - S 109

15. Rösen P, Eckel J, and Reinauer H. Influence of bradykinin on glucose
    uptake and metabolism studied in isolated cardiac myocytes and isolated
    perfused rat hearts. Hoppe Seylers Z Physiol Chem (1983); 364:431-438

EFFECTS OF BRADYKININ AND DES-ARG^9-BRADYKININ ON THE ISCHEMIC RAT HEART

R. Chahine, A. Adam, N. Yamaguchi, R. Nadeau

Research Center, Hôpital du Sacré-Coeur, 5400 boul. Gouin W., H4J 1C5 Montréal and Faculties of Medicine and Pharmacy, Université de Montréal, Québec, Canada

SUMMARY: In isolated rat hearts subjected to 30 min stop flow followed by 5 min reperfusion, Bradykinin (Bk), 1 and 0.1 μM, significantly decreased ventricular tachycardia and fibrillation, noradrenaline (NA) output at reperfusion as well as myocardial NA loss and lipid peroxidation, and preserved creatine kinase activity. Des-Arg^9-Bk 0.1 μM had also similar effects. Hence, the active metabolite des-Arg^9-Bk may be involved in the protective action of Bk, partly mediated by its inhibitory effect on NA liberation.

INTRODUCTION

Several mechanisms have been proposed to explain the genesis of ischemia/reperfusion arrhythmias. NA liberation into the coronary effluent at reperfusion may be a potential cause of ventricular arrhythmias (1). Oxygen free radicals have been also implicated in reperfusion-induced myocardial damage and ventricular fibrillation via peroxidation of membrane phospholipids (2). Pharmacological control of ventricular arrhythmias is of particular importance as they occur in a number of pathological circumstances. Bk, an autacoid nonapeptide, which acts locally on B$_2$ receptors (3) producing vasodilation and increasing vascular permeability, may be effective against ischemia/reperfusion injury. In isolated hearts, an abolition of reperfusion arrhythmias (4) and a reduction of NA overflow (5) have been attributed to Bk. Des-Arg^9-Bk, the active metabolite of Bk, has also significant cardiovascular effects. Acting via the B$_1$ receptor which is not present in normal tissues, it can often be expressed in pathological conditions (8).

The present study was undertaken in isolated rat hearts submitted to global ischemia/ reperfusion. The aim was to investigate whether exogenous Bk by reducing NA overflow and lipid peroxidation may abolish reperfusion arrhythmias and preserve enzyme loss from the damaged tissue and whether des-Arg^9-Bk may contribute to the effects of Bk.

MATERIALS AND METHODS

Hearts from Male Wistar rats were removed rapidly and perfused retrogradely at a constant pressure of 10 Kpa (7) with a modified Krebs-Henseleit (KH) solution, pH 7.4, saturated with O_2/CO_2 (95:5) at 37°C. After 20 min of equilibration, aortic flow was interrupted for 30 min followed by 5 min of reperfusion (ISC group, n=10). In the treated groups (n=6), KH solution containing Bk (1 or 0.1 μM) or des-Arg^9-Bk (0.1 μM) was perfused 10 min before stop flow and throughout the experimental period. In control group (CTL, n=5) hearts were perfused for 35 min without any intervention. Electro-cardiogram (ECG) was monitored continuously, the effluent of perfusate was collected before stop flow and at reperfusion for the determination of coronary flow (CF) and NA output by radioenzymatic assay (8). At the end of the experiments hearts were frozen at -80°C until assay. Myocardial NA content was determined by HPLC using about 100 mg of from heart apex tissue (7). Myocardial Malondialdehyde (MDA) content was measured in terms of thiobarbituric acid (9) from the supernatant of approximately 200 mg of ventricular homogenate centrifuged at 4000 rpm. Myocardial creatine kinase (CK) activity was determined in the same supernatant by spectrophotometry using a Beckman assay kit. Proteins were measured by the method of Lowry et al (10). Statistical significance between groups was determined by student t test at the level of $p < 0.05$, after an analysis of variance. All values are expressed as means $\pm$ SEM.

RESULTS

In the totality of ischemic untreated hearts, ventricular fibrillation (VF) and/or tachycardia (VT) were observed at reperfusion (table 1). Bk (0.1 μM) decreased the incidence of arrhythmias, and abolished completely VF at 1 μM. In the same experimental conditions, des-Arg^9-Bk (0.1

μM) also significantly decreased the incidence of arrhythmias. In parallel, Bk (1 μM) abolished almost completely the ischemia-induced release of NA at reperfusion and des-Arg^9-Bk reduced also NA overflow to 70 % of the control value (figure 1, left). Such results were corroborated by the determination of myocardial NA content which was preserved under Bk and des-Arg^9-Bk perfusions (figure 1, right). CK activity was reduced (-27 %) in hearts submitted to global ischemia comparatively to control hearts (fig 2, left). In hearts pretreated with Bk and des-Arg^9-Bk, enzyme activity was close to control values. At the doses used, lipid peroxidation evaluated by the MDA assay, in the presence of Bk (1 μM) and its active metabolite (0.1 μM), was only reduced to 11. 3 and 17 % respectively, comparatively to 32 % in ischemic untreated hearts (fig 2, right).

DISCUSSION

In the present study the effects of Bk and its active metabolite des-Arg^9-Bk were examined in an experimental model of ischemia/reperfusion injury. Owing to the non reproducibility of a sampling in areas affected by coronary ligation (regional ischemia), total ischemia was chosen to induce a global attack throughout the heart and thus arbitrary areas need not to be selected. In our experimental conditions, we found a causal relation between NA levels and reperfusion arrhythmias. In fact, neuronal activity from central origin do not lead to substantial accumulation of NA within the ischemic myocardium, therefore intramyocardial NA stocks are expected to play the major role in modulating transmitter release (11). It has been demonstrated that NA overflow during reperfusion represents a washout of the transmitter released from sympathetic nerve endings previously accumulated in the tissue during the ischemic period and released via an inversion of the carrier system normally responsible for uptake-I, an indirect consequence of the metabolic imbalance within the nerve endings (12). This extracellular NA accumulation should be involved in cell injury and the development of malignant arrhythmias. Since Bk and des-Arg^9-Bk considerably limited both NA output and the incidence of reperfusion-induced ventricular dysrhythmias, Bk and des-Arg^9-Bk could act as blockers of NA liberation by affecting the uptake-I mechanism, limiting thereby reperfusion damages.

Table 1. Incidence of ventricular tachycardia (VT) and fibrillation (VF) at reperfusion in the absence (ISC) and presence of bradykinin (Bk) or des-Arg^9-Bk.

	VT	VF
ISC	10/10	8/10
Bk 1 μM	4/6	0/6
Bk 0.1 μM	5/8	4/8
Des-Arg^9-Bk 0.1 μM	5/6	2/6

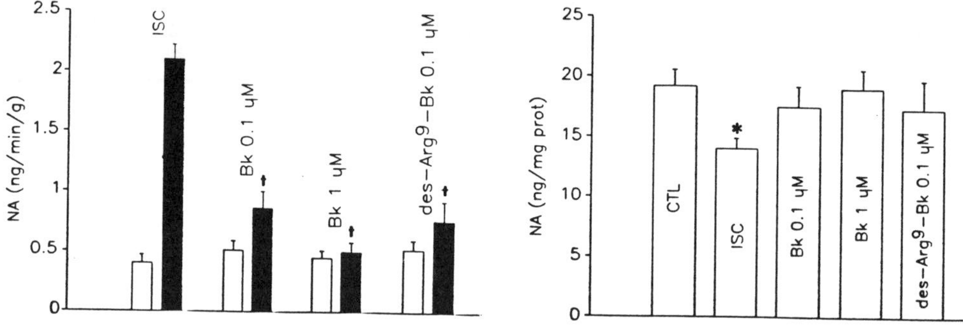

Figure 1. Left panel: Cumulative noradrenaline (NA) overflow (ng/min/g of heart) during 5 min reperfusion (filled bars) versus preischemic values (empty bars) in the ischemic untreated hearts (ISC) and in the treated hearts by different concentrations of bradykinin (Bk) or des-Arg^9-Bk. n = 5 - 8, + p < 0.05 vs ISC. **Right panel**: Myocardial NA content (ng/mg protein). CTL = control hearts. n = 5, * p < 0.05 vs CTL.

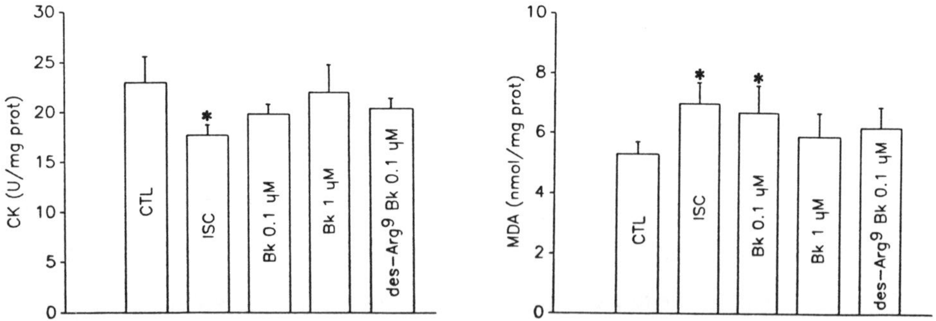

Figure 2. Bradykinin (Bk) and des-Arg^9-Bk effects on the myocardial creatine kinase (CK) expressed in U/mg protein **(left panel)** and myocardial malondialdehyde (MDA) content in nmol/mg protein **(right panel)**. CTL = control hearts. ISC = hearts submitted to 30 min ischemia followed by 5 min reperfusion. n = 5, * p < 0.05 vs CTL.

Catecholamines and their degradation byproduct may generate free radicals by autooxydation (13) and may contribute to the mosaic of reperfusion damages including arrhythmias. Lipid peroxidation is the primary mechanism for free radical-mediated cell and tissue injury (2). Once fatty acids begin to be released from membranes, they may bring more Ca^{2+} into the cell resulting from the disruption of membrane phospholipids increasing intracellular osmotic pressure and membrane permeability (13), leading to ionic alteration which may change conduction properties, and in the presence of catecholamines may predispose the tissue to dysrhythmias. The exact mechanism by which Bk and des-Arg^9-Bk attenuated reperfusion-induced arrhythmias has not been resolved. However, the improved CK activity used as a marker of myocyte injury, decreased lipid peroxidation and decreased NA loss from tissue are witnesses to its beneficial effect.

Several mechanisms for the cardioprotective effects of Bk have been reported. Thus, in isolated perfused hearts Bk increases the rate of glucose uptake and oxidation as well as the formation of lactate, it enhances the nutritional flow across the capillary wall and indirectly glucose metabolism (14). Bk causes also a dose-dependent release of nitric oxide (15) from endothelial cells and of prostacyclin (16) which is known as blocker of NA release and possesses several pharmacological properties which protect the ischemic myocardium.

Bk is metabolized by three main carboxypeptidases. Kininase I is known to generate the active metabolite des-Arg^9-Bk. Bk is also degraded by angiotensin converting enzyme (ACE). Using a specific carboxiterminal antiserum for Bk quantification by RIA, we could not recover in the perfusate effluent the initial concentration added to the fluid perfusion (data not shown). The measured concentration plead for an important metabolism of Bk in the heart tissue. It should be noted that we have used in this study Bk at the concentrations of 0.1 and 1 μM (considered to be within the pharmacological range) without ACE inhibitors and near those used by Linz et al (4) in the presence of ACE inhibitors. Nevertheless, we can not exclude that part of the effect observed with Bk are due to a partial metabolism in des-Arg^9-Bk. In fact we were able to reproduce an inhibition of NA release and a decrease of VT and VF when ischemic hearts were perfused with des-Arg^9-Bk. These observations lead to the hypothesis that B_1 receptors are involved in the ischemia/reperfusion injury, and are compatible with those of Marceau and Regoli (6) and Bouthillier et al (17) who noticed an induction of B_1 receptor in pathological conditions such anoxia.

CONCLUSION

These results suggest that the cardioprotective action of Bk may be partly mediated by its active metabolite des-Arg^9-Bk, which constitute a basis to study the effects of ischemia on the B_1 receptors expressed at the myocardial level and its characterisation using specific pharmacological antagonist.

REFERENCES

1.   Godin D, Campeau N, Nadeau R. Cardinal R, de Champlain J. Catecholamine release and ventricular arrhythmias during coronary occlusion and reperfusion in the dog. Can J Physiol Pharmacol 1985; 63: 1088 - 1095.

2.   Simpson PJ, Lucchesi BR. Free radicals and myocardial ischemia and reperfusion. J Lab Clin Med 1987; 10: 13 - 30.

3.   Regoli D, Barabe J. Pharmacology of bradykinin and related kinins. Pharmacol Rev 1980; 32: 1 - 46.

4.   Linz W, Martorana PA, Grotsch H, Qi Bei-Yin, Scholkens BA. Antagonizing bradykinin obliterates the cardioprotective effects of bradykinin and angiotensin-converting enzyme inhibitors in ischemic hearts. Drug Development Research 1990; 19: 393 - 408.

5.   Carlsson L, Abrahamsson T. Ramiplirat attenuates the local release of noradrenaline in the ischemic myocardium. Eur J Pharmacol 1989; 166: 157 - 164.

6.   Marceau F, Regoli D. Kinins receptors of the B_1 type and their antagonists. In: Bradykinin Antagonists: Basic and Clinical Research, Burch RM editor. New York: Marcel Dekker, 1991: 33 - 49.

7.   Chahine R, Chen X, Yamaguchi N, de Champlain J, Nadeau R. Myocardial dysfunction and norepinephrine release in isolated rat heart injured by electrolysis-induced oxygen free radicals. J Mol Cell Cardiol 1991; 23: 279 - 286.

8   Peuler JD, Johnson JA. Simultaneous single isotope radioenzymatic assay of plasma norepinephrine, epinephrine an dopamine. Life Sci 1977; 21: 625 - 636.

9.   Ohkawa H, Ohishi N, Yagi K. Assay for lipid peroxide in animal tissue by thiobarbituric acid reaction. Analyt Biochem 1979; 95: 351 - 358.

10. Lowry OH, Rosenbrough NJ, Fara AL, Randall RJ. Protein measurement with the Folin phenol reagent. J Biol Chem 1951; 193: 265 - 275.

11. Abrahamsson T, Almgren O, Carlsson L, Svensson L. Antiarrhythmic effect of reducing myocardial noradrenaline stores with alpha methyl-meta tyrosine. J Cardiovasc Pharmacol 1985; 7(suppl 5): 81 - 85.

12. Schomig A, Dart A, Dietz R, Kubler W, Mayer E. Paradoxical role of neuronal uptake for the locally mediated release of endogenous noradrenaline in the ischemic myocardium. J Cardivasc Pharmacol 1985; 5 (suppl 5): S40 - 44.

13. Hess ML, Manson NH. Molecular oxygen: Friend and foe. The role of oxygen free radical system in the calcium paradox and ischemia/reperfusion injury. J Mol Cell Cardiol 16, 969 - 985.

14. Rosen P, Eckel J, Reinauer H. Influence of bradykinin on glucose uptake and metabolism studied in isolated cardiac myocytes and isolated perfused rat hearts. Hope-Seyler's Z Physiol Chem 1983; 364: 1431 - 1438.

15. Palmer RMJ, Ferrige AG, Moncada S. Nitric oxide release accounts for the biological activity of endothelium-derived relaxing factor. Nature 1987; 327: 524 -526.

16. Needleman P, Bronson SD, Wyche A, Sivakoff M, Nicolaou KC. Cardiac and renal prostaglandin I$_2$. J Clin Invest 1978; 61: 839 - 849.

17. Bouthillier J, Deblois D, Marceau F. Studies on the induction of pharmacological responses to des-Arg^9 Bk in vitro and in vivo. Br J Pharmacol 1987; 92: 257 - 264.

AAS 38/III
Recent Progress on Kinins
© 1992 Birkhäuser Verlag Basel

THE EFFECT OF BRADYKININ ON CORONARY FLOW AND ITS POTENTIATION BY SH-CONTAINING ACE-INHIBITORS

P.A. de Graeff, W.H. van Gilst

Department of Pharmacology/Clinical Pharmacology, University of Groningen, Bloemsingel 1, 9713 BZ Groningen, The Netherlands

SUMMARY: Bradykinin plays a role in the regulation of coronary blood flow. Under basic conditions the vasodilating effect is primarily mediated by stimulation of the endogenous nitrovasodilator endothelium-derived relaxing factor (EDRF). This effect is shortlasting but can be increased and prolonged by sulfhydryl(SH)-containing agents. ACE-inhibitors may cause coronary vasorelaxation by a bradykinin-mediated release of EDRF. This can be potentiated by the presence of SH-groups, as was shown with captopril and zofenoprilat. As a consequence, SH-containing ACE-inhibitors may potentiate nitrates, because they act as exogenous nitrovasodilators, and reverse tolerance to their therapeutic effect.

INTRODUCTION

Interest in the role of bradykinin in the regulation of coronary blood flow has increased when it became known as an endothelium-dependent vasodilator, stimulating local production of prostacyclin and endothelium-derived relaxing factor (EDRF), also identified as nitric oxide (1). Although this role appears limited under basic conditions this may change under pathophysiological circumstances, especially myocardial ischaemia. It may also contribute to the local coronary vasodilatory effects of ACE-inhibitors, as has been demonstrated both *in vitro* and *in vivo*, because these drugs prevent bradykinin breakdown by inhibition of kininase II, which is identical to the ACE (2).

Evidence that ACE-inhibitors may potentiate the effects of bradykinin on coronary flow was obtained by *in vitro* experiments with a specific bradykinin antagonist (3,4). The increase in coronary flow by enalaprilat and ramiprilat, given both *in vitro* and *ex vivo*, was completely abolished by this antagonist. Similarly, *in vivo*, in rat experiments using a microsphere technique, the presence of a bradykinin antagonist reduced the increase in local conductance in the heart by enalaprilat (5).

However, a more puzzling picture arose when captopril and zofenoprilat were studied in these models. These sulfhydryl containing ACE-inhibitors induced a much more pronounced effect than the non-sulfhydryl containing ACE-inhibitors enalaprilat and ramiprilat (3,5,6,7). Evidence that this was due to the presence of a sulfhydryl group was obtained by other experiments with the inactive (R,S)-isomer of captopril and other SH-containing agents, like cysteine and glutathione in the isolated rat heart under normoxic conditions, which demonstrated similar effects (6,7). We also knew from studies with nitrates, which as exogenous nitrovasodilators act by the formation of intermediate metabolites, among which nitric oxide (NO), that their vasodilating effects can be enhanced by SH-containing agents (8). Thus, it was postulated that captopril and zofenoprilat potentiated the effect of endothelial NO (=EDRF) by their thiol moiety. This led to the following hypothesis:

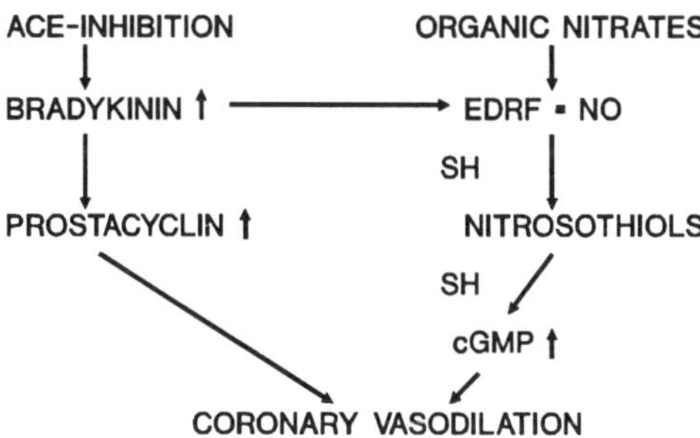

Fig. 1. Hypothesis for the interaction between sulfhydryl (SH)-containing angiotensin-converting enzyme (ACE) inhibitors and organic nitrates. EDRF, endothelium derived relaxing factor;NO, nitrous oxide;cGMP,cyclic GMP.

Subsequently, a number of experiments were performed to prove this hypothesis.

EFFECTS OF BRADYKININ ON CORONARY FLOW AND POTENTIATION BY ACE-INHIBITORS

2.1 Interference with prostacyclin. It has been demonstrated conclusively, both in cultured endothelial cells and *in vitro* and *in vivo* heart preparations that bradykinin can stimulate arachidonic release and subsequent prostacyclin formation in the endothelial cell of the coronary vasculature (9,10,11). In our *in vitro* rat heart preparation under normoxic conditions, bradykinin did induce transient coronary vasodilation but a significant increase in overflow of 6-keto-PGF1α, the stable metabolite of prostacyclin, was only found after high concentrations of bradykinin suggesting that other mechanisms might be involved (3). This was further corroborated by a number of experiments with cyclooxygenase inhibitors like indomethacin and acetylsalicylic acid, both by our and by other groups, which demonstrated no effect on the bradykinin-mediated increase in coronary flow (3,12).

Similar effects on coronary flow were obtained with ACE-inhibitors. Although a number of reports have shown that ACE-inhibitors may stimulate prostacyclin formation (10) and a slow increase in 6-keto-PGF1α in the coronary effluent was found following ramiprilat (6,7), the coronary vasodilating effects were not reduced following cyclooxygenase inhibition, despite effective blockage of the outflow of prostacyclin (3,6,7,13). Furthermore, the differential effects of the ACE-inhibitors with a thiol moiety remained unexplained.

All these results suggested that under stable, normoxic conditions the role of a bradykinin-mediated effect on prostacyclin, potentiated by ACE-inhibitors is limited. However, this may change under ischemic conditions as has been shown in patients with coronary artery disease, in whom indomethacin significantly reduced coronary sinus blood flow (14). Notwithstanding, other underlying mechanisms must be involved in the regulation of coronary flow by bradykinin and attention has recently concentrated on the release of EDRF.

2.2 Interference with EDRF. Kinins is one group of mediators which stimulate the release of EDRF by endothelial cells as has been demonstrated in cultured bovine and human cells (8). EDRF, or nitric oxide, stimulates the production of cGMP, a second messenger, by autocrine and paracrine mechanisms, leading to vasorelaxation and inhibition of platelet adhesion and aggregation (1). The importance of this mechanism has been shown *in vitro* by complete inhibition of nitric oxide release when bradykinin was administered in the presence of a bradykinin antagonist (8). Further evidence of a contribution of bradykinin-mediated release of EDRF to coronary flow was demonstrated *in vivo* in the canine coronary circulation by attenuation of the effects of bradykinin by concomitant administration of inhibitors of EDRF, especially L-N^G-monomethylarginine (L-NMMA)(11). Similar effects have also been shown in human coronary arteries although a reduced response to bradykinin was found in atherosclerotic arteries, indicating damage to local vascular control mechanisms (15).

The effects of ACE-inhibitors on the release of EDRF may mimic those of bradykinin, suggesting a contribution of locally generated kinins in the vascular wall and subsequent release of EDRF to the vasodilatory effects of these drugs. Studies in cultured bovine and human endothelial cells with ramiprilat and enalaprilat have shown that ACE-inhibitors are able to unmask a release of bradykinin with a subsequent increase in EDRF (8,16). *In vivo* studies in

SHR-rats have shown that co-administration of the inhibitor of nitric oxide synthesis L-NMMA attenuated the hypotensive effect of several ACE-inhibitors (17).

We investigated in the isolated rat heart under normoxic conditions:

1. The effect of 10 and 30 mM L-NMMA on the cumulative dose-response curve of captopril on coronary flow, and

2. the effect of the "true" precursor L-arginine on coronary flow in the presence and absence of captopril.

The following results were obtained (18, printed with permission of the authors):

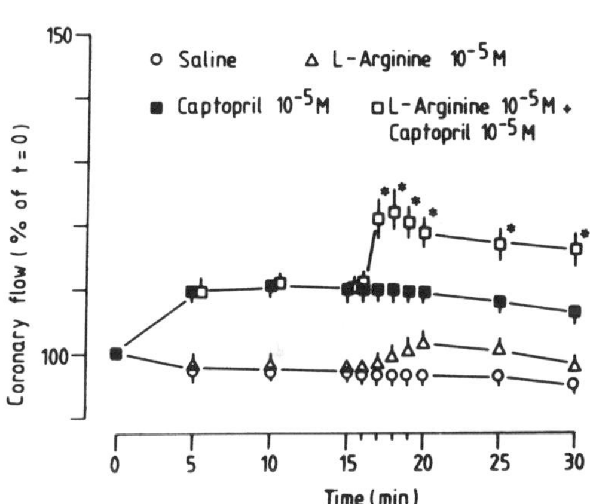

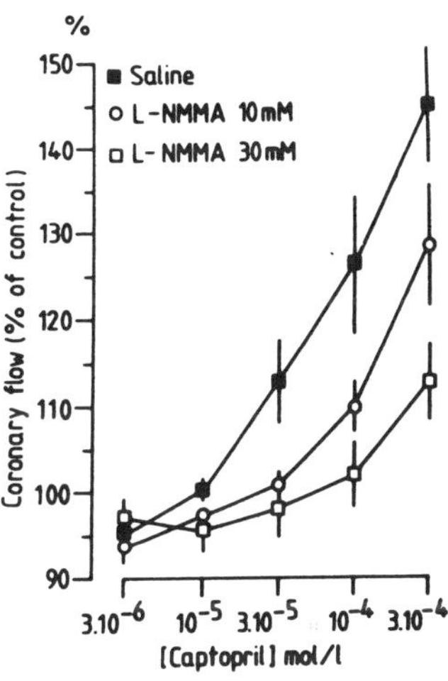

Fig. 2A. Effect of captopril, arginine and the combination on coronary flow.

Fig. 2B. Effect of L-NMMA on the cumulative dose-response curve of captopril

These results strongly suggest that a bradykinin-mediated release of EDRF indeed plays a role in the coronary vasodilating effects of ACE-inhibitors. As both the "true" and the "false" precursor already affected coronary flow in the absence of the ACE-inhibitor, it is suggested that EDRF is produced under normal conditions and participates in the regulation of coronary flow.

POTENTIATION OF THE EFFECT OF BRADYKININ ON THE RELEASE OF EDRF IN THE PRE-
SENCE OF SULFHYDRYL GROUPS

The bradykinin-induced release of prostaglandins and EDRF is a transient response that that does not persist in the continuous presence of bradykinin or during repeated administration of the compound (10). This tachyphylaxis occurs at the receptor level and is due to a homologous down-regulation of the kinin receptors, occurring *in vitro* within a few minutes. However, another mechanism may also be involved. Already in the early 1960s, potentiation of the effect of bradykinin by sulfhydryl(SH)-containing compounds was observed (18). We know from studies with organic nitrates that tissue sulfhydryl groups in vascular smooth muscle are involved in their mechanism, supposedly by activation of guanylate cyclase through the formation S-nitrosothiols from intermediate metabolites, among which nitric oxide (8). It has been shown repeatedly that administration of SH-containing agents like acetylcysteine, enhances nitrate vascular effects (19). Furthermore, EDRF is destroyed by free oxygen radicals which may be scavenged by sulfhydryl containing agents, including the ACE-inhibitor captopril, acting as antioxidants, thus leading to a potentiation of its effect (20).

The interaction of SH-containing agents on the effects of bradykinin on coronary flow was studied by us in a number of experiments in the isolated rat heart under normoxic conditions (18). The effect of 5 min of bradykinin infusion showed een biphasic pattern with a maximum after 1-2 min followed by a complete loss of its effect, suggesting the development of tolerance. Concomitant treatment with the sulfhydryl donor 10 μM of cysteine increased and prolonged this effect. This effect was also reflected in the dose-response curves of bradykinin on coronary flow in the presence of different concentrations of SH-containing ACE-inhibitors captopril and zofenoprilat and the non-SH containing ACE-inhibitor enalaprilat (Fig. 3, derived from 18, printed with permission of the authors).

Cysteine also potentiated the cumulative dose-response curve of bradykinin though to a lesser extent than captopril (18). These results clearly suggest that SH-containing agents can potentiate the effect of bradykinin on coronary flow. This effect is independent of ACE-inhibition and is probably related to a synergism with EDRF. Further proof could be obtained by another set of experiments showing a reduced response of bradykinin and an unchanged response to captopril on coronary flow when endogenous SH-depletion was evoked by pretreatment with high dosages of nitrates (18).

Apparently, thiol-groups are important for the vasorelaxing action of EDRF. One should be cautious however, to extrapolate our findings to the *in vivo* situation, because high levels of free radicals under basic conditions are only found *in vitro*, and not *in vivo*. Furthermore, the *in vivo* capacity of SH-containing agents to interact with locally generated EDRF needs to be studied further. Some data are available already on the interaction with exogenous nitrovasodilators, as will be discussed below.

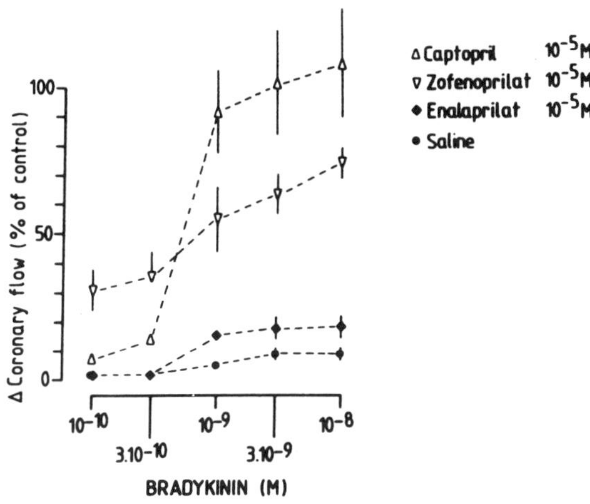

Fig. 3. Effect of captopril, zofenoprilat, and enalaprilat on the cumulative dose-response curve of bradykinin on coronary flow.

POTENTIATION OF NITRATES AND REVERSAL OF TOLERANCE BY SH-CONTAINING ACE-INHIBITORS

During the past century nitrates have become the most widely used agents in angina pectoris and congestive heart failure. The main action appears to be a consequence of their vasodilating effects on the systemic circulation, predominantly the capacitance veins. As described above, nitrate-induced vasodilator responses may be modified by changes in SH-availability (8). A deficiency of reduced SH-groups during continued administration of nitrates may explain why their vascular effect are rapidly attenuated during sustained therapy, a phenomenon which is known as tolerance (19). The clinical relevance of this phenomenon is shown by the almost complete loss of anti-anginal efficacy during continuous treatment with transdermal nitroglycerin patches. Further evidence for this hypothesis was derived from studies with N-acetylcysteine, which reverses the loss of vascular responsiveness (19). By combining bradykinin-mediated release of endogenous EDRF and donation of SH-groups, captopril appears to be a very attractive agent to prevent the development of tolerance to the effect of nitrates.

Furthermore, neurohumoral responses, especially noradrenaline and angiotensin II, also play an important role in the attenuation of nitrates, which can be effectively blocked by ACE-inhibition (21).

Experimental evidence was obtained by experiments in our *in vitro* heart model under normoxic conditions, where it was shown that captopril can potentiate the coronary vascular response to isosorbide dinitrate, comparable to cysteine, whereas ramiprilat, lacking a SH-moiety, had no significant effect (22). In rat aortic rings both acetylcysteine and captopril, but not enalaprilat, potentiated the vasorelaxant effects of nitroglycerin (23). Clinically this interaction was tested in 10 patients with chronic stable angina, already treated with isosorbide dinitrate, who received either captopril or placebo 1 hour before exercise testing in a randomized, double-blind, cross over design (18). Compared with placebo, captopril significantly improved efficacy parameters during exercise testing. Interestingly, no differences in pressure rate index were present at rest and during exercise, suggesting that other (SH-group?) mechanisms than hemodynamic changes might explain this beneficial effect. These effects were recently corroborated (24). Other clinical studies show however that other, non SH-containing ACE-inhibitors may also prevent nitrate tolerance by other mechanisms, especially by a reduction of counter-regulatory mechanisms (25). Thus, comparative studies between SH-containing and non-SH containing ACE-inhibitors are needed to elucidate the relative importance of the SH group of captopril.

REFERENCES

1.   Vanhoutte PM. Endothelium and control of vascular function: state of the art lecture. Hypertension 1989; 13: 658-67.

2.   Ertl G. Angiotensin converting enzyme inhibition and ischaemic heart disease. Eur Heart J 1988; 9: 716-27.

3.   Tio RA, Gilst WH van, Rett K, Wolters K, Dietze GJ, Wesseling H. The effects of bradykinin and the ischemic isolated rat heart. Horm Metabol Res 1990; 22: 85-9.

4.   Schölkens BA, Linz W, König W. Effects of the angiotensin converting enzyme inhibitor, ramipril, in isolated ischaemic rat heart are abolished by a bradykinin antagonist. J Hypertension 1988; 6(suppl 4): S25-S28

5.   Tio RA, Heiligers J, Langen CDJ de, Saxena PR, Wesseling H. Systemic and regional effects of the ACE-inhibitors zofenoprilat and enalaprilat in the rat, with and without treatment with a bradykinin antagonist. PhD Thesis 1990: pp. 119-32.

6.   Gilst WH van, Wijngaarden J van, Scholtens E, Graeff PA de, Langen CDJ de, Wesseling H. Captopril-induced increase in coronary flow; An SH-dependent effect on arachidonic acid metabolism? J Cardiovasc Pharmacol 1987; 9(suppl 2): S31-S36.

7.   Gilst WH van, Scholtens E, Graeff PA de, Langen CDJ de, Wesseling H. Differential influences of angiotensin converting-enzyme inhibitors on the coronary circulation. Circulation 1988; 77(suppl I):I-24 - I-29.

8.   Ignarro LJ, Lippton H, Edwards JC, Baricos WH, Hyman AL, Kadowitz PJ, Gruetter CA. Mechanisms of vascular smooth muscle relaxation by organic nitrates, nitrites, nitroprusside and nitric oxide: evidence for the involvement of S-nitrosothiols as active intermediates. J Pharmacol Ther 1981; 218: 739-49.

9.   Wiemer G, Schölkens BA, Becker RHA, Busse R. Ramiprilat enhances endothelial autocoid formation by inhibiting breakdown of endothelium-derived bradykinin. Hypertension 1991; 18: 558-63.

10. Schrör K. Converting enzyme inhibitors and the interaction between kinins and eicosanoids. J Cardiovasc Pharmacol 1990; 15(suppl 6): S60-S68

11. Pelc LR, Gross GJ, Warltier DC. Mechanism of coronary vasodilation produced by bradykinin. Circulation 1991; 83: 2048-56.

12. Baydoun AR, Woodward B. Effects of bradykinin in the rat isolated perfused heart: role of kinin receptors and endothelium-derived relaxing factor. Br J Pharmacol 1991; 123: 1829-33.

13. Wijngaarden J van, Tio RA, Gilst WH van, Graeff, PA de, Langen CDJ de, Wesseling H. Coronary vasodilation induced by captopril and zofenoprilat: evidence for a prostaglandin-independent mechanism. Naunyn-Schmiedeberg's Arch Pharmacol 1991; 343: 491-5.

14. Friedman PL, Brown EJ, Gunther S, Alexander RW, Barry WH, Mudge GH, Grossman W. Coronary vasoconstrictor effect of indomethacin in patients with coronary-artery disease. N Engl J Med 1981; 305 1171-5.

15. Chester AH, O'Neil G, Moncada S, Tadjkarimi S, Yacoub MH. Low basal and stimulated release of nitric oxide in atherosclerotic epicardial coronary arteries. Lancet 1990; ii: 897-900.

16. Busse R, Lamontagne D. Endothelium-derived bradykinin is responsible for the increase in calcium produced by angiotensin-converting enzyme inhibitors in human endothelial cells. Naunyn-Schmiedeberg's Arch Pharmacol 1991; 344: 126-9.

17. Cachofeiro V, Sakakibara T, Nasjletti A. Kinins, nitric oxide, and the hypotensive effect of captopril and ramiprilat in hypertension. Hypertension 1992; 19: 138-45.

18. Gilst WH van, Graeff PA de, Leeuw, MJ de, Scholtens E, Wesseling H. Converting enzyme inhibitors and the role of the sulfhydryl group in the potentiation of exo- and endogenous nitrovasodilators. J Cardiovasc Pharmacol 1991; 18: 429-36.

19. Elkayam U. Tolerance to organic nitrates: evidence, mechanisms, clinical relevance, and strategies for prevention. Ann Int Med 1991; 114: 667-77.

20. Goldschmidt JE, Tallarida RJ. Pharmacological evidence that captopril possesses an endothelium-mediated component of vasodilation: effect of sulfhydryl groups on endothelium-derived relaxing factor. J Pharmacol Exp Ther 1991; 257: 1136-45.

21. Parker JD, Farrell B, Fenton T, Cohanim M, Parker JO. Counter-regulatory responses to continuous and intermittent therapy with nitroglycerin. Circulation 1991; 84: 2336-45.

22. Gilst WH van, Graeff PA de, Scholtens E, Langen CDJ de, Wesseling H. Potentiation of isosorbide dinitrate-induced coronary dilatation by captopril. J Cardiovasc Pharmacol 1987; 9: 254-5.

23. Lawson DL, Nichols WW, Mehta P, Mehta JL. Captopril-induced reversal of nitroglycerin tolerance: role of sulfhydryl group vs. ACE-inhibitory activity. J Cardiovasc Pharmacol 1991; 17: 411-8.

24. Metelitsa VI, Martsevevich SY, Kozyreva MP, Slastnikova ID. Enhancement of the efficacy of isosorbide dinitrate by captopril in stable angina pectoris. Am J Cardiol 1992; 69: 291-6.

25. Katz RJ, Levy WS, Buff L, Wasserman AG. Prevention of nitrate tolerance with angiotensin converting enzyme inhibitors. Circulation 1991; 83: 1271-7.

Circulation
Kidney

CARDIAC ANGIOTENSIN CONVERTING ENZYME AND DIASTOLIC FUNCTION OF THE HEART

H. SCHUNKERT and M. PAUL

Medizinische Klinik II, Universität Regensburg, Franz-Josef Strauß Allee, D-8400 Regensburg, FRG; and German Insitute for High Blood Pressure Research, Pharmakologisches Institut, Universität Heidelberg, Im Neuenheimer Feld 366, D-6900 Heidelberg, FRG

SUMMARY: It has been known for a long time that systemic infusion of angiotensin II in patients with coronary artery disease or normal control subjects causes a marked increase in left ventricular end diastolic pressure (LVEDP) and systolic pressure (LVP) (1,2). In this setting angiotensin II produces a marked increase in afterload that makes it difficult to acknowledge possible local myocardial effects of the peptide. The studies (3-8) summarized in the present paper were designed to examine the physiological role of local cardiac angiotensin II generation and local bradykinin degradation on cardiac function in the normal and hypertrophied rat heart. Angiotensin I and angiotensin II, infused in isolated, well oxygenated, buffer perfused normal rat hearts, produced a mild increase in LVEDP with no change in systolic function (3). In contrast, in hypertrophied rat hearts, angiotensin I and angiotensin II caused a marked deterioration of diastolic function, increasing LVEDP from 10 to 25-37 mmHg on average (3,5). Preliminary evidence suggests that angiotensin II effects on diastolic function are mediated via a protein kinase C dependent pathway that might involve Na^+/H^+ exchange (4,5). When cardiac angiotensin converting enzyme was blocked by infusion of an ACE inhibitor prior and in parallel to angiotensin I infusion no changes in diastolic function were noted (6). Furthermore, ACE inhibition blunted the diastolic dysfunction during low flow ischemia in isolated hypertrophied rat hearts (7). This effect of ACE inhibition was even more remarkeable, since no exogenous angiotensin was infused in this experiment. Bradykinin infusion or prevention of bradykinin breakdown by ACE inhibition, on the other hand, had no effect on diastolic function in isolated normal oxygenated rat or rabbit hearts, despite causing a mild vasodilatation (6,8).

INTRODUCTION

Angiotensin converting enzyme (ACE) is a dipeptidase that is intimately involved in the metabolism of a whole group of peptides and substances regulating cardiac and vascular function (9). First, ACE facilitates the cleavage of angiotensin I to generate the active metabolite angiotensin II, which acts as a regulator of cardiovascular homeostasis. In addition, angiotensin II controls the activity of the renin angiotensin system by feedback regulation (10) and mediates norepinephrine release (11). However, ACE is also involved in the metabolism of bradykinin which, in part, is acting through formation of prostacyclin and nitric oxide (EDRF) (12). The enzyme is identical

with kininase II which is involved in bradykinin metabolism by liberating the C-terminal dipetide
phenylalanyl-arginine (9).

The complexicity of systemic and local functions of ACE makes it difficult to dissect out the
mechanisms responseable for the long scale of effects seen after pharmacological inhibition of the
enzyme. The aim of the present studies (3-8) was to examine the role of ACE in the regulation of
the diastolic function of the heart.

MATERIALS AND METHODS

Weanling male Wistar rats underwent ligation of the ascending aorta or open chest operation
(sham) using standard methods (3). Under pentobarbital anesthesia the anterior thorax was
dissected, the aorta was localized and a metal clip with an internal diameter of 0.6 mm was placed
on the ascending aorta.

Perfusion Technique: Eight weeks after the operation, rats were injected intraperitoneally with 1.5
ml pentobarbital and the thorax was rapidly opened. The heart was removed, placed in a water-
jacketed constant temperature chamber (37^0 C). Orthograde perfusion of the coronary arteries was
restored through a short cannula inserted in the aortic root, as previously described (3). The
perfusate consisted of modified Krebs-Henseleit buffer: 118 mM NaCl, 4.7 mM KCl, 2.0 mM
$CaCl_2$, 1.2 mM KH_2PO_4 , 1.2 mM $MgSO_4$,25mM Na HCO_3, 5.5 mM glucose and 1.0 mM
lactate. The oxygenated perfusate was equilibrated with a 5% CO_2 -95% O_2 gas mixture such that
the perfusate PO_2 was approximately 550 mm Hg.

After coronary perfusion was initiated, the left ventricle was decompressed by apical puncture
and insertion of a short apical drain to vent left ventricular Thebesian drainage. A collapsed latex
balloon, slightly larger than the left ventricular chamber, was inserted into the left ventricle via a
left artial incision. The balloon was filled with bubble-free saline and attached to a Statham P23Db
pressure transducer (Staham Instruments, Inc., Puerto Rico) via a 15 cm lenght of polyethylene
tubing. Coronary perfusion pressure was measured via a Y-shaped connecter attached to the aortic
infusion cannula. A thermistor and a bipolar pacing electrode were inserted in the right ventricle
and the heart was stimulated by a Grass S5 stimulator (Grass Instrument Co., Waltham, Ma).

Experimental Protocols: The performance of each heart stabilized by 15 min buffer perfusion and
constant pacing at a heart rate of 4 Hz. At the end of the stabilisation period left ventricular
pressures, and coronary flow e.g. coronary resistance were measured.

1.) Protocol (3). Angiotensin I or angiotensin II (Sigma Chemical Co., St.Louis, Mo.) in 0.5% pasteurized bovine serum albumin (Sigma Chemical Co.) was added to the system by a Harvard pump (flow 0.5ml/min) to achieve a final concentration of 3.8×10^{-8} M (16-25 minute) and 3.8×10^{-7} M (26-35 minute) in the Krebs-Henseleit buffer. Ventricular function and coronary perfusion of LVH and sham hearts were recorded at baseline and at the end of each dosage period (minutes 25 and 35).

2.) Protocol (6). LVH and sham hearts received enalaprilat infusion (4×10^{-6} M) for 15 min prior and together with 30 min 3×10^{-7} M angiotensin I infusion.

3.) Protocol (7). One group of LVH and sham hearts was perfused with enalaprilat, whereas respective controls received vehicle. After a stabilisation period coronary perfusion pressure was reduced from 75-100 mmHg to 12 mmHg such that LVH and sham hearts were perfused at low flow ischemia for 30 min. Diastolic and systolic function were recorded via a LV balloon in 5 min intervals.

4.) Protocol (8). The effects of 10^{-5} M bradykinin infusion on cardiac function and vascular resistance were studied in isolated perfused, normal rabbit hearts.

<u>Biochemical analysis:</u> The perfusates were heated to boiling and precipitated proteins were removed by centrifugation. The supernatent was partially purified with RP 18 SEP PAK cartridges (Waters Associates, Milford, Ma.) as previously described (3). After washing with 0.01 M trifluoroacetic acid, the SEP PAK cartridges were eluted with 80% acetonitrile in 0.01 M trifluoroacetic acid. The eluate was lyophilized, then redissolved in 0.02 M acetic acid and subjected to high performance liquid chromatography (HPLC). HPLC was performed employing a Varian 5000 solvent delivery system combiend wifh a Spectroflow 757 variable UV monitor (LKB Instruments, Paramus, NY) tuned to 216 nm. The data handling and processing was performed using an Apple IIe PC, loaded with Chromatochart software (Interactive Microware,Inc., State College, PA.). The angiotensins were separated on a LKB ultraspect column (Spherisorb ODS-2,3 μm;4.6 x 50 mm). The solvent consisted of 40% methanol in 10 mM Na acetate, pH 5.6 (A) and 80% methanol in Na acetate 10 mM, pH 5.6 (B). The gradient was B=0% at 0 to B=100% at 30 minutes. The flow rate was 1 ml/min. The fractions were collected in polypropylene test tubes; the collection time for a fraction was 30 seconds. Synthetic angiotensins were used for calibration of the HPLC column. The recovery was 83 % for Ang I and 94 % for Ang II %.

The fractions corresponding to the retention times of synthetic angiotensin I, II and III were pooled, respectively. After lyophilization, samples were redissolved in 0.2 ml of 0.02 M acetic acid, diluted with radioimmunoassay buffer (1 mM Tris, pH 7.4 and 1 mg/ml crystalline BSA), and RIA performed as previously described in detail (3). The sensitivity is 0.1-1.2 ng / tube. Since RIA was always performed after HPLC, the crossreactivity of the angiotensin II antibody (a kind gift of Dr. Arthur Freedlander) with angiotensins III or nonangiotensin peptides was prevented (3). The conversion rate was calculated as Ang II / Ang I + Ang II [M] x100= %.

Statistical analysis: Results are expressed as mean +/- standard errow. Differences observed between groups were analysed by Student's t-test for unpaired data. Statistical significance was accepted at the 95% confidence limit.

RESULTS

Ligation of the ascending aorta resulted in slightly lower body weights at sacrifice as compared with sham operated controls (218 g vs 243 g ; p=0.16). Left ventricular weight in relation to body weight was significantly higher in clipped animals (LVH) as compared with shams (3.6 vs 2.6 g/kg, p<0.005).

1.) Protocol (3). As soon as angiotensin I was added, angiotensin II appeared in the effluent. The intracardiac conversion rate of angiotensin I to angiotensin II during perfusion with 3.5×10^{-8} M angiotensin I was 17.3 +/- 4.1% in LVH (n=9) and 6.8 +/-1.3% in shams (n=13; p<0.01) (3). The conversion per gram cardiac tissue (left + right ventricle) was 15.3 % vs 6.3 %, respectively (p<0.05) (3). In response to angiotensin I infusion in isolated buffer perfused rat hearts, there was a dose-related increase in coronary perfusion pressure in both normal and hypertrophied hearts as compared to baseline (p<0.05) (3). The increase of coronary perfusion pressure was similar in LVH and control hearts. Angiotensin I had no effect on systolic developed pressure in the isolated rat hearts. In contrast, there was a dose-related increase in LVEDP in hypertrophied rat hearts perfused with angiotensin I (Fig.1) (3). To study whether the increase in LVEDP is related to the increase in coronary perfusion pressure, five LVH hearts were perfused at increasing levels of coronary flow to gradually increase coronary perfusion pressure. The increase of coronary perfusion pressure achieved by increasing flow was associated with a small increase of LVEDP (3 mmHg) (6). This increase of LVEDP was significantly lower as that seen at identical coronary perfusion pressure after angiotensin I infusion (15-25 mmHg; p<0.05), thus suggesting that the effects of angiotensin on LVEDP are only partially mediated by changes in coronary perfusion pressure (6).

2.) Protocol (6). Infusion of enalaprilat alone had no effect on cardiac function under well oxygenated conditions. When enalaprilat was infused in parallel to angiotensin I the intracardiac angiotensin I conversion rate of the LVH hearts was inhibited by 70% (p<0.05) as compared with vehicle plus angiotensin I perfused LVH hearts. Furthermore, coronary resistance as well as diastolic function remained at baseline levels and were sigificantly lower in enalaprilat plus angiotensin I as compared with vehicle plus angiotensin I perfused LVH hearts. Indeed, LVEDP and coronary resistance in enalaprilat plus angiotensin I perfused LVH hearts were similar to that in hearts infused with vehicle only (Fig. 2) (6).

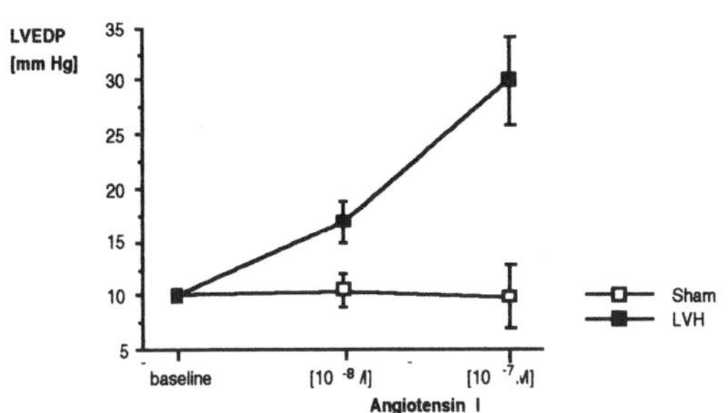

Figure 1. The effect of increasing doses of angiotensin I on left ventricular end diastolic pressure in normal (sham) and hypertrophied (LVH) rat hearts. (From:Schunkert H, Dzau VJ, Tang SS, Hirsch AT, Apstein C, Lorell BH33 (1990). Increased rat cardiac angiotensin converting enzyme activity and mRNA levels in pressure overload left ventricular hypertrophy. J Clin Invest 86: 1913-1920; with permission)

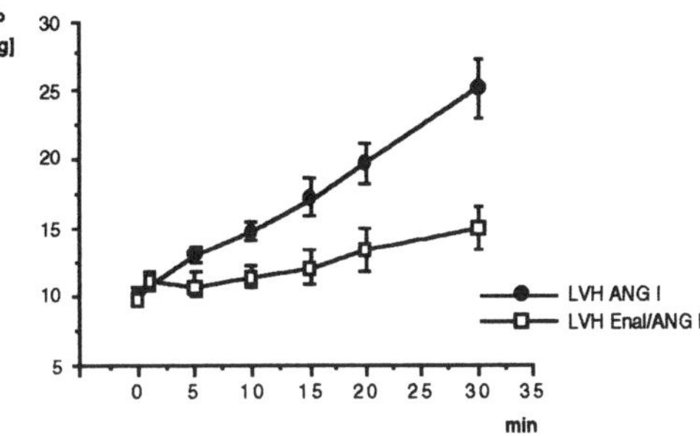

Figure 2. The effect of angiotensin I alone (LVH ANG I) as compared with coadministration of the ACE-inhibitor enalaprilat with angiotensin I (LVH Enal/ANG) on left ventricular end diastolic pressure in hypertrophied (LVH) rat hearts. (Adapted from: Schunkert H, Jackson B, Tang SS, Lorell BH (1991). Localization and functional significance of cardiac angiotensin converting enzyme (ACE) in hypertrophied rat hearts. Circulation 84 (Suppl II): II-307.)

3.) Protocol (7). At baseline, left ventricular systolic developed pressure/gram (LV develop P/g) was higher in LVH, whereas during ischemia LV develop P/g was lower in LVH as compared with sham hearts. Enalaprilat did not affect LV develop P/g under either condition. The ischemic rise of LVEDP was more pronounced in vehicle treated LVII as compared with vehicle treated sham hearts (38 vs 20 mmHg, p< 0.05), however, inhibition of ACE diminished the rise of LVEDP in LVH hearts from 38 to 28 mmHg, such that LVEDP in LVH plus enalaprilat was no longer different from sham.

4.) Protocol (8). Bradykinin was infused in isolated beating rabbit hearts at a concentration of 10^{-5} M. At this dose a slight decrease in coronary perfusion pressure was observed (about 10% of baseline) (8). However, there was neither a change in systolic or diastolic pressures nor in dP/dt (8).

DISCUSSION

Taken together, these observations indicate that angiotensin I to angiotensin II conversion takes place in the coronary circulation (3,6). The angiotensin I conversion rate is increased in the myocardium of hearts with left ventricular hypertrophy, extending recent findings of increased cardiac ACE activity and mRNA levels in this model of pressure overload LVH in the rat (3, 6, 13). Blockade of the enzyme by an ACE inhibitor can decrease intracardiac angiotensin I conversion rate in isolated beating LVH hearts (6). The data suggest that ACE is the main enzyme involved in conversion of angiotensin I to angiotensin II in the intact, isolated hypertrophied rat heart.

Although the physiological significance of the cardiac ACE is still not clear, an increasing number of studies suggest that the inhibition of the tissue enzyme accounts for numerous benefits of ACE-inhibitors. The decrease of coronary flow in angiotensin I perfused, isolated hearts can be prevented by pretreatment with ACE-inhibitors (11) resulting in improved energy metabolism and reduced tissue injury (14). In the present studies coronary flow was kept constant. Thus, ACE inhibition in angiotensin I perfused rat hearts caused a decrease in coronary perfusion pressure (6). The positive inotropic effects of angiotensin II are species-specific and cannot be studied in the rat. In sheep and cats, cardiac ACE inhibition or angiotensin receptor blockade prevented angiotensin I related positiv inotropy (15). Cardiac ACE inhibition may also affect the electrophysiological actions in prevention of ventricular arrhythmias during ischemia or reperfusion (14, 16). Linz et al. showed that cardiac ACE inhibition during reperfusion of rat hearts can protect against ventricular fibrillation and tissue injury. Co-administration of angiotensin II, but not angiotensin I, reversed these effects (14). This finding might be explained by the high density of angiotensin II receptors along the conduction system in this species (17). Another effect of tissue ACE-inhibition, demonstrated in the isolated hearts, is the deceleration of bradykinin metabolism, which improved the coronary hemodynamics (18). Both mechanisms may contribute for the observation, that treatment with ACE-inhibtors reduced significantly infarct size in dogs (19).

Further evidence for the functional relevance of cardiac ACE was provided by experiments studying the effects of angiotensin I on diastolic function in isolated normal and hypertrophied hearts (3). Angiotensin I infusion in rat hearts with LVH caused a dramatic increase of LVEDP (3). The deterioration of diastolic function could not be explained by the concomitant increase of coronary perfusion pressure, since a flow mediated increase of coronary perfusion pressure had only a mild effect on LVEDP (6). Rather, direct myocardial effects of angiotensin II might explain the reduction of lusitropy of the hypertrophied left ventricle. Preliminary studies of Mochizuki et al. provide evidence that angiotensin II acts via stimulation of protein kinase C (4). Inhibition of protein kinase C abolished the effects of angiotensin II on relaxation, whereas stimulation of the enzyme caused physiological changes similar to that of angiotensin II infusion, including a sharp rise in LVEDP (4). In addition to protein kinase C activation (or as a result of protein kinase C activation), angiotensin also stimulates Na^+/H^+ exchange. In fact, inhibition of the Na^+/H^+

antiporter by amiloride has also been shown to prevent the deterioration of diastolic function in isolated hypertrophied rat hearts (5).

ACE inhibition may also improve diastolic function of hypertrophied hearts during ischemia. Eberli et al. recently demonstrated in isolated hearts that enalaprilat can attenuate the increase of LVEDP in LVH during low flow ischemia even in the absence of exogenous angiotensin I (7). As to whether the effects of ACE inhibition on bradykinin degradation are involved in these changes remains to be elucidated. Under oxygenated conditions, enalaprilat infusion alone (without angiotensin I) had no effect on cardiac function (6,7). Furthermore, bradykinin infusion in normal rabbit hearts did not change cardiac function (8). However, the effects of bradykinin on myocardial relaxation in the presence of LVH have not been investigated, yet.

The data from experimental models of left ventricular hypertrophy also expand to the clinical arena. Short term oral treatment with the ACE inhibitor cilazapril in hypertensive patients with left ventricular hypertrophy resulted in markedly improved diastolic function (20). Furthermore, acute intravenous enalaprilat administration improved left ventricular filling in patients with myocardial ischemia, especially during exercise (21). In the same studies, angiotensin I and angiotensin II content was measured in the coronary effluent and the arterial circulation demonstrating a similar reduction of conversion rate during ACE inhibition (21). These data suggest that in humans both systemic and cardiac ACE can be effectively inhibited by enalaprilat. Only limited experience is available on intracoronary ACE inhibition. Foult et al. has shown in patients with dilated cardiomyopathy that cardiac ACE inhibition may improve coronary hemodynamics with only minor changes in cardiac function (22).

CONCLUSION

Angiotensin converting enzyme is the major enzyme activating angiotensin I in the coronary circulation of the rat. The product, angiotensin II, has mulitple local actions including vasoconstriction and positive inotropy in some species. Furthermore, angiotensin II can deteriorate diastolic filling of the ventricle, in particular when pressure overload left ventricular hypertrophy is evident. Inhibition of cardiac ACE may reverse the functional changes seen after angiotensin I adminstration. Further clinical studies are needed to test the efficacy of ACE inhibitors in treatment of diastolic dysfunction in patients with left ventricular hypertrophy.

REFERENCES

1. Ross J, Braunwald E. The study of left ventricular function in man by increasing resistance to ventricular ejection with angiotensin. Circulation 1964; 29:739.

2. Ronan JA, Steelman RB, Schrank JP, Cochran PT. The angiotensin infusion test as a method evaluating left ventricular function. Am Heart J 1975; 89: 554-560.

3. Schunkert H, Dzau VJ, Tang SS, Hirsch AT, Apstein C, Lorell B. Increased rat cardiac angiotensin converting enzyme activity and mRNA levels in pressure overload left ventricular hypertrophy: effects on coronary resistance, contractility and relaxation. J Clin Invest 1990; 86: 1913-1920.

4. Mochisuki T, Schunkert H, Ngoy S, Apstein CS, Lorell BH. The effects of angiotensin in pressure overload hypertrophy are simulated by protein kinase C activation. Circulation 1991; 84 (Suppl II): II-308.

5. Mochisuki T, Weinberg E, Apstein CS, Schunkert H, Lorell BH. Impairment of diastolic function by angiotensin II in pressure overload hypertrophy: evidence for Na^+/H^+ exchange. Circulation 1991; 84 (Suppl II): II-280.

6. Schunkert H, Jackson B, Tang SS, Lorell BH. Localization and functional significance of cardiac angiotensin converting enzyme (ACE) in hypertrophied rat hearts. Circulation 1991; 84 (Suppl II): II-307.

7. Eberli FR, Apstein CS, Ngoy S, Lorell BH: Enalaprilat decreases ischemic diastolic dysfunction in pressure overload hypertrophy. Circulation 1990;82 (Suppl III): III 161

8. Lorell BH, Apstein CS, Cunningham MJ, Schoen FJ, Weinberg E, Peeters GA, Barry WH. Contribution of endothelial cells to calcium dependent fluorescence transients in rabbit hearts loaded with Indo 1. Circ Res 1990; 67: 415-425.

9. Erdös EG. Angiotensin I converting enzyme and the changes in our concepts through the years. Hypertension 1990; 16: 363-370.

10. Schunkert H, Hirsch AT, Pinto Y, Pelletier P, Jacob H, Remme WJ, Ingelfinger JR, Dzau VJ. Feedback regulation of angiotensin converting enzyme mRNA and activity by angiotensin II. Circulation 1990; 82 (Suppl III): III-230.

11. Xiang J, Linz W, Becker H, Ganten D, Lang RE, Schoelkens B, Unger T. Effects of converting enzyme inhibitors:ramipril and enalapril on peptide action and sympathetic neurotransmission in the isolated rat heart. Eur J Pharmacol 1984; 113:215-223

12. Wiemer G, Schölkens BA, Becker RHA, Busse R. Ramiprilat enhances endothelial autacoid formation by inhibiting breakdown of endothelium derived bradykinin. Hypertension 1991; 18: 558-563.

13. Paul M, Schunkert H. Detection of angiotensin converting enzyme mRNA in the rat heart by use of the polymerase chain reaction (PCR). Agents and Actions 1992 (in present suppl).

14. Linz W, Scholkens BA, Han YF. Beneficial effects of the converting enzyme inhibitor, ramipril, in ischemic rat hearts. J Cardiovasc Pharm 1986; 8(suppl 10):S91-S99

15. Cross RB, Dallemagne C, Beswick E, Khosla MC, Bumpus FM. A cardiac specific analog of angiotensin I potentiated by converting enzyme inhibition. Life Sci 1985; 36: 613-618

16. vanGilst WH, deGraeff PA, Wessling H, deLangen CDJ. Reduction of reperfusion
    arrhythmias in the ischemic isolated rat heart by angiotensin converting enzyme inhibitors: a
    comparison of captopril, enalapril and HOE 498. J Cardiovasc Pharm 1987; 9:254-255

17. Saito K, Gutkind JS, Saavedra JM. Angiotensin II binding sites in the conduction system of
    rat hearts. Am J Physiol 1987;253: H1618-H1622

18. Schölkens BA, Linz W, Lindpaintner K, Ganten D. Angiotensin deteriorates but bradykinin
    improves cardiac function following ischemia in isolated rat hearts. J Hypertens 1987, 5
    (suppl): S7-S9

19. Ertl G, Kloner RA, Alexander RA, Braunwald E: Limitation of experimental infarct size by an
    angiotensin converting enzyme inhibitor. Circulation 1982;65: 40-48.

20. Marmor A, Green T, Krakuer J, Szucs T, Schneeweis A. A single dose of cilazapril improves
    diastolic function in hypertensive patients. Am J Med 1989; 87: 61S-63S

21. Remme WJ, Kruyssen HACM, Look MP, Bootsma M, de Leeuw PW. Enalaprilat acutely
    reduces myocardial ischemia in normotensive patients with coronary artery disease through
    neurohumoral modulation. In Remme WJ (ed.): Metabolic and neurohumoral aspects of acute
    myocardial ischemia in man, 1990; CASPARIE (Publ.), Heerhugowaard, 2:346-362

22. Foult JM, Tavolaro O, Antony I, Nitenberg A (1988) Direct myocardial and coronary effects
    of enalaprilat in patients with dilated cardiomyopathy: assessment by a bilateral intracoronary
    infusion technique. Circulation 77: 337-344

AAS 38/III
Recent Progress on Kinins
© 1992 Birkhäuser Verlag Basel

DIRECT EFFECTS OF BRADYKININ ON GLOMERULAR FILTRATION AND PROXIMAL TUBULE REABSORPTION IN RAT KIDNEY

J.L. Ader, F. Praddaude, T. Tran-Van, C. Emond, I. Tack, D. Regoli* and J.P Girolami

Physiology Department and INSERM Unit 133, School of Medicine, Paul Sabatier University,
Toulouse F-31062, France
*Pharmacology Department, Sherbrooke University, PQ, Canada

SUMMARY : Intrarenal infusion of a low dose of bradykinin devoid of extrarenal action exerts direct vasodilator and diuretic effects on rat kidney. It does not significantly alter glomerular filtration rate and absolute proximal reabsorption but increases the delivery of fluid from the proximal tubule evaluated by the clearance of lithium. None of these effects are observed in kidneys infused with both bradykinin and a specific bradykinin antagonist.

INTRODUCTION

Although many previous studies have demonstrated that intravenous or intrarenal infusion of bradykinin increases renal blood flow and urinary excretion of sodium and water, several questions regarding the renal effect of bradykinin remain unanswered. The dose of bradykinin capable of eliciting effects on the kidney is one of these questions. In this regard, and only considering the previous studies conducted on anaesthetised rats with bradykinin infused into the suprarenal aorta or into the renal artery, the effective doses used have been 0.5 µg/min (1), 1 µg kg^{-1} min^{-1} (2), and 2.5 µg kg^{-1} min^{-1} (3). In addition, these doses have been reported to induce either an increase (1) or a decrease (2, 3) in renal vascular resistance. Certain authors have reported that they induce an increase (1), whereas others have reported a decrease (2, 3) in arterial pressure. Another question concerns the effects of bradykinin on glomerular filtration and proximal tubule reabsorption. Bradykinin has little effect on the whole-kidney or single-nephron glomerular filtration rate in spite of large measured increases in renal or single-nephron plasma flow (2, 3, 4, 5). Despite the decreased filtration fraction and the likely subsequent reduction in peritubular colloid osmotic pressure, the absolute rate of fluid reabsorption along the proximal tubule was not changed by bradykinin infusion when evaluated in micropuncture experiments on rat and dog kidneys (2, 3, 4). In contrast, a free water clearance study during water diuresis in dogs indicated earlier that proximal tubular reabsorption was decreased by intrarenal bradykinin infusion (6).

This study was thus conducted to examine (i) the in vivo renal action of a low dose of bradykinin in the absence of any systemic effect, and (ii) the effect of bradykinin infusion on the proximal tubule by using the clearance of lithium as a marker of overall proximal fluid delivery to the distal nephron.

METHODS

Clearance experiments were performed on 50 male Wistar-Kyoto rats aged 10-11 weeks. All rats were fed a normal sodium diet and fasted the night before an experiment. After anaesthesia with pentobarbital sodium (50 mg/kg body wt IP), the animals were placed on a servo-controlled heating table and the trachea was cannulated. One polyethylene tubing cannula was inserted into the left carotid artery for blood sampling and direct AP measurement using a Statham P10 EZ pressure transducer connected to a Gould TA 4000 recorder. Two cannulae were inserted into the right jugular vein, one for infusion of Ringer-lactate solution containing 3% gelatin to replace fluid losses, and the other for infusion of normal saline (10 µl/min) to deliver pentobarbital, inulin, PAH and lithium (2.86 µg/min). A ventral midline incision was made and both ureters were catheterized for urine collection. The right suprarenal artery was carefully cannulated, as previously described (7), under the lens of a microscope with a very thin catheter made of a stretched PE-10 tubing, and infused with normal saline at the rate of 10 µl/min. The rats were allowed to equilibrate for 1 h after completion of surgical preparation.

Right kidney and left kidney haemodynamics and functions were then evaluated separately during two 20-min periods. During the second period, the saline infused into the right renal artery was either pure (15 control rats : C) or conveying 20 ng/min of bradykinin (21 rats : BK) or conveying 20 ng/min of bradykinin and 200 ng/min of the specific antagonist D-Arg[Hyp^3, D-Phe^7]-bradykinin (14 rats : BK+Antag). The infused doses were determined in pilot experiments using a slotted periarterial sensor connected to an electromagnetic flowmeter (Skalar MDL 1401). The dose of bradykinin was determined as the lowest dose that increased renal blood flow by at least 0.5 ml/min every time. The dose of antagonist was determined as the lowest dose that suppressed this effect of bradykinin every time.

Glomerular filtration rate (GFR) was equated with the clearance of inulin (C In) and effective renal plasma flow (RPF) with the clearance of PAH (C PAH). Renal blood flow (RBF) was calculated as the ratio of RPF to 1 - haematocrit. The filtration fraction (FF) was calculated as the ratio of C In to C PAH. Urine and plasma concentration of lithium was determined with a flame emission spectrophotometer (Instrumentation Laboratory 943). The delivery of fluid from the proximal tubules to distal nephron segments, that is to say absolute proximal excretion (APE), was equated with the clearance of lithium (C Li). Fractional proximal excretion (FPE) was calculated as the ratio of APE to GFR. Absolute proximal reabsorption (APR) was calculated as GFR minus APE. All rates of filtration, flow, excretion and reabsorption were expressed per 1 g of kidney weight (g kw) and fractional data as percentages.

The values are reported as means ± SD. The Mann-Whitney U test and the Wilcoxon signed rank test were performed and results with P<0.05 were considered significant.

RESULTS

Mean AP and haematocrit remained unchanged during the experiments. Left contralateral kidney RBF, GFR, FF, urine flow rate and C Li did not differ between and did not significantly change during vehicle, BK or BK+Antag infusion at any time.

BK infusion significantly increased right kidney RBF by 1197 ± 898 μl min^{-1} g kw^{-1} and urine flow rate by 1.2 ± 0.7 μl min^{-1} g kw^{-1} (p<0.001) whereas GFR was not altered significantly (Table 1). Consequently, FF was decreased by 3.0 ± 3.6 % (p<0.001).

Table 1. Right kidney clearances and excretory function in anaesthetised rats.
      1 : during intrarenal infusion of 10 μl/min of saline.
      2 : during intrarenal infusion of 10 μl/min of saline either pure (Control) or conveying
          bradykinin (BK) or conveying bradykinin and bradykinin antagonist (BK+Antag).
Values represent means ± SD. Statistical significance : *** p < 0.001 for differences between period 1 and period 2.

		Control	BK	BK + Antag
Renal blood flow	1	8237 ± 1583	7088 ± 1484	7061 ± 2169
μl min^{-1} g kw^{-1}	2	7633 ± 1473	8285 ± 1405***	6650 ± 1520
Glomerular filtration rate	1	1027 ± 258	925 ± 180	1014 ± 243
μl min^{-1} g kw^{-1}	2	1002 ± 211	968 ± 161	1007 ± 174
Filtration fraction	1	22.6 ± 6.9	23.9 ± 6.1	26.8 ± 5.5
%	2	$23.5 \pm 6,5$	$20.9 \pm 4,2$***	27.6 ± 5.9
Urine flow rate	1	4.8 ± 1.2	4.2 ± 0.8	4.0 ± 1.0
μl min^{-1} g kw^{-1}	2	4.8 ± 1.3	5.5 ± 1.0***	4.0 ± 1.2
Lithium clearance	1	285 ± 116	236 ± 65	249 ± 50
μl min^{-1} g kw^{-1}	2	291 ± 148	321 ± 101***	259 ± 59

Bradykinin increased right kidney C Li by 85 ± 54 μl min^{-1} g kw^{-1} (p<0.001) (Table 1). Proximal excretion was also significantly increased by 7.4 ± 5.0 % when expressed as a fractional value (p<0.001). Nevertheless, the decrease in absolute proximal reabsorption (42 ± 133 μl min^{-1} g kw^{-1}) was not significant (Fig. 1).

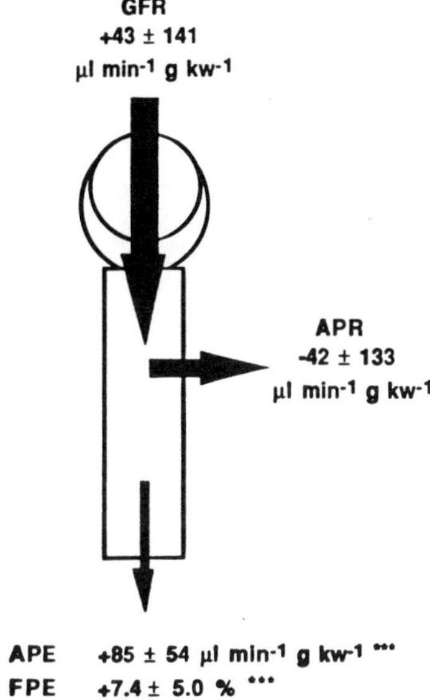

Figure 1. Schematic glomerulus and proximal tubule showing the alterations in glomerular filtration rate (GFR), absolute proximal reabsorption (APR), and absolute and fractional proximal excretion (APE and FPE respectively) induced by intrarenal bradykinin infusion. *** : p < 0.001

DISCUSSION

This study documents the effects of an acute infusion of a low dose of bradykinin into one renal artery on glomerular filtration and proximal tubule reabsorption in anaesthetised rats. Our observations demonstrate that a low dose of bradykinin directly increases renal blood flow without significantly altering glomerular filtration rate and absolute proximal reabsorption.

An important consideration in the conception of the study was to observe the direct renal action of bradykinin. It appears that this objective has been achieved for the following three reasons : (i) the low dose of bradykinin used caused no extrarenal alteration, namely changes in arterial pressure and heamatocrit, that could affect the renal response ; (ii) no alteration in the contralateral kidney heamodynamics and function was detected (which also supports the interpretation of an action restricted to the infused kidney) ; (iii) no effect was observed in kidneys infused with both bradykinin and bradykinin antagonist, thus allowing the reported alterations to be ascribed to bradykinin. Low doses of BK incontestably elicit renal effects. In this regard, it has recently been demonstrated that an intravenous dose as low as 25 ng kg^{-1} min^{-1} was able to displace the intrarenal tubulo-glomerular feedback in rats (8).

In agreement with previous experimental evidence (6, 9, 10, 11), our results indicate that intrarenal bradykinin infusion induces an increase in renal blood flow with no significant change in GFR. Contrary to these findings, it has been reported that intrarenal bradykinin may decrease renal blood flow and increase renal vascular resistance in both conscious and anaesthetised Wistar rats (1, 12). In these studies, the pressor effect of intrarenal bradykinin was only observed with high doses (at least 300 ng/min) that resulted not only in an increase in the infused kidney resistance but also in increases in arterial pressure and in contralateral kidney resistance. In contrast, these doses had no significant effect when infused intravenously, neither on arterial pressure nor on renal resistance (12). Thus, as suggested by the authors, high intrarenal doses do not provide an indication on the direct renal effect of bradykinin but probably evoke a reflex pattern of haemodynamic changes depending on activation of local sensory nerves and, in return, of renal sympathetic efferent nerves.

To our knowledge, the clearance of lithium has never been used to examine the effect of bradykinin on proximal tubule. Interpretation of the tubular function data is based on the assumption that lithium is reabsorbed like sodium and water in the proximal tubule and then passes through the distal segments without significant reabsorption, ie that the C Li is an accurate marker of overall proximal fluid delivery to the distal nephron. (13). This has been conclusively demonstrated in sodium-replete rats (14). Thus, the C Li can be used with some degree of confidence in this study since the rats were on a normal sodium diet prior to the experiments and were infused with 0.9 % NaCl all in all at the rate of 30 µl/min during the experiments.

Contrary to a free water clearance study during water diuresis in dogs (6), our results indicate that absolute proximal reabsorption calculated from GFR and C Li is not altered by intrarenal infusion of BK in rats. To our knowledge, the effect of BK on APR has been previously examined in micropuncture studies performed on dogs (4) and on rats (2, 3). All three studies also failed to detect any alteration in single nephron proximal reabsorption. Taken together, the results indicate that APR remains stable in spite of the generally agreed decrease in whole kidney filtration fraction induced by BK (4, 10, 11). This stability probably results from opposite changes in the Starling forces governing peritubular uptake of reabsorbate (2, 3). In contrast with the stable absolute proximal reabsorption, the delivery of fluid out of the proximal tubule, whether expressed in absolute or in fractional value, is increased by BK. This increase results from the conjunction of both non significant alterations in GFR and APR. It therefore cannot be excluded that the 36 % increase in APE participates in the increased urine flow rate during BK infusion.

To summarize, intrarenal infusion of a low dose of bradykinin devoid of extrarenal action exerts direct vasodilator and diuretic effects on the kidney. It does not significantly alter GFR and APR but increases the delivery of fluid from the proximal tubule evaluated by lithium clearance. None of these effects are observed in kidneys infused with both bradykinin and bradykinin antagonist.

REFERENCES

1. Smits JFM, Brody MJ. Activation of afferent renal nerves by intrarenal bradykinin in conscious rats. Am J Physiol 1984 ; 247 : R1003-R1008.

2. Baylis C, Deen WM, Myers BD, Brenner BM. Effects of some vasodilator drugs on transcapillary fluid exchange in renal cortex. Am J Physiol 1976 ; 230 : 1148-1158.

3. Mertz JI, Haas JA, Berndt TJ, Burnett Jr JC, Knox FG. Effects of bradykinin on renal interstitial pressures and proximal tubule reabsorption. Am J Physiol 1984 ; 247 : F82-F85.

4. Stein JH, Congbalay RC, Karsh DL, Osgood RW, Ferris TF. The effects of bradykinin on proximal tubular sodium reabsorption in the dog : evidence for functional nephron heterogeneity. J C Lin Invest 1972 ; 51 : 1709-1721.

5. Thomas CE, Bell PD, Navar LG. Influence of bradykinin and papaverine on renal and glomerular hemodynamics in dogs. Renal Physiol 1982 ; 5 : 197-205.

6. Willis LR, Ludens JH, Hook JB, Williamson HE. Mechanism of natriuretic action of bradykinin. Am J Physiol 1969 ; 217 : 1-5.

7. Smits JFM, Kasbergen CM, Van Essen H, Kleinjans JC, Struyker-Boudier HAJ. Chronic local infusion into the renal artery of unrestrained rats. Am J Physiol 1983 ; 244 : H304-H307.

8. Morsing P., Persson EG. Kinin and tubuloglomerular feedback in normal and hydronephrotic rats. Am J Physiol 1991 ; 260 : F868-F873.

9. Stein JH, Ferris TF, Huprich JE, Smith TC, Osgood RW. Effect of renal vasodilatation on the distribution of cortical blood flow in the kideny of the dog. J C Lin Invest 1971 ; 50 : 1429-1438.

10. Flamenbaum W, Gagnon J, Ramwell P. Bradykinin-induced renal hemodynamic alterations : renin and prostaglandin relationships. Am J Physiol 1979 ; 237 : F433-F440.

11. Granger JP, Hall JE. Acute and chronic actions of bradykinin on renal function and arterial pressure. Am J Physiol 1985 ; 248 : F87-F92.

12. Janssen BJA, Van Essen H, Struyker Boudier HAJ, Smits JFM. Hemodynamic effects of activation of renal and mesenteric sensory nerves in rats. Am J Physiol 1989 ; 257 : R29-R36.

13. Thomsen K. Lithium clearance : a new method for determining proximal and distal tubular reabsorption of sodium and water. Nephron 1984 ; 37 : 217-223.

14. Thomsen K, Holstein-Rathlou NH, Leyssac PP. Comparison of three measures of proximal tubular reabsorption : lithium clearance, occlusion time and micropuncture. Am J Physiol 1981 ; 241 : F348-F355.

AAS 38/III
Recent Progress on Kinins
© 1992 Birkhäuser Verlag Basel

KALLIKREIN AND KININ RELEASE USING SUPERFUSED DISAGGREGATED CORTICAL CELLS FROM RAT AND HUMAN KIDNEY: EFFECTS OF AVP AND DOPAMINE

K.G. Marshall, [1]D.G. Waller and J.D.M. Albano

Department of Renal and Endocrine Medicine, University of Southampton, St. Mary's Hospital, Portsmouth PO3 6AD and [1]Clinical Pharmacology Group, Southampton General Hospital, Southampton SO9 4XY, UK

SUMMARY: Potential regulators of the renal Kallikrein-Kinin System are poorly defined. We have therefore examined the effect of arginine vasopressin and dopamine on the release of kallikrein and kinin from collagenase-dispersed rat and human renal cortical cells.

INTRODUCTION

The renal Kallikrein-Kinin System (KKS) is one of several intra-renal hormone systems involved in the physiological control of renal blood flow, water and electrolyte transport. Within the kidney, the KKS has been implicated in the regulation of local blood pressure, and a deficiency in the production of components of the system may play an important role in the pathogenesis of hypertension (1,2). However, control of the KKS is poorly understood and potential regulators of the KKS components remain ill-defined.

The vasoconstrictor hormone arginine vasopressin (AVP), has been shown to perform distinct intra-renal roles such as its hydro-osmotic action on the collecting duct and pressor effects on vascular smooth muscle. An interaction between AVP and kallikrein release has been proposed from both in vivo and in vitro studies (3,4) and there is evidence that kinins are involved in the modulation of AVP-mediated anti-diuresis in the collecting duct (5,6). However, AVP-stimulated release of both kallikrein and kinin, independent of humoral, neural and haemodynamic factors has not been shown.

Over the last decade, dopamine (DA) has become established as a major natriuretic hormone which is synthesized and secreted by the kidney (7). In normotensive man, intra-renal DA generation is increased after dietary salt loading. By contrast, some salt-sensitive essential hypertensive patients fail to mobilize DA after an acute sodium load (8). Studies in dogs and man have led to the proposal that DA and the KKS may act together as components in a natriuretic cascade (9, 10). Moreover, the localization of cortical DA receptors at sites close to the location of renal KKS components (11) suggests that part of the vasodilatatory and natriuretic actions of intra-renal DA may be mediated through the KKS. However, no clear relationship between dopaminergic activity and the KKS has been reported.

For a detailed investigation of the factors regulating the release of KKS components at the cellular level, a preparation devoid of haemodynamic and neuronal influences is required. Superfusion of enzymically dispersed rat renal cortical cells has facilitated the delineation of regulatory stimuli and intra-cellular signal transduction systems responsible for renal renin (12) and prostaglandin (13) release. This suggests that this preparation may be applied to the in vitro study of the factors underlying the regulation of the renal KKS. In the experiments to be described, the column superfusion system was adapted for investigating kallikrein and kinin release from disaggregated renal cortical cells in response to AVP or DA.

MATERIALS AND METHODS

Polytetrafluoroethylene (PTFE) superfusion chambers were constructed by Mr. B. Horne, Department of Medical Physics, Middlesex Hospital Medical School (London, UK). ^{125}I Tyr^8 BK was from NEN, Du Pont Limited (Hertfordshire, UK). Bradykinin (BK) was obtained from Biogenesis (Bournemouth, UK) and Worthington collagenase (CLS) was from Lorne Laboratories (Berkshire, UK). AVP, DA and other reagents were obtained from SIGMA Chemical Company (Poole, Dorset, UK) and BDH (Poole, Dorset, UK).

Superfused disaggregated rat and human renal cortical cells were prepared from female Wistar-Kyoto rat kidney tissue and macroscopically normal, surgically removed human kidney tissue using a modification of a method, previously described (14). Briefly, finely minced renal cortical slices were digested in collagenase solution [0.7 mg/ml in Krebs-Ringer bicarbonate buffer (KRBGA) with 0.2% (w/v) glucose and 2% (w/v) bovine serum albumin (BSA)] for three 25 min periods at 37°C. The pooled cell harvest was then spun for 15 min at 200g, the supernatant discarded, the pellet resuspended and then spun again. The washing procedure was repeated a further two times, and the final pellet resuspended in KRBGA containing 0.2% (w/v) BSA after filtering through a 50 μm nylon gauze. Cell numbers were estimated using a haemocytometer. Cell viabilty was assessed by Trypan blue exclusion.

Prepared cells were placed on an inert matrix of polyacrylamide gel in parallel PTFE superfusion chambers designed to display good flow characteristics with minimal adsorption (Figure 1).

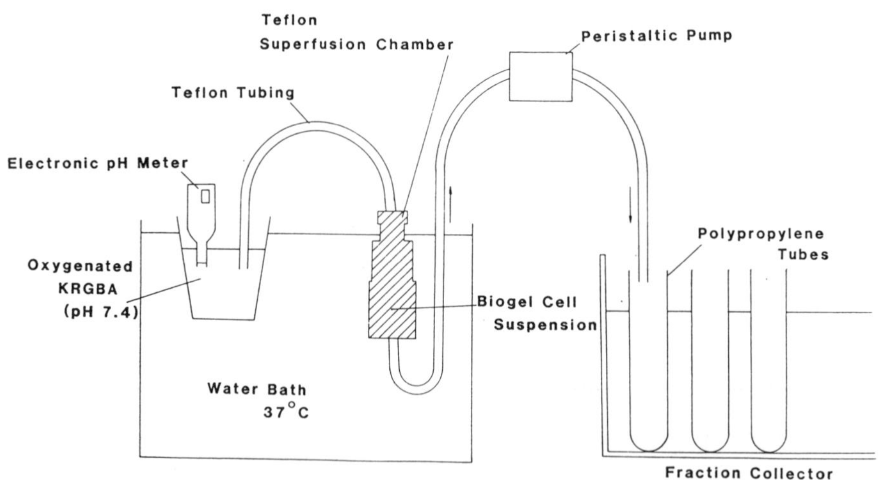

Fig. 1. Components of the column superfusion system

After 90 min equilibration, involving superfusion with KRBGA alone, superfusate samples were collected over 5 min periods. Cells were challenged with 10 min pulses of either AVP or DA, covering the range 10^{-10}-10^{-6}M, in ascending order of concentration, with 30 min recovery periods allowed between doses.

Collected samples were assayed for kallikrein activity and kinin concentration. Kallikrein activity was determined in a kininogenase assay using single donor, human heat-inactivated plasma as a source of kininogen substrate. Generated kinins and superfusate kinin release were then quantitated in a kinin radioimmunoassay using a monoclonal antibody.

Analysis of data

Responses for all superfusion experiments were expressed as integrated Response Ratios (RR). For both secreta, RR was calculated by comparing the mean basal release in the three fractions immediately before and after stimulation, with the mean stimulated response. Each agonist response was therefore compared with an independent set of control measurements. Two-way analysis of variance (ANOVA) and a paired Students t-test was used to compare response values with basal values. A $P \leq 0.05$ was taken to indicate a statistically significant increase. The results are the means $\pm$ SEM.

RESULTS

The effect of AVP

Neither 10^{-10} nor 10^{-9}M AVP significantly increased kininogenase release from rat renal cortical cells. However, a statistically significant increase was obtained after 10^{-8}M AVP (RR= 228.2 $\pm$ 68.4, P< 0.05, n=7), with similar increases at concentrations of 10^{-7}M AVP, and higher. Mean kinin responses (n=10) to AVP were significantly increased by 10^{-7}M AVP and 10^{-6}M AVP (Figure 2).

Using human renal cortical cells (n=6), AVP produced a significant kininogenase response after 10^{-9}M AVP (RR= 172.5 $\pm$ 14.7, P< 0.005) and 10^{-7}M AVP. Significant kinin responses (n=6)were observed at a concentration of 10^{-8}M AVP, and above.

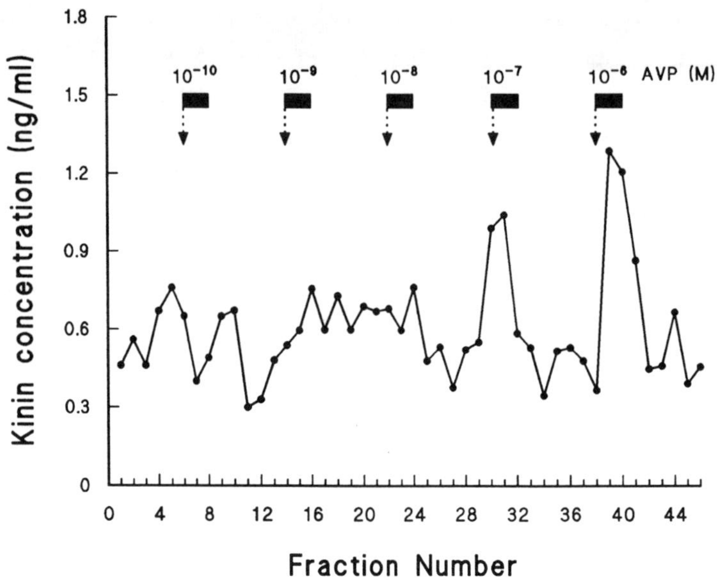

Fig. 2. Sample trace showing the effect of increasing AVP concentrations on kinin release from superfused disaggregated rat renal cortical cells. Dashed arrows represent the beginning of AVP challenge; solid bars the period of AVP superfusion.

The effect of DA

Using rat renal cortical cells (n=5), an increased kininogenase response was demonstrated after 10^{-9}M DA (RR= 176.6 ± 27.3, P< 0.05), and at higher concentrations. A similar pattern was shown for kinin responses (n=11) with a statistically significant kinin response at 10^{-9}M DA (RR= 206.4 ± 16.8, P< 0.001) and above.

Using human renal cortical cells (n=5) 10^{-8}M DA was the threshold concentration required for a statistically significant maximal kinin response (RR= 197.3 ± 32.6, P< 0.05). Statistically significant increments were also demonstrated at 10^{-7}M and 10^{-6}M DA.

DISCUSSION

The distal nephron of the mammalian kidney is responsible for the final regulation of salt and water excretion. In this part of the kidney, the transport of water, electrolytes and urea is adjusted to the specific needs of the whole organism. AVP is capable of acting on the connecting tubule (15), the intra-renal site of kallikrein localization. Moreover, it has been postulated that AVP-mediated stimulation of the KKS may serve to buffer excessive or prolonged hydro-osmotic effects of this hormone, allowing for a precise adjustment of AVP-regulated water and solute permeabilites.

In the experiments described, it was observed that superfused disaggregated rat renal cortical cells challenged with ascending concentrations of AVP, respond with dose-related increases in kallikrein and kinin release. These results provide evidence consistent with the suggestion that the KKS, via increased kinin generation, is capable of responding to AVP-stimulated water reabsorption, by opposing its actions on the collecting duct. However, in the latter regions of the nephron, species differences are evident with respect to the actions of AVP on adenylate cyclase activity or salt transport in hormone responsive cells (16). Extrapolations of AVP-mediated actions from one species to another may therefore be inappropriate and independent observations from human tissue are desirable. It was, therefore, of interest to note that human renal cortical cells responded in a similar manner to rat renal cortical cells, with enhanced kallikrein and kinin release following AVP administration. Thus, the mechanisms controlling kallikrein and kinin responses to AVP appear to operate similarly in both the rat and human distal nephron.

Increasing concentrations of DA significantly enhanced kallikrein and kinin release from superfused dispersed rat renal cortical cells. We are unaware of previous reports of a direct action of DA on kidney KKS activity. These results support the proposal that DA and the KKS may be functionally coupled in the integrated regulation of intra-renal salt and water handling (9). Moreover, the pattern of kinin release from human renal cortical cells in response to DA challenge closely parallels that shown using the rat cell model. Complementary data on the kallikrein response to DA using cells prepared from human kidney tissue was not obtained, since the kininogenase assay was still under development. We conclude that the superfusion system for the study of disaggregated renal cortical cells is applicable to the investigation of the KKS in both human and rat kidney.

ACKNOWLEDGEMENTS

Financial support from the National Kidney Research Fund and the Foundation for Nephrology is acknowledged. We thank Dr. M. Webb (Sandoz Institute for Research) for the gift of the monoclonal antibody.

REFERENCES

1. Fuller PJ, Funder JW. The cellular physiology of glandular kallikrein. Kidney Int 1986; 29:953-964.

2. Scicli AG, Carretero OA The renal kallikrein-kinin system. Kidney Int 1986; 29:120-130.

3. Yamada K, Hasunuma K, Sniina T, Ito K, Tamura Y, Yoshida S. Inter-relationship between urinary kallikrein-kinins and arginine vasopressin in man. Clin Sci 1989; 76:13-18.

4. Chapman ID, Leach JT, Bhoola KD. Studies on renal and urinary kallikrein in spontaneously hypertensive rats. Adv Exp Med Biol 1985; 198A:219-224.

5. Schuster VL, Kokko JP, Jacobsen, HR. Interactions of lysyl bradykinin and anti-diuretic hormone in the rabbit cortical collecting tubule. J Clin Invest 1984; 73:1659-1607

6. Tomita K, Pisano JJ, Burg MB, Knepper MA. Effects of vasopressin and bradykinin on anion transport by the rat cortical collecting duct. J Clin Invest 1986; 77:136-141.

7. Lee MR. Dopamine and the kidney. Clin Sci 1982; 71:439-448.

8. Lee MR. Dopamine, the kidney and essential hypertension. Clin Exp Hypertens 1989; A11 (Suppl 1):149-158.

9. Mills I, Newport P and Obika L. Kallikrein and kinins in the control of blood pressure. In: Secondary forms of

Hypertension. Blaufox MD and Bianchi O, editors. Grune and Stratton, New York: 1981: 195-203.

10. Iimura O, Shimammoto K, Ura N, Nakagawa M, Nishimiya T, Ando T, Yamguchi Y, Masuda A, Ogata H, Saito S, Yamaji I, Fukuyama S. The pathophysiological role of renal dopamine, kallikrein-kinin and prostaglandin systems in essential hypertension. Agents Actions 1987; (Suppl 22) 247-256.

11. Felder RA, Jose PA. Dopamine receptors in rat kidneys identified with ^{125}I-SCH-23982. Am J Physiol 1988; 255:F790-F797.

12. Drury PL, Williams BC, Edwards CRW, Oddie CJ, Horne B. Development and application of a superfusion technique for the study of renin secretion in rat renal cortical cells. Clin Sci 1986; 62:581-587.

13. Wuthrich RP, Loup R, Favre L, Vallotton MB. Dynamic response of PG synthesis to peptide hormones and osmolality in renal tubular cells. Am J Physiol 1986; 250:F709-F797.

14. Williams BC, Duncan FM, Drury PL, Train LMC, Edwards CRW. Dopamine stimulates renin release in isolated renal cortical cells by activation of specific dopaminergic receptors. J Hypertens 1983; 1(Suppl 2):177-179.

15. Imai M. The connecting tubule: A functional subdivision of the rabbit distal nephron segments. Kidney Int 1979; 15:346-356.

16. Abramow M, Beauwens R, Cogan E. Cellular events in vasopressin action. Kidney Int 1987; 15(Suppl 6):S25-S29.

AAS 38/III
Recent Progress on Kinins
© 1992 Birkhäuser Verlag Basel

RENAL HYPERFILTRATION STATES: RELATIONSHIP TO KALLIKREIN AND KININS

R.K. Mayfield[1], A.A. Jaffa[1], A.W. Edmundson[1] and J.N. Harvey[2]

[1]Departments of Medicine and Pharmacology, Medical University of
South Carolina and V.A. Medical Center, Charleston, S.C. 29425,
and [2]University Department of Medicine, The General Infirmary,
Leeds LS1 3EX, United Kingdom

SUMMARY: Animal models and humans with glomerular hyperfiltration and
hyperperfusion secondary to diabetes or high protein diet show increased
renal production of kallikrein and kinins.   Acute aminoacid infusion or
ingestion also raises GFR, RPF and urinary kinins.   Treatment with
aprotinin or a kinin receptor antagonist reverses or prevents
hyperfiltration in these rat models.

INTRODUCTION

Nephropathy occurs in approximately 40% of patients with insulin-
dependent diabetes mellitus (IDDM).   One recent study has suggested that
IDDM patients with glomerular hyperfiltration are at increased risk to
subsequently develop nephropathy (1).   Animal model studies also provide
evidence that glomerular hemodynamic abnormalities that lead to increased
glomerular filtration rate (GFR) are pathogenetically important in the
development of glomerulopathy.

In streptozotocin (STZ)-diabetic rats treated with enough insulin to
maintain normal growth and moderate hyperglycemia, GFR is chronically
increased (2).   These rats develop glomerulosclerosis over months.
Micropuncture measurements show that resistance of the glomerular
arterioles is reduced, afferent more than efferent, resulting in increased
glomerular plasma flow and intracapillary pressure (2).   If STZ-diabetic
rats are not treated with insulin, hyperglycemia is severe, glomerular
arteriolar resistances are increased, and glomerular plasma flow and
filtration are reduced (3).

Altering dietary protein intake has similar effects on glomerular
hemodynamics.   Like severe hyperglycemia, restricting protein intake

increases afferent and efferent arteriolar resistances and reduces glomerular flow and filtration (4). Chronic high protein intake or acute parenteral aminoacid infusion vasodilates the glomerular circulation and raises GFR, similar to the effects of moderate hyperglycemia (5,6).

The factors that mediate these changes in renal vascular resistance are not clearly defined. Although converting enzyme inhibitors lower intraglomerular pressure in hyperfiltering diabetic rats, they do not reverse the glomerular vasodilation and hyperfiltration (2). We have investigated the role of renal kallikrein and its vasodilatory kinin product in diabetes-induced and protein-induced hyperfiltration.

METHODS

Two groups of diabetic rats underwent separate study. The first group (n = 11) received no insulin and displayed severe hyperglycemia (>400 mg/dl) throughout 3 wk of study (SD rats). A second group (n = 18) was treated daily with 1.5-1.75 U of protamine zinc insulin (Lilly, Indianapolis, IN). Insulin was administered within 48 h after STZ administration, and plasma glucose levels were maintained between 200-300 mg/dl over 5-6 wk (MD rats). Separate groups of age-matched control rats were studied with each diabetic group (n = 16 and 17, respectively). All groups were fed 25% protein diets ad lib.

To study effects of dietary protein, normal rats were fed ad lib diets containing 9%, 25% or 50% protein for 8 to 13 days (U.S. Biochemical Corp., Cleveland, Ohio, USA). The dietary protein source was casein. The change in protein content was offset by a change in carbohydrate content (dextrose and cornstarch). Vitamins and electrolytes were constant in the three diets; however, the 9% diet was supplemented with methionine.

At the end of each study, rats were placed individually in metabolic cages, and 24-h urine collections were obtained. Aliquots were stored at -20°C for measurement of glucose, active kallikrein and prokallikrein (7). Immediately after completing urine collections, GFR and renal plasma flow (RPF) were measured by single-injection methods with ^{51}Cr-EDTA and ^{125}I-orthoiodohippuran, respectively (8).

The effects of a kallikrein inhibitor, aprotinin (Bayer A.G., Wuppertal, FRG), on GFR, RPF, and renal and urinary kallikrein activity were studied in MD rats fed 25% protein and normal rats fed 50% dietary protein for 12 days. During the last three days of these studies, rats received twice-daily subcutaneous injections of aprotinin (35,000 KIU) or vehicle. Normal rats fed 25% protein and vehicle-treated were controls for both studies. Left kidney GFR and RPF were measured at the end of these studies by direct clearance methods (8). Thirty minutes prior to measuring GFR and RPF, aprotinin-treated rats were given an intravenous dose of aprotinin (20,000 KIU/kg) followed by an infusion of 2,000-5,000 KIU/kg/min, throughout the measurement period. Immediately thereafter, kidneys were excised, perfused and stored for measurement of tissue kallikrein-like esterase activity and protein (8).

RESULTS

Table 1. Characteristics of untreated severely hyperglycemic diabetic (SD) rats and insulin-treated moderately hyperglycemic diabetic (MD) rats

	Body Wt. (g)	Kidney Wt. (g)	Urine Vol. (ml/day)	Urine Glucose (g/day)
SD	268 ± 10^a	3.25 ± 0.08^a	209 ± 18^a	16.6 ± 1.1^a
Control	373 ± 5	2.83 ± 0.05	21 ± 2	0.0040 ± 0.0001
MD	379 ± 6	2.93 ± 0.08	$42 \pm 4^{b,c}$	$1.7 \pm 0.4^{a,c}$
Control	386 ± 4	2.77 ± 0.07	27 ± 2	0.0040 ± 0.0003

Urine volume and glucose were measured the day before kidney-function studies. Body and kidney weights were measured on the day of kidney studies.
a P <0.001, b P<0.005, vs respective control group.
c P <0.001 vs. SD rats
Reprinted from: Harvey et al., Reference 7.

In SD rats, RPF was significantly reduced, compared to controls (Figure 1A). Although GFR tended to be lower, the reduction was not significant. Excretion of active kallikrein was reduced 30% in SD, compared to age-matched control rats. In contrast to these abnormalities in SD rats, MD rats showed significant elevations in GFR, RPF and active kallikrein (Figure 1B). Active kallikrein excretion in MD was increased nearly 50%

above the excretion rate in controls. Prokallikrein excretion rate was not altered in either SD or MD rats.

Treatment of MD rats with aprotinin, a kallikrein inhibitor, reduced left kidney GFR to 1.59 $\pm$ 0.05 ml/min, a level significantly lower than in vehicle-treated MD rats (2.01 $\pm$ 0.10 ml/min, P < 0.02), and not different from the level in nondiabetic control rats (1.59 $\pm$ 0.13 ml/min). RPF in these respective groups showed similar changes: 4.98 $\pm$ 0.28, 5.95 $\pm$ 0.32 and 4.51 $\pm$ 0.36 ml/min (P < 0.05, aprotinin, vehicle and control). In vehicle-treated MD rats, renal vascular resistance was lower than in normal control rats. Aprotinin treatment raised resistance. Aprotinin also reduced urinary and renal kallikrein-like esterase activity more than 80%, compared to vehicle-treated MD, but did not alter plasma or urinary glucose levels, blood pressure, plasma protein concentration or hematocrit.

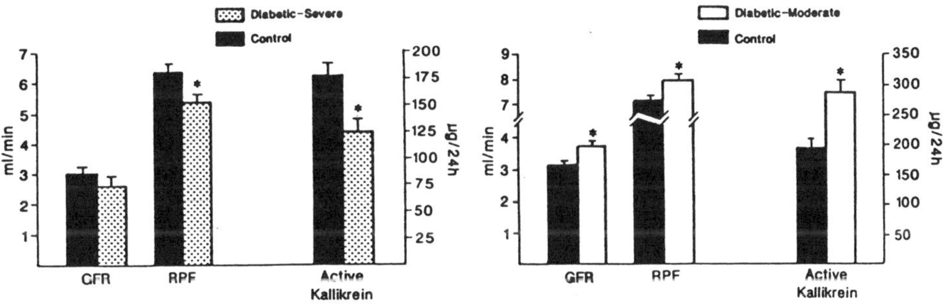

Figures 1A and 1B. GFR, RPF, and active kallikrein excretion in SD and MD rats. * P < 0.005 vs control rats. Reprinted from reference 7.

In response to increasing dietary protein content, GFR showed a progressive increase in normal rats (Figure 2A). RPF was reduced in 9% protein-fed rats compared to 25% protein fed rats. However, RPF did not show an increase in 50% compared to 25% protein-fed rats. Active and prokallikrein excretion progressively increased with increasing protein intake (Figure 2B). In these normal rats fed different protein diets, and in SD and MD diabetic rats, both GFR and RPF correlated directly with excretion rate of active kallikrein. Similar to MD rats, 50% protein-fed normal rats treated with aprotinin showed a reduction in GFR and RPF. Left kidney GFR and RPF in aprotinin-treated 50% rats were 1.54 ± 0.15 and 4.86 ± 0.38 ml/min, respectively, compared to 1.89 ± 0.10 and 5.93 ± 0.22 ml/min in vehicle-treated 50% rats (P < 0.05 or less), and 1.56 ± 0.05 and 5.07 ± 0.15 ml/min in untreated 25% protein-fed rats (P < 0.02 or less vs vehicle-treated 50%).

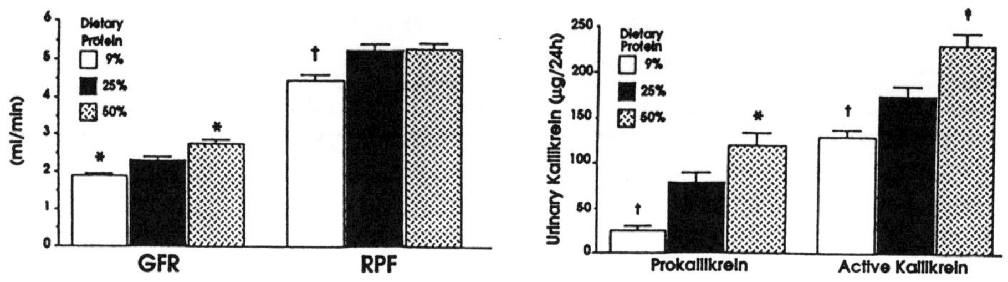

Figures 2A and 2B.   GFR, RPF and excretion rates of active and prokallikrein in normal rats fed 9%, 25%, or 50% protein diets. * P < 0.05;   P < 0.02;   P < 0.005. Reprinted from: Jaffa et al., Reference 8.

DISCUSSION

These data show that changes in renal vascular resistance produced by diabetes or altered protein intake are associated with changes in renal kallikrein excretion, and treatment with a kallikrein inhibitor reverses hyperperfusion and hyperfiltration characteristic of moderate hyperglycemia and high protein intake.

Several observations support a role for renal kallikrein in regulating glomerular hemodynamics. Kallikrein-containing cells of the kidney reside in the distal connecting tubule, which makes a second nephron loop after the macula densa and passes within a few microns of the afferent arteriole (9). Kallikrein traffics in these cells to the basolateral membrane, which is adjacent to the extralumenal surface of the afferent arteriole (9). Kininogen substrate is produced in the principal cells that are interspersed amongst kallikrein-containing connecting tubule cells (10).

These anatomical locations are made more relevant by the finding that addition of kinins to the extralumenal surface of isolated perfused afferent arterioles reduces resistance (11). This effect may be a direct action of kinins or may be mediated via prostacyclin or endothelial derived relaxing factor. The production of both are stimulated by kinins, and there is evidence that both can modulate afferent vascular tone (12, 13).

In a recently completed but unpublished study, we found that type I diabetic patients with hyperfiltration excrete more active kallikrein than patients with normal GFR or normal control subjects (14). We have also reported that normal humans show a rapid increase in renal excretion of active kallikrein, prokallikrein and kinins when protein intake is increased from 40 to 140 g/day (15). These changes precede a rise in GFR. Moreover, acute ingestion of aminoacids in humans or intravenous infusion of aminoacids in rats raises urinary kinins within 1-2 h, and this is accompanied by an increase in RPF and GFR (15, 16). In rats the responses to aminoacids can be blocked by pre-treatment with aprotinin or a B_2 kinin receptor antagonist (16). The same antagonist reverses hyperfiltration and hyperperfusion in MD rats (unpublished data). Jaffa et al. have reported that when MD rats are fed a protein restricted (9%) diet, a diet previously shown to reverse hyperfiltration, renal excretion of active kallikrein is reduced nearly 50% (17).

Finally, our most recent findings indicate that the effects of diabetes and dietary protein on renal tissue kallikrein occur at the level of kallikrein gene transcription (18). Studies are now exploring what signals mediate these gene effects.

ACKNOWLEDGEMENTS

This work was supported by the Research Service of the Department of Veterans Affairs. Dr. Jaffa is the recipient of a Research Development Award from the American Diabetes Association.

REFERENCES

1. Mogensen CE, Christiansen CK. Predicting diabetic nephropathy in insulin-dependent patients. N Engl J Med 1984; 311:89-93.
2. Zatz R, Dunn BR, Meyer TW, Anderson S, Rennke HG, Brenner BM. Prevention of diabetic glomerulopathy by pharmacologic amelioration of glomerular capillary hypertension. J Clin Invest 1986; 77:1925-1930.
3. Hostetter TH, Troy JL, Brenner BM. Glomerular hemodynamics in experimental diabetes mellitus. Kidney Int 1981; 19:410-415.
4. Ichikawa I, Purkerson ML, Klahr S, Troy JL, Martinez-Maldonado M, Brenner BM. Mechanism of reduced glomerular filtration rate in chronic malnutrition. J Clin Invest 1980; 65:982-988.
5. Zatz R, Meyer TW, Rennke HG, Brenner BM. Predominance of hemodynamic rather than metabolic factors in the pathogenesis of diabetic glomerulopathy. Proc Natl Acad Sci USA 1985; 82: 5963-5967.
6. Meyer TW, Ichikawa I, Zatz R, Brenner BM. The renal hemodynamic response to aminoacid infusion in the rat. Trans Assoc Am Phys 1983; XCVI: 76-83.
7. Harvey JN, Jaffa AA, Margolius HS, Mayfield RK. Renal kallikrein and the hemodynamic abnormalities of the diabetic kidney. Diabetes 1990; 39:299-304.
8. Jaffa AA, Harvey JN, Sutherland SE, Margolius HS, Mayfield RK. Renal kallikrein response to dietary protein: A possible mediator of hyperfiltration. Kidney Int 1989; 36: 1003-1010.
9. Vio CP, Figueroa CD, Caorsi I. Anatomical relationship between kallikrein-containing tubules and the juxtaglomerular apparatus in the human kidney. Am J Hypertension 1988; 1:269-271.
10. Figueroa CD, MacIver AG, Mackenzie JC, Bhoola KD. Immunocytochemical localisation of low molecular weight kinogen (LMWK) in the human kidney. Histochemistry 1988; 89:437-442.
11. Edwards RM. Response of isolated renal arterioles to acetylcholine, dopamine, and bradykinin. Am J Physiol 1985; 248:F183-F189.
12. Hura´ CE, Kunau RT. Angiotensin II-stimulated prostaglandin production by canine afferent arterioles. Am J Physiol 1988; 254:F734-F738
13. Ito S, Johnson CS, Carretero OA. Modulation of angiotensin II-induced vasoconstriction by EDRF in the isolated microperfused rabbit afferent arteriole. J Clin Invest 1991; 87:1656-1663.
14. Harvey JN, Edmundson AW, Mayfield RK. Renal kallikrein in IDDM patients with glomerular hyperfiltation. Diabetes 1990; 39:72A.
15. Bolin P, Jaffa AA, Rust PF, Mayfield RK. Acute and chronic responses of human renal kallikrein and kinins to dietary protein. Am J Physiol 1989; 257:F718-F723.
16. Jaffa AA, Vio CP, Silva RH, Vavrek RJ, Stewart JM, Rust PF, Mayfield RK. Evidence for renal kinins as a mediator of aminoacid-induced hyperperfusion and hyperfiltration. J Clin Invest 1992; 89:1460-1468.
17. Jaffa AA, Silva RH, Mayfield RK. Modulation of renal kallikrein production by dietary protein in streptozotocin-diabetic rats with hyperfiltration. Diabetes 1991; 50: 127A.
18. Jaffa AA, Chai KX, Chao J, Chao L, Mayfield RK. Effects of diabetes and insulin on expression of kallikrein and renin genes in the kidney. Kidney Int 1992; 41:789-795.

REGIONAL HEMODYNAMIC EFFECTS OF A KININ ANTAGONIST IN AWAKE NORMOTENSIVE RATS

P. Madeddu, V. Anania, P. Pinna Parpaglia, M. P. Demontis,
M.V. Varoni, M.C. Fattaccio, N. Glorioso

Clinica Medica and Farmacologia, Sassari University,
Viale S. Pietro 8, 07100 Sassari, Italy

SUMMARY: We evaluated the effects of the B_2-receptor antagonist Ac-D-Arg[Hyp^3,D-Phe^7,Leu^8]-bradykinin on mean blood pressure (MBP) and Doppler mesenteric blood flow (DMBF). The antagonist, at a dose able to completely block the hemodynamic effects of exogenous bradykinin, increased MBP by $3 \pm 1\%$ and decreased DMBF by $12 \pm 3\%$. These responses persisted after pharmacological neuro-hormonal blockade. Our results suggest that the kallikrein-kinin system plays a role in the regulation of mesenteric blood flow in rats.

INTRODUCTION

Kinins are endothelium-dependent vasodilators which may be involved as autacoids in the regulation of blood pressure and regional blood flow (1,2). Indirect evidence for a physiological role of endogenous kinins is derived from the cardiovascular effects induced by the administration of exogenous bradykinin into the circulation (3,4). In 1985, five years after the proposal that the biological effects of kinins are mediated by two different receptor types, named B_1 and B_2 (1), the first effective competitive antagonists of bradykinin were discovered (5). Evidence from this experimental approach indicated that B_2 receptors may subserve most of the hemodynamic effects of endogenous kinins.

In the present study, the B_2-receptor antagonist Ac-D-Arg[Hyp^3,D-Phe^7,Leu^8]-bradykinin was used *in vivo* to verify the hypothesis that endogenous kinins play a role in the regulation of mesenteric blood flow in conscious normotensive rats.

MATERIALS AND METHODS

Male Wistar rats (270-330 g) were anesthetized with pentobarbital sodium (50 mg/kg i.p.) and, through a vertical mid-abdominal incision, a miniaturized pulsed Doppler flowmeter probe (Crystal Biotech, Holliston, MA, USA) was placed and sutured around the superior mesenteric artery; two PE-10 tubings (Clay-Adams, Parsippany, NJ, USA) were inserted into the femoral arteries and passed to the abdominal aorta above the origin of the superior mesenteric artery. Doppler probe wires and femoral catheters were tunneled under the skin and then exteriorized at the back of the neck. Five days afterfore, mean blood pressure (MBP) and Doppler mesenteric blood flow (DMBF) were measured with a Statham tranducer (Gould, Oxford, CA, USA) and a three Channel pulsed Doppler instrument (Pulse-Doppler, University of Iowa, Iowa City, IA, USA) connected to a Quartet recorder (Basile, Varese, Italy).

Continuous recordings of MBP (from one of the two arterial catheters) and DMBF were made before (at least 30 min), during and after pharmacological interventions. Drugs were administered via the second femoral catheter with a 100-μl Hamilton syringe (Hamilton, Reno, NV, USA).

Experiment 1: rats were given intra-arterial boluses of 30 μl saline (vehicle), 300 ng/kg bradykinin (Peninsula Laboratories, Belmont, CA, USA) in 30 μl saline, or 300 ng/kg bradykinin combined with 100 μg/kg Ac-D-Arg[Hyp^3,D-Phe^7,Leu^8]-bradykinin (Novabiochem, Laufelfingen, Switzerland) in 30 μl saline. Each group consisted of 8 rats.

Experiment 2: rats were given bolus injection of 30 μl saline (vehicle) or 100 μg/kg Ac-D-Arg[Hyp^3,D-Phe^7,Leu^8]-bradykinin in 30 μl saline.

The same experiment was performed in rats premedicated with 25 mg/kg hexamethonium (Sigma Chemical Company, St. Louis, MO, USA), 4 mg/kg phentolamine (CIBA, Basel, Switzerland), 2 mg/kg propranolol (Sigma Chemical Company, St. Louis, MO, USA), 240 μg/kg chlorpheniramine (Shering, Kenilworth, NJ, USA), 16 mg/kg cimetidine (SK&F Laboratories, Rome, Italy), 30 μg/kg of the vasopressin V_1-receptor antagonist d[CH$_2$]$_5$T[Me]-vasopressin (Bachem, Torrance, CA, USA) and 500 µg/kg enalaprilat (Merck, Sharp & Dohme, Rome, Italy) 5 min prior to the i.a. injection of kinin antagonist or vehicle. The above doses proved to be effective for completeness of blockade in previous studies (6). Each group consisted of 8 rats.

All data were represented as mean ± SE. MBP and DMBF were expressed in millimeters of Hg (mm Hg) and kilohertz (kHz), respectively. Multivariate repeated measures analysis of variance (ANOVA) and profile analysis were performed. Differences between groups were determined using Tukey and paired *t*-tests with Bonferroni multiple comparison adjustment. Mathematical and statistical analyses were performed with a Statview II package, on a Macintosh IICX computer.

RESULTS

No significant hemodynamic change was observed in rats given vehicle (MBP, from 116 ± 4 to 115 ± 3 mm Hg; DMBF, from 4.2 ± 0.3 to 4.2 ± 0.2 kHz). Bradykinin decreased MBP (from 116 ± 3 to 104 ± 2 mm Hg, $p < 0.01$) and increased DMBF (from 4.2 ± 0.3 to 5.7 ± 0.5 kHz, $p < 0.01$). Bradykinin-induced hemodynamic effects were nullified by simultaneous injection of the antagonist (MBP, from 114 ± 4 to 115 ± 2 mm Hg; DMBF, from 4.3 ± 0.2 to 4.3 ± 0.2 kHz).

As shown in Table 1, the antagonist given alone increased MBP by 3 ± 1% ($p < 0.05$) and decreased DMBF by 12 ± 3 % ($p < 0.05$). The hemodynamic parameters came back to baseline within 1 min. The injection of vehicle did not cause any effect.

MBP was lower in rats with neurohormonal blockade compared with that of intact rats. The antagonist caused a mild but significant increase in MBP (from 63 ± 7 to 67 ± 5, $p < 0.05$) and decreased DMBF by 6 ± 1 % (from 3.4 ± 0.4 to 3.2 ± 4 kHz, $p < 0.05$). The latter effect corresponded to increased mesenteric vascular resistances (from 21.6 ± 4.2 to 23.9 ± 4.7 RU, 10 ± 2 % $p < 0.01$). The injection of vehicle did not cause any hemodynamic effect.

Table 1. Effect of the intra-aortic bolus injection of vehicle or kinin antagonist on mean blood pressure and mesenteric blood flow in awake rats.

	Vehicle (n=8)			K.ant. (n=8)		
	BAS	EXP	%	BAS	EXP	%
MBP (mmHg)	115±2	115±3	0±1	114±2	118±2[*]	3±1[#]
DMBF (kHz)	4.6±0.3	4.6±0.4	0±2	4.8±0.6	4.2±0.5[*]	-12±3[#]
DMVR (RU)	29.4±3.5	29.8±3.6	0±2	27.2±4.1	31.8±4.4[*]	20±9[#]

Values are mean ± SE. BAS, baseline value; EXP, experimental value at the peak of response; MBP, mean blood pressure; DMBF, Doppler mesenteric blood flow; DMVR, Doppler mesenteric vascular resistances; vascular resistances are calculated by the formula: mean blood pressure/blood flow (RU, arbitrary vascular resistance unit).
[*] $p < 0.05$, (experimental vs basal value)
[#] $p < 0.05$, (kinin antagonist vs vehicle)

DISCUSSION

In the present study we have found that intra-aortic bolus injection of a bradykinin B_2-receptor antagonist causes a mild increase in blood pressure associated with mesenteric vasoconstriction. These hemodynamic effects persisted following pharmacological neurohormonal blockade.

Bradykinin reportedly causes renal vasodilation in dogs (7) while renal vasoconstriction prevails in anesthetized rats and in rabbit isolated kidney (8,9). In the rat mesenteric vascular bed, bradykinin relaxes the arteries and constricts the veins (10). The diversity of hemodynamic responses to exogenous bradykinin might be due to differences in regional vascular receptor sensitivity in various animal species.

The occurrence of vasoconstriction following the administration of a kinin receptor antagonist may provide more direct evidence for a vasorelaxant action of endogenous kinins. The antagonist Ac-D-Arg[Hyp^3,D-Phe^7,Leu^8]-bradykinin, at a dose able to prevent the cardiovascular effect of exogenous bradykinin, caused a modest increase in MBP. Similar pressor effects have been found to occur with the antagonist D-Arg$[Hyp^3,Thi^{5,8},D-Phe^7]$-bradykinin at doses of 4 mg/kg (6) and 50 µg/rat (11), while lower doses did not alter blood pressure (12). It is conceivable that high doses are needed to antagonize kinins formed in the arterial wall.

In the present study, the antagonist caused a reduction in mesenteric blood flow (a response opposite to that induced by exogenous bradykinin). The finding that the cardiovascular effects of acetylcholine, dopamine or prostaglandin E_2 are not affected by the antagonist (Paolo Madeddu, unpublished results) discounts the possibility that mesenteric vasoconstriction is caused by interaction of the antagonist with vascular receptors of unrelated vasodilators. Vasoconstriction induced by an antagonist of the first generation, D-Arg$[Hyp^3,Thi^{5,8},D-Phe^7]$-bradykinin, in the renal vascular compartment has been attributed to stimulation of the release of catecholamines and renin due to residual agonistic activity (13,14). The antagonist Ac-D-Arg$[Hyp^3,D-Phe^7,Leu^8]$-bradykinin belongs to a new series of pure B_2-receptor antagonists whose agonistic activity was strongly reduced by replacing the aminoacid Phe^8 with Leu (15). In addition, we found that mesenteric vasoconstriction induced by the antagonist was not prevented by pharmacological blockade of baroreceptors and major, rapidly acting hormone systems. All together these data speak against the possibility that the hemodynamic effect of the antagonist was caused by reflex neurohormonal activation.

Responses to kinin antagonist were short in duration possibly because of its susceptibility to fast enzymatic degradation. Due to this limitation, the present study cannot say whether endogenous kinins exert a long-term control of local blood flow.

The gastrointestinal tract and related exocrine glands are rich in kallikrein which, through the release of kinins from kininogen, may participate to the regulation of regional blood flow in that district (16). In particular, our results indicate that endogenous kinins may play an important role in the regulation of mesenteric blood flow in conscious normotensive rats.

REFERENCES

1. Regoli D, Barabe' J. Pharmacology of bradykinin and related kinins. Pharmacol Rev 1980; 32: 1-46.

2. Carretero OA, Scicli AG. Kinins, paracrine hormones, in the regulation of blood flow, renal function, and blood pressure. in Laragh JH, Brenner BM, Kaplan NM (Eds) Endocrine Mechanisms in Hypertension, New York, Raven Press Publishers, 1988; pp 219-239.

3. Goldberg LI, Dollery CT, Pentecost BL. Effect of intrarenal infusions of bradykinin and acetylcholine on renal blood flow in man. J Clin Invest 1965; 44: 1052-1056.

4. Flamenbaum W, Gagnon J, Ramwell P. Bradykinin induced renal hemodynamic alterations: renin and prostaglandin relationships. Am J Physiol 1979; 237: F433-F440.

5. Vavrek RJ, Stewart JW. Competitive antagonists of bradykinin. Peptides 1985; 6: 161-164.

6. Carbonell LF, Carretero OA, Madeddu P, Scicli AG. Effect of a kinin antagonist on mean blood pressure. Hypertension 1988; 11 (Suppl.I): I-84-I-88.

7. Regoli D. Neurohumoral regulation of precapillary vessels. The kallikrein-kinin system. J Cardiovasc Pharmacol 1984; 6 (Suppl 2): S401-S412.

8. Wong PYK, Terragno DA, Terragno NA, McGiff JC. Dual effects of bradykinin on prostaglandin metabolism: relationship to the dissimilar vascular actions of kinins. Prostaglandins 1977; 13: 1113-1125.

9. Barabe' J, Marceau F, Theriault B, Drouin JN, Regoli D. Cardiovascular actions of kinins in rabbits. Can J Physiol Pharmacol 1979; 57: 78-91.

10. Northover RM, Northover BJ. The effect of vasoactive substances on rat mesenteric blood vessels. J Pathol 1970; 101: 99-108.

11. Seino M, Abe K, Nushiro N, Omata K, Kasai Y, Yoshinaga K. Contribution of bradykinin to maintainance of blood pressure and renal blood flow in anaesthetized spontaneously hypertensive rats. J Hyperten 1988; 6, (Suppl. 4) S401-S403.

12. Madeddu P, Glorioso N, Soro A, Manunta P, Troffa C, Tonolo GC, Melis MG, Pazzola A. Effect of a kinin antagonist on renal function and haemodynamics during alterations in sodium balance in conscious normotensive rats. Clin Sci 1990; 78: 165-168.

13. Mulinari A, Benetos A, Gavras I, Gavras H. Interaction of bradykinin with catecholamines in blood pressure regulation. Hypertension 1987; 10: 374-377.

14. Beierwaltes WH, Carretero OA, Scicli AG. Renal hemodynamics in
    response to a kinin analogue antagonist. Am J Physiol 1988;
    255: F408-F414.

15. Rhaleb NE, Telemaque S, Rouissi N, Dion S, Jukic D, Drapeau G,
    Regoli D. Structure-activity studies of bradykinin and related
    peptides: B_2-receptor antagonists. Hypertension 1991; 17: 107-
    115.

16. Berg T, Carretero OA, Scicli AG, Tilley B, Stewart J. Role of
    kinins in the regulation of rat submandibulary gland blood
    flow.   Hypertension 1989; 14: 73-80.

AAS 38/III
Recent Progress on Kinins
© 1992 Birkhäuser Verlag Basel

KININ ANTAGONIST BLUNTS THE DIURETIC EFFECT OF FUROSEMIDE IN DEOXYCORTICOSTERONE-TREATED RATS

P. Madeddu, N. Glorioso, P. Pinna Parpaglia, M.P. Demontis,
M.V. Varoni, M.C. Fattaccio, G. Sabino, G. Pisanu,
G. Patteri and V. Anania

Clinica Medica and Farmacologia, Sassari University,
Viale S. Pietro 8, 07100 Sassari, Italy

SUMMARY: Furosemide at a dose of 3 mg/kg body wt increased urinary volume (vehicle: 12.8 ± 0.6; furosemide: 42.4 ± 2.6 ml/8h, p < 0.01) and urinary sodium excretion (vehicle: 0.9 ± 0.1; furosemide: 5.0 ± 0.4 mM /8h, p < 0.01) in deoxycorticosterone-treated rats. These effects were associated to a decrease in mean blood pressure (from 122 ± 4 to 113 ± 3 mmHg, p < 0.01) and renal vascular resistances (from 15.6 ± 0.6 to 14.3 ± 0.7 RU, p < 0.05). The B_2-receptor antagonist D-Arg[Hyp^3,Thi^5,D-Tic^7,Oic^8]-bradykinin significantly blunted the diuretic and natriuretic effect of furosemide and completely prevented the decrease in blood pressure and renal vascular resistances. The renal kallikrein-kinin system may modulate the diuretic and hemodynamic effects of furosemide in conditions of increased mineralcorticoid activity.

INTRODUCTION

Furosemide increases urinary kallikrein excretion in rats and in man (1,2) either by depleting the enzyme content (3,4) or by activating its synthesis at the level of the connecting tubule (5). As adrenelectomy blunts the urinary kallikrein response to furosemide and deoxycorticosterone treatment of adrenalectomized rats restores it to normal levels, mineralcorticoid activity might modulate the release of kallikrein induced by the furosemide (5). In addition, the kallikrein-kinin system may contribute to the the renal effects induced by furosemide. Aprotinin, an inhibitor of the enzymatic activity of kallikrein, prevents the increase in plasma renin activity induced by furosemide (6). Hypertensive patients responding with systemic and renal vasodilation to diuretic treatment show an increase in urinary kallikrein excretion significantly greater than others without such an hemodynamic effect (7).

It is well known that inhibition of Na^+-Cl^--cotransport on the luminal side of the ascending limb of Henle's loop is the main mechanism of the diuretic action of furosemide. Prostaglandin-induced renal vasodilation may also contribute to the increase in urinary volume and sodium excretion (8). Kinins might be involved in renal effects of furosemide by stimulating prostaglandin release (9,10).

The aim of the present study was to determine if inhibition of endogenous kinins by a long-acting B_2-receptor antagonist of bradykinin can alter the diuretic and natriuretic action of furosemide in rats with or without deoxycorticosterone pretreatment.

MATERIALS AND METHODS

Male Wistar rats (Morini, Como, Italy) weighing 270-330 g were housed at constant room temperature with a 12-hour light-dark cycle and had free access to tap water and rat chow.

Experiment 1: Rats received weekly subcutaneous injections of deoxycorticosterone enantate (DOC, Shering Company, Milan, Italy) at a dose of 25 mg/kg or vehicle (sesame oil) for two weeks. Then, a PE-10 tubing catheter (Clay Adams, Parsipanny, NJ, USA) was inserted into the left femoral vein under light ether anesthesia, advanced into the vena cava, tunneled under the skin and exteriorized at the back of the neck. Experiments were conducted twenty four hours after catheter implantation. Rats were given 300 µg/kg of the antagonist D-Arg[Hyp^3,Thi^5,D-Tic^7,Oic^8]-bradykinin (Hoe 140, Hoechst AG, Frankfurt, Germany) or vehicle (saline, 100 µl) subcutaneously. Ten minutes later, rats (n = 8, each group) received intravenous boluses of: 1) vehicle (100 µl normal saline), 2) furosemide (0.3 mg/kg in 100 µl saline) and 3) furosemide (3 mg/kg in 100 µl saline). Immediately after injection, rats were placed in individual metabolic cages to obtain eigth-hour urine collections. During this period of time, they had free access to tap water but they were deprived of food. Urinary volume (UV) was determined gravimetrically. Urinary sodium ($U_{Na}V$) was determined by flame photometry.

Experiment 2:   Rats received weekly injections of DOC (25 mg/kg, s.c.). After two weeks, they were anesthetized with pentobarbital sodium (50 mg/kg, i.p.) and, through a vertical mid-abdominal incision, miniaturized pulsed Doppler flowmeter probes (Crystal Biotech, Holliston, MA, USA) were placed and sutured around the left renal artery. The probe wires were tunneled out of the back of the abdominal cavity, exteriorized and sutured on the back of the neck. PE-10 tubing catheters were inserted into the left femoral artery and vein, advanced into the abdominal aorta and vena cava, respectively, and exteriorized at the back of the neck. Following closure of the incisions, rats were treated with ampicillin (7 mg/kg, s.c.) and allowed to recover. Experiments were performed 5 days thereafter. The probe wires were connected to the alligator clips of Doppler cables whose terminations were plugged into a three-channel pulsed Doppler instrument (Pulsed-Doppler, University of Iowa, Iowa City, IA, USA) and the arterial catheter was connected to a Statham transducer (Gould, Oxford, CA, USA). Mean Doppler shift which corresponds to renal blood flow and mean blood pressure (MBP) were recorded on a Quartet recorder (Basile, Varese, Italy). On the day of the experiment, rats were placed in cylindrical restrainers. After 30 min of stabilization, they were given 300 µg/kg of Hoe 140 (n = 6) or vehicle (100 µl, n = 6) subcutaneously. Ten minutes later, both groups received furosemide (3 mg/kg in 100 µl saline i.v.). MBP and Doppler renal blood flow (DRBF) were expressed in millimeters of Hg (mm Hg) and kilohertz (kHz), respectively. Renal vascular resistances were calculated by the following formula: MBP/DRBF and expressed in arbitrary resistance units (RU).

All data are expressed as mean ± standard error of the mean. Analysis of variance was performed to test for differences among groups and over time. Differences between groups were determined using paired or unpaired *t*-tests with Bonferroni multiple comparison adjustment.

RESULTS

<u>Experiment 1:</u> as shown in Table 1, furosemide induced a dose-dependent increase in UV and $U_{Na}V$ in the control group given DOC-vehicle. Furosemide-induced diuretic and natriuretic effects were not altered by pretreatment with Hoe 140. In DOC-treated rats, furosemide increased UV and $U_{Na}V$ significantly. Both effects were significantly blunted by pretreatment with Hoe 140.

Table 1. Effects of Hoe 140 on the diuretic and natriuretic action of furosemide in rats pretreated with deoxycorticosterone or vehicle.

GROUP		VEHICLE-TREATED			DOC-TREATED		
Furosemide (mg/kg)		0	0.3	3	0	0.3	3
CONTROL	UV	4.5± 0.5	8.2±[#] 0.8	44.8±[#] 1.4	12.8± 0.6	36.0±[#] 0.5	42.4±[#] 2.6
	$U_{Na}V$	0.3± 0.1	1.5±[#] 0.2	5.4±[#] 0.4	0.9± 0.1	3.5±[#] 0.3	5.0±[#] 0.4
EXPERIM.	UV	3.7± 0.4	11.0±[#] 0.9	39.3±[#] 1.3	15.1± 0.9	21.0±[#o] 1.2	21.8±[#o] 1.4
	$U_{Na}V$	0.4± 0.1	1.6±[#] 0.2	5.3±[#] 0.3	0.8± 0.1	2.5±[#] 0.3	3.2±[#o] 0.3

Values are mean ± SE. Control group received saline and experimental group received 300 µg/kg of antagonist 10 min prior to the intravenous injection of furosemide.
UV, urinary volume (ml/8h); $U_{Na}V$, urinary sodium excretion (mM/8h).
# = p < 0.01 vs vehicle (0 mg/kg furosemide)
o = p < 0.01 vs control

<u>Experiment 2:</u> in DOC-treated rats, furosemide decreased MBP by 7% (from 122 ± to 113 ± 3 mmHg, p < 0.01) and RVR by 8% (from 15.6 ± 0.6 to 14.3 ± 0.7 RU, p < 0.01). DRBF remained unchanged (from 7.9 ± 0.4 to 8.0 ± 0.5 kHz, N.S.). No significant hemodynamic change was induced by furosemide in DOC-treated rats

given Hoe 140 (MBP, from 121 ± 4 to 120 ± 5 mmHg; DRBF, from 8.0 ± 0.3 to 7.9 ± 0.3 kHz; RVR, from 15.2 ± 0.4 to 15.2 ± 0.4 RU).

DISCUSSION

We found that a B_2-receptor bradykinin antagonist blunts the diuretic effect and prevents the decrese in mean blood pressure and renal vascular resistances caused by a single intravenous dose of furosemide in rats given deoxycorticosterone.

Furosemide increases diuresis and natriuresis by inhibiting Na^+-Cl^--cotransport at the luminal side of the ascending limb of the loop of Henle. There are contradictory reports on the existence of a relationship between urinary kallikrein excretion and changes in urinary volume and urinary excretion of sodium after furosemide administration. Abe *et al* (11) and Croxatto *et al* (12) found that urinary kallikrein excretion is closely related to the diuretic and natriuretic action of furosemide, whereas Cinotti *et al* (3) did not find such a correlation and interpreted the increase in urinary kallikrein excretion as a 'wash-out' phenomenon. Bonner *et al* (4) showed that furosemide enhances the urinary excretion of kinins; this effect correlated directly with the changes in urinary volume and sodium excretion. The possibility that kinins generated within the kidney might modulate the pharmacological action of furosemide is supported by our finding that a long-acting and potent B_2-receptor bradykinin antagonist blunts the acute diuretic and natriuretic action of furosemide in DOC-treated rats. Among various factors influencing renal kallikrein gene expression, mineralcorticoids seem to play a central role. Obika *et al* (5) showed that adrenelectomy decreases and DOC treatment enhances urinary kallikrein excretion after furosemide administration. Thus, mineralcorticoid activity may modulate the response of the kallikrein-kinin system to diuretics.

Specific bradykinin-binding sites have been identified in various parts of the nephron such as cortical epithelial cells, collecting tubules, and glomerular membranes (12,13). As far as we know, receptors for bradykinin were not demonstrated at the level of the ascending limb of Henle's loop where furosemide exerts its pharmacological effect. Inhibition of electrolyte reabsorption by furosemide increases the delivery of sodium and water at the level of the distal tubule and stimulates the kallikrein-kinin system. It is conceivable therefore that Hoe 140 blunts the increase in sodium and water excretion induced by furosemide by inhibiting the action of kinins on tubular electrolyte transport at the level of the distal nephron.

Part of the diuretic effect of furosemide has been attributed to renal vasodilation which may be mediated by enhanced formation of renal prostaglandins (8). We observed a reduction in blood pressure in DOC-treated rats; renal vascular resistances also decreased, maintaining renal blood flow constant. The finding that bradykinin-antagonist completely inhibits these effects suggests that endogenous kinins are responsible for the changes in renal vascular tone caused by furosemide. Kallikrein is located not only at the luminal but also at the basolateral side of the distal tubular cells. From the latter location, kallikrein releases kinins into the interstitial fluid and these substances can modulate renal vascular resistances by stimulating the release of arachidonic acid and the synthesis of prostaglandin E_2. Since indomethacin, a cyclooxygenase inhibitor, inhibits the renal vasodilation and blunts the diuretic effect caused by furosemide (8), it is conceivable that kinins initiate but prostaglandins are the final mediators of the renal effects of this diuretic.

In conclusion, our data indicate that the renal kallikrein-kinin system, when activated by high levels of corticosteroids, may contribute to the renal effets of furosemide.

RERERENCES

1.   Croxatto HR, Roblero J, Garcia R, Corthorn J, San Martin ML.
      Effect of furosemide upon urinary kallikrein excretion. Agents
      and Actions 1973; 3/5: 267-274.

2   Olshan AR, O'Connor DT, Preston RA, Frigon RP, Stone RA.
      Involvement of kallikrein in the antihypertensive response to

furosemide in essential hypertension. J. Cardiovasc. Pharmacol. 981; 3: 161-167.

3. Cinotti GA, Stirati G, Taggi F, Ronci R, Simonetti BM, Pierucci R. Relationship between kallikrein and PRA after intravenous furosemide. J. Endocrinol. Invest. 1979: 2: 147-150.

4. Bonner G, Beck D, Deeg M, Marin-Grez M, Gross F. Effects of frusemide on the kallikrein-kinin system of the rat. Clin. Science 1982; 63: 447-453.

5. Obika LFO, Marin-Grez M. Urinary kallikrein response to repeated frusemide injections in rats: effect of adrenalectomy and deoxycorticosterone acetate treatment. Clin. Science 1986; 71: 497-503.

6. Seito S, Kher V, Scicli AG, Beierwaltes WH, Carretero OA. The effect of aprotinin (a serine protease inhibitor) on renal function and renin release. Hypertension 1983; 5: 893-899.

7. O'Connor DT. Response of the renal kallikrein-kinin system, intravascular volume, and renal hemodynamics to sodium restriction and diuretic treatment in essential hypertension. Hypertension 1982; (suppl III): III-72-III-78.

8. Nies AS, Gal J, Fadul S, Gerber JG. Indomethacin-furosemide interaction: the importance of renal blood flow. J. Pharmacol. Exp. Ther. 1983; 226: 27-32.

9. Blasingham MC, Nasjletti A. Contribution of renal prostaglandins to the natriuretic action of bradykinin in the dog. Am. J. Physiol. 1979; 237: F182-F187.

10. Needleman P, Wyche A, Bronson SD, Holmberg S, Morrison AR. Specific regulation of peptide-induced renal prostaglandin synthesis. J. Biol. Chem. 1979; 254: 9772-9777.

11. Abe K, Irokawa N, Yasujima K, Saito T. The kallikrein-kinin system and prostaglandins in the kidney; their relation to furosemide induced diuresis and to the renin-angiotensin-aldosterone system in man. Circ. Res. 1978; 43: 254-260.

12. Regoli D, Barabe' J. Pharmacology of bradykinin and related kinins. Pharmacol. Rev. 1980; 32; 1-46.

13. Bascands JL, Pecher C, Cabos G, Gerolami PG. B2-kinin receptor like binding in rat glomerular membranes. Biochem. Biophys. Res. Communic. 1989; 158: 99-104.

Circulation

Effects of ACE Inhibitors

STIMULATION OF ENDOTHELIAL AUTACOID FORMATION BY INHIBITORS OF ANGIOTENSIN-CONVERTING ENZYME

Markus Hecker, Thomas Benzing and Rudi Busse

Department of Applied Physiology, University of Freiburg, Hermann-Herder-Str. 7, D-7800 Freiburg i.Br., Germany

SUMMARY: We have investigated in human endothelial cells in culture the effects of angiotensin-converting enzyme (ACE) inhibitors on the concentration of intracellular free Ca^{2+} ($[Ca^{2+}]_i$) and the formation of nitric oxide (NO) and prostacyclin (PGI_2). Enalaprilat, moexiprilat and ramiprilat similarly potentiated the increase in $[Ca^{2+}]_i$ elicited by bradykinin and caused an increase in resting $[Ca^{2+}]_i$ when given alone. The latter effect was long-lasting and accompanied by an increased formation of NO and PGI_2. All of these effects were inhibited by the B_2-kinin receptor antagonist Hoe 140, suggesting that the endogenous synthesis/release of bradykinin represents an autocrine mechanism for the stimulation of endothelial autacoid formation. Thus these findings strongly support the concept that ACE inhibitors promote local vasodilation by increasing the level of bradykinin generated in subthreshold concentrations by the endothelium.

INTRODUCTION

Although the primary action of angiotensin-converting enzyme (ACE) inhibitors is thought to be the inhibition of systemic and local angiotensin II formation, a number of experimental and clinical data suggest that other dilator mechanisms may be involved in the hypotensive effect of these compounds (1). One hypothesis is that the blood pressure-lowering effect of ACE inhibitors is to a significant extent due to the accumulation of vasoactive kinins (2). Studies on plasma kinin levels in hypertensive patients receiving ACE inhibitors (3) do not suggest a major role for circulating kinins in this response, so that a local generation of kinins, e.g., in the vessel wall, appears to be the more likely cause. Indeed, ACE inhibitors prevent and reverse the functional and morphological alterations of the endothelium in spontaneously hypertensive rats (4) and hypercholesterolemic rabbits (5).

The potential sources of intravascular kinins remain to be elucidated. ACE, which is identical to the kininase II of the kallikrein-kinin system, however, inactivates bradykinin by liberating the C-terminal dipeptide L-Phe-L-Arg (6), and activation of B_2-kinin receptors on endothelial cells results in the concomitant release of the potent vasodilators prostacyclin (PGI_2) and nitric oxide (NO) (7). Moreover, there is some evidence that stimulators of endothelial NO formation, such as acetylcholine, ATP or substance P, can be released from endothelial cells themselves, thus representing an effective paracrine dilator mechanism (8).

Consequently, we have studied the effects of the ACE inhibitor ramiprilat on the formation of PGI_2 and NO by cultured endothelial cells (9). We have also investigated its effect on endothelial Ca^{2+} levels (10) and found that ramiprilat enhances both PGI_2 and NO biosynthesis by inhibiting the breakdown of endothelium-derived bradykinin. Herein we demonstrate that two other ACE inhibitors, enalaprilat and moexiprilat, mimick the effect of ramiprilat and that the bradykinin-mediated increase in endothelial autacoid formation is based on an elevation of the intracellular free Ca^{2+} concentration ($[Ca^{2+}]_i$).

MATERIALS AND METHODS

MATERIALS. 3-Isobutyl-1-methyl-xanthine (IBMX) and N^G-nitro-L-arginine were purchased from Serva (Heidelberg, FRG) and Indo-1/AM from Boehringer (Mannheim, FRG). Hippuryl-L-histidyl-L-leucine was obtained from Sigma (Deisenhofen, FRG) and [glycine-1-^{14}C] hippuryl-L-histidyl-L-leucine (specific activity 2.97 Ci/mol) from DuPont de Nemours (Dreieich, FRG). The B_2-kinin receptor antagonist Hoe 140 (D-Arg-[Hyp^3, Thi^5, D-Tic^7, Oic^8]bradykinin) and ramiprilat were kindly provided by Hoechst (Frankfurt a.M., FRG) and enalaprilat and moexiprilat by MSD Sharp & Dohme (Munich, FRG) and Pharma Schwarz (Monheim, FRG), respectively.

CELL CULTURE. Human umbilical vein endothelial cells (HUVECs) were isolated from umbilical cords, seeded on 6-well or 24-well plates precoated with 25 μg/ml fibronectin or on quartz coverslips and grown to confluence in M-199 culture medium as described previously (9).

DETERMINATION OF ACE ACTIVITY. Cells grown in 6-well plates were incubated with 1.1 mM [glycine-1-^{14}C]-hippuryl-L-histidyl-L-leucine (specific activity 0.1 Ci/mol) in 1 ml of 10 mM potassium phosphate buffer, pH 8.3, containing 300 mM sodium chloride (11), for 60 min. Incubations were terminated by adding 1 ml 0.1 N hydrochloric acid followed by a 10 min centrifugation at 2,500xg. Labelled hippurate was extracted from the deproteinated samples with 1 ml ethyl acetate followed by another 10 min centrifugation at 2,500xg to facilitate separation of the aqueous and organic layer. An aliquot of the organic layer (400 μl) was mixed with 10 ml scintillation fluid (Biofluor, DuPont de Nemours) and counted for radioactivity in a ß-liquid scintillation counter (Kontron MR 300).

DETERMINATION OF ENDOTHELIAL CYCLIC GMP LEVELS AND PROSTACYCLIN RELEASE. After aspiration of the culture medium, the cell monolayers were washed twice with 2 ml HEPES/Tyrode's solution and then incubated for 15 min with 100 μM IBMX at 37°C. Drugs and solvents were added to the cells at the concentrations and times indicated in the results section. Incubations were terminated by removing the supernatant and extracting the cells with 6% (v/v) ice-cold trichloroacetic acid. The concentration of 6-keto-prostaglandin $F_{1\alpha}$, the stable hydrolysis product of prostacyclin, in the supernatant was determined by using a specific radioimmunoassay. The cells were then extracted, processed and cyclic GMP levels determined by radioimmunoassay as described (9). Sample protein concentrations were determined by using Peterson's modification of the micro-*Lowry* assay (12).

MEASUREMENT OF INTRACELLULAR CALCIUM. Cells on coverslips were loaded with the fluorescent Ca^{2+} indicator dye Indo-1/AM and changes in $[Ca^{2+}]_i$ were monitored essentially as described previously (10).

STATISTICAL ANALYSIS. All data in the figures are expressed as mean±SEM. Statistical evaluation was performed by Student's t test or one-way analysis of variance (followed by a Bonferroni t test), where appropriate, with a p-value <0.05 considered statistically significant.

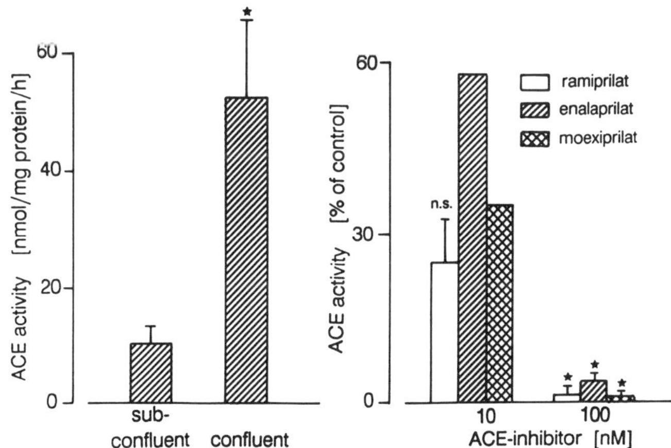

Figure 1. ACE activity (nmol [1-^{14}C]hippurate/mg protein/h) of sub-confluent (70% density) or confluent HUVECs in the absence or presence of ACE inhibitors ($n=3$ except enalaprilat (10^{-8}M, $n=1$) and moexiprilat (10^{-8}M, $n=1$); *$p<0.05$ vs. control, i.e. sub-confluent or confluent cells; n.s., not significant).

RESULTS

ACE ACTIVITY. ACE activity was detected in both sub-confluent and confluent primary cultures of HUVECs (Fig. 1) with the confluent cells being significantly more active than sub-confluent cells (5.3-fold, $p < 0.05$). Thus the expression of ACE activity in primary cultures of human endothelial cells seems to be cell cycle-dependent. Ramiprilat, enalaprilat or moexiprilat almost completely abrogated the ACE activity in confluent HUVECs at $10^{-7}M$ (93-98% inhibition, $p < 0.05$, Fig. 1).

CA^{2+} HOMEOSTASIS. As with ramiprilat (10), incubations of HUVECs with enalaprilat $(3x10^{-7}M)$ for 90 min caused a small but significant increase in resting $[Ca^{2+}]_i$ (1.5-fold, $p < 0.05$, Fig. 2) and potentiated (2.4-2.9-fold, $p < 0.05$) the dose-dependent mobilization of $[Ca^{2+}]_i$ elicited by bradykinin (Fig. 2). Moexiprilat $(3x10^{-7}M)$ also caused an increase in resting Ca^{2+} and this effect and that of the other ACE inhibitors was abolished by the B_2-kinin receptor antagonist Hoe 140 $(10^{-7}M)$. Interestingly the increase in basal $[Ca^{2+}]_i$ induced, e.g., by ramiprilat, was maintained for several hours (Fig. 4).

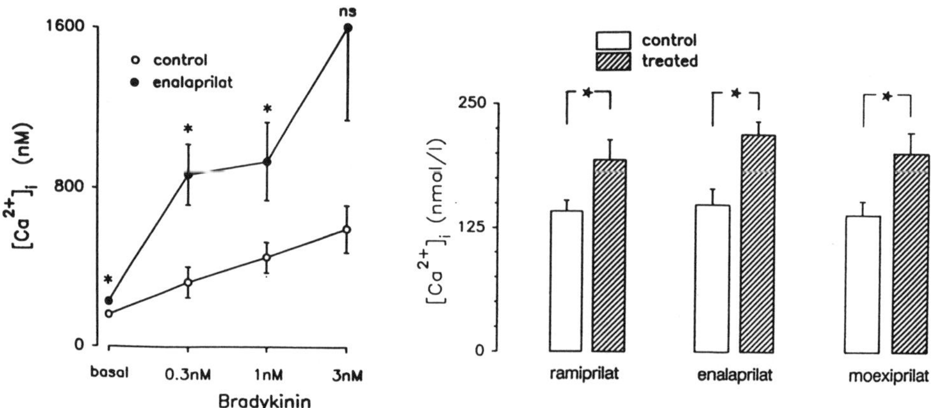

Figure 2. Effect of enalaprilat $(3x10^{-7}M)$ on resting (basal) and bradykinin-stimulated $[Ca^{2+}]_i$. HUVECs on coverslips were loaded with Indo-1/AM, in the presence or absence of enalaprilat, for 90 min at 37°C, and changes in $[Ca^{2+}]_i$ were monitored with or without stimulation by bradykinin. The figure $(n = 3-5)$ shows the peak increase in $[Ca^{2+}]_i$ (*$p < 0.05$ vs. control).

Figure 3. Changes in resting $[Ca^{2+}]_i$ in HUVECs after 90 min in the presence of ramiprilat $(3x10^{-7}M, n = 11)$, enalaprilat $(3x10^{-7}M, n = 5)$ or moexiprilat $(3x10^{-7}M, n = 4$; *$p < 0.05$ vs. control).

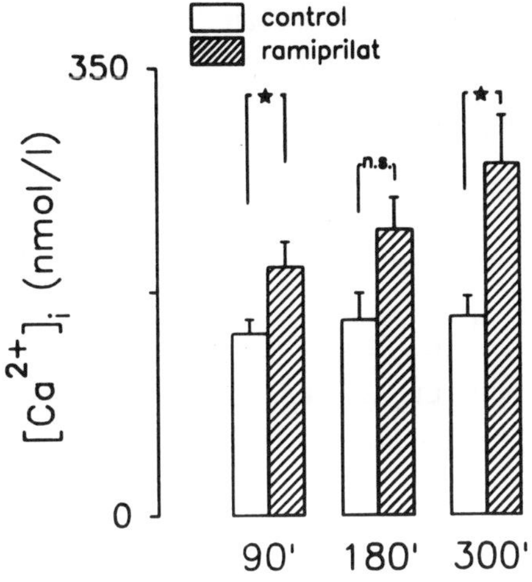

Figure 4. Time course of the increase in resting $[Ca^{2+}]_i$ elicited by ramiprilat (3×10^{-7}M, $n=5$). HUVECs on coverslips were incubated, with or without ramiprilat, for 0, 90 or 210 min at 37°C followed by 90 min loading with Indo-1/AM, in the presence or absence of ramiprilat, and subsequent determination of $[Ca^{2+}]_i$ (*$p<0.05$ vs. control; n.s., not significant).

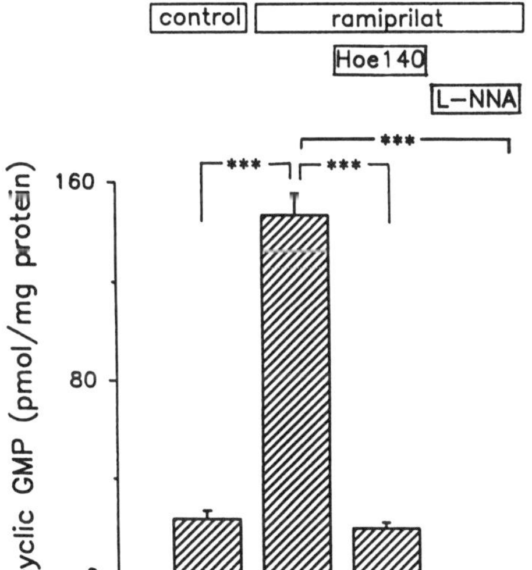

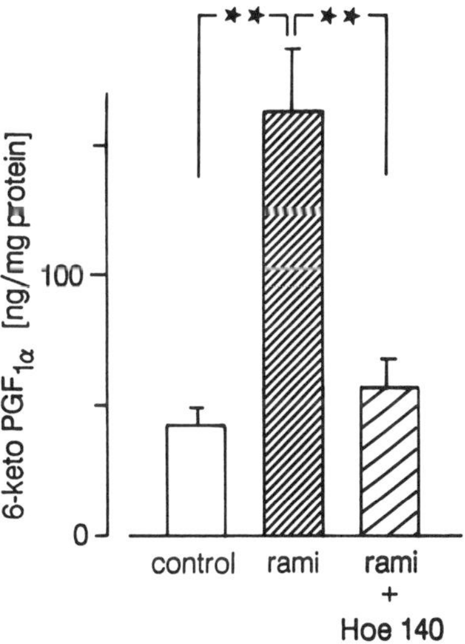

Figure 5. Cyclic GMP levels in HUVECs after 30 min at 37°C in the absence (control, $n=9$) or presence of ramiprilat (3×10^{-7}M, $n=5$). Note the inhibition by Hoe 140 (10^{-7}M, $n=5$) or N^G-nitro-L-arginine (L-NNA, 3×10^{-5}M, $n=3$) of the increase in intracellular cyclic GMP content evoked by ramiprilat (***$p<0.001$ vs. control or ramiprilat, respectively).

Figure 6. Ramiprilat (10^{-7}M) potentiates the basal release of PGI_2 from HUVECs. The cells were incubated with ramiprilat for 15 min at 37°C; Hoe 140 (10^{-7}M) was added 5 min before the addition of ramiprilat ($n=5$, **$p<0.01$ vs. control or ramiprilat, respectively).

INTRACELLULAR CYCLIC GMP. Incubations of ramiprilat (3×10^{-7}M) with HUVECs for 30 min substantially elevated (6-fold, $p < 0.001$, Fig. 5) their basal cyclic GMP level (24 ± 2 pmol/mg protein, $n = 9$), and this increase, as with the Ca^{2+} signal, was maintained for several hours. Co-incubations with Hoe 140 (10^{-7}M) abolished the effect of the ACE inhibitor and no increase in cyclic GMP was observed in the presence of the NO synthase inhibitor N^G-nitro-L-arginine (L-NNA, 3×10^{-5}M, Fig. 5), whereas N^G-nitro-D-arginine had no effect ($n = 3$).

PGI$_2$ RELEASE. When incubated with monolayers of HUVECs for 15 min, ramiprilat (10^{-7}M) significantly potentiated (3.8-fold, $p < 0.05$, Fig. 6) the accumulation of 6-keto-PGF$_{1\alpha}$ in the supernatant, and this effect was prevented by an equimolar dose of Hoe 140 (88% inhibition, $p < 0.05$).

DISCUSSION

This study demonstrates that incubation of human endothelial cells with ACE inhibitors such as ramiprilat, enalaprilat or moexiprilat leads to a significant increase in resting $[Ca^{2+}]_i$ and, as a consequence, in NO and PGI$_2$ formation. As all of these effects were abolished by the potent B$_2$-kinin receptor antagonist Hoe 140, these findings suggest that bradykinin (or a related peptide), formed in and released by endothelial cells, can act in an autocrine manner, provided that kinin degradation is prevented by ACE inhibition. This hypothesis is supported further by the finding that primary cultures of human endothelial cells contain substantial ACE activity which is blocked by the ACE inhibitors at the concentrations tested in this study.

The sustained elevation of endothelial NO and PGI$_2$ production after ACE inhibition may have important physiological implications and could explain the cytoprotective effects of ACE inhibitors in the vascular wall (4,5). Besides their potent vasodilator effects, NO and PGI$_2$ act synergistically to inhibit platelet aggregation and adhesion (13). Both endothelial autacoids can also modulate monocyte-vessel wall interactions with NO preventing the adhesion of monocytes to endothelial cells, whereas PGI$_2$ inhibits their chemotactic response (14). As the adherence of monocytes and their subsequent migration into the arterial wall are thought to be early events in the development of atherosclerosis, it is conceivable that an enhanced endothelial autacoid formation may help to prevent the initiation of atherosclerosis. PGI$_2$ may have additional anti-atherosclerotic effects resulting from a cyclic AMP-mediated reduction in cholesterol accumulation in arterial smooth muscle cells (15). Finally, NO and PGI$_2$, via their respective cyclic nucleotide second messenger systems, may inhibit smooth muscle mitogenesis and proliferation (16,17).

The accumulation of kinins in the culture medium of endothelial cells grown in dishes under no-flow conditions is not comparable with the physiological situation *in vivo* where kinins released from the endothelium may be considerably diluted by the streaming blood. However, the blood flow itself, by generating shear stress (viscous drag) on the luminal surface of the endothelium, may greatly enhance kinin release. Indeed compounds known to stimulate NO biosynthesis such as ATP, acetylcholine or substance P are released from cultured human endothelial cells *in vitro* by increased flow (8). Substance P is also released from the rat hind limb vasculature *in vivo* when the endothelium is intact (18), and an endothelium-dependent relaxation by ACE inhbibitors has been documented in perfused canine carotid arteries (19,20). Thus it is conceivable that flow also facilitates the release of bradykinin from the endothelium.

In conclusion our findings suggest that the protection by ACE inhibitors of the morphological and functional integrity of the vessel wall *in vivo* may be due to the local accumulation of endothelium-derived bradykinin which, in an autocrine manner and by elevating $[Ca^{2+}]_i$, acts as a potent stimulus for endothelial autacoid formation.

ACKNOWLEDGMENTS

This study was supported by the Deutsche Forschungsgemeinschaft (Bu 436/4-2, He 1587/2-1) and in part by MSD Sharp & Dohme GmbH, Germany. The authors are indebted to Isabel Winter, Christine Kircher and Olaf Sommer for expert technical assistance and the midwifes of the St. Josephs Hospital Freiburg for providing umbilical cords.

REFERENCES

1. Kramer HJ, Glänzer K, Meyer-Lehnert H, Mohaupt M, Predel HG. Kinin- and non-kinin-mediated interactions of converting-enzyme inhibitors with vasoactive hormones. J Cardiovasc Pharmacol 1990; 15(Suppl.6):S91-S98.

2. Schrör K. Converting-enzyme inhibitors and the interaction between kinins and eicosanoids. J Cardiovasc Pharmacol 1990; 15(Suppl.6):S60-S68.

3. Iimura O, Shimamoto K. Role of kallikrein-kinin system in the hypotensive mechanisms of converting-enzyme inhibitors in essential hypertension. J Cardiovasc Pharmacol 1989; 13(Suppl.3):S63-S66.

4. Clozel M, Kuhn H, Hefti F. Effects of angiotensin converting enzyme inhibitors and of hydralazine on endothelial function in hypertensive rats. Hypertension 1990; 16:532-540.

5.   Wiemer G, Becker RHA, Linz W. Diminished endothelial dysfunction in long term hypercholesterolemic rabbits by inhibition with ramipril. Arch Int Pharmacodyn Ther 1990 (abstract); 305:293.

6.   Erdös EG. Some old and some new ideas on kinin metabolism. J Cardiovasc Pharmacol 1990; 15(Suppl.6):S20-S24.

7.   Lückhoff A, Pohl U, Mülsch A, Busse R. Differential role of extra- and intracellular calcium in the release of EDRF and prostacyclin from cultured endothelial cells. Br J Pharmacol 1988; 95:189-196.

8.   Milner P, Kirkpatrick KA, Ralevic V, Toothill V, Pearson J, Burnstock G. endothelial cells cultured from human umbilical vein release ATP, substance P and acetylcholine in response to increased flow. Proc R Soc Lond B 1990; 241:245-248.

9.   Wiemer G, Schölkens BA, Becker RMA, Busse R. Ramiprilat enhances endothelial autacoid formation by inhibiting breakdown of endothelium-derived bradykinin. Hypertension 1991; 18:558-563.

10.   Busse R, Lamontagne D. Endothelium-derived bradykinin is responsible for the increase in calcium produced by angiotensin-converting enzyme inhibitors in human endothelial cells. Naunyn-Schmiedebergs Arch Pharmacol 1991; 344:126-129.

11.   Okamura T, Miyazaki M, Inagami T, Toda N. Vascular renin-angiotensin system in two-kidney, one clip hypertensive rats. Hypertension 1986; 8:560-565.

12.   Lowry OH, Rosebrough NJ, Farr AL, Randall RJ. Protein measurement with phenol reagent. J Biol Chem 1951; 193:265-275.

13.   Radomski MW, Palmer RMJ, Moncada S. The anti-aggregating properties of vascular endothelium: interactions between prostacyclin and nitric oxide. Br J Pharmacol 1987; 92:639-646.

14.   Bath PMW, Hassall DG, Gladwin A, Palmer RMJ, Martin JF. Nitric Oxide and Prostacyclin. Divergence of inhibitory effects on monocyte chemotaxis and adhesion to endothelium in vitro. Arteriosclerosis Thrombosis 1991; 11:254-260.

15.   Hajjar DP. Prostaglandins and cyclic nucleotides - Modulators of arterial cholesterol metabolism. Biochem Pharmacol 1985; 34:295-300.

16.   Garg UC, Hassid A. Nitric oxide-generating vasodilators and 8-bromo-cyclic guanosine monophosphate inhibit mitogenesis and proliferation of cultured rat vascular smooth muscle cells. J Clin Invest 1989; 83:1774-1777.

17.   Moncada S, Higgs AE. Prostaglandins in the pathogenesis and prevention of vascular disease. Blood Rev 1987; 1:141-145.

18.   Ralevic V, Milner P, Hudlicka O, Kristek F, Burnstock G. Substance P is released from the endothelium of normal and capsaicin-treated rat hind-limb vasculature, invivo, by increased flow. Circ Res 1990; 66:1178-1183.

19.   Vanhoutte PM, Auchschwelk W, Biondi ML, Lorenz RR, Schini VB, Vidal MJ. Why are converting enzyme inhibitors vasodilators. Br J Clin Pharmacol 1989; 28:S95-S104.

20.   Mombouli JV, Vanhoutte PM. Kinins and endothelium-dependent relaxation to converting enzyme inhibitors in perfused canine arteries. J Cardiovasc Pharmacol 1991; 18:926-927.

EFFECTS OF CONVERTING ENZYME INHIBITION ON ENDOTHELIAL BRADYKININ METABOLISM AND ENDOTHELIUM-DEPENDENT VASCULAR RELAXATION

Claus Bossaller, Wolfgang Auch-Schwelk, Michael Gräfe,
Kristof Graf, Claus Baumgarten* and Eckart Fleck.

Department of Cardiology, German Heart Institute Berlin,
Augustenburger Platz 1, 1000 Berlin 65, Germany
*Department of Clinical Immunology, Free University, Berlin

SUMMARY: The effects of ACE-inhibitors on bradykinin metabolism and bradykinin-induced endothelium-dependent relaxtion were studied in isolated coronary arteries and endothelial cells in culture. The results suggest that ACE-inhibitors affect coronary vascular tone by at least two endothelium-dependent and bradykinin-mediated mechanisms: First, ACE-inhibitors decrease endothelial bradykinin degredation which is accomponied by an augmented bradykinin mediated endothelium-dependent relaxation. Second, ACE-inhibitors evoke endothelium-dependent relaxations in coronary arteries stimulated with threshold concentrations of bradykinin, which cannot be attributed to an inhibition of bradykinin degradation. The effect appears to represent a new mechanism which may be based on an interaction of the bradykinin receptor and the angiotensin converting enzym on the cellular level.

INTRODUCTION

The vascular effects of ACE-inhibitors are partially attributed to bradykinin mediated effects (1,2). Bradykinin is a potent vasodilator in vivo and relaxes isolated arteries in the presence of an intact endothelium. The compound is inactivated by angiotensin converting enzyme (3). Thus, ACE-inhibitors may increase bradykinin concentrations and thereby induce vaso-dilation. Although measurements of plasma bradykinin during ACE-inhibitor treatment gave inconsistent results, which may be attributed to methodological difficulties in the measurement of bradykinin in blood (4), ACE-inhibitors may modify bradykinin meta-bolism at the local vascular tissue level not reflected by changes in circulating bradykinin. In addition, ACE-inhibitors may modulate bradykinin mediated vasodilation by other mechanism, including an interaction with bradykinin receptors. In order to investigate the effects of ACE-inhibitors on the local endothelial bradykinin metabolism and endothelium-dependent vascular relaxation, the effects of these drugs were studied in isolated coronary arteries and endothelial cells in culture.

METHODS

MEASUREMENTS OF ISOMETRIC TENSINON IN ISOLATED CORONARY ARTERIES: Bovine and human coronary arteries (from heart transplantation patients) were placed in modified Krebs Henseleit bicarbonat solution (equilibrated at 37°C with 95%O_2-5%CO_2) of the following composition (mM): NaCl 118.0; KCl 4.7; $CaCl_2$ 2.5; $MgSO_4$ 1.2; $NaHCO_3$ 25.0; KH_2PO_4 1.2; Glucose 11.1, and EDTA 0.025. Rings were mounted in organ chambers at their optimal passive tension for the measurement of isometric force. Indomethacin (10^{-5}M) was present throughout the experiments to prevent synthesis of vasoactive prostaglandins. In a first set of experiments, relaxations in response to brady-kinin or substance P were studied in rings contracted with $PGF_{2\alpha}$ (1-5 x 10^{-6}M) after one hour of preincubation with or without ACE-inhibitors. In a second set of experiments, the rings were contracted with $PGF_{2\alpha}$ and relaxed with threshold concentrations of bradykinin or substance P followed by the cummulative addition of ACE-inhibitors.

BRADYKININ RIA: Bradykinin concentrations in the incubation medium were deter-mined using a specific radioimmunoassay (5). The RIA employed Tyr-8-Bk which was iodinated with ^{125}J by the Method of Hunter and Grennwood (6). The specific activity of the labelled bradykinin obtained by this method was 1.3-1.4 Ci/umol. Rabbit antiserum to bradykinin was produced by multiple injections of bradykinin, which had been coupled to chicken serum albumin by the method of Goodfriend et al. (7). Results were calculated using a spline-fit computer programme. The minimum detection level of this method is 20 pg/ml bradykinin. The crossreactivity is <2% to highly purified human kininogen and <1% to des-arg9-bradykinin.

MEASUREMENT OF BRADYKININ BREAKDOWN IN ISOLATED CORONARY ARTERIES: Bradykinin (10^{-8}M) was added to the coronary arteries (approximately 5 grams) after a preincubation period of 2 h in 30 ml oxygenated Krebs buffer at 37° C in the presence or absence of ACE-inhibitors. Bradykinin concentrations were measured by RIA at 10, 30, 60 and 120 minutes after the addition of bradykinin.

MEASUREMENT OF BRADYKININ BREAKDOWN IN ENDOTHELIAL CELL CULTURE: Human umbilical vein endothelial cells (HUVEC) were prepared by standard methods (8) with slight modifications (9). Confluent cultured cells from the second passage were incubated in serum-free medium (RPMI 1640), containing 10 mM HEPES, antibiotics and peptidase inhibitors (amastatin, 10^{-5}M; phosphoramidone, 10^{-8}M; MGTA, 10^{-4}M) in the presence or absence of ACE-inhibitors. After the addition of 10^{-8}M bradykinin, samples were taken at various times. The samples were mixed with 0.4ml 0.2mM Tris + 0.01 M EDTA buffer (pH 6.4), frozen immediately in liquid nitrogen and stored at -70° C for further use.

RESULTS

Bradykinin (10^{-12}-10^{-9}M) and substance P (10^{-9}-10^{-8}M) induced endothelium-dependent relaxations in bovine coronary arteries. The relaxations in response to bradykinin (Fig. 1) but not to substance P were augmented after preincubation with the ACE-inhibitor lisinopril (10^{-6}M). In human coronary arteries, bradykinin induced smaller endothelium-dependent relaxations compared to the bovine tissue. The threshold effect was at 10^{-9}M and the maximal relaxation was 32±12% in response to 10^{-6}M bradykinin (n=6). These relaxations were significantly augmented in the presence of lisinopril (+10^{-6}M: 52±13%, n=6, p<0.01). Similar results were obtained with enalaprilat and captopril.The effects of the ACE-inhibitor on the degradation of bradykinin by bovine coronary arteries is shown in Figure 2. Lisinopril significantly reduced the bradykinin degradation.

In endothelial cell monolayers the bradykinin breakdown was inhibited by various ACE-inhibitors including enalaprilat, captopril and lisinopril. Concentrations as little as 10^{-10}M already partially inhibited the degradation. Almost complete inhibition was achieved with ACE-inhibitor concentrations of 10^{-6}M (enalaprilat, lisinopril) and at 10^{-5}M captopril (Fig. 3).

The results presented so far demonstrate that ACE-inhibitors inhibit bradykinin degradation in coronary arteries and endothelial cells in culture. This effect is accompanied by (and may be the cause of) the augmentation of bradykinin mediated endothelium-dependent relaxation observed in the presence of ACE-inhibitors.

In further experiments coronary arteries were relaxed with threshold concentrations of bradykinin followed by the cumulative addition of ACE-inhibitors. In bovine coronary arteries with intact endothelium, 100pM bradykinin caused 15-20% relaxation. The cumulative addition of different ACE-inhibitors (e.g. captopril, lisinopril, enalaprilat) resulted in further relaxations. Figure 4 illustrates the effects of enalaprilat on arteries stimulated with threshold concentrations of bradykinin (10^{-10}M) which caused 18±9% relaxation. The accumulative addition of enalaprilat resulted in further relaxations with a maximum of 86±5% at 10^{-7}M enalaprilat. Similar results were obtained in human coronary arteries although the tissue revealed a lower sensitivity to bradykinin. Measurements of bradykinin concentrations in the organ bath showed no significant changes. ACE-inhibitors did not relax the arteries in the presence of threshold concentrations of other endothelium-independent and endothelium-dependent vasodilators including substance P and acetylcholine.

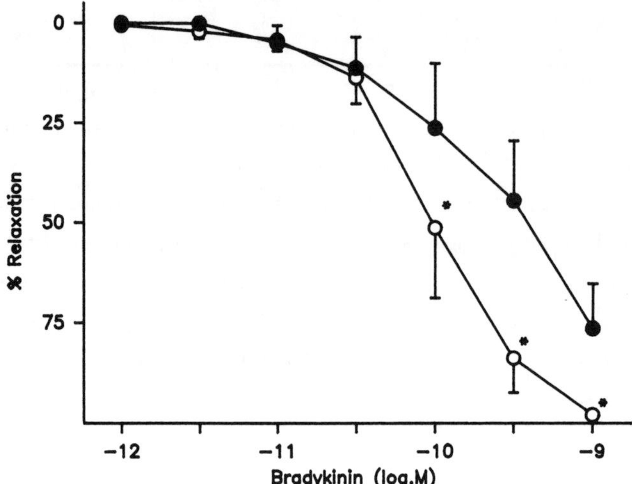

Figure 1.

Effect of lisinopril on endothelium-dependent relaxations induced with bradykinin in bovine coronary artery.

The relaxations (control = ●) are augmented in the presence of 10^{-6}M lisinopril (o).

Data are shown as means ±SEM, n=6, *=p<0.05.

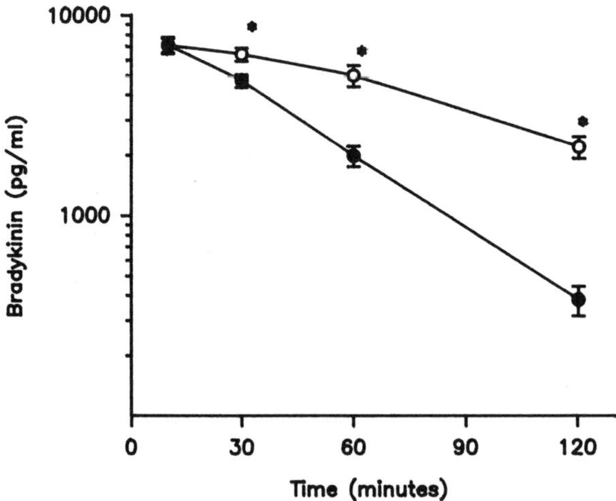

Figure 2.

Effect of lisiniopril on the degradation of bradykinin in bovine coronary artery.

10^{-6}M Lisinopril (o) significantly reduced the decline of bradykinin (● = controls).

Data are shown as means ± SEM, n=6, *p<0.01.

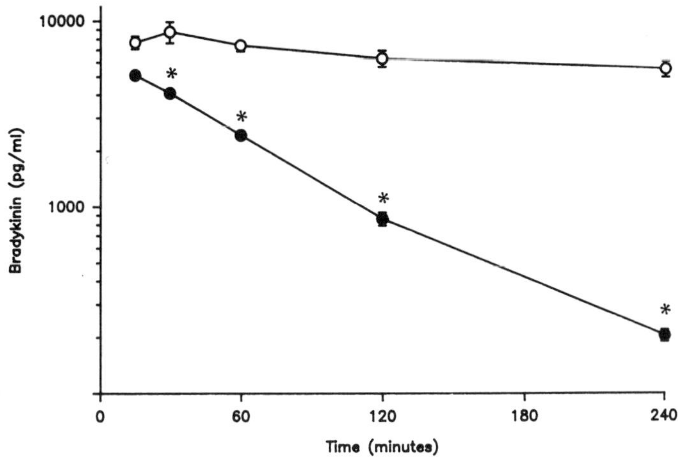

Figure 3.

Effect of captopril on bradykinin degradation in endothelial cell monolayers.
Values are means ± SEM (n=9) in the absence (• =control) or presence of
10^{-6}M captopril (o).

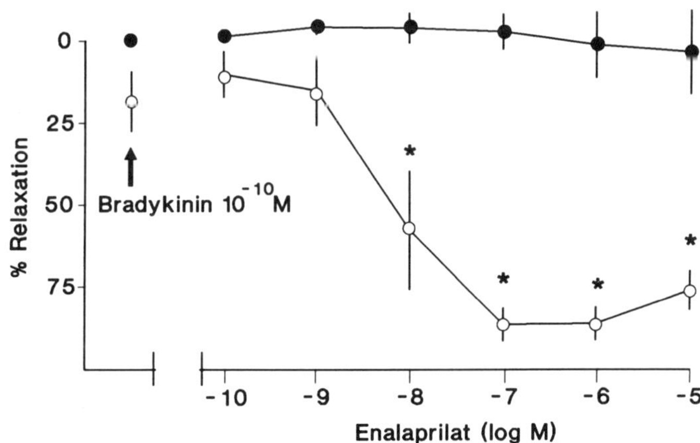

Figure 4.

Effects of enalaprilat (10^{-10}-10^{-5}M) on bovine coronary artery with intact endothelium.
o = in the presence of 10^{-10}M bradykinin, • = in the absence of bradykinin.
Data are shown as means ± SEM of 5 experiments.

DISCUSSION

The results of the experiments provide evidence that the endothelium-dependent relaxation in response to bradykinin can be affected by ACE-inhibitors via at least two different mechanisms.

Incubation of human and bovine coronary arteries with ACE-inhibitors augments relaxations in response to subsequently added bradykinin. Since bradykinin is known to be metabolized by kinase II, which is identical to angiotensin converting enzyme, one explanation for the observed augmented bradykinin effect could be the inhibition of the degradation of bradykinin by the ACE-inhibitors (10). Evidence for that hypothesis is provided by the experiments showing a reduced bradykinin degradation in the organ bath of the incubated arteries in the presence of ACE-inhibitors. In addition, the bradykinin degradation was almost completely blocked by ACE-inhibitors in intact endothelial cells in culture. Concentrations as low as 10^{-10}M of the ACE-inhibitors already induced a significant inhibition (10). However, the endothelium-dependent relaxations in response to bradykinin were completed within minutes whereas the metabolism of bradykinin and the effect of ACE-inhibitors on it in isolated arteries and in cell cultures revealed much slower kinetics.

A second effect of ACE-inhibitors on bradykinin mediated endothelium-dependent relaxations was observed after stimulation of the arteries with threshold concentrations of bradykinin. In these experiments the arteries were relaxed with bradykinin and the ACE-inhibitors were added after the relaxations with bradykinin were completed. The observed increase in relaxation in response to the ACE-inhibitors in these experiments can hardly be attributed to an inhibition of bradykinin metabolism since the relaxations to the ACE-inhibitors clearly exceed the initial responses to bradykinin. Indeed, measurements of bradykinin concentrations before and after the addition of ACE-inhibitors showed no change. Therefore, a mechanism different of an inhibition of the degradation of bradykinin is proposed. Further experiments of our group show that these relaxations are mediated by B_2-receptors and nitric oxide since they are blocked by B_2-receptor antagonists and abolished after preincubation with nitro-1-arginine (11). They may be based upon an interaction of the ACE-enzyme and the B_2-receptor (12).

REFERENCES

1. Ishida H., Scicli AG, Carretero OA. Role of angiotensin converting enzyme and other peptidases in vivo metabolism of kinins. J Hypertension 1989; 14:322-327.

2. Carbonell LF, Carretero OA, Stewart JM Scicli AG. Effect of a kinin antagonist on acute antihypertensive activity of enalaprilat in severe hypertension. J Hypertension 1988; 11:239-243.

3. Yang HYT, Erdös EG, Levin Y. A dipeptidyl carbooxypeptidase that converts angiotensin I and inactives bradykinin. Biochem Biophys Acta 1970; 214:374-376.

4. Bönner G, Iwersen D. The analytical value for kinin concentratrion in blood depends on the antiserum used in the bradykinin radioimmunoassay. Journ Clin Chem 1987; 25:29-43.

5. Baumgarten CR, Togias AG, Naclerio RM et al. Influx of kininogens into nasal secretions after antigen challenge of allergic individuals. J Clin Invest 1985; 76:191-197.

6. Hunter WM, Greenwood FC. Preparation of 131I labelled human growth hormone of high specific activity. Nature 1962; 194:495-496.

7. Goodfriend TL, Levine L, Fasman GD. Antibodies to bradykinin and angiotensin: a use of carbodiimides in immunology. Science 1962; 144:1344-1346.

8. Jaffe EA, Nachman RL,Becker CG, Minick CR. Culture of human endothelial cells derived from umbilical veins. Identification by morphologic and immunologic criteria. J Clin Invest 1973; 52:2745-2756.

9. Franke RP, Gräfe M, Schnittler H, Seiffge D, Mittermayer C, Drenckhahn D Induction of human vascular endothelial stress fibres by fluid shear stress. Nature 1984; 307:648-649.

10. Gräfe M, Bossaller C, Graf K, Auch-Schwelk W, Baumgarten CR, Hildebrand A, Fleck E. Effect of angiotensin converting enzyme inhibition on bradykinin metabolism by human vascular endothelial cells.Am J Physiol, submitted 1991.

11. Auch-Schwelk W, Bossaller C, Claus M, Graf K, Gräfe M, Fleck E. ACE-inhibitors are endothelium-dependent vasodilators in coronary arteries stimulated with threshold concentrations of bradykinin. Cardiovasc Res submitted 1992.

12. Roscher AA, Klier C, Dengler R, Faußner A, Müller-Esterl W. Regulation of bradykinin action at the receptor level. J Cardiovasc Pharmacol 1990; 15(suppl.6):39-43.

AAS 38/III
Recent Progress on Kinins
© 1992 Birkhäuser Verlag Basel

ACE INHIBITOR EFFECT ON BRADYKININ METABOLISM IN THE VASCULAR WALL

Peter Gohlke, *Peter Bünning, **Gerd Bönner and Thomas Unger

Department of Pharmacology and German Institute for High Blood Pressure Research, University of Heidelberg, *Hoechst AG Frankfurt/M and **Department of Internal Medicine II, Cologne, Germany

SUMMARY: In the isolated rabbit thoracic aorta the ACE inhibitor ramiprilat attenuated bradykinin degradation by enzymes localized on vascular endothelial cells as well as on vascular smooth muscle cells by 26% and 32%, respectively. We conclude that the ACE inhibitor can attenuate vascular bradykinin degradation not only by inhibition of endothelial ACE but also by inhibition of other bradykinin degrading enzymes in deeper layers of the vascular wall.

INTRODUCTION

The antihypertensive effect of angiotensin converting enzyme (ACE) inhibitors is determined in part by the blockade of kinin degradation, since ACE, also known as kininase II, is one of the most important bradykinin (BK) degrading enzymes.

This has been shown by a number of studies indicating that the acute effects of ACE inhibitors on blood pressure were partially blocked by inhibitors of the kallikrein-kinin-system (KKS) (1,2).

The vascular bed of the lung has been considered to be one of the most important inactivation sites for systemic BK. BK administered intravenously (i.v.) was inactivated by more than 90% due to enzymatic degradation in the lung (3). Inactivation of locally produced BK in different vascular beds by the action of BK degrading vascular enzymes may be an important determinator of BK related effects on blood pressure and regional blood flow. Besides ACE, a number of other vascular enzymes including carboxypeptidases, aminopeptidases and endopeptidases have been implicated in vascular BK metabolism (4)

In the present study we used an in vitro model, the isolated rabbit thoracic aorta, to investigate the effect of an ACE inhibitor on the metabolism of BK by enzymes localized on endothelial and vascular smooth muscle cells (VSMC) of the vascular wall.

MATERIALS AND METHODS

Male Chinchilla rabbits (Möllegard, Skensved, Denmark) weighing 2-3 kg were sacrificed by cervical dislocation. The thoracic aorta was removed and rinsed with a modified Krebs-Henseleit buffer pH 7,4 (118 mM NaCl; 4.75 mM KCL; 1.2 mM KH_2PO_4; 1.7 mM $CaCl_2$; 1.2 mM $MgSO_4$; 25 mM $NaHCO_3$; 10 mM HEPES; 11 mM glucose; gasses with a 95% O_2-5% CO_2 mixture).

After dissection of perivascular tissue all side branches were tied off. The aorta was cannulated at both ends and suspended in an organ chamber containing the modified Krebs-Henseleit buffer pH 7,4. The incubation bath was kept at $37^{\circ}C$. The aorta was instilled with 200 to $300\mu l$ buffer, closed at both ends and preincubated for one hour before BK incubations were started.

In a first experiment, intact (n=3) or endothelially denuded (n=3) aortic vessels were filled with 1nM BK and incubated for 5, 10, 15, 30, 45 and 60 minutes. Each incubation period was followed by a washout period of 10 minutes with buffer.

In a second experiment, intact (n=3) or endothelially denuded (n=3) aortic vessels were filled with 1 nM BK together with the ACE inhibitor ramiprilat ($10^{-7}M$) and incubated for 5, 10, 15, 30, 45 and 60 minutes. Each incubation period was followed by a washout period of 10 minutes with buffer containing $10^{-7}M$ ramiprilat. The integrity of the endothelium as well as the completeness of endothelial denudation was verified histologically by silver nitrate staining.

Incubations were terminated by sampling of the luminal fluid. Aliquots of $200\mu l$ were added to $50\mu l$ 0.5M HCl and stored at $-20^{\circ}C$. The concentration of BK in the luminal fluid was determined by radioimmunoassay as desribed previously (5). The major degradation products of BK, des-Arg^9-BK, des-Phe^8-Arg^9-BK and des-Arg^1-BK did not crossreact with the BK antibody (< 0.01%).

In a third experiment the time-dependent distribution of the ACE inhibitor ramiprilat in the vascular wall was investigated by administration of tritium labelled ramiprilat into the aortic lumen. The aorta was incubated with $10^{-7}M$ ^3H-ramiprilat (0.108 mCi/ml; 55.33 Ci/mmol, Hoechst AG, Frankfurt) for 5, 15, 30, 60 and 90 minutes. ^3H-ramiprilat was diluted with Krebs-Henseleit buffer to a final concentration of $10^{-7}M$ prior to incubation.

One aorta was used for each time period. At the end of an incubation period radioactivity was measured in the luminal fluid, the endothelial, medial and adventitial layer as well as in the incubation bath. The vascular layers were separated machanically and transferred into porcelain vessels for catalytic combustion in the O_2-stream of a combustion furnace (OX 300, Zinser Analytics, Frankfurt, Germany). The procedure resulted in the generation of ^3H_2O which was rinsed together with scintillation fluid into a vial. The recovery was determined using a known amount of radioactivity with and without catalytic combustion. Radioactivity was measured in a ß-scintillation counter (2000 CA Tricarb, Packard Instruments) and expressed as decays per minutes (dpm).

RESULTS

The initial BK concentrations measured intraluminally before the start of aortic incubations ranged from 96 pg/100μl to 146 pg/100μl (mean ± SD: 123.6 ± 13.1 pg/100μl).

The intraluminal BK concentration decreased time-dependently after instillation of BK into an aorta with intact endothelium (Figure 1). The breakdown of luminal BK by endothelial enzymes was almost complete (96%) after one hour of incubation. The addition of the ACE inhibitor ramiprilat to the incubation medium reduced BK degradation by endothelial enzymes to 70% after one hour of incubation (Figure 1).

After removal of the endothelium, BK was also effectively degraded by enzymes localized on VSMC (82% after one hour of incubation) (Figure 1). Addition of the ACE inhibitor ramiprilat to the incubation medium still reduced BK degradation to 50% after one hour of incubation (Figure 1).

The time-dependend distribution of radioactivity in the aortic lumen, the vascular layers and in the incubation bath after aortic incubation of ^3H-ramiprilat ($10^{-7}M$) is presented in figure 2. After intraluminal incubation the ACE inhibitor quickly diffused into the vascular wall, reaching a distribution equilibrium in the endothelial, medial and adventitial layer within 5 minutes. Thereafter, the percentage amount of radioactivity in the three vascular layers remained constant during the 90 minute incubation period. The percentage of radioactivity in the incubation bath increased in a linear fashion with time to 33.7% after 90 minute incubation.

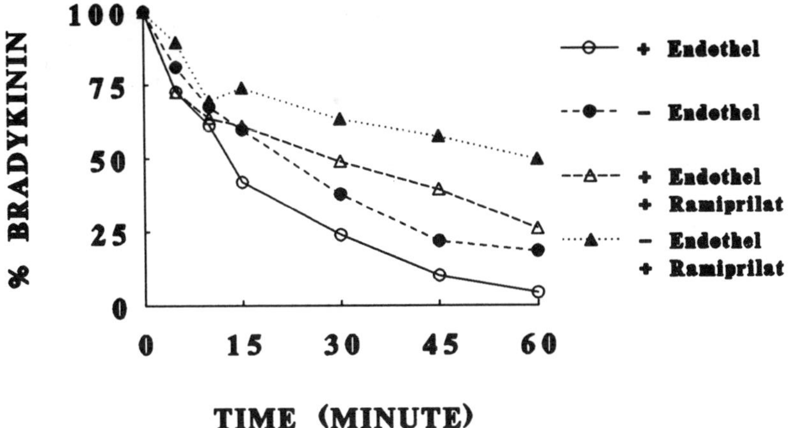

Figure 1. Percentage amount of intraluminal bradykinin after intraluminal bradykinin $(10^{-9}M)$ in rabbit thoracic aorta.

DISCUSSION

The isolated rabbit thoracic aorta represents a valuable tool for studies on the metabolism of peptides, including BK, in the vascular wall. In contrast to aortic ring preparations peptides can be administered intraluminally to be exposed exclusively to enzymes localized on endothelial cells or - after removal of the endothelium - on VSMC. The study of a time-dependent degradation of BK by vascular enzymes is facilitated by the long half-life of BK in our preparation (10-15 minutes in an intact aorta). An explanation for this prolonged half-life resides in the limited amount of endothelial cells or VSMC and thus, the limited amount of BK degrading enzymes in this preparation compared to e.g. whole organ preparations or to in vivo conditions (half-life <30 sec). Furthermore, histologically controlled studies with and without the endothelium are possible. On the other hand, the lack of pulsatile perfusion inherent to this model has to be acknowleged.

The administration of BK into an aorta with intact endothelium resulted in a nearly complete breakdown of the peptide during an incubation period of one hour. BK

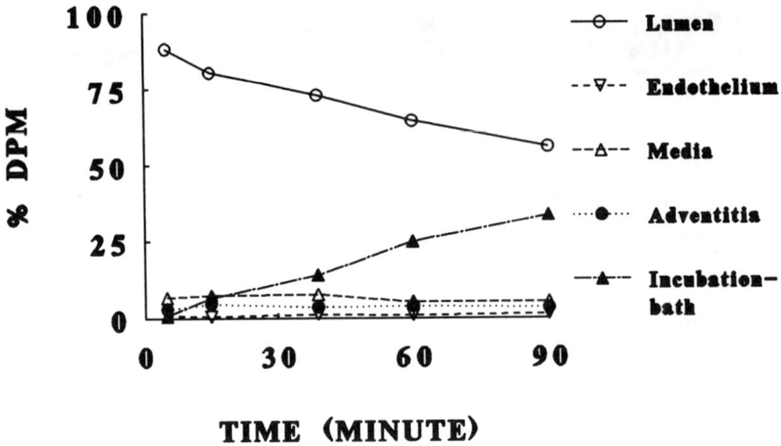

Figure 2. Time-dependent distribution of radioactivity in the vascular wall after administration of ^{3}H-ramiprilat (10^{-7}M) into the lumen of the rabbit thoracic aorta. The radioactivity at the start of each incubation period was taken as 100%, and the results were expressed as the percentage amount of radioactivity.

degradation by endothelial enzymes was reduced by 26% after the addition of the ACE inhibitor ramiprilat, most likely by inhibition of endothelial ACE.

Since BK is a potent vasodepressor peptide this finding clearly supports the hypothesis that the antihypertensive actions of ACE inhibitors are not only determined by the inhibition of ANG II generation but also by inhibition of BK degradation. Evidence for a role of endogenous BK in the antihypertensive action of ACE inhibitors in vivo was provided by several studies showing that the acute antihypertensive effect of ACE inihibitors were attenuated after BK receptor blockade (1,2). Furthermore, in kinin deficient Brown-Norway rats with renovascular hypertension, the blood pressure decrease caused by i.v. injection of an ACE inhibitor was significantly reduced (2).

More recently we have demonstrated that in ACE inhibitor-pretreated Wistar rats with renovascular hypertension chronic blockade of BK B_2-receptors for 6 weeks partially reversed the antihypertensive effect of the ACE inhibitor. Similarly, co-administration of the BK B_2-receptor antagonist Hoe 140 and the ACE inhibitor ramipril for 6 weeks resulted in a marked attenuation of the antihypertensive effect of the ACE inhibitor (6).

Removal of the aortic endothelium allows the study of BK metabolism in deeper layers of the vascular wall. After administration of BK into the lumen of a denuded aorta the peptide was effectively degraded by 82% within one hour of incubation. Again, the addition of the ACE inhibitor ramiprilat attenuated BK degradation by VSMC enzymes by 32%.

On first glance this finding suggests that ACE is localized not only in the endothelial but also in the medial layer of the vascular wall where it contributes to local BK inactivation. However, in a previous study using the same in vitro model, we have demonstrated a complete blockade of angiotensin I conversion to angiotensin II after removal of the endothelium demonstrating that ACE was absent in the endothelially denuded aortic preparation (7).

The absence of ACE in the vascular media has been shown by a number of previous studies investigating the localization of ACE in the vascular wall by means of biochemical, immunohistochemical and autoradiographical methods (for review see 8). All these studies failed to detect any ACE activity in the vascular medial layer.

In view of these findings, the ACE inhibitor-induced attenuation of BK degradation by VSMC enzymes seems to be largely independent of ACE. Thus, it is most likely that the ACE inhibitor attenuated BK breakdown by inhibiting BK degrading enzymes other than ACE on VSMC.

A number of enzymes have been implicated in kinin degradation in addition to ACE including carboxypeptidases (e.g. carboxypeptidase N) aminopeptidases (e.g. aminopeptidase P) and endopeptidases (e.g. neutral endopeptidase 24.11 and endopeptidase 24.15) (4,10). Only few studies have addressed the interaction of ACE inhibitors with BK degrading enzymes other than ACE. In one these, Drummer et al. (10) observed a 50% inhibition of the endopeptidase 24.11. and a 42% inhibition of an amastatin sensitive aminopeptidase in rat kidney fractions after chronic oral enalapril treatment, while other enzymes like cathepsin B or carboxypeptidase B were not affected. However, the role of these enzymes in the regulation of BK levels in VSMC of the vasculature has not yet been investigated.

Oza et al (11) reported the presence of a kallikrein-like-enzyme, kininogen and kininases in rat VSMC in culture. Although the individual kininases were not further characterized in this study, these findings suggest a local KKS which may act in a paracrine/autocrine manner in blood vessels to regulate vascular tone.

The distribution of ramiprilat in the vascular wall was investigated by aortic incubation of ^{3}H-ramiprilat. Our results demonstrate a very quick diffusion of the ACE inhibitor into and through the blood vessel wall, indicating a lack of major diffusion barriers which could have prevented access of the ACE inhibitor to possible extracellular target sites within the different layers of the vascular wall. Based on this finding, we can assume that ACE inhibitors can generally access target enzymes in deeper layers of the vascular wall, despite the fact that their major target is endothelial ACE.

CONCLUSION

We have demonstrated that BK breakdown by enzymes localized not only on vascular endothelial cells but also on VSMC is attenuated by the ACE inhibitor ramiprilat. Since ACE has not been found in VSMC, this finding suggests that the ACE inhibitor is able to inhibit VSMC enzymes other than ACE that are relevant for vascular BK breakdown.

REFERENCES

1. Carbonell LF, Carretero OA, Stewart JM, Scicli AG. Effect of a kinin antagonist on the acute antihypertensive activity of enalaprilat in severe hypertension, Hypertension 1988; 11:239.

2. Danckwardt L, Shimizu I, Bönner G, Rettig R Unger Th. Converting enzyme inhibition in kinin-deficient Brown Norway rats, Hypertension 1990; 16:429.

3. Baker Jr CRF, Little AD, Little GH Canizaro PC Behal FJ. Kinin metabolism in the perfused ventilated rat lung. I: Bradykinin metabolism in a system modeling the normal, uninjured lung, Circulatory Shock 1991; 33: 37.

4. Ryan JW. Peptidase enzymes of the pulmonary vascular surface, Am. J. Physiol. 1989; 257:L53.

5. Bönner G. Das renale Kallikrein-Kinin-System (Springer Verlag, Berlin, Heidelberg, New York, London, Paris, Tokyo), 1988.

6. Bao G, Gohlke P, Qadri F, Unger Th. Chronic kinin receptor blockade attenuates the antihypertensive effect of ramipril, Hypertension 1992; in press.

7. Gohlke P, Bünning P, Unger Th. Distribution and metabolism of angiotensin I and II in the blood vessel wall, Hypertension 1992; in press.

8. Unger Th, Gohlke P, Paul M, Rettig R. Tissue renin-angiotensin systems: Fact or fiction?, J. Cardiovasc. Pharmacol. 1991; 18 (Suppl.2):S20.

9. Carretero OA, and Scicli AG. Zinc metallopeptidase inhibitors: a novel antihypertensive treatment, Hypertension 1991; 68:366.

10. Drummer OH, Kourtis S, Johnson H. Effect of chronic enalapril treatment on enzymes responsible for the catabolism of angiotensin I and formation of angiotensin II, Biochem. Pharmacol. 1990; 39: 513.

11. Oza NB, Schwartz JH, Goud HD, Levinsky NG. Rat aortic muscle cells in culture express kallikrein, kininogen and bradykininase activity, J. Clin. Invest. 1990; 85: 597.

AAS 38/III
Recent Progress on Kinins
© 1992 Birkhäuser Verlag Basel

LOCAL PRODUCTION OF KININS CONTRIBUTES TO THE ENDOTHELIUM DEPENDENT RELAXATIONS EVOKED BY CONVERTING ENZYME INHIBITORS IN ISOLATED ARTERIES

Jean-Vivien Mombouli, Stephane Illiano, and Paul M. Vanhoutte

Center For Experimental Therapeutics, Baylor College of Medicine, One Baylor Plaza, Houston Texas 77030, USA.

SUMMARY: The angiotensin converting enzyme (ACE) inhibitor perindoprilat evokes endothelium-dependent relaxations in perfused isolated canine arteries. Kininogens, the precursors of bradykinin, elicit endothelium-dependent relaxations which are potentiated by perindoprilat, inhibited by B_2-kinin antagonists and partially impaired after inhibition of NO synthase. These observations suggest that locally produced kinins may stimulate the production of NO and endothelium-derived hyperpolarizing factor, and that this action is potentiated by ACE inhibitors.

INTRODUCTION

In isolated canine arteries bradykinin causes endothelium-dependent relaxations that are potentiated by ACE inhibitors[1] and are mediated by both nitric oxide (NO) and endothelium-derived hyperpolarizing factor (EDHF), following the activation of B_2-kinin receptors.[2] Bradykinin is inactivated rapidly by a variety of proteases, of which the most active is angiotensin I converting enzyme (ACE) present both in the vascular wall and in the plasma.[3] Thus, it has been suggested that prevention of the degradation of bradykinin may play a role in the antihypertensive actions of ACE inhibitors.[4] In perfused canine carotid arteries, the non-sulfhydryl ACE inhibitor perindoprilat[5] evokes endothelium-dependent relaxations, that are antagonized by B_2-kinin receptor antagonists.[6] In the present study, the contributions of NO and EDHF in the relaxation evoked by ACE inhibitors were investigated.[6,7] Furthermore, since kininogens are processed into kinins by enzymes present in isolated arteries,[8,9] including the canine coronary,[10] experiments were designed to determine whether or not these precursors of bradykinin mimic the endothelium-dependent effects of the peptide.

MATERIALS AND METHODS

Carotid and coronary arteries were removed from anesthetized (30 mg/ml pentobarbital intravenously followed by exsanguination) mongrel dogs (15-30 kg), and placed in cold modified Krebs-Ringer bicarbonate solution [composition in mM; NaCl 118.3; KCl 4.7; $CaCl_2$ 2.5; $MgSO_4$ 1.2; KH_2PO_4 1.2; $NaHCO_3$ 25; glucose 11.1; calcium disodium edetate 0.026, pH 7.4 (control solution)]. The arteries were cleaned of connective tissue and cut into segments (5 cm) or rings (3-4 mm); in some preparations, the endothelium was removed mechanically. All experiments were carried out in the presence of indomethacin (10^{-5} M; to prevent the production of vasoactive prostaglandins).

Perfusion cascade bioassays. Segments of carotid arteries were cannulated and suspended horizontally in a organ chamber filled with warm and aerated control solution between two stirrups, one of which was connected to a force transducer (Grass, FTC103, Cleveland, OH, USA) for isometric recording of changes in tension.[1] The arteries were perfused (5 ml/min) with warm and aerated control solution and used as a source of endothelium-derived mediators. In some experiments a second segment without endothelium was placed in series with the donor and served as a detector of endothelium-derived mediators (Figure 1). Alternatively a ring of canine coronary artery without endothelium (detector ring) was suspended directly below the perfusion lines by means of two stainless steel stirrups for isometric recording of changes in tension.The preparations were stretched progressively to the optimal point of their length-tension relationship as determined by the response to KCl (60 mM) obtained at each level of stretch.

Organ chambers. Rings were suspended between two stirrups in organ chambers (20 ml) filled with control solution (gassed with 95% O_2-5% CO_2 and maintained at 37°C). One of the stirrups was anchored inside the organ chamber and the other was connected to a force transducer. The rings were stretched to the optimal point of their length-active tension relationship (8-10 g) with KCl (60 mM). The responses to low and high molecular weight kininogens (LMWK and HMWK, respectively) were assessed in rings with and without endothelium contracted with prostaglandin $F_{2\alpha}$ ($4x10^{-6}$ M). The effects of enzyme-inhibitors or receptor-antagonists were studied in paired rings following 30 minute incubation.

Drugs and chemicals. The following drugs were used; D-Arg^0[Hyp^3,$DPhe^7$]-bradykinin, indomethacin, prostaglandin $F_{2\alpha}$, sodium nitroprusside (all from Sigma, St-Louis, MO, USA), [Leu^8]-desArg^9-bradykinin (Peninsula, Belmont, CA, USA), high and low molecular weight kininogens, DL-2-mercaptomethyl-3-guanidinoethylpropanoic acid (Calbiochem, La Jolla, CA, USA), N^G-nitro-L-arginine (Aldrich, Milwaukee, WI, USA), perindoprilat (S9490-3, Servier, Neuilly sur seine, France). Drugs were prepared in water except for indomethacin (dissolved in water and Na_2CO_3, which had no effect at the final bath concentration of $5x10^{-6}$ M).

Statistical analysis. In each experimental group, n refers to the number of animals from which blood vessels were taken. Results are presented as means±SEM. Relaxations are expressed as percentage changes of tension from the plateau contractions to constrictor agonists. Statistical comparisons were performed by means of a two way analysis of variance or Student's t test for paired observations. Differences were considered to be significant when P was less than 0.05.

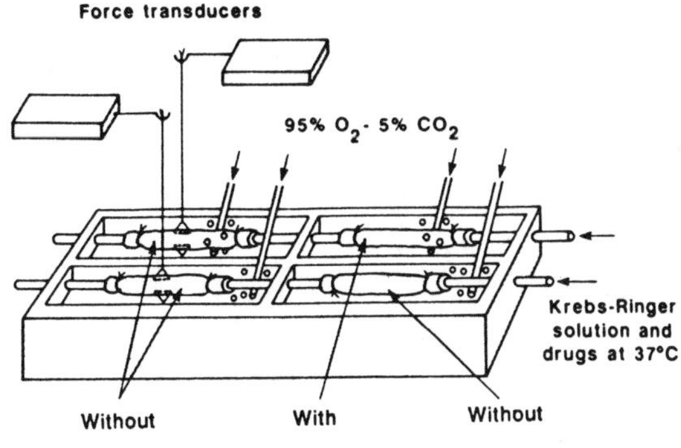

Figure 1. *Double vessel perfusion bioassay apparatus.* Isometric tension is monitored in vessels (with or without endothelium) with force transducers (up to four) connected to a chart recorder. Drugs are administered through infusion sites placed between, or upstream of, these preparations. Under these conditions, the effects of drugs on the abluminal and luminal release of vasoactive substances by blood vessels can be examined.

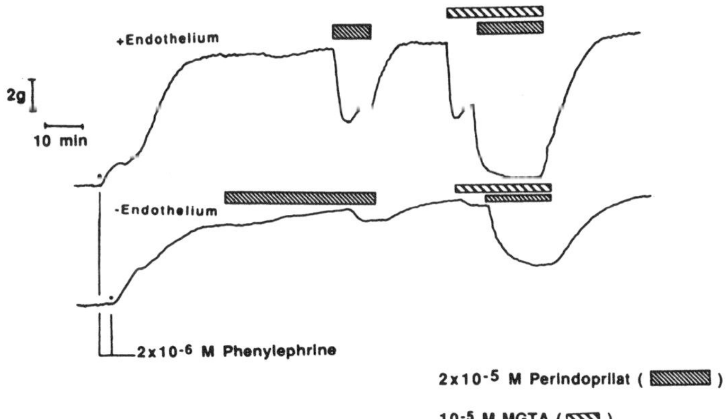

Figure 2. *Relaxations to perindoprilat.* Two segments of carotid arteries were placed in series and contracted with phenylephrine. Treatment of the vessel downstream (without endothelium; see figure 1) with perindoprilat did not alter the contraction to phenylephrine. Addition of the ACE inhibitor upstream evoked a relaxation in both the vessel with and the one without endothelium. Infusions of MGTA and perindoprilat augmented the relaxation.

RESULTS

Perindoprilat. In bioassay experiments, using two carotid arteries in series (Figure 1), infusions of perindoprilat in the segment without endothelium did not alter the contraction to phenylephrine (Figure 2, bottom trace). However, addition of perindoprilat (2×10^{-5} M) upstream of the artery with endothelium, elicited relaxations in both preparations (Figure 2, top trace). After washout, addition of the carboxypeptidase-inhibitor MGTA caused a relaxation, that was enhanced further by perindoprilat (Figure 2). In experiments where MGTA was added prior to perindoprilat it did not evoke relaxations; MGTA like perindoprilat did not possess direct effects on arteries without endothelium (data not shown). In another series of experiments in which a coronary artery ring without endothelium was used as a detector, perindoprilat elicited relaxations in the donor arteries which was transmitted to the detector preparation (Figure 3). The relaxations were inhibited by the inhibitor of NO synthase N^{G}-nitro-L-arginine (10^{-4} M; Figure 3).

Kininogens. In rings without endothelium, cumulative additions of either LMWK or HMWK (1 ng/ml to 1µg/ml) during sustained contractions to prostaglandin $F_{2\alpha}$ (4×10^{-6} M) did not cause significant changes in tension. In rings with endothelium, additions of either HMWK (Figure 4) or LMWK (Figure 5A) elicited concentration-dependent relaxations. These relaxations were not affected by MGTA (10^{-5} M; a carboxypeptidase-inhibitor; Figure 3 and 4B). The relaxations evoked by HMWK (Figure 4) or LMWK (Figure 5B) were potentiated by perindoprilat (10^{-6} M); the combination of perindoprilat plus MGTA did not cause a significant further potentiation than perindoprilat alone (Figure 4 and 5B). In rings with endothelium incubated with both perindoprilat and MGTA, the relaxations to either HMWK (Figure 4) or LMWK (data not shown) were not affected by the B_1-kinin receptor antagonist [Leu[8]]-desArg[9]-bradykinin (10^{-5} M). The B_2-kinin receptor antagonist D-

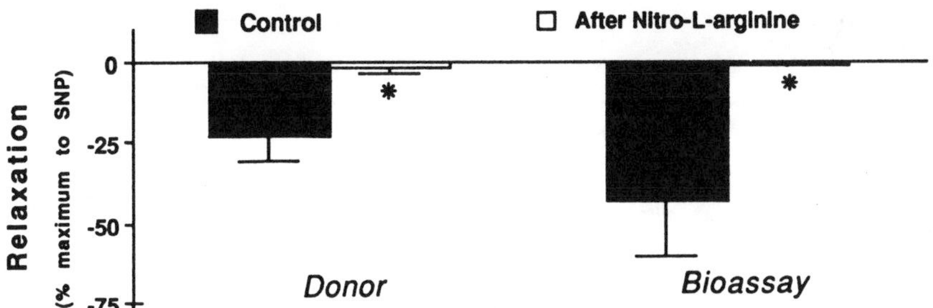

Figure 3. *Effects of N^G-nitro-L-arginine.* Relaxations to perindoprilat were determined in segments of carotid (donor) and rings of coronary (bioassay) arteries, contracted with prostaglandin $F_{2\alpha}$ (10^{-6} M). Results are shown as means±SEM and expressed as a percentage of the relaxation induced by sodium nitroprusside (10^{-6} M), added at the end of the experiment. Asterisks denote statistically significant differences with responses obtained under control conditions.

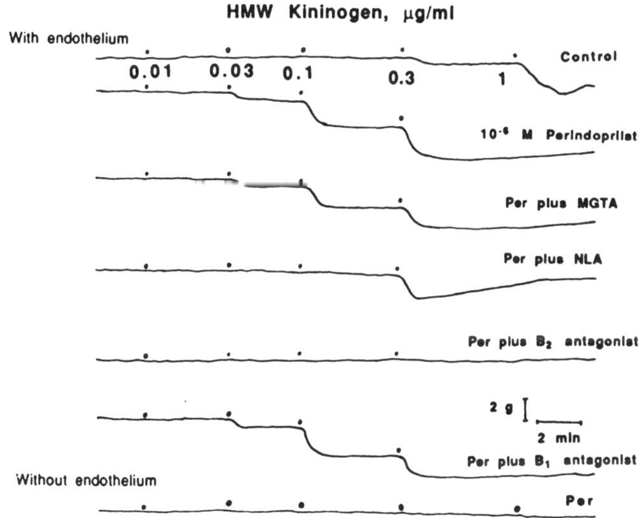

Figure 4. *Relaxations to High molecular weight kininogen.* The kininogen was added cumulatively (from 0.01 to 1 µg/ml), during plateau contractions to prostaglandin $F_{2\alpha}$ (4×10^{-6} M), to rings with (upper six traces) and without (bottom trace) endothelium. The effects of perindoprilat (Per) either alone or in combination with the carboxypeptidase inhibitor MGTA (10^{-5} M), N^G-nitro-L-arginine (NLA; 10-4 M), a B₂ (D-Arg⁰[Hyp³,D-Phe⁷]-bradykinin; 3×10^{-7} M), or a B₁ ([Leu8]-desArg9-bradykinin; 10^{-5} M) kinin receptor antagonists were examined.

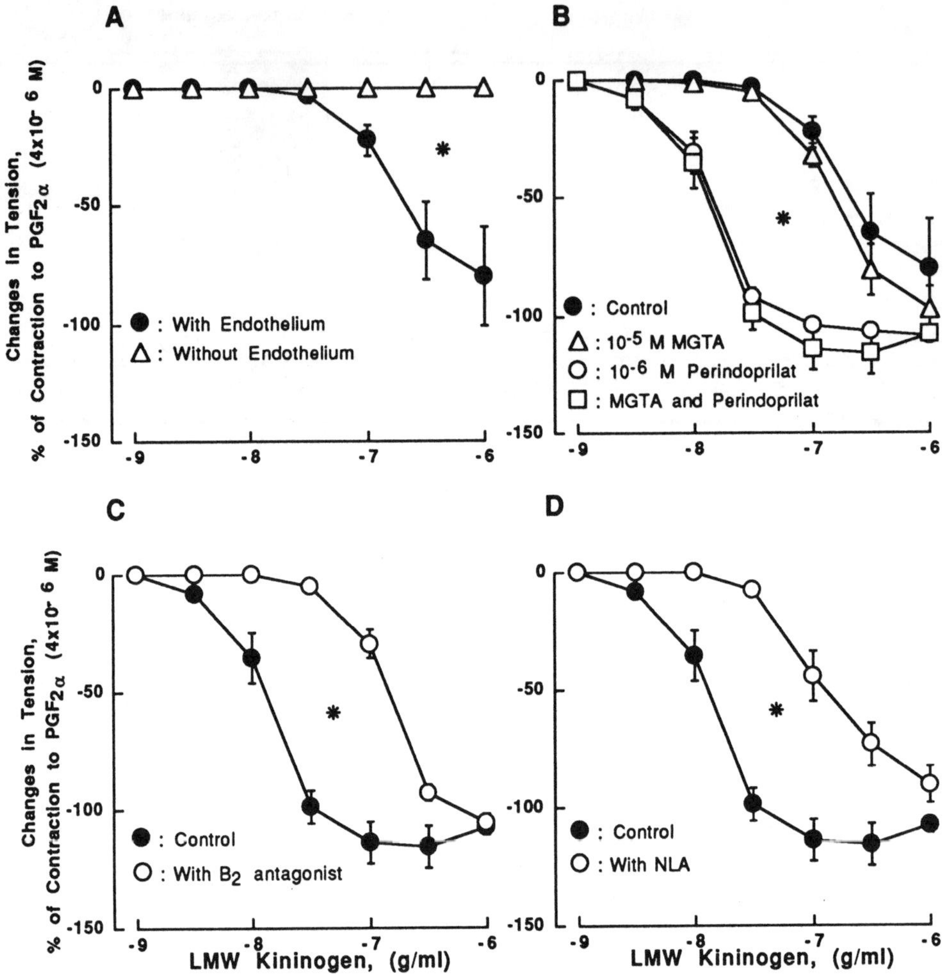

Figure 5. *Relaxations to low molecular weight (LMW) kininogen.* Results are shown as means±SEM and expressed as a percentage of the contraction to prostaglandin $F_{2\alpha}$ ($PGF_{2\alpha}$). Effects of endothelium removal (A), of the ACE inhibitor perindoprilat and the carboxypeptidase inhibitor MGTA (B), of the B_2-kinin antagonist D-Arg^0[Hyp^3,DPhe^7]-bradykinin (3×10^{-7} M; C) and N^Gnitro-L-arginine (NLA; 10^{-4} M; D). Asteriks indicate statistically significant differences of areas under the curves (P<0.05) with control conditions (closed circles).

Arg[0][Hyp[3],DPhe[7]]-bradykinin (3×10^{-7} M) caused a significant shift to the right of the concentration-relaxation curves to either kininogens (Figure 4 and 5C, respectively).

In rings with endothelium treated with N[G]nitro-L-arginine (10^{-4} M), the response to HMWK (Figure 4) or LMWK (data not shown) were transient. The sensitivity of the tissues to either kininogens was reduced significantly (Figure 4 and 5D).

DISCUSSION

The present findings confirm that perindoprilat induces relaxations in perfused canine carotid arteries with, but not in those without, endothelium.[6] Since both the donor and the detector preparations responded, it can be concluded that perindoprilat elicited the release of relaxing factors both in the luminal and abluminal directions. Infusion of both MGTA and perindoprilat enhanced the relaxation, suggesting that under flow conditions vascular carboxypeptidase activities[3] may contribute significantly to the degradation of bradykinin. Since the data were collected in the presence the cyclooxygenase inhibitor indomethacin, vasodilator prostanoids do not contribute to the response evoked by perindoprilat. However, in the presence of the nitric oxide synthase inhibitor N[G]-nitro-L-arginine the response induced by perindoprilat in both the donor vessel and the bioassay preparation is abolished. This indicates that NO is the diffusible factor detected downstream. A basal production of EDHF has been documented in the carotid artery of the dog.[11] In canine coronary arteries the release of EDHF by bradykinin is less efficiently coupled to B$_2$-kinin receptors.[2] Therefore, it is possible that the levels of endogenous kinins achieved under the present experimental conditions are insufficient to trigger the production of EDHF.

Kininogens elicit endothelium-dependent relaxations, which are potentiated by perindoprilat; this suggests the involvement of a substrate of ACE. Furthermore, the relaxations are inhibited significantly by B_2-kinin receptor antagonist. This indicates that kininogens are converted into kinins. Like for bradykinin[2] N^G-nitro-L-arginine inhibits partially the response; this suggests that at higher concentrations kininogens elicit the production of EDHF, the mediator of these NO-independent relaxations[2]. Thus it can be can be concluded that kinin-generating enzymes in canine arteries can process kininogens into kinins which in turn stimulate the production of NO and EDHF.

The present experiments suggest that a vascular kallikrein-kinin system operates in parallel with the renin-angiotensin system.[12] Since ACE inhibitors augment endothelium-dependent relaxations to ADP and aggregating platelets,[1] endothelium-derived vasodilator mediators may play an important role in the therapeutic actions of ACE-inhibitors.[7]

ACKNOWLEDGEMENTS. This study was supported in part by grant No HL31183 from the National Institute of Health. The authors are grateful to Dr E. Scalbert (Servier) for the generous gift of perindoprilat.

REFERENCES

1. Mombouli, J.V., Nephtali, M., and Vanhoutte, P.M. Effects of the non-sulfhydryl angiotensin I converting enzyme inhibitor cilazaprilat on endothelium-dependent responses in isolated canine arteries. Hypertension 1991, 18 (suppl. II): II22-II29.

2. Mombouli, J.V. Illiano, S., Nagao, T. and Vanhoutte, P.M. The potentiation of bradykinin-induced relaxations by perindoprilat in canine coronary arteries involves both nitric oxide and endothelium-derived hyperpolarizing factor. Circulation 1991, 84 (suppl. II): 2482.

3. Erdös, E.G. Angiotensin I converting enzyme and the changes in our concepts through the years. Lewis. K. Dahl Memorial lecture. Hypertension 1991, 16: 363-370.

4. Carretero, O.A. and Scicli, A.G. Local hormonal factors (intracrine, autocrine, and paracrine) in hypertension? Hypertension 1991, 18 (suppl. I): I58-I69.

5. Laubie, M., Schiavi, P., Vincent, M. Schmitt, H. Inhibition of angiotensin converting enzyme with S9490: biochemical effects, interspecies differences and role of sodium diet in hemodynamic effects. J Pharmacol 1984, 6: 1076-1082.

6. Mombouli, J.V. and Vanhoutte, P.M. Kinins and endothelium-dependent relaxations to converting enzyme inhibitors in perfused canine arteries. J Cardiovasc Pharmacol 1991, 18: 926-927.

7. Vanhoutte, P.M., Auch-Schwelk, W., Biondi, M.L., Lorenz, R.R., Schini, V.B. and Vidal, M.J. Why are converting enzyme inhibitors vasodilators? Br J Clin Pharmacol 1989, 28: 95S-104S.

8. Nolly, H., Scicli, A.G. Scicli,G. and Carretero, O.A. Characterization of a kininogenase from rat vascular tissue resembling tissue kallikrein. Circ Res 1985, 56: 816-821.

9. Oza, N.B., Schwartz, J.H., Goud, H.D. and Levinsky, N.G. Rat aortic smooth muscle cells in culture express kallikrein, kininogen, and bradykininase activity. J Clin Invest 1990, 85: 597-600.

10. Zeitlin, I.J., Fagbemi, S.O. and Parratt, J.R. Enzymes in normally perfused and ischemic dog hearts which release a substance with kinin like activity. Cardiovasc Res 1989, 23: 91-97.

11. Siegel, G., Grote, J., Schnalke, F. and Zimmer, K. The significance of endothelium for hypoxic vasodilatation. Z Kardiol 1989, 78 (suppl 6): 124-131.

12. Kifor, I. and Dzau, V.J. Endothelial renin-angiotensin pathway: evidence for intracellular synthesis and secretion of angiotensins. Circ Res 1987, 60: 422-428.

AAS 38/III
Recent Progress on Kinins
© 1992 Birkhäuser Verlag Basel

CONVERTING ENZYME INHIBITOR-STIMULATED FORMATION OF NITRIC OXIDE AND PROSTACYCLIN IN ENDOTHELIAL CELLS FROM BOVINE AORTA IS MEDIATED BY ENDOTHELIUM-DERIVED BRADYKININ

G. Wiemer, B.A. Schölkens, R.H.A. Becker

Hoechst AG, SBU Cardiovascular Agents; 6230 Frankfurt / M. 80, FRG

SUMMERY : Like bradykinin the converting enzyme inhibitor ramiprilat concentration-dependently enhances the formation of nitric oxide and prostacyclin assessed by intracellular cyclic GMP accumulation and 6-keto prostaglandin $F_{1\alpha}$ resp. Both ramiprilat-induced effects are completely suppressed by the specific kinin receptor antagonist Hoe 140. The ramiprilat-induced cyclic GMP increase is totally blocked by the stereospecific inhibitor of nitric oxide synthase, N^G-nitro-L-arginine.

INTRODUCTION

From a number of experimental and clinical data it has been proposed that the antihypertensive (1, 2) and also the cardioprotective (3) and antiarteriosclerotic (4) effects of converting enzyme (CE) inhibitors in vivo are not only the consequence of reducing the formation of angiotensin II but also reducing the degradation of locally generated kinins in the vascular wall. However, up to now the potential sources of intravascular kinins have not yet been identified. Since confluent endothelial cells (EC) possess considerable amounts of the bradykinin (BK) precursor kininogen (5) as well as CE activity at the luminal cell surface we studied the effect of ramiprilat (RT) in primary cultured EC from bovine aorta. In detail we investigated whether these cells are able to synthetize and release BK and related kinins which in the presence of CE inhibitors may accumulate in the extracellular space in significant amounts that stimulate via kinin receptor activation and increase in cytosolic calcium (6) the formation of the vasodilators nitric oxide (NO) and prostacyclin.

MATERIAL AND METHODS

EC from bovine aorta were isolated by digestion with dispase and were cultered according to the method of Lückhoff et al. (7). Incubation of primary cultered cells and measurement of intracellular cyclic GMP and the stable hydrolysis product of prostacyclin, 6-keto prostaglandin $F_{1\alpha}$ (6-KPGF$_{1\alpha}$) in the supernatants were carried out as recently described (8).

RESULTS

In cultured bovine aortic EC addition of BK (10^{-7} mol/l) led to a fast and transient increase in cyclic GMP within 1 min. This short effect of a high concentration of BK reflects a desensitization of B_2 kinin receptors (6). The concentration-response relation of BK on cyclic GMP content revealed a threshold concentration for BK of about 3×10^{-10} mol/l and a maximum at about 10^{-8} mol/l. Preincubation of the cells with 10^{-7} mol/l of the specific kinin receptor antagonist Hoe 140 (9) totally suppressed BK-induced responses in EC. A complete inhibition of BK-induced cyclic GMP formation was also observed in monolayers that were preincubated with the inhibitor of NO-synthase, N^G - nitro-L-arginine (L-NNA). Concentration- and time-dependent increases in cyclic GMP production were also induced after incubation of EC with the CE-inhibitor RT (Fig. 1 A and B). Maximal increases in cyclic GMP content were obtained between 10^{-8} - 10^{-7} mol/l at threshold concentrations of about 10^{-9} mol/l (Fig. 1 B).

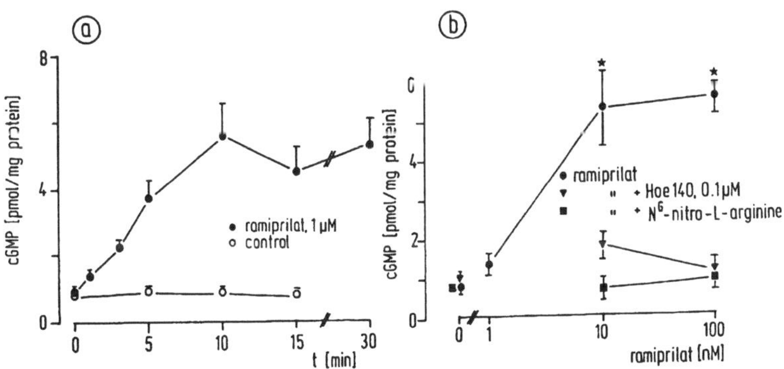

Fig. 1 Effects of ramiprilat on the accumulation of cyclic GMP (cGMP) in cultured bovine aortic endothelial cells as a function of time (panel a) and as a function of concentration (panel b) 10 min incubation. Hoe (140 10^{-7} mol/l) and N^G-nitro-L-arginine (10^{-5} mol/l) were added 5 min before addition of ramiprilat.

Preincubation of the cells with either L-NNA or Hoe 140 abolished the increases in cyclic GMP content (Fig. 1 B). Hoe 140 concentration-dependently inhibited the BK and RT-induced cyclic GMP formation with IC_{50} - values of about 3×10^{-8} and 10^{-10} mol/l, resp. (Fig. 2).

                                                                    G. Wiemer et al.

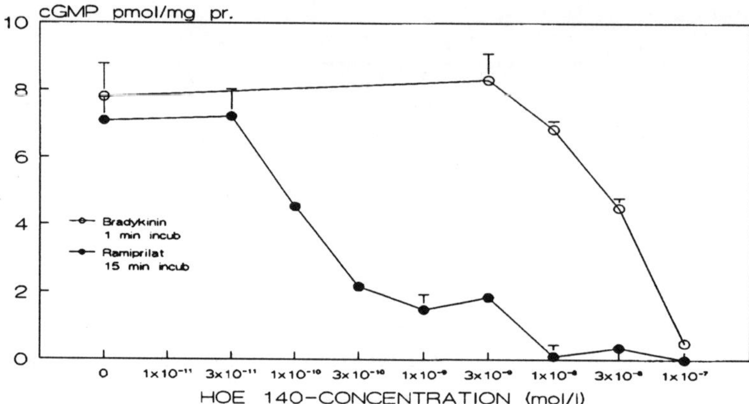

Fig. 2 Inhibition of bradykinin (10^{-8} mol/l) -and ramiprilat (10^{-8} mol/l) -induced formation of cyclic GMP (cGMP) in bovine aortic endothelial cells by Hoe 140. Bradykinin was incubated for 1 min, ramiprilat for 10 min.

In contrast to the fast and transient cyclic GMP kinetic after stimulation with BK, the RT-induced increase in cyclic GMP content developed slowly, reached a plateau level after 10 min, and remained stable for several hours (Fig. 1 A and Fig. 3).

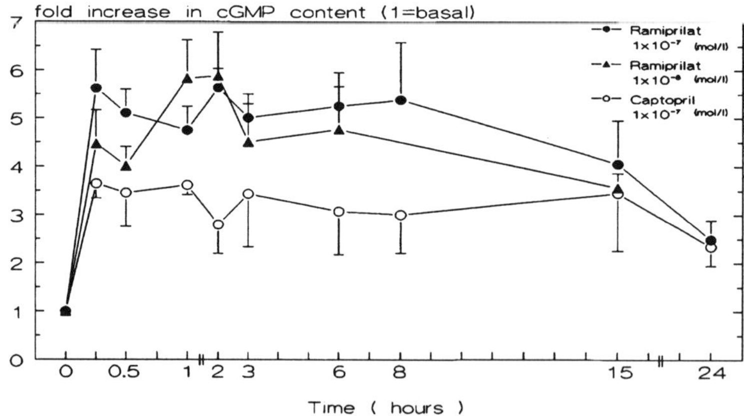

Fig. 3 Effects of ramiprilat (10^{-8} and 10^{-7} mol /l) and captopril (10^{-7} mol/l) on the accumulation of cyclic GMP (cGMP) in cultured bovine aortic endothelial cells as a function of time. The x-fold increases in cGMP over basal values after incubation in HEPES-buffer for the times indicated in the presence of ramiprilat or captopril are depicted.

BK also stimulated the release of prostacyclin in EC in a concentration- and time-dependent manner. Threshold concentrations were in the range of about 10^{-8} mol/l and maximal increases were observed between 10^{-7} - 10^{-6} mol/l. Near maximal increases were reached after 1 min of BK-stimulation and thereafter declined despite continuous stimulation. Similar findings have already been reported in a number of previous studies (10, 11, 12). Like BK RT elicited the range between 10^{-8} - 10^{-6} mol/l an increase in prostacyclin synthesis and release that was totally inhibited by Hoe 140 (10^{-7} mol/l). The time course of prostacyclin formation was quite parallel to that obtained for RT-induced cyclic GMP accumulation. Compared to the rapid and transient BK-induced increase in prostacyclin release, significant increases were only observed after 10 min.

DISSCUSION

The presented results provide circumstantial evidence of the formation of BK by cultured EC. It was shown that CE inhibitors stimulate the formation of NO and prostacyclin in EC by inhibiting the breakdown of endothelium-derived BK. This was supported by using the specific kinin receptor antagonist Hoe 140, that abolished the enhanced NO and prostacyclin formation observed after CE inhibition. The complete inhibition of BK- and RT-induced NO- synthesis by the stereospecific inhibitor of NO-synthase, N^G -nitro-L-arginine (L-NNA), indicates that an increased formation of NO, which in turn stimulates endothelial soluble guanylyl cyclase is responsible for the observed cyclic GMP increases. In contrast to the fast and transient increase in cyclic GMP content by exogenously added BK, the increase in response to RT developed slowly, peaking only after 10-15 min and was maintained for several hours. The rapid development of homologous desensitization after exposure to a high concentration of BK has been described (6). The question why the endogenously formed kinins do not lead to receptor desensitization cannot be satisfactorily answered by the performed experiments. It is likely, however, that desensitization is a process occurring only at higher concentrations of exogenously added BK ($> 3 \times 10^{-9}$ mol/l). The endogenous concentration of BK induced by RT is about 10^{-10} mol/l estimated from the concentration-dependent inhibition of the RT-induced cyclic GMP increase by Hoe 140 (Fig.2). This deduction is allowed since either BK or Hoe 140 exhibit a nearly same affinity to the BK-receptor as shown by the concentration-dependent inhibition of the BK-induced cyclic GMP increase by Hoe 140 (Fig.2). The cyclic GMP increase by BK (10^{-8} mol/l)is inhibited by Hoe 140 with an IC_{50} -value of about 3×10^{-8} mol/l.

The long-lasting stimulation of NO formation after CE inhibition mediated by the inhibition of the breakdown of endothelium-derived BK might have important beneficial effects in certain vascular diseases such as hypertension and arteriosclerosis.

REFERENCES

1. Kramer HJ, Glänzer K, Meyer-Lehnert H, Mohaupt M, Predel HG: Kinin - and non-kinin - mediated interactions of converting-enzyme inhibitors with vasoactive hormones. J Cardiovasc Pharmacol 1990; 15 (suppl 6): S91 - S98.

2. Scherf H, Pietsch R, Landsberg G, Kramer HJ, Düsing R: Converting-enzyme inhibitor ramipril stimulates prostacyclin synthesis by isolated rat aorta: Evidence for a kinin -dependent mechanism. Klin Wochenschr 1986; 64: 742 - 745.

3. Linz W, Martorana PA, Schölkens BA: Local inhibition of bradykinin degradation in ischemic hearts. J Cardiovasc Pharmacol 1990; 15 (suppl 6): S99 - S109.

4. Becker RHA , Wiemer G, Linz W: Preservation of endothelial function by ramipril in rabbits on a long-term atherogenic diet. J Cardiovasc Pharmacol 1991; 18 (suppl 2): S110 - S115.

5. Van Iwaarden F, de Groot PG, Sixma JJ , Berrettini M, Bouma BN: High-molecular weight kininogen is present in cultured human endothelial cells: Localization, isolation, and characterization.Blood 1988; 71: 1268 - 1276.

6. Lückhoff A, Zeh R, Busse R: Desensitization of the bradykinin-induced rise in intracellular free calcium in cultured endothelial cells. Pflügers Arch 1988; 412: 654 - 658.

7. Lückhoff A, Busse R, Winter I, Bassenge E: Characterization of vascular relaxant factor released from cultured endothelial cells. Hypertension 1987; 9: 295 - 303.

8. Wiemer G, Schölkens BA , Becker RHA , Busse R: Ramiprilat enhances endothelial autacoid formation by inhibiting breakdown of endothelium-derived bradykinin.Hypertension 1991; 18: 558 - 563.

9. Hock FJ , Wirth K, Albus U, Linz W, Gerhards HJ , Wiemer G, Henke S, Breipohl G, König W, Knolle J, Schölkens BA: Hoe 140 a new potent and long acting bradykinin antagonist: In vitro studies. Br J Pharmacol 1991; 102: 769 - 773.

10. Lückhoff A, Pohl U, Mülsch A, Busse R: Differential role of extra- and intracellular calcium in the release of EDRF and prostacyclin from cultured endothelial cells.Br J Pharmacol 1988; 95: 189 - 196.

11. Brotherton AFA: Induction of prostacyclin biosynthesis is closely associated with increased guanosine 3', 5' - cyclic monophosphate accumulation in cultured human endothelium.J Clin Invest 1986; 78: 1253 - 1260.

12. Alhenc-Gelas F, Tsai SJ, Callahan KS, Campbell WB, Johnson AR: Stimulation of prostaglandin formation by vasoactive mediators in cultered human endothelial cells. Prostaglandins 1982; 24: 723-742.

CAPTOPRIL AS A MODIFIER OF THE ARACHIDONATE CASCADE OF RAT PLATELETES

Á. Gecse, G. Telegdy

Department of Pathophysiology Albert Szent-Györgyi University Medical School of Szeged, Hungary

SUMMARY: The lipoxygenase pathway of arachidonate cascade in the platelets of spontaneously hypertensive rats has been found to be significantly higher than in normotensive animals. Repeated oral administration of Captopril (in drinking water - 200 mg/100 ml) for 14 days resulted in an elevation in the activity of arachidonate cascade in the platelets of treated rats. At the same time the Captopril treatment induced the formation of 12-hydroxy-heptadecatrienoic acid (12-HHT), which molecule is known to be a potent prostacyclin (PGI_2) releaser and/or synthesis inducer. PGI_2 is one of the most potent vasodilatator molecule in living organisms. The in vitro experiments in rat platelets suggest, that very low doses of Captopril (10^{-12} to 10^{-10} M) result in a significantly elevated 12-HHT synthesis. Captopril migth act through the 12-HHT - PGI_2 mechanism, resulting in blood pressure reduction. The lipoxygenase pathway of platelets, the formation of 12-hydroxyeicosatetraenoic acid (12-HETE), was significantly elevated in vitro in the presence of low dose of Captopril (10^{-11} and 10^{-10} M).

INTRODUCTION

Evidences indicate that various eicosanoids may participate in the vascular and renal mechanisms controlling blood pressure. Prostaglandins (PGE_2, PGD_2, and PGI_2) subsurve antihypertensive processes by opposing pressor mechanisms that bring about vasoconstriction and conservation of salt and water. On the other hand, thromboxane A_2 (TXA_2) causes vasoconstriction, PGF_{2alpha} reduces venous complience, and PGE_2 and PGI_2 stimulate renin secretion. This dual nature of prostanoids in blood pressure control seems to be very important mechanism (1, 2).

Arachidonate metabolites are cell-to-cell messengers produced by eukaryotic cells, often in response to extracellular stimuli (3). The metabolites of arachidonate cascade are not stored to any appreciable extent in any tissue, and there presence and actions therefore reflect *de novo* synthesis and release (4). It is the availability of free arachidonic acid, the substrate for eicosanoids - PG-s, Tx, PGI_2, hydroperoxy-eicosatetraenoic acids (HPETE-s), hydroxyeicosatetraenoic acids (HETE-s) and for peptide containing lipids, leukotrienes (LT-s) - that limits arachidonate cascade. The precursor fatty acid is liberated by a receptor-mediated activation of phospholipase A_2, phospholipase C and diacylglycerol kinase from phospholipids followed by the conversion of this precursor to eicosanoids.

The mechanisms by which various stimuli induce the deacylation and/or reacylation of arachidonic acid esterified in position 2 of phospholipids, and the synthesis of arachidonate metabolites, have not been unequivocally demonstrated. The evidences suggest that peptides may exert regulatory role through activation of calcium channel and/or phospholipid-arachidonate metabolism.

The platelets have been studied specifically, because platelet-vessel wall interactions might occur, and result inthe generation of endothelium and platelet derived factors (including arachidonate metabolites), playing important role in the pathogenesis of hypertension.

The locally synthesized and released arachidonate metabolites (either cyclooxygenase or lipoxygenase products) might modify both the endothelial cell and circulating blood cell functions, resulting in an altered blood vessel response and reactivity.

Studies were undertaken to exlore the possible role played in the regulation of the cyclooxygenase and lipoxygenase pathways of aracidonate cascade of platelets by Captopril, which blood pressure regulating action has only been known partly.

MATERIALS AND METHODS

Chemicals

Arachidonic acid (grade I) was purchased from Sigma Chemical Co, St. Louis, Mo (USA). 1-[14]-Arachidonic acid (2035 MBq/mM specific activity) was obtained from Amersham (England). TC Medium 199 was purchased from DIFCO Laboratories Detroit, Mich. USA. Prostaglandins (PGE_2, PGD_2, PGF_{2alpha}, and 6-oxo-PGF_{1alpha}), and tromboxanes (TxA_2, TxB_2) were generously provided by Dr J.E. Pike, Upjohn Co, Kalamazoo (USA). 12-hydroxyeicosatetraenoic acid (12-HETE) and 12-hydroxyheptadecatrienoic acid (12-HHT) standards were purchased from Calbiochem, La Jolla, Ca (USA).

Chemicals were of analytical grade and obtained commercially, unless otherwise stated.

Isolation of platelets

Blood was drawn from the abdominal aorta of male rats of the Wistar strain (body weight: 180 _+ 20g) under light ether anaesthesia. Samples were diluted with phosphate buffer (pH 7.4) which contained 5.8 mM EDTA and 5.55 mM glucose. The platelet rich plasma was collected after the whole blood had been centrifuged at 200 g for 10 min at room temperature. The platelets were sedimented from the supernatant by means of centrifugation at 2000 g for 10 min. The pellet contained the platelets (2×10^9 ml^{-1}) and the red blood cells (1×10^8 ml^{-1}). Erythrocytes are capable to metabolize arachidonic acid by the lipoxygenase pathway and release 12-HETE, dihydroxyeicosatetraenoic acid and 6-sulfidopeptide-containing leukotrienes (5). Therefore erythrocytes were lysed with hyposmic ammonium chloride (0.83 %, 9 parts) containing EDTA (0.02 %, 1 part) at 4 $^{\circ}$C for 15 min. According to our preliminary experiments this treatment did not modify the arachidonate cascade of platelets. The platelets were washed 3 times with phosphate buffer (pH 7.4, containing 5.8 mM EDTA and 5.55 mM glucose). After centrifugation at 2000 g for 10 min at room temperature, the platelets were resuspended in TC Medium 199. The TC Medium 199 was applied, because the arachidonate cascade of platelets was most active in this medium and less active or inactive in Ca^{2+} depleted or Ca^{2+} free incubation mixtures.

Analysis of eicosanoids

The platelets (10^8 ml^{-1} in each sample) were preicubated at 37 ^{o}C for 5 min, then 1-^{14}C-arachidonic acid was given to the incubation mixture. Ten minutes later the samples were centrifuged at 2000 g for 5 minutes. The pellets were resuspended in fresh, prewarmed ($37^{o}C$) TC Medium 199 and the Captopril (10^-^{12} to 10^{-6} M) was introduced to the incubation mixture, and 10 min later the enzyme reaction was quenched by bringing the pH to 3 with formic acid. In the control samples no Captopril was given to the incubation tubes. According to Dahl and Uotila (6) and our unpublished results 5 - 10 min is an adequate time period for labelling platelets with ^{14}C-arachidonic acid. The samples were extracted with ethyl acetate (2 x 3 ml). The organic phases were pooled and evaporated to dryness under nitrogen. The samples were stored under nitrogen at -36 ^{o}C for 24 h without any loss of activity. Each sample was subjected to HPLC within 24 h.

High-performance liquid chromatography

Reversed phase high-performance liquid chromatography (HPLC) was performed on a column (4.6 x 250 mm) packed with LiChrosorb^R C-18 (7 µm particles). The solvent used for the separation of 12-HETE, 12-HHT and TxB_2 was - acetonitril (700 ml): water (300 ml): concentrated phosphoric acid (10 µl; 2.5 mM) - and the pH was adjusted to 4 with NH_4OH. The solvent was passing through the column with a speed of 1 ml/min at room temperature. The samples were dissolved in 15 µl solvent and loaded into the 10 µl microcapillary of the ISCO (Mod. 2350) HPLC pump. 12-HETE was monitored at 235 nm with UV detector, integration took place with a Hewlett-Packard integrator (HP 3396A).Four samples were collected in each minute for 10 min in liquid scintillation vials. From the 11th min, samples were collected in each min for 50 min, to detect the radioactive cyclooxygenase metabolites. The solvent was diluted in each scintillation vial with 2 ml metanol, and the radioactivity was determined in a Beckman LS 1800 liquid scintillation counter, using 3,5 ml toluene containing 0.4 % w/v PPO, 0.02 % w/v POPOP and 10 % v/v ethanol. The radiolabelled products of arachidonic acid were identified with unlabelled authentic standards.

Statistical analysis was performed via *Student*s* t test.

Captopril treatment of rats

The effects of Captopril treatment on the arachidonate cascade of rat platalets were tested both in male and female spontaneously hypertensive rats (SHR) and in male and female normotensive Wistar-Kyoto rats in 19 - 20 weeks of age (body weight: 180 _+ 15 g), as well. The Captopril was dissolved in the drinking water (200 mg/100 ml), and was freshly prepared in each day. The oral Captopril treatment was maintained for 14 days. The platelets of control and treated rats were separated according to the above described method.

RESULTS

The lipoxygenase pathway dominated the arachidonate cascade of rat platelets. The lipoxygenase pathway was slightly, but significantly elevated in the platelets of spontaneously hypertensive rats (Fig.1.).

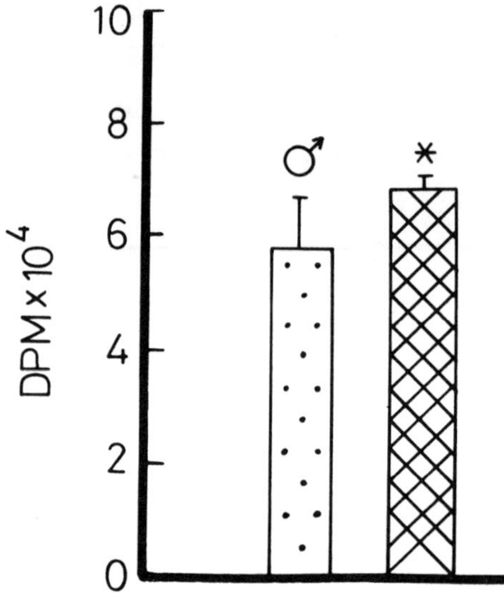

Figure 1. The lipoxygenase pathway of arachidonate cascade in the platelets of spontaneously hypertensive and normotensive rats.

The released radioactive lipoxygenase metabolites (12-HETE, 12-HPETE) of arachidonate cacade were separated by HPLC, and quantitatively determined in Beckman LS 1800 counter.
Each column represents the mean $+$ SEM of the data on 10 rats
normotensive rats
spontaneously hypertensive rats
* $p < 0.05$

Both pathways (cyclooxygenase and lipoxygenase) of the arachidonate cascade of platelets separated from either male or female spontaneously hypertensive rats was slightly, but significantly stimulated in those groups of animals, which were treated with Captopril (data are not shown).

The 12-HHT - the most potent endogenous prostacyclin liberator - generation was measured in the platelets of spontaneously hypertensive Captopril treated an untreated male rats. The formation of 12-HHT in the platelets of rats subjected to Captopril treatment was significantly induced (Fig. 2.).

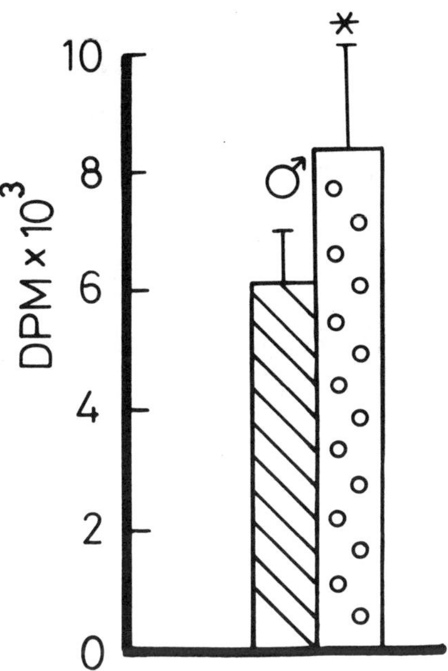

Figure 2. The effect of oral (14 days) administration Captopril on the 12-hydroxyheptadecatrienoic acid (12-HHT) generation in platelets of spontaneously hypertensive male rats.

The released radiolabelled 12-HHT was separated by HPLC and quantitatively determined by Beckman LS 1800 liquid scintillation counter.
SHR untreated group ▨
SHR Captopril treated group ☐
Each column represents the mean + SEM of the data on 10 animals.
* = p < 0.05

The platelets were isolated from normotensive rats, and the radiolabelled 12-HHT generation was measured after preincubation of platelets with different concentrations of Captopril (10^{-12} to 10^{-6} M). Even in the presence of 10^{-12} M Captopril, the synthesis of 12-HHT elevated twice of the control. The most potent Captopril dose was found to be 10^{-11} M, when the 12-HHT generation has almost reached three times value of the untreated animals (Fig.3.).

The formation of 12-HHT was falling gradually, increasing the dose of Captopril to 10^{-7} M, where the second most effective peak was observed. This dose of Captopril (10^{7} M) was almost as active as the 10^{-11} M (Fig. 3.).

The results were similar when the platelets of spontaneously hypertensive rats were subjected to Captopril action in vitro (data are not shown).

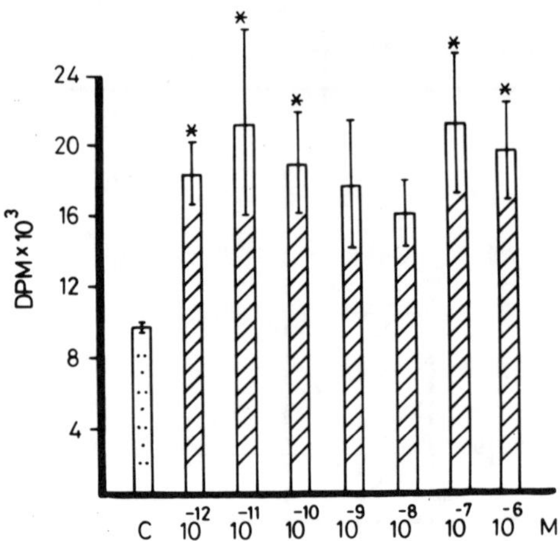

Figure 3. The in vitro effect of Captopril on the generation of radiolabelled 12-HHT from ^{14}C-arachidonic acid in the platelets of normotensive rats.

The doses of Captopril is listed on the abscissa, and the released 12-HHT is registered on the ordinate, in DPM. The 12-HHT was separated with HPLC and quantitatively determined in Beckman LS 1800 scintillation counter.
Each column represents the mean + SEM of the data on 10 animals.
Control group - ⬚
Captopril treated group - ▨
* = $p < 0.05$

The lipoxygenase pathway (12-HETE and 12-HPETE) of arachidonate cascade in the platelets of normotensive and spontaneously hypertensive rats were also investigated in vitro in the presence of Captopril. Three doses if Captopril resulted in a significant elevation of 12-HETE formation in the platelets of normotensive rats, but the increase was not so pronounced than in case of 12-HHT synthesis. The most effective dose of Captopril was found to be 10^{-7} M, and less effective were the 10^{-11} M and 10^{-10} M, respectively.

Much less 12-HPETE could be detected in the incubation mixture of platelets, but the effect of Captopril was similar as it was found with 12-HETE (data are not shown).

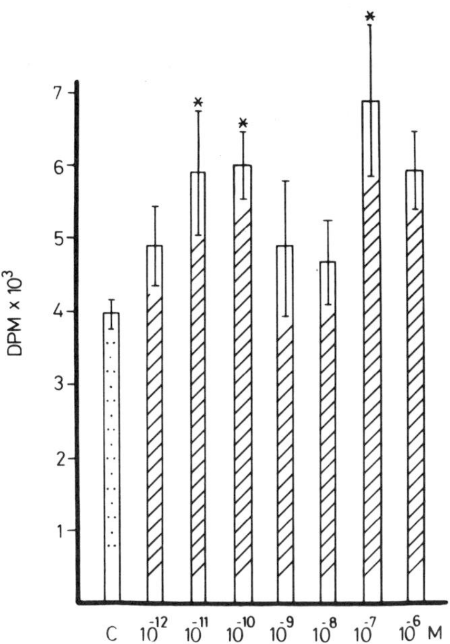

Fig. 4. The in vitro effect of Captopril on the radiolabelled 12-hydroxyeicosatetraenoic acid (12-HETE) synthesis from ^{14}-C-arachidonic acid in the platelets of normotensive rats.

The 12-HETE was separated on a HPLC column, the radioactivity was quantitatively measured in a Beckman LS 1800 liquid scintillation counter.
Each column represents the mean + SEM of the data on 10 animals.
Control group - ⬚
Captopril treated group - ⬚
* = $p < 0.05$

DISCUSSION

It is known that the lifetime Captopril treatment of spontaneously hypertensive rats results in a blood pressure of animals in the normal range (7), through several mechanisms (8, 9, 10). Much less experiments were assessed to detect any action of Captopril on SH rats after a short period of treatment.

Our results suggest that some of the antihypertensive actions of Captopril might be elicited by the 12-HHT - prostacyclin mechanism, because the Captopril (both in vivo and in vitro) induced the generation and release of 12-HHT from the platelets of SH rats. The 12-HHT is an endogenous regulator of endothelial PGI_2 synthesis (11). The PGI_2 might inhibit either the aggregation and release of platelets or induce vasodilation, increasing the sensitivity of baroreceptors, which then leads to the normalisation of blood pressure (9).

Our second important finding was that the 12-HETE generation was induced in the platelets treated with Captopril either in vitro or in vivo. The 12-HETE the end-lipoxygenase metabolite of arachidonic acid in platelets has been previously shown to prevent PGH_2/TxA_2-induced platelet activation, aggregation (12), which might be beneficial in hypertensive states.

ACKNOWLEDGEMENT

This work was partly supported by the grant (No 2683) of the National Scientific Research Foundation Hungary.

REFERENCES

1. Lin L., Nasljetti A. Role of endothelium-derived prostanoid in angiotensin-induced vasoconstriction. Hypertension. 1991; 18: 158-164.

2. Levens N. R., Cote D., Ksander G. Inhibiton of thromboxane synthetase potentiates the antihypertensive actiom of an angiotensin-converting enzyme inhibitor by prostaglandin-dependent but kinin-independent mechanism. J. Pharm. Exp. Ther. 1991; 259: 219-227.

3. Harris R. H., Ramwell P. W., Gilmer P. J. Cellular mechanism of prostaglandin action. Ann. Rev. Physiol. 1979; 41: 653-665.

4. Piper P., Vane J. The release of prostaglandins from lung and other tissue. Ann. N. Y. Acad. Sci. 1971; 183: 363-385.

5. Kobayashi T., Levine L. Arachidonic acid metabolism by erythrocytes. J. Biol. Chem. 1983; 258: 9116-9119.

6. Dahl M. L., Uotila P. Verapamil decreases the formation of thromboxane from exogenous [14]-C-arachidonic acid in human platelets in vitro. Prostaglandins Leukotrienes Med. 1983; 17: 191-202.

7. Berecek K. H., Wyss J. M., Swords B. H. Alterations in vasopressin mechanisms in Captopril-treated spontaneously hypertensive rats. Clin. Exp. Hypert. Part A - Theory and Practice 1991; 13: 1019-1031.

8. Busse R., Lamontagne D. Endothelium-derived bradykinin is responsible for the increase in calcium produced by angiotensin-converting enzyme inhibitors in human endothelial cells. Naunyn Schmiedebergs Arch. Pharmacol. 1991; 344: 126-129.

9. Chapleau M. W., Hajduczok G., Abboud F. M. Paracrine role of prostanoids in activation of arterial baroreceptors - An owerview. Clin. Exp. Hypert. Part A - Theory and Practice 1991; 13: 817-824.

10.Codde J. P. Beilin L. J. Prostaglandins and experimental hypertension: A review with special emphasis on the effect of dietary lipids. J. Hypertension. 1986; 4: 675-686.

11.Sadowitz P. D., Yamaja Setty B. N., Stuart M. 12-HHT as an endogenous regulator of prostacyclin synthesis. Prostaglandins. 1987; 34: 749-759.

12.Fonlupt P., Croset M., Lagarde M. 12-HETE inhibits the binding of PGH_2/TXA_2 receptor ligands in human platelets. Thromb. Res. 1991; 63: 239-248.

Recent Progress on Kinins
© 1992 Birkhäuser Verlag Basel

RAMIPRILAT PREVENTS PAF-INDUCED MYOCELLULAR AND ENDOTHELIAL INJURY IN A NEUTROPHIL-PERFUSED HEART PREPARATION

K. Schrör and A. Felsch

Institut für Pharmakologie, Heinrich-Heine-Universität Düsseldorf, D-4000 Düsseldorf 1, F.R.G.

SUMMARY: This study investigates the action of PAF-stimulated human polymorphonuclear leukocytes (PMN) on myocardial integrity and function in Langendorff-perfused guinea-pig hearts. Infusion of 10^6 PMN/ml resulted in a negative inotropic effect without larger biochemical evidence for myocardial tissue injury while infusion of PAF (1 μM) did not cause any permanent effect at all. However, the combined administration of PAF-stimulated PMN resulted in severely depressed myocardial contractile function and biochemical evidence for myocardial tissue injury. This was probably due to an enhanced uptake of PMN from the coronary perfusate and accumulation within the myocardial tissue. Ramiprilat, (10 μM) significantly improved left ventricular function and myocardial cell integrity. Similar results were obtained with bradykinin (1 nM). The data suggest a PAF-induced, PMN-mediated myocardial tissue injury as well as cardioprotective actions of ACE inhibition which are possibly related to stimulation of the kinin/prostacyclin axis.

INTRODUCTION

In living cells and tissue, any inflammatory stimulus such as activated neutrophils (PMN) will also result in the activation of defense mechanisms (1). These tend to reduce the noxious reaction and to focus it on the site of injury. Angiotensin converting enzyme (ACE) inhibitors are a group of compounds, capable of stimulation of endogenous prostacyclin formation (2). This is associated with tissue protective actions in experimental set-ups (3, 4). It is currently unknown whether the cardioprotective effect of ACE inhibitors also involves antineutrophil actions. Clearly, stimulation of prostaglandin generation might be valuable in this respect.

The present study was designed to investigate the possible cardioprotective actions of ACE inhibition in a model of the PMN-perfused isolated heart preparation (5). To elucidate the possible contribution of prostaglandins, control experiments were performed with the same experimental set-up, using bradykinin.

MATERIALS AND METHODS

Langendorff hearts were prepared from guinea pigs (270-350 g body weight) and perfused at 35°C and pH 7.4 with oxygenated (5% CO_2 in O_2) Tyrode solution, supplemented with glucose (10.0 mM) and EDTA (0.05 mM). The flow rate was kept constant at 10 ml/min and the heart rate at 240 beats/min by electrical pacing.

Left ventricular peak actively developed pressure (LVP), left ventricular enddiastolic pressure (LVEDP) and coronary perfusion pressure (CPP) were measured continuously. PMN were prepared from fresh human citrated blood as previously described (6). PMN uptake into myocardial tissue was determined by the myeloperoxidase (MPO) assay. Superoxide dismutase-(SOD) and creatinphosphokinase-specific activities were determined in left ventricular homogenates at the end of the experiment. Cardiocoronary prostacyclin generation was measured in the coronary effluent in terms of 6-oxo-$PGF_{1\alpha}$ by radioimmunoassay.

At time 0, infusion was started of either PMN, PMN + ramiprilat (10 μM), ramiprilat alone (10 μM) or vehicle alone for a total time of 40 min. Human PMN were infused into the aortic inflow tract from a separate syringe at a constant rate of 0.2 ml/min. This was equivalent to a final concentration of 10^6 cells/ml in the Tyrode buffer.

If not otherwise indicated, infusion of PAF (1 μM) was started at time 20 min and was maintained until time 40 min. In another set of experiments, bradykinin (Sigma, Deisenhofen, FRG) (1 nM) was infused for 20 min, i.e. from times 20 min to 40 min. All infusions were stopped at time 40 min. After a 20 min washout-period, the heart was removed and subjected to the biochemical assays as described above.

The data are mean ± SEM (x ± SEM) of n observations. Statistical analysis was made by the two-tailed Student's t-test. P values of $\leq$ 0.05 were considered significant.

RESULTS

There were no significant differences in LVP, LVEDP and CPP between the several groups of hearts at the beginning of the experiment (time 0). Infusion with PMN alone for 40 min resulted in a significant drop in LVP while LVEDP and CPP remained unchanged. PAF alone did not modify LVP or LVEDP. There were also no significant alterations in these parameters by ramiprilat alone or in combination with PAF. However, ramiprilat prevented the reduction of LVP in PMN-perfused hearts.

Infusion of PAF-stimulated PMN resulted in a considerable increase in LVEDP from 0 mmHg to 14 ± 2 mmHg (P < 0.01) at 40 min and of the CPP from 67 ± 5 mmHg to 75 ± 6 mmHg ($P < 0.05$) while the LVP was reduced from 51 ± 2 to 39 ± 2 mmHg ($P < 0.01$). Coadministration of PMN and ramiprilat significantly attenuated the decrease in LVP and the increase in LVEDP. Table 1 summarizes the data on LVP.

Table 1. Alterations in left ventricular actively developed pressure (LVP) as compared to vehicle-treated controls (control).

Group	n	LVP [mmHg] at time [min]			
		0	20	40	60
Control	8	58 ± 2	55 ± 3	55 ± 5	51 ± 4
PAF	6	59 ± 1	59 ± 1	55 ± 5	$55 \pm 5_*$
PMN	6	59 ± 2	51 ± 4	$41 \pm 1_*$	$40 \pm 1_*$
RAM + PAF	3	57 ± 4	62 ± 5	63 ± 5	61 ± 8
PMN + PAF	8	55 ± 1	51 ± 2	$39 \pm 2^{*+}$	$32 \pm 2^{*+}$
PMN + PAF + RAM	8	57 ± 1	51 ± 1	49 ± 6	41 ± 3

Data are mean $\pm$ SEM of n experiments
$_*$): $P < 0.05$ (PMN or PMN + PAF vs. Control at the same time)
$^+$): $P < 0.05$ (PMN + PAF vs. PMN + PAF + RAM)

The total 6-oxo-PGF$_{1\alpha}$ release during the first 20 min amounted to 46 ± 1 pmoles in the absence (n = 3) and 45 ± 7 pmoles in the presence of PMN ($P > 0.05$). PAF alone did not cause any significant increase in 6-oxo-PGF$_{1\alpha}$ release in the absence of PMN. In contrast, there was a marked 7-fold increase in 6-oxo-PGF$_{1\alpha}$ levels after infusion of PAF- stimulated PMN (n = 8; $P < 0.05$).

Treatment with ramiprilat resulted in a significant, about twofold, increase of 6-oxo-PGF$_{1\alpha}$ release both in the presence and absence of PMN prior to PAF stimulation. During the first 5 min of PMN infusion, according to the "area under the curve", $2.3 \pm 0.4 \times 10^7$ PMN were taken up in vehicle-treated hearts but only $1.3 \pm 0.2 \times 10^7$ cells in ramiprilat-treated preparations ($P < 0.05$). This was equivalent to a 50% inhibition of PMN uptake by ramiprilat during initial PMN infusion ($P < 0.05$; n = 8).

After initiation of PAF stimulation, there was another uptake of PMN by the heart. A total number of $1.8 \pm 0.4 \times 10^7$ cells was taken up during PAF stimulation in the absence but only $1.0 \pm 0.1 \times 10^7$ cells in the presence of ramiprilat (n = 8; $P < 0.05$). Thus, *in toto* 4.1×10^7 PMN were taken up by the heart in the absence and 2.3×10^7 PMN in the presence of ramiprilat. This was equivalent to a 50% reduction of PMN uptake by ramiprilat. This agrees well with the 50% reduction in myocardial MPO activity by ramiprilat treatment (see below) and suggests the reduced uptake of PMN as major explanation for cardioprotective actions of ramiprilat.

A 3-4-fold increase in myocardial MPO activity was obtained after PAF stimulation and reduced by 55% in the presence of ramiprilat ($P < 0.05$). Similarly, infusion of PAF-stimulated PMN resulted in a significant, 36% reduction of CK which was significantly attenuated by ramiprilat treatment. A more than 60% loss of cytosolic SOD was obtained after administration of PAF-stimulated PMN which was significantly attenuated by ramiprilat treatment ($P < 0.05$). Ramiprilat itself caused no significant reduction in any of these parameters of myocardial biochemistry (Table 2).

Table 2. Actions of ramiprilat (10 μM) on biochemical parameters in PAF-stimulated PMN perfused guinea-pig hearts

	n	Myocardial SOD [IU/mg protein]	Myocardial MPO [mU/gww]	Myocardial CK [IU/mg protein]
CON	8	28.4 ± 2.0	0	13.6 ± 0.3
PAF	6	25.0 ± 0.6	0	11.5 ± 0.3
PMN	6	26.6 ± 0.6	6.4 ± 0.2	12.3 ± 0.7
RAM + PAF	3	25.5 ± 1.0	0	11.0 ± 1.7
PMN + PAF	8	11.0 ± 0.5	18.6 ± 2.8	8.9 ± 0.5
PMN + PAF + RAM	8	17.8 ± 1.7	10.8 ± 2.4	10.9 ± 0.5

In order to establish the possible involvement of kinins in the beneficial effects of ACE inhibition, two additional series of experiments were carried out where bradykinin (1 nM) was infused instead of ramiprilat into the coronary circulation of PAF + PMN-treated hearts. Infusion of bradykinin resulted in an almost complete inhibition of PMN uptake during PAF stimulation ($P < 0.05$; n = 6). This was accompanied by a considerable increase in 6-oxo-PGF$_{1\alpha}$ levels in the coronary effluent, reaching peak levels that were 50-60-fold above basal. Bradykinin also significantly attenuated the loss of CK-, SOD- and MPO-specific activities. Table 3 summarizes these results. This was accompanied by a considerably improved myocardial and coronary function, including a significant reduction of enhanced LVEDP: 9 ± 2 vs. 14 ± 2 mmHg and increase in LVP: 39 ± 2 to 70 ± 8 mmHg at time 40 min (n = 6; $P < 0.05$) (Table 3).

Table 3. Actions of bradykinin (1 nM) on biochemical parameters in PAF-stimulated PMN perfused guinea-pig hearts

Parameter	n	no bradykinin	n	bradykinin	P
Myocardial SOD [IU/mg protein]	8	11.0 ± 0.5	6	17.3 ± 1.5	< 0.05
Myocardial MPO [mU/gww]	8	16.4 ± 2.5	6	9.9 ± 2.1	< 0.05
Myocardial CK [IU/mg protein]	8	8.8 ± 0.8	6	12.2 ± 0.8	< 0.05

DISCUSSION

Continuous infusion of human polymorphonuclear neutrophils (PMN) into the coronary circulation of an isolated Langendorff heart is associated with a considerable loss of PMN. These cells probably migrated from the vessel lumen into the myocardial tissue. Indeed, histological examination of myocardial tissue confirmed accumulation of PMN in the extravascular space after stimulation by PAF. This was not seen in the absence of PAF and was prevented by both ramprilat and bradykinin treatment (not shown). These changes were paralleled by alterations in MPO activity in myocardial homogenates.

Infusion of PAF through the preparation resulted in some transient negative inotropic effect, previously described by others (7). Otherwise, PAF did not significantly change myocardial tissue integrity as seen from unchanged levels of CK and SOD and normal ventricular function.

A completely different picture was obtained when PAF-stimulated PMN were infused. PAF is known to be a potent stimulus for PMN degranulation, chemotaxis, aggregation and superoxide anion generation (1, 8). There was a considerable loss of myocardial cytosolic enzymes, such as SOD and CK, by 50-60% of control, suggesting severe myocardial tissue injury. This injury was probably related to immigrating PMN, as seen from the immediate drop in PMN recovery from the coronary effluent after starting PAF infusion and an about 3-fold increase in tissue MPO activity. These biochemical alterations were accompanied by a severe impairment of cardiac and coronary function.

Piper and Stewart (9) have shown that about 4% of PAF, administered to guinea-pig hearts is taken up by myocardial tissue. In our experimental conditions, these 4% are equivalent to about 2 nmoles PAF per g wet weight. This is probably sufficient to stimulate PMN chemotaxis and degranulation (8). Thus, PAF induces myocardial injury in this preparation primarily by acting as local chemoattractant for PMN accumulation within the cardiac tissue and in addition stimulates PMN secretion and generation of toxic oxygen species.

Administration of ramiprilat at a concentration sufficient to block ACE activity in isolated heart preparations (10) did not cause any changes in left ventricular function, coronary perfusion pressure or myocardial biochemistry (not shown). When ramiprilat was administered together with PAF-stimulated PMN, there was a marked improvement of all parameters measured, including preservation of cardiac CK and SOD levels and improvement of cardiac and coronary function (cf. Table 2). This suggests that ramiprilat protects the myocardium from PMN-induced injury and is in line with earlier reports from others, dealing with protective actions of ramiprilat in ischemic myocardial injury (3, 4). In separate control experiments, ramiprilat did not affect PMN functions in vitro (not shown). Thus, the PMN-antagonistic, i.e. tissue-protective properties of ramiprilat appear to be indirect in nature.

In this respect, it is interesting to note that ramiprilat significantly enhanced basal PGI_2 release. Although the stimulation was small, it resulted in the generation of functionally active amounts of prostacyclin as seen from the reduced PMN uptake during early infusion and the reduced myocardial MPO activity. This finding confirms the existence of an active angiotensin converting enzyme in the vasculature of the isolated guinea-pig heart. The enzyme may become activated by adhering PMN (11), possibly associated with reduced kinin degradation and PGI_2 release (12).

In order to establish that stimulation of the kinin/prostaglandin axis is involved in the beneficial effects of ramiprilat in this model, additional experiments were performed using bradykinin as a positive control. We have originally described that the bradykinin-induced coronary vasodilatation in guinea-pig hearts involves release of prostacyclin and another indomethacin-insensitive vasodilator (13), later shown to be probably NO (14). Bradykinin clearly protected the heart from PMN-induced myocardial injury in a way very similar to ramiprilat (cf. Table 3). The more pronounced response to bradykinin might be due to the considerably higher PGI_2 release of or by additional actions of bradykinin. This includes inhibition of PMN by release of nitric oxide, since bradykinin stimulates this pathway in the guinea-pig heart (14) and both cAMP- and cGMP-dependent pathways of human neutrophil inactivation act synergistically (15).

CONCLUSION

In conclusion, we have shown that the ACE inhibitor ramiprilat protects the PMN-perfused Langendorff heart of the guinea pig from tissue injury by inhibition of neutrophil accumulation. These effects are similar to those, obtained with administration of bradykinin and suggest that the protective effect of ramiprilat on PMN-induced myocardial damage is related to stimulation of cardiocoronary PGI_2 release. The possible involvement of additional protective factors, such as NO (10) can not be excluded. In any case, the data clearly suggest that cardioprotective actions of ACE inhibitors may also be due to inhibition of injury-induced PMN accumulation. This is a new aspect in the biological profile of ACE inhibitors that requires further studies.

ACKNOWLEDGEMENTS

This study was supported in part by the Deutsche Forschungsgemeinschaft (Schr 194/7-4) and a grant of the Ministerium für Wissenschaft und Forschung des Landes Nordrhein-Westfalen. The authors are grateful to Dipl.-Chem. Hans Strobach for performing the radioimmunological determinations, to Marie Palmér for expert technical assistance and to Erika Lohmann for competent secretarial help.

REFERENCES

1.	Hallett B. The significance of stimulus-response coupling in the neutrophil for physiology and pathology. In: Hallett B, editor. The neutrophil: Cellular Biochemistry and Physiology. Boca Raton: CRC-Press 1989; 1-22.

2.	Schrör K. Converting enzyme inhibitors and the interaction between kinins and eicosanoids. J Cardiovasc Pharmacol 1990;15 Suppl 6:S60-8.

3.	Schölkens BA, Linz W, König W. Effects of the angiotensin converting enzyme inhibitor, ramipril, in isolated ischaemic rat heart are abolished by a bradykinin antagonist. J Hypertension 1988;6 Suppl 4:S25-8.

4.	Van Gilst WH, de Graeff PA, Wesseling H, de Langen CDJ. Reduction of reperfusion arrhythmias in the ischemic isolated rat heart by angiotensin converting enzyme inhibitors: A comparison of captopril, enalapril, and HOE 498. J Cardiovasc Pharmacol 1986;8:722-28.

5.	Gillespie MN, Kojima S, Kunitomo M, Jay M. Coronary and myocardial effects of activated neutrophils in perfused rabbit hearts. J Pharmacol Exp Ther 1986;239:836-40.

6.	Hecker G, Ney P, Schrör K. Cytotoxic enzyme release amd oxygen centered radical formation in human neutrophils are selectively inhibited by E-type prostaglandins but not by PGI_2. Naunyn-Schmiedeberg's Arch Pharmacol 1990;341:308-15.

7.   Levi R, Burke JA, Guo Z-G, Hattori Y, Hoppens CM, McManus LM, Hanahan DJ, Pinckard RN. Acetyl glyceryl ether phosphorylcholine (AGEPC). A putative mediator of cardiac anaphylaxis in the guinea pig. Circ Res 1984;54:117-24.

8.   Braquet P, Touquil L, Shen TY, Vargaftig BB. Perspectives in platelet-activating factor research. Pharmacol Rev 1987;39:97-145.

9.   Piper PJ, Stewart AG. Antagonism of vasoconstriction induced by platelet-activating factor in guinea-pig perfused hearts by selective platelet-activating factor receptor antagonists. Br J Pharmacol 1987;90:771-83.

10. Wiemer G, Schölkens BA, Becker RHA, Busse R. Ramiprilat enhances endothelial autacoid formation by inhibiting breakdown of endothelium-derived bradykinin. Hypertension 1991;18:558-63.

11. Miller DK, Sadowski S, Soderman DD, Kuehl FA Jr. Endothelial cell prostacyclin production induced by activated neutrophils. J Biol Chem 1985;260:1006-14.

12. Ward PA, Varani J. Mechanisms of neutrophil-mediated killing of endothelial cells. J Leukocyte Biol 1990;48:97-102.

13. Schrör K, Metz U, Krebs R. The bradykinin-induced coronary vasodilatation. Evidence for an additional prostacyclin-independent mechanism. Naunyn-Schmiedeberg's Arch Pharmacol 1979;307:213-21.

14. Kelm M, Schrader J. Nitric oxide release from the isolated guinea pig heart. Eur J Pharmacol 1988;155:317-21.

15. Schröder H, Ney P, Woditsch I, Schrör K. Cyclic GMP mediates SIN-1 induced inhibition of human polymorphonuclear leukocytes. Eur J Pharmacol 1990;182:211-18.

CARDIOPROTECTION OF ACE INHIBITOR IN ISCHEMIC HEART IS NOT DEPENDENT ON THE LOCAL ANGIOTENSIN II FORMATION

Keita Noda, Manabu Sasaguri, Munehito Ideishi, Masaharu Ikeda and Kikuo Arakawa

Department of Internal Medicine of Fukuoka University
45-1, 7-chome Nanakuma, Jonan-ku, Fukuoka city, Japan 814-01

SUMMARY: Angiotensin II (AII) and bradykinin (BK) release into anterior interventricular vein (AIV) increased significantly 30 minutes after left anterior descending artery (LAD) occlusion in the absence of kidneys. Captopril enhanced BK release, but did not suppress the increase of AII release. Nafamostat suppressed both releases. Infarct size was significantly reduced by captopril but not by nafamostat. These results suggest that cardioprotective effect of captopril might be dependent on local BK accumulation, but not on suppression of local AII generation.

INTRODUCTION

Captopril, an angiotensin converting enzyme (ACE) inhibitor, has been shown to have cardioprotective effects, and most important effect is the reduction of myocardial infarct size (1-5). The underlyng mechanisms of this effect seem to be related not only to their improvement of systemic hemodynamics but also to local effects on ischemic tissue. Increases of prostaglandin synthesis (6,7), the formation of endothelium-derived relaxation factor (8), direct vasodilation due to reduced BK degradation (9-11), and scavenging of free radicals by their sulfhydryl groups (9,12) could all contribute to the diminution of infarct size by these agents in addition to the reduction of AII formation. However, there has been no direct in vivo evidence of the suppression of local AII formation and BK accumulation after treatment of ACE inhibitor during myocardial ischemia.

Present study investigated the local AII and BK formation in the ischemic heart and those effects on progression of myocardial infarction. To estimate AII and BK which are generated in ischemic area and released to circulating blood more precisely than previous experiment (13), anterior interventricular venous blood was selectively collected because the blood perfused in ischemic area, after LAD occlusion, circulated into AIV. To identify each influence of local AII or

BK production in ischemic area on the size of myocardial infarction, ACE inhibitor ; captopril and tissue kallikrein inhibitor ; nafamostat mesilate (6-amidino-2-naphthyl p-guanidinobenzoate dimethanesulfonate), were chosen. Nafamostat is known to inhibit trypsin, tissue kallikrein, plasmin, Cl_r, Cl_s and thrombin, but not to inhibit acid proteases such as papain and cathepsin D (14,15).

MATERIALS AND METHODS

All animals received humane care in compliance with guidelines formulated in Fukuoka university. Forty-seven male mongrel dogs weighing 8-17 kg were anesthetized with sodium pentobarbital (25 mg/kg), and bilaterally nephrectomized to eliminate the influence of the renal renin-angiotensin system. After 24 hours, dogs were re-anesthetized with sodium pentobarbital (15 mg/kg), and ventilated with a Harvard respirator (Model 607E). Halothane was used for maintaining anesthesia in 0.5-1.0 % v/v with pure oxygen.

Arterial blood pressure, left atrial pressure and central venous pressure were monitored by calibrated pressure transducers (MP-24T Nihon Kohden, Japan). To measure cardiac output, a thermodilution catheter (5F Criti Cath Gould, U.S.A.) was placed in the main pulmonary artery via the left femoral vein. A polyethylene catheter was inserted into the right femoral vein to administer drugs. The electrocardiogram (lead II) was monitored continuously.

The heart was exposed by left thoracotomy through the fifth intercostal space and was placed in a pericardial cradle. To determine the local production of AII and BK, an 18-gage polyethylene catheter (Medicut, Argyle, Japan) was inserted into the AIV, placing the tip of the catheter 2-3 cm from the origin of LAD. When all systemic hemodynamic parameters became stable, each inhibitor were administered. After all systemic parameters reached the steady state, LAD was occluded for 90 minutes at 2-3 cm distal to its origin. The hematocrit was measured every 30 minute.

The following 4 groups of dogs were used in this study:
Sham operation (SHAM) group:
 Dogs underwent sham operation and the LAD remained unoccluded.
Control (CONT) group:
 Dogs were subjected to LAD occlusion for 90 minutes with no drug administration.
Captopril (CAPT) group:
 Dogs received an intravenous bolus of captopril (0.25 mg/kg) in 5 ml of saline 30 minutes
 before LAD occlusion followed by infusion of captopril (0.5 mg/kg/hour).
Nafamostat mesilate (NAFA) group:
 Dogs received an continuous infusion of nafamostat mesilate (3.0 mg/kg/hour) from 30
 minutes before LAD occlusion. (0.25 mg/kg/min)

Blood gas analysis

Blood was collected from AIV and aorta before occlusion and 15, 30, and 90 minutes after occlusion. Blood was drawn into a heparinized polyethylene syringe and kept on ice until analysis. The pH and Po_2 were measured using a blood gas analyzer (ABL 2, Acid-Base Laboratory, Denmark).

Coronary circulation

To calculate the anterior interventricular venous blood flow (AIVBF), 6.5 ml of blood was collected from the 18-gage catheter inserted into AIV at the level of the coronary sinus before occlusion and 5, 15, 30, 60, and 90 minutes after occlusion. The time (T) required for this collection was measured. AIVBF was calculated as follows; AIVBF = 6.5 / T (ml/min)

Measurement of Angiotensin II and Bradykinin

AII and BK were measured by radioimmunoassay (RIA). Antisera against AII and BK were raised in male white rabbits. Each peptide was coupled to thyroglobulin using carbodiimide (16), and 3 mg of the preparation was injected into the backs of rabbits every 3 weeks for 4 months. The Antisera against AII and BK were used at 1:200,000 and 1:80,000 dilutions, respectively. The Cross-reactivity of anti-AII antiserum with angiotensin I and angiotensin III was 0.3 % and 50 %, respectively. The Cross-reactivity of anti-bradykinin antiserum with Lys-bradykinin and Met-Lys-bradykinin was 53 % and 70 %, respectively.

Blood was collected from AIV and aorta before drug administration and occlusion as well as 5, 15, 30, 60, and 90 minutes after occlusion. To 6.5 ml of blood added was 1 ml of protease inhibitor mixture containing 12 mg of sodium ethylenediamine tetraacetate, 2.4 mg of polybrene, 6 mg of 1, 10-phenanthroline, 6,000 KIE of aprotinin and 3 mg of soybean trypsin inhibitor. The samples were centrifuged at 3,000 rpm for 10 minutes at 4 ^oC. Two milliliters of the plasma was applied to a C18 cartridge (Sep-Pak, Waters Associates, U.S.A.) pretreated with 5 ml of 100 % methanol and 5 ml of distilled water. The cartridge was washed twice with 5 ml of distilled water and then eluted with 2 ml of 80 % methanol containing 0.1 % trifluoroacetic acid. The eluate was evaporated to dryness and stored at -20 ^oC. Samples were dissolved in 2 ml of RIA buffer which was composed of 0.02 M potassium phosphate pH 7.4, 0.1 M NaCl, 1 mg/ml of bovine serum albumin, 2.6 mM sodium ethylenediamine tetraacetate, 3 mM 1,10-phenanthroline, and 1 mg/ml of NaN_3. For the measurement of BK, a mixture consisting of 100 µl of sample, 200 µl of diluted antiserum, 400 µl of RIA buffer and 100 µl of ^{125}I-labeled bradykinin (Du Pont/NEN Research Product, U.S.A.) was incubated for 24 hours at 4 ^oC. For the measurement of AII, a mixture consisting of 100 µl of sample, 200 µl of diluted antiserum and 400 µl RIA buffer was incubated for 20 hours at 4 ^oC, and ^{125}I-labeled angiotensin II (Du Pont/NEN Research Product,

U.S.A.) was added to the mixture and incubated for 4 hours at 4 °C. Antibody bound radioactivity was precipitated with 100 µl of 2 mg/ml bovine gamma globulin dissolved in saline and 1 ml of 25 % polyethylene glycol. The radioactivity of the precipitate was measured by a gamma counter (Autowell Gamma System ARC 600, Aloka, Japan).

The plasma concentration (C) of AII and BK was calculated as follows because the plasma had been diluted by 1 ml of protease inhibitor mixture:

$$C = Co \times \{6.5 \times (1\text{-Ht} / 100) + 1\} / 6.5 \times (1\text{-Ht} / 100) \text{ (pg/ml)}$$

where Co = the concentration measured by RIA, Ht = the hematocrit determined at the same time.

To evaluate local production in the ischemic zone, each release into AIV was calculated as follows:

$$\text{Release} = (C_{AIV} - C_{Ao}) \times \text{AIVBF} \times (1 - \text{Ht} / 100) \text{ (pg/min)}$$

where C_{AIV} = the concentration in the anterior interventricular venous plasma, C_{Ao} = the concentration in the aortic plasma, AIVBF = the anterior interventricular venous blood flow.

Infarct size

The extent of infarction in the CONT (N=5), CAPT (N=5), and NAFA (N=5) groups was determined by a macroscopic enzyme technique (17). After 90 minutes of occlusion, the heart was arrested by bolus injection of saturated KCl solution and excised. Then, 1 % 2,3,5-triphenyltetrazolium chloride (Sigma Chemical U.S.A) dissolved in 6 % Dextran T70 in 5 % glucose was infused into LAD just distal to the occlusion, while 1 % monastral blue dissolved in 6 % dextran T70 in 5 % glucose was perfused via left and right coronary arteries from the ascending aorta at the same and constant pressure (100cm H2O). Each heart was cut into 6 slices transversely and weighed. Uninfarcted myocardium containing dehydrogenase in the zone at risk area was stained red and infarcted myocardium remained unstained. The normal zone was stained with blue. The infarct and risk zones were traced on an acetate sheet and the percentage of infarcted area in relation to the zone at risk and to the whole left ventricle were calculated planimetrically.

Statistical analysis

All data are expressed as the mean ± SE. Means of all values except for infarct size was compared by two-way analysis of variance and, where appropriate, by Duncan's multiple range test. The data for the infarct size were compared with the control group by one-way analysis of variance and, where appropriate, by Duncan's multiple range test. Pearson correlation coefficient was calculated for some pairs of variables. Probability values of less than 0.05 were considered as significant. Statistical Analysis System computer program was used for statistcal processing.

RESULTS

Systemic hemodynamics

In the SHAM group, systemic hemodynamics were stable throughout the experimental period, although cardiac output decreased and total peripheral resistance increased at 60 and 90 minutes after occlusion. In the CONT group, mean aortic pressure (101±5.7 to 98± 5.4 mmHg) and cardiac output (2.25±0.09 to 2.12±0.11 l/min) were decreased and left atrial pressure (5.8±0.4 to 7.3±0.5 mmHg) was increased after occlusion. In the CAPT group, mean aortic pressure (97±2.5 to 83±2.5 mmHg), left atrial pressure (5.8±0.4 to 5.2±0.4 mmHg) and total peripheral resistance (3396±220 to 2850±226 dyne.sec/cm^5) significantly decreased 30 min after the drug administration. But, aortic pressure, cardiac output, left atrial pressure and total peripheral resistance did not change during occlusion. Heart rate was significantly reduced after occlusion. Nafamostat increased heart rate (127±5.2 to 141±6.2 mmHg), and decreased left atrial pressure (6.2±0.4 to 5.5±0.4 mmHg) and central venous pressure (2.41±0.16 to 2.36±0.14 mmHg). Aortic pressure, cardiac output and total peripheral resistance showed the same time course changes as in the CONT group after occlusion. The pressure rate product was significantly decreased after coronary artery occlusion in the CONT, CAPT and NAFA groups and it was significantly smaller in the CAPT group than in the NAFA group.

Coronary circulation

Anterior interventricular venous blood flow was significantly decreased in the CONT, CAPT and NAFA groups, in contrast to the SHAM group (Table 1).

Table 1. Change of anterior interventricular venous blood flow

Anterior interventricular venous blood was collected from the AIV at the level of the coronary sinus before occlusion and 5, 15, 30, 60, and 90 minutes after occlusion. Anterior interventricular venous blood flow (ml/minute) was calculated from 6.5 ml of anterior interventricular venous blood divided by the time required for each collection. Values are means±SE. SHAM:sham operation group, CONT:control group, CAPT:captopril group, NAFA:nafamostat group. PRE-CAO:before coronary artery occlusion, POST-CAO:after coronary artery occlusion. *:significantly different from PRE-CAO value

Group	PRE-CAO	POST-CAO				
		5 min	15 min	30 min	60 min	90 min
SHAM	0.78±0.05	0.79±0.05	0.80±0.06	0.78±0.06	0.76±0.05	0.72±0.06*
CONT	0.79±0.04	0.61±0.04*	0.63±0.04*	0.63±0.04*	0.62±0.04*	0.61±0.05*
CAPT	0.83±0.06	0.66±0.04*	0.68±0.04*	0.71±0.05*	0.69±0.04*	0.68±0.04*
NAFA	0.86±0.07	0.62±0.05*	0.66±0.05*	0.67±0.05*	0.66±0.04*	0.66±0.05*

                                                                K. Noda et al.

Table 2. Change in pH of anterior interventricular venous and aortic blood in the CONT group
The pH in blood collected from AIV and aorta before occlusion and 15, 30, and 90 minutes after occlusion was measured using a blood gas analyzer (ABL 2, Acid-Base Laboratory, Denmark). PRE-CAO:before coronary artery occlusion, POST-CAO:after coronary artery occlusion. AIV: anterior interventricular venous blood, Aorta:aortic blood. *:significantly different from PRE-CAO pH, †:significantly different from aortic pH determined at the same time.

	PRE-CAO	POST-CAO		
		15 min	30 min	90 min
AIV	7.26 ± 0.015	7.15 ± 0.024*†	7.15 ± 0.021*†	7.18 ± 0.018*†
Aorta	7.30 ± 0.015	7.29 ± 0.017	7.29 ± 0.016	7.28 ± 0.015*

The pH of anterior interventricular venous blood after occlusion was significantly lower than before occlusion as was that of aortic blood in the CONT (Table 2), and the CAPT and NAFA groups also showed the similar pH change, while the SHAM group showed no change. The pH difference correlated significantly with the percent change of the anterior interventricular venous blood flow;in the CONT group (n=48, r=0.52, p=0.0001), in the CAPT group (n=48, r=0.54, p=0.0001) and in the NAFA group (n=48, r=0.63, p=0.0001). There was no significant correlation in the SHAM group.

Changes of hormonal parameters

Local AII and BK releases did not change significantly in the SHAM group. However, in the CONT group, AII release increased significantly at the 30th minute after occlusion in comparison to both before and 90 minutes after occlusion and BK release increased significantly at 30 minutes after occlusion in comparison to that before occlusion. AII release negatively correlated with pH in the anterior interventricular venous blood, (r=-0.29, p=0.031) and BK release showed a correlation with the pH difference between anterior interventricular venous and aortic blood, (r=0.28, p=0.037). Local AII release did not change significantly after captopril administration and during occlusion, but AII release at 30 minutes after occlusion did not show the significant decrease compared with the CONT group. Local BK release markedly increased at 5 and 15 minutes after occlusion. BK release showed a correlation with the pH difference between anterior interventricular venous and aortic blood (r=0.36, p=0.012). Local AII and BK release did not change significantly after nafamostat administration or during occlusion.

Changes of AII and BK release after occlusion were compared among the CONT, CAPT and NAFA (Fig.1). The increment of AII release in the NAFA group was significantly smaller than that in the CONT group at 30 minutes after occlusion and that in the CAPT group at 15 minutes after occlusion. The increments of BK in the CAPT group were greater than in all other groups at 5 and 15 minutes after occlusion. In the NAFA group, it was smaller than in the CONT group at 30 minutes after occlusion.

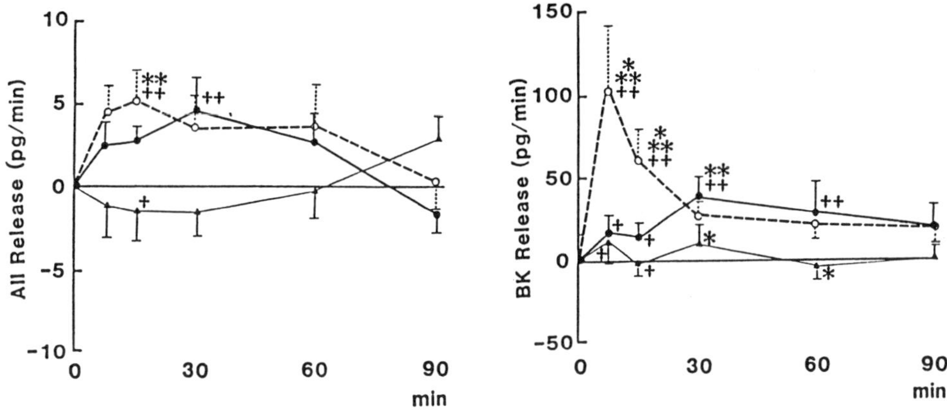

Figure 1. Changes of angiotensin II and bradykinin release during 90 minutes of coronary artery occlusion in 3 groups. Each point shows the mean ± SE. CONT (●), CAPT (○),NAFA (▲)
** ; significantly different from pre-occlusion   (p<0.05),   * ; significantly different from CONT   (p<0.05)
† ; significantly different from CAPT (p<0.05),   †† ; significantly different from NAFA   (p<0.05)

Infarct size

Infarct size expressed as a percentage of wet weight relative to the zone at risk and to the left ventricle were 15.6±3.71 and 6.38±1.57 % in the CONT group (N=5), 3.91±1.91 and 0.86±0.44 % in the CAPT group (N=5) and 8.90±5.49 and 2.33±1.52 % in the NAFA group (N=5). Infarct size in the CAPT group alone was significantly smaller than in the CONT group.

DISCUSSION

The characterization of anterior interventricular venous blood

Anterior interventricular venous blood is essentially derived from the blood supplied by LAD. Vinten-Johansen et al. (18) reported that about 97 % of anterior interventricular venous blood was supplied by LAD and that it was supplied from collateral blood during LAD occlusion.

In a preliminary study, we compared the blood in the great cardiac vein and AIV. The pH of anterior interventricular venous blood decreased significantly from 7.27±0.015 to 7.15±0.021 (N=14) with a reduction of flow after occlusion. But the pH in the great cardiac venous blood did not change significantly (7.29±0.009 to 7.26±0.006, N=10) and the great cardiac venous blood flow was more than anterior interventricular venous blood, but did not decrease after occlusion because the great cardiac venous blood was also supplied from the circumflex artery. Thus, the blood from AIV appears to be the unique blood perfusing the zone at risk and it should provide us the most available method to evaluate the local hormonal changes by ischemia introduced by left anterior descending arterial occlusion.

Change of Angiotensin II and Bradykinin Release in the Ischemic Heart of Nephrectomized Dogs.

The renin-angiotensin system in ischemic heart is not clear, but a few studies have reported an increase of AII levels a few days after an episode of myocardial ischemia without congestive heart failure (19), and a decrease of myocardial renin-like activity or an increase of renin-like activity in coronary arterial endothelium during myocardial ischemia (20).

To estimate local AII production precisely, the renal renin-angiotensin system was eliminated and AII release from the heart was calculated from the difference of AII concentrations between blood in AIV and aorta multiplied by anterior interventricular plasma flow. AII release rose significantly at 30 minutes after occlusion. Captopril did not suppress the increase of local AII release and the time course change was similar to that in the CONT group, but nafamostat mesilate, a serine protease inhibitor which has a strong inhibitory effect on kallikrein, plasmin and thrombin but a weak inhibitory effect on acid proteases suppressed AII release only during occlusion. This increase might relate to activation of angiotensin generating serine protease like tissue kallikrein (13,21,22) and/or to angiotensin generating serine protease secreted from inflammatory cells (23,24).

In experimental canine myocardial ischemia model, BK concentrations in the coronary sinus have been reported to reach a maximum within 10 minutes after coronary artery occlusion and then decreased rapidly (25-28). In our experiment, BK release increased generally at 30 minutes after occlusion, and nafamostat inhibited the the increase of BK after myocardial ischemia but captopril enhanced the BK release from 5 minutes after occlusion. The reason why BK did not increase immediately after occlusion despite the blood collected from specifically the ischemic zone may be that locally formed BK was degraded rapidly by kininase II. Ishida et al. (29) have reported that BK levels in the arterial blood fell rapidly even when BK was being infused continuously via the inferior vena cava.

These results suggest that increased local AII generation in the ischemic heart of the nephrectomized dogs mainly dependson serine protease(s), and tissue renin-angiotensin system might not play an major role in AII generation in the early stage of myocardial infarction. However, a great amount of BK is generated in ischemic myocardium and captopril has a great capacity of inhibiting the degradation of BK.

Infarct size

ACE inhibitors have been reported to reduce the size of myocardial ischemia (1-5). In addition to the beneficial changes of systemic hemodynamics and scavenging of free radicals by their sulfhydryl groups (9,12), improvement of coronary circulation in jeopardized area might play an important role in the reduction of infarct size (1,2). The mechanisms of increase of coronary flow were reported to be the increases of prostaglandin synthesis (6,7), the formation of endothelium-derived relaxation factor (8), direct vasodilation due to reduced BK degradation (9-11) in addition to the reduction of AII formation. Treatment with captopril was effective in the reduction of myocardial infarct size even during a short ischemic period (90 minutes). During 90 minutes of ischemia, BK release in ischemic area was enhanced, while increased AII release remained unchanged. Though it is not still clear that the local AII formation by ACE was not inhibited by captopril, captopril itself could not suppressed the increase of AII generation even if enough captopril is administered to reduce the blood pressure and the size of myocardial infarction. The major effect of captopril on protecting the myocardium from ischemia may have been due to the marked local accumulation of BK and other related vasodilators, such as prostaglandins and endothelium-derived relaxation factor. The results of treatment of nafamostat also showed that the suppression of local AII formation did not contribute the reduction of the infarct size in the absence of BK generation. Martorana (30) have also reported that BK antagonist, HOE 140, eliminated the cardioprotective effect of ACE inhibitor, ramiprilat, on the reduction of infarct size.

These results suggest that the local BK accumulation may play an important role in the reduction of myocardial infarction. However, the suppression of local AII formation alone could not diminish the extent of myocardial infarction in this nephrectomized dog model. Additional effects produced by ACE inhibitor such as vasodilation by BK accumulation might also effectively protect the ischemic myocardium.

ACKNOWLEDGMENTS

This work was supported in the part by a Grant-in-Aid from the Ministry of Education, Science and Culture of Japan (No 62480221). We gratefully acknowledge M. Ishibashi. MD, Department of Internal Medicine, Fukuoka University for his technical advice.

REFERENCES

1.   Ertl G., Alexander R.A. and Kloner R.A.: Interaction between coronary occlusion and the Renin-Angiotensin System.*Basic Res Cardiol* 1983;78:518-533.
2.   Ertl G., Kloner R.A., Alexander R.W. and Braunwald E.: Limitation of experimental infarct size by an angiotensin-converting enzyme inhibitor. *Circulation* 1982;65:40-48.
3.   Carl E., Hoch C., Ribeiro L.G.T. and Lefer A.M.: Preservation of ischemic myocardium by a new converting enzyme inhibitor, Enalaprilic acid, in acute myocardial infarction. *Am Heart J* 1986;109:222-228.
4.   Lefer D.J. and Lefer A.M.: Coronary vascular actions of the converting enzyme inhibitor, enarapril (41790). *Proc Soc Exp Biol* 1984;175:211-214.
5.   Lefer A.M. and Rock R.C.: Cardioprotective effects of enarapril in acute myocardial ischemia. *Pharmacology* 1984; 29:61-69.
6.   Dusing., Scherhag R., Landsberg G., Glanzer K. and Kramer H.J.: The converting enzyme inhibitor captopril stimulates prostacyclin synthesis by isolated rat heart. *Eur J Pharmacol* 1983;92:501-504.

7.   Noguchi K., Kato T., Ito H., Aniya Y. and Sakanashi M.: Effect of intracoronary captopril on coronary blood flow and regional myocardial function in dogs. *Eur J Pharmacol* 1985;110:11-19.

8.   Radomski M.W., Palmer R.M.J. and Moncada S.: Endogenous nitric oxide inhibits human platelet adhesion to vascular endothelium. *Lancet* 1987;2:1057-1058.

9.   von Gilst W.H., Scholtens E., Graeff P.A., Langer C.D. and Weseling H.: Differential influences of angiotensin converting-enzyme inhibitors on coronary circulation. *Circulation* 1988;77(Suppl. I):I-24-I-29.

10.   Scholkens B.A. and Linz W.: Local inhibition on angiotensin II formation and bradykinin degradation in isolated heart. *Clin Exp Hypertension* 1988;A10(6):1259-1270.

11.   Toda N. and Okunishi T.: Endothelium-dependent and -independent response to vasoactive substances of isolated human coronary arteries. *Am J Physiol* 1989;257:H988-H995.

12.   Westlin W. and Mullane K.: Does captopril attenuate reperfusion-induced myocardial dysfunction by scavenging free radicals? *Circulation* 1988;77(Suppl. I):I30-I39.

13.   Gondo M., Maruta H and Arakawa K.: Direct formation of angiotensin II without renin or converting enzyme in the ischemic dog heart. *Jpn Heart J* 1989;30:219-229.

14.   Aoyama T., Ino Y., Ozeki M., Oda M., Satoh T., Koshiyama Y., Suzuki S. and Fuzita M.: Pharmacological studies of FUT-175, Nafamostat mesilate I. Inhibition of protease activity in vitro and vivo experiments. *Jpn J Pharmacol* 1984;35:203-227.

15.   Iwasaki M., Ino Y., Motoyoshi A., Ozeki M., Satoh.T, Kurumi M. and Aoyama T.: Pharmacological studies of FUT-175, nafamostat mesilate V. Effects of the pancreatic enzymes and experimental acute pancreatitis in rats. *Jpn J Pharmacol* 1986;41:155-162.

16.   Goodfriend T.L. and Odya C.E.: Bradykinin in methods of hormone radioimmuno-assay. edited by Jaffe B.M and Behrman H.R. New York Academic Press. 1979;909-923.

17.   Kloner R.A., Darsee J.R., DeBoer L.W.V. and Carlson N.: Early pathologic detection of acute myocardial infarction. *Arch Pathol Lab Med* 1981;105:403-406.

18.   Vinten-Johansen J., Johnson W.E., Crtstal G.J., Mills S.A., Santamore W.P. and Cordell A.R.: Validation of local venous sampling within the at risk left anterior descending artery vascular bed in the canine left ventricle. *Cardiovasc Res.*1987;21:646-651.

19.   McAlpine H.M. and Cobbe S.M.: Neuroendocrine change in acute myocardial infarction. *Am J Med* 1988;84 (Suppl. 3A):61-66.

20.   Haulica I., Ncamtu C., Slatincanu S., Prcdcscu D., Popcscu G., Stratonc A.,Rosca V., Rusu G. and Inonescu G: A comparative study of the renin-like activity in the heart and vascular system under various experimental conditions. *Rev Roum Morphol Embryol Physiol* 1984;21:3-11.

21.   Arakawa K. and Maruta H.:Ability of kallikrein to generate angiotensin-like presser substance and a proposed "Kinin-Tensin enzyme system". *Nature* 1980288:705-706

22.   Maruta H. and Arakawa K.: Confirmation of direct angiotensin formation by kallikrein. *Biochem J* 1983;213:193-200.

23.   Wintroub B.U., Klickstein L.B. and Watt K.W.K.: A human neutrophil-dependent pathway for generation of angiotensin II. *J Clin Invest* 1981;68:484-490.

24.   Wintroub B.U, Kaempfer C.E., Schechter N.M. and Proud D.: A human lung mast cell chymotrypsin-like enzyme. *J Clin Inv* 1986;77:196-201.

25.   Kimura E., Hashimoto K., Furukawa S. and Hayakawa H.: Changes in bradykinin level in coronary sinus blood after the experimental occlusion of a coronary artery. *Am Heart J* 1973;85:635-647.

26.   Hashimoto K., Hirose M., Fukukawa S., Hayakawa H. and Kimura E.: Changes in hemodynamics and bradykinin concentration in coronary sinus blood in experimental coronary artery occlusion. *Jpn Heart J* 1977;18:679-689.

27.   Matsuki T., Shoji T.,Yoshida S., Kudoh Y., Motoe M., Inoue M., Nakata T.,Hosoda S.,Shimamoto K.,Yellon D. and Iimura O. : Sympathetically induced myocardial ischemia causes the heart to release plasma bradykinin. *Cardiovasc Res* 1987; 21:428-432.

28. Hashimoto K., Hamamoto H., Handa Y., Hirose M., Fukukawa S. and Kimura E.: Changes in components on bradykinin system and hemodynamics in acute myocardial infarction. *Am Heart J* 1987;95:619-629.
29. Ishida H., Scicli A.G. and Carretero O.A.: Role of angiotensin converting enzyme and other peptidases in in vivo metabolism of kinins. *Hypertension* 1989;14:322-327.
30. Martorana P.A., Kettenbach B., Breipohl G., Linz W.and Scholkens B.A. : Reduction of infarct size by local angiotensin-converting enzyme inhibition is abolished by a bradykinin antagonist. *Eur.J.Pharmacol* 1990;182:395-396.

Hypertension

AAS 38/III
Recent Progress on Kinins
© 1992 Birkhäuser Verlag Basel

SALT SENSITIVITY AND BRADYKININ ACTIVITY IN RATS

Athanase Benetos, Hervé Bouaziz, Michel Safar

INSERM U.337, Hôpital Broussais, Paris, FR

INTRODUCTION

Salt sensitivity is associated with a number of neuro-hormonal alterations. Exposure to a high salt diet increases blood pressure and modifies the activity of different vasoconstrictor or vasodilator systems. Chronic elevation of blood pressure in Dahl sensitive rats, on a high salt diet, is related to an increase in vascular resistance, whereas blood volume, after a transient increase, returns to normal values (1). Different mechanisms have been proposed to explain this arteriolar vasoconstriction. Several studies suggest that the effects of chronic NaCl loading on sympathetic activity are different in Dal sensitive (DS) and Dahl resistant (DR) rats (2,3). Other mechanisms, such as changes in baroreceptor activity (4), renal function (5) or responsiveness to the vasoconstrictors and the vasodilators (6,7) have been proposed. Recent studies show that NaCl loading can suppress, in some experimental models, the bradykinin (BK) plasma or tissue levels (7,8). However the hemodynamic impact of such a suppression in salt sensitive or resistance experimental models is not clear. The aim of the present study was to investigate the role of endogenous bradykinin on blood pressure regulation in DS , DR and spontaneous hypertensive rats (SHR), fed with a low or high salt diet for a period of 5 weeks. The bradykinin activity was assessed with the use of a peptide analog which is a specific antagonist of bradykinin(9).

MATERIAL AND METHODS

Thirty nine male Dahl sensitive and Dahl resistant rats (Mollegaard Breeding Center, Skensved, Denmark) and 19 SHR (Laboratoires Janvier France) were used in these experiments. The rats were delivered to our animal house at 7 weeks of age. Four days after arrival, half of the DS and DR rats were placed on 7% NaCl diet (U.A.R., Epinay sur Orge, France) while the other half, remained on the standard rat diet (.4% NaCl). Both diets contained the same amount of potassium (.73%). Tap water was provided at libitum throughout the study. All rats remained on these diets for 5 weeks, housed at constant temperature (24 ± 1°C) and humidity, with a 12h light-dark cycle. On the day of the experiment, the rats were anesthetized with intraperitoneal pentobarbital(60 mg/kg). The right femoral artery and the right carotid artery were cannulated, for direct arterial pressure recording and drug infusions respectively. BP and HR were continuously recorded with Statham P23XL pressure transducer, Gould Brush pressure computer and Gould Brush 2400 recorder, (Gould Cleveland, Ohio).

After 20 minutes of stabilization a continuous infusion of .9% NaCl (control) was administrated (0.1 ml/min) for a period of 10 minutes. Blood pressure and heart rate were measured throughout the infusion. Twenty minutes later, an infusion of the BK antagonist B4146 - (Hyp^3,Thi ^{5,8}, D-Phe^7)-BK (NOVACHEM France)(40 μg/ml/mn) was performed for a period of 10 minutes. Thirty minutes after the end of the B4146 infusion, bolus injections of bradykinin (SIGMA France) (.1, .5, 1, 5 μg) were performed. In pilot experiments, we observed that the activity of the B4146 had ended completely by 10 minutes after its administration. Animal handling and experimental procedures which followed, were in accordance with institutional regulations.

STATISTICS

All results were expressed as mean ± SEM. Basal values of the different groups were compared by one way ANOVA. The effects of NaCl diet on the response to the BK antagonist infusion and the BK injections were performed in each group by a multiple variance analysis. When F was significant differences were tested by non parametric Kulls Newman test. A P < .05 was considered significant.

RESULTS

Table I shows the weight and basal values of blood pressure and heart rate in the 6 groups of rats, on the day of the experiments. SHR on both diets and DS on high salt (DSH) had significantly higher MBP values as compared to the 3 other groups. High salt fed SHR (SHRH) showed relatively higher MBP levels as compared to the SHR on low salt diet (SHRL), but this difference was not significant. Heart rate was the same in all studied groups.

Table I. Basal values in the 6 studied groups

Groupe	n	Weight (gr)	MBP (mmHg)	HR (b/min)
DSL	11	381 ± 7	143 ± 5	382 ± 11
DSH	9	316 ± 16	185 ± 7*	336 ± 17
DRL	10	341 ± 14	134 ± 6	333 ± 9
DRH	9	338 ± 10	132 ± 10	352 ± 8
SHRL	10	352 ± 8	178 ± 6*	335 ± 10
SHRH	9	360 ± 9	190 ± 12*	325 ± 11

*p<0.001 vs other groups

Fig.1 shows maximumm mean blood pressure changes following bradykinin antagonist infusion in the different groups. In the DS rats, NaCl diet significantly modified the response to the bradykinin antagonist (p < 0.001). Thus in DSL mean blood pressure increased by 13.2 ± 1.5 mmHg whereas in DSH no change in blood pressure was observed (1.1 ± 0.8 mmHg) .In DR and SHR Nacl diet did not modify the response to the bradykinin antagonist. Fig.2 shows the dose-response curve to the BK in the 6 groups. Salt diet had no significant effect on BP response to the bradykinin in any of the studied rat strains.

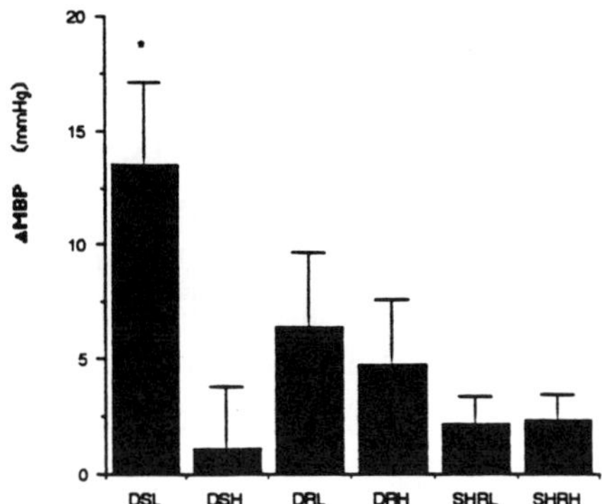

Fig. 1. Changes in MBP during infusion of the Bradykinin antagonist * p < 0.01 vs the other groups.

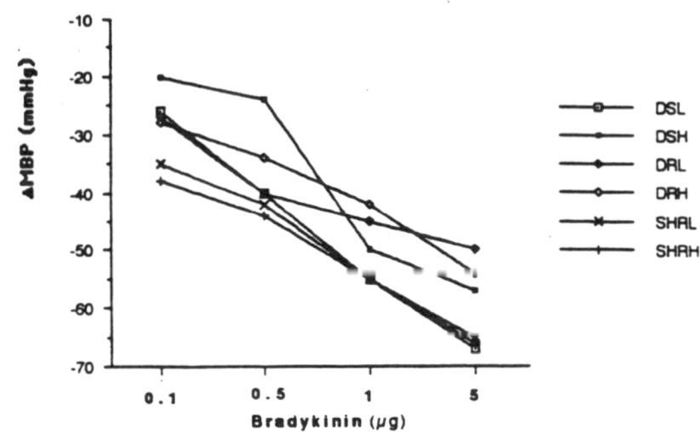

Fig. 2. Changes in MBP following Bradykinin injections

DISCUSSION

Recently, several BK analogs with antagonistic properties have been utilized in an attempt to define the role of the Kallikrein - Kinin system in cardiovascular homeostasis (9,10). With the use of these antagonists, different studies have shown the participation of the bradykinin in the antihypertensive effects of converting enzyme inhibitors (10,11,12).

Moreover using the same antagonists, several investigators have attempted to study the basal activity of the endogenous bradykinin in different experimental models. The results obtained from these studies have not been consistent. In some experiments, acute administration of a bradykinin antagonist raised blood pressure in normotensive rats (13). However, in other studies, endogenous bradykinin did not seem to participate in normal blood pressure control unless rats were pretreated with different vasoconstrictors (14,15). These conflicting results, obtained with the different BK antagonists, could be at least partially explained by non specific effects of these compounds (15,16).

In our study we assessed the effects of endogenous bradykinin inhibition in Dahl sensitive or resistant rats and in SHR. The main result of this study is that in DS rats fed a low salt diet endogenous bradykinin plays a significant role in blood pressure regulation, since its inhibition with the B4146 increases blood pressure. In this strain of rats NaCl loading abolishes the response to the BK antagonist. In the salt resistant groups (DR rats and SHR) the NaCl diet does not influence the response to the BK antagonist.

Blood pressure-increase during kinin receptor antagonism, in DSL rats suggests a basal vasodepressor action of endogenous bradykinin in this model. However, before reaching such a conclusion, different considerations must be taken into account. Previous studies have shown that BK antagonists may have a pressor effect, independent of BK inhibition (15,16). Thus activation of the sympathetic system by some of these antagonists has been reported. However this activation occured when very large amounts of these antagonists were used (4 mg/kg), which is about 40 times higher than the concentration used in our study. Moreover, if this hypothesis was correct, the pressor effects of BK antagonist should be present in all groups of rats, which was not the case in our study.

The lack of increase in MBP during the BKA infusion in DSH suggests an interaction between salt loading and the bradykinin system in the NaCl sensitive rat model. This interaction can be either related to the suppression of endogenous bradykinin, or to a decreased vascular responsiveness to the BK. Our results can eliminate the latter hypothesis, since dose-response curves to the exogenous bradykinin were not altered by the salt diet. Lüscher et al (7), using arterial rings, observed a decreased responsiveness in endothelium-dependent vasorelaxation in DS rats fed a high salt diet, whereas the endothelium independent vasorelaxation was much less affected. The authors of this study concluded that NaCl could either decrease endothelium-derived relaxing factors in Dahl rats fed a high salt diet, or increase the release of endothelium derived constrictor factors. Our results suggest that NaCl loading is responsible for a decrease in the endogenous bradykinin activity but has no effect on the responsiveness to the exogenous BK.

The results of the present study suggest that endogenous bradykinin does not participate in the basal blood pressure homeostasis in salt resistant rats, whereas it plays an important role in the blood pressure regulation in DS rats. Indeed DS rats present a number of alterations of the vasoactive systems with a possible failure of different vasodilators (5,7). Therefore the relative role of bradykinin in the maintenance of blood pressure is more important in DS rats. When a high salt diet is applied, endogenous bradykinin is suppressed, which could be one of the mechanisms of blood pressure elevation in these rats. In a previous study Waeber et al (14) observed that bradykinin participates in the blood pressure regulation when other vasodilating systems were inhibited. These data support the hypothesis that endogenous bradykinin is a "second line" vasodilator which plays a major role when other vasodilating systems are impaired as is the case in Dahl sensitive rats.

Hence, we suggest that NaCl-included suppression of endogenous bradykinin can participate in the initiation and/or maintenance of salt dependent hypertension in Dahl rats.

REFERENCES

1.   Ganguli M, Tobian L, Iwai J. Cardiac output and peripheral resistance in strains of rats sensitive and resistant to NaCl hypertension. Hypertension 1979; 1: 3-7.

2.   Racz K, Kuchel O, Buu NT. Abnormal adrenal catecholamine synthesis in salt-sensitive dahl rats. Hypertension 1987; 9: 76-80.

3.   Chen YF, Meng Q, Wyss JM, Jin H, Rogers CF, Oparil S. NaCl does not affect hypothalamic noradrenergic input in deoxycorticosterone acetate/NaCl and dahl salt-sensitive rats. Hypertension 1990; 16: 55-62.

4.   Ferrarie AU, Mark AL. Sensitization of aortic baroreceptors by high salt diet in dahl salt-resistant rats. Hypertension 1987; 10: 55-60.

5. Rapp JP. Dahl salt-susceptible and salt-resistant rats. Hypertension 1982; 4: 753-63.

6. Kong JQ, Taylor DA, Fleming WW, Kotchen TA. Specific supersensitivity of the mesenteric vascular bed of dahl salt-sensitive rats. Hypertension 1991; 17: 349-56.

7. Lüscher TF, Raij L, Vanhoutte PM. Endothelium-dependent vascular responses in normotensive and hypertensive dahl rats. Hypertension 1987; 9: 157-63.

8. Madeddu P, Glorioso N, Soro A, Manunta P, Troffa C, Tonolo G, Melis MG, Pazzola A. Effect of a kinin antagonist on renal function and haemodynamics during alterations in sodium balance in conscious normotensive rats. Clinical Science 1990; 78: 165-8.

9. Vavrek RJ, Stewart JM. Competitive antagonists of bradykinin. Peptides 1985; 6: 161-4.

10. Benetos A, Gavras H, Stewart JM, Vavrek RJ, Hatinoglou S, Gavras I. Vasodepressor role of endogenous bradykinin assessed by a bradykinin antagonist. Hypertension 1986; 8: 971-4.

11. Danckwardt L, Shimizu I, Bönner G, Rettig R, Unger T. Converting enzyme inhibition in kinin-deficient brown norway rats Hypertension 1990; 16: 429-3

12. Carbonell LF, Carretero OA, Stewart JM, Scicli AG. Effect of a kinin antagonist on the acute antihypertensive activity of enalaprilat in severe hypertension. Hypertension 1988; 11: 239-43.

13. Carbonnel LF, Carretero OA, Madeddu P, Scicli AG. Effects of a kinin antagonist on mean blood pressure. Hypertension 1988; 11(suppl I): 84-8.

14. Waeber B, Niederberger M, Gavras H, Nussberger J, Brunner HR. Hemodynamic effects of a kinin antagonist. J Cardiovasc Pharmacol 1990; 15(Suppl.6): S78-S82

15. Aubert JF, Waeber B, Nussberger J, Vavrek RJ, Stewart JM, Brunner HR. Influence of endogenous bradykinin on acute blood pressure response to vasopressors in normotensive rats assessed with a bradykinin antagonist. J Cardiovasc Pharmacol 1988; 11: 51-5.

16. Mulinari R, Benetos A, Gavras I, Gavras H. Vascular and sympathoadrenal responses to bradykinin and bradykinin analogue. Hypertension 1988; 11: 754-7.

ESSENTIAL ROLE OF KALLIKREIN-KININ SYSTEM IN SUPPRESSION OF BLOOD PRESSURE RISE DURING THE DEVELOPMENTAL STAGE OF HYPERTENSION INDUCED BY DEOXYCORTICOSTERONE ACETATE-SALT IN RATS

M. Katori[1], M. Majima[1], S.S.J. Mohsin[1], M. Hanazuka[2], S. Mizogami[2] and S. Oh-ishi[3]

Department of Pharmacology, Kitasato University School of Medicine, Sagamihara, Kanagawa [1], Pharmaceutical Laboratory, Research Center, Mitsubishi Kasei Corp., Yokohama[2], and Department of Pharmacology, Kitasato University School of Pharmaceutical Sciences, Tokyo[3], Japan

SUMMARY: Deoxycorticosterone acetate (DOCA)-salt hypertension was induced in Brown Norway (BN) kininogen-deficient rats (BN-Ka) and normal rats from the same strain (BN-Ki) after nephrectomy. Systolic blood pressure, which was determined by the tail-cuff method, of BN-Ki increased gradually during this treatment. In contrast, the blood pressure of mutant BN-Ka increased rapidly 2 weeks after the onset of the treatment. Urinary excretion of active kallikrein and prokallikrein increased at the same degree in rats of both strains during this treatment. Significant increase in urinary sodium excretion was observed with a tendency to increase in urine volume during the treatment in normal BN-Ki rats, whereas both parameters were essentially not increased in mutant BN-Ka rats, which could not generate urinary kinin. Aprotinin infusion by osmotic minipump to normal BN-Ki rats during the DOCA-salt treatment resulted in significant further increase in the systolic blood pressure.

INTRODUCTION

Roles of kallikrein-kinin system in hypertension have been claimed (1-4), but the conclusive evidence has not been provided. In order to clarify roles of this system in hypertension, we used mutant kininogen-deficient rats, which are deficient in both high molecular weight (HMW) and low molecular weight (LMW) kininogens in plasma (Brown Norway-Katholiek, BN-Ka) and were compared with normal rats of the same strain (Brown Norway-Kitasato, BN-Ki). In experimental hypertension induced by deoxycorticosterone acetate (DOCA)-salt in uni-nephrectomized rats, normal BN-Ki rats increased the blood pressure gradually, whereas mutant BN-Ka rats showed rapid increase in the systolic blood pressure (5). The present paper describes the time course of urinary kallikrein during the developmental stage of hypertension in BN-strain rats.

MATERIALS AND METHODS

BN-Ka rats (Rattus norvegicus, BN/fMai) were initially obtained from the Katholiek Universiteit of Leuven, Belgium. Normal rats of the same strain (BN-Ki) have been kept in Kitasato University. Male rats of 5-20 weeks of age were used.

Systolic blood pressure of unanesthetized rats was determined using a tail cuff plethysmography method.

DOCA-salt hypertension was induced by weekly subcutaneous injection of DOCA solution (5 mg/kg, Sigma Chemical Co., St. Louis, Mo., USA) and by giving 1 % NaCl solution in drinking water after removal of the left kidney of rats of both strains under light ether anesthesia at 7 weeks of age.

The activity of the active kallikrein in urine collected 24 hrs was measured using a peptidyl fluorogenic substrate selective for glandular kallikrein, Pro-Phe-Arg-methylcoumarinylamide (Peptide Institute, Minoh, Osaka, Japan). One arbitrary unit (AU) was defined as the amount of urinary kallikrein that released 1×10^{-10} mol AMC (aminomethyl-coumarin) in 1 μl urine for 10 min. Active kallikrein activity was calculated as the difference between amidase activity in the presence of soya bean trypsin inhibitor (0.5 μg/μl urine) and that in the presence of aprotinin (0.5 μg/μl urine). The activity of urinary prokallikrein was determined by the increase in kallikrein activity after treatment of urine with trypsin, which was determined by the same method for urinary active kallikrein. The urine for measurement of the kallikrein activity was collected over 24 hours on the fourth day, during keeping rats in metabolic cages for five days each week.

The plasma levels of HMW and LMW kininogens were determined by Uchida's method reported previously (6), in which kininogens were converted to bradykinin and the amounts of kinin generated were measured by a bradykinin enzyme immunoassay kit. The level of plasma prekallikrein was measured by a previously reported method (7).

Urine from individual rats was collected in a metabolic cage for 24 hours and urine volume was measured. Sodium and potassium were determined with coated wire electrodes selective for sodium or potassium.

Free kinin was measured in the urine collected via catheter inserted into both ureters of untreated rats of both strain under pentobarbital anesthesia. The kinin levels were determined by the bradykinin enzyme immunoassay kit after separation with Sep-Pak C18 column (Waters Associates, Milford, Ms., USA).

Continuous administration of aprotinin was performed by osmotic minipump (model 2001, Alza Corp., Palo Alto, Calif., USA) implanted subcutaneously in the back.

All data were expressed as mean ± SEM. Significance of difference was evaluated by Student's t.

RESULTS

The plasma level of HMW kininogen of mutant BN-Ka at 10 weeks of age was below the detection limit and that of LMW kininogen was also very low, whereas the level of both kininogens in normal BN-Ki rats were approximately the same as those in Sprague-Dawley strain rats. The plasma prekallikrein level of BN-Ka rats was also lower than in normal BN-Ki rats. Free kinin from ureters could not be detected in mutant BN-Ka rats, but normal BN-Ki rats excreted 9.5 ng of kinin for 120 min. These results indicate that kallikrein-kinin system is abnormal in mutant BN-Ki rats.

The systolic blood pressure of mutant BN-Ka rats at 5 weeks of age (106 ± 0.4 mmHg) was not different from that of normal BN-Ki rats (104 ± 1.0 mmHg) and the rate of the spontaneous increase of the systolic blood pressure with age did not differ between the strains. Under the DOCA-salt treatment from 7 weeks of age after uninephrectomy, the systolic blood pressure of normal BN-Ki rats increased gradually with time and reached a plateau at 18 weeks of age, as shown in Table 1. In contrast,

Table 1. Changes in the systolic blood pressure of normal BN-Ki and mutant BN-Ka rats during DOCA-salt treatment.

Age (weeks)	systolic blood pressure (mmHg)	
	BN-Ki (n=7-16)	BN-Ka (n=5-14)
7	121 ± 3	118 ± 3
9	133 ± 5	$158 \pm 6**$
11	147 ± 4	$171 \pm 5**$
13	161 ± 6	177 ± 8
15	170 ± 7	174 ± 8
18	180 ± 9	172 ± 12

**, $p<0.01$. The treatment started at 7 weeks of age.

that of mutant BN-Ka rats rose rapidly 2 weeks after the onset of the treatment and continued to increase at a slower rate to reach a plateau at 13 weeks of age. The difference of the systolic pressure between both strains was statistically significant at 9 and 11 weeks of age.

Figure 1 indicates the time course of the urinary levels of active kallikrein and prokallikrein during the DOCA-salt treatment. The activities of urinary active kallikrein and prokallikrein were not changed with age either in mutant BN-Ka or normal BN-Ki without the DOCA-salt treatment. In contrast, DOCA-salt treatment induced the significant increase in active kallikrein and prokallikrein with peaks at 10 weeks of age (the third week under the treatment). The activities of both kallikrein under the treatment were significantly higher than those in untreated groups between 7 and 12 weeks of age. The activities, however, were nearly restored to those of untreated groups at 15 weeks of age, when the systolic blood pressure reached the plateau level.

Although there was no difference in the activities of active kallikrein and prokallikrein in the urine between mutant BN-Ka and normal BN-Ki rats during the DOCA-salt treatment, urine volume in mutant BN-Ka did not increase during the treatment and was not different from the untreated groups, whereas that in normal BN-Ki tended to increase (Figure 2). Urinary excretion of sodium in BN-Ka was kept much lower during the study period, except at 10 and 15 weeks of age, at which the excretion rate was slightly, but significantly, increased, whereas the

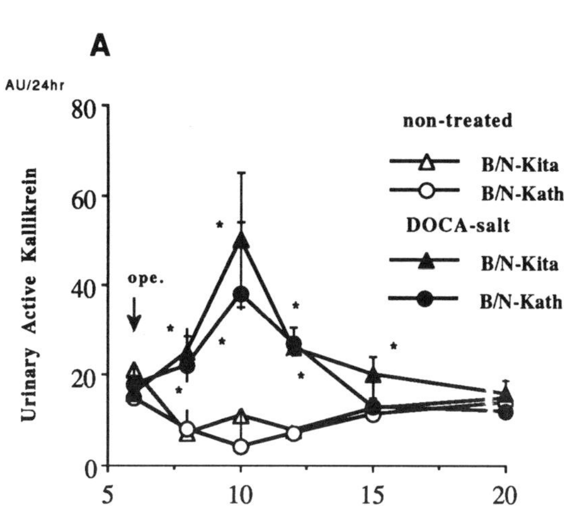

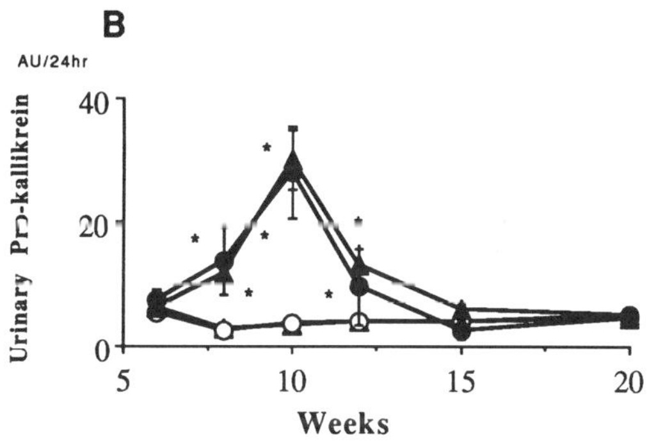

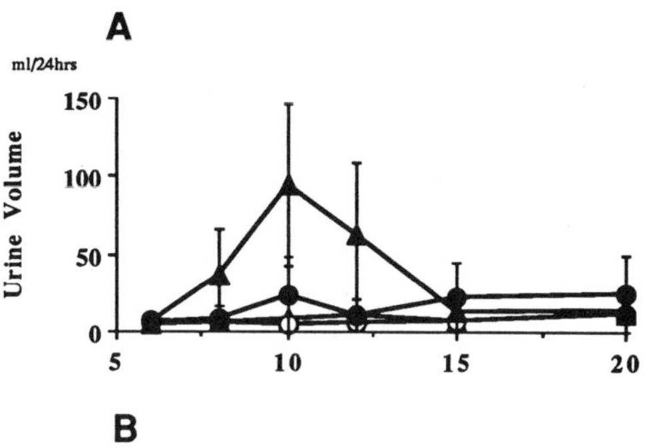

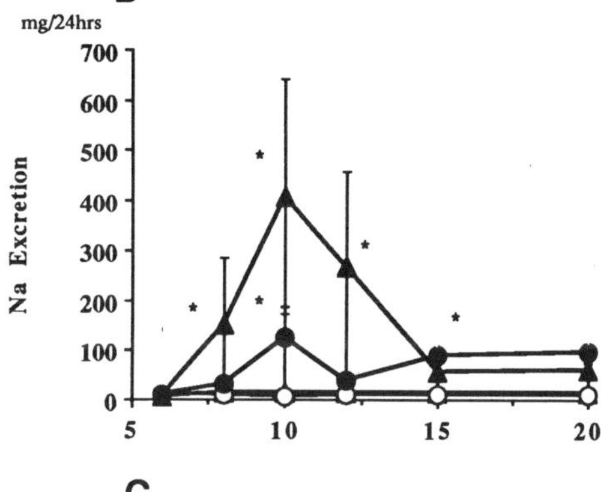

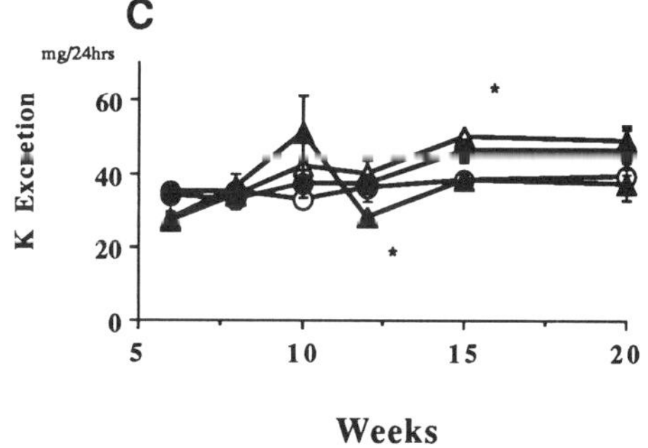

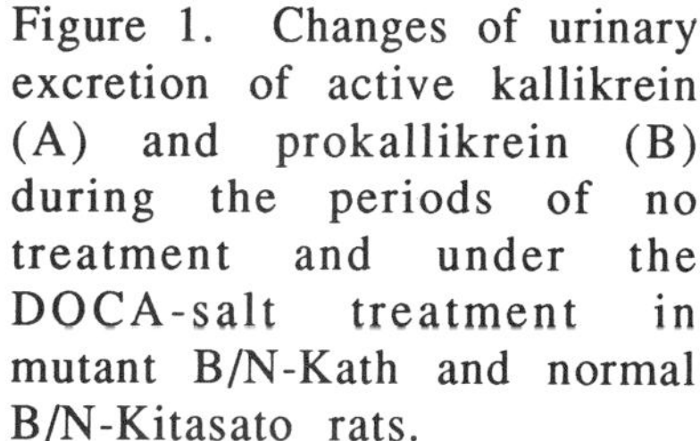

Figure 1. Changes of urinary excretion of active kallikrein (A) and prokallikrein (B) during the periods of no treatment and under the DOCA-salt treatment in mutant B/N-Kath and normal B/N-Kitasato rats.

Values indicate mean (± SEM) from 4-8 rats and those under the treatment were compared with those of non-treated groups.

* p< 0.05, ** p< 0.01

Figure 2. Changes of urine volume (A), sodium excretion (B) and potassium excretion (C) during periods of no treatment or under the DOCA-salt treatment.

Values indicate mean (± SEM) from 4-8 rats and those under the treatment were compared with those of non-treated groups.

*p< 0.05

sodium excretion rate in normal BN-Ki rats increased markedly by the treatment and the time course agreed with that of the activities of urinary active kallikrein and prokallikrein. The potassium excretion levels were essentially not different between both strains in the treated and non-treated groups, although statistically significant changes were observed during the time course.

Continuous infusion of aprotinin, a kallikrein inhibitor (10^5 KIU/day), by osmotic minipump implanted subcutaneously in the back of normal BN-Ki over one week from sixth day after the onset of the treatment caused the significant increase in the systolic blood pressure from 141 to 158 mmHg.

DISCUSSION

It has been well known that DOCA-salt treatment caused gradual, but marked increase in the systolic blood pressure. This is true in normal BN-Ki rats. The gradual increase in the systolic blood pressure was replaced by the rapid increase to the nearly maximum level within two weeks of the treatment, when the same treatment was made in mutant BN-Ka rats, in which plasma HMW and LMW kininogens were essentially deficient and urinary kinin was unable to be generated (5). This fact provides reliable evidence that kallikrein-kinin system takes a definite role in suppression of the early developmental stage of this hypertension. This hypothesis is further strengthened by significant increase in the systolic pressure of normal BN-Ki rats during continuous infusion of a kallikrein inhibitor, aprotinin, subcutaneously for 7 days from the fourteenth day after the onset of the DOCA-salt treatment.

The essential suppressive role of kallikrein-kinin system may be played in spontaneously hypertensive rats (SHR), which increased the systolic blood pressure gradually from 4 weeks of age and reached a plateau at 8 weeks of age (8). The urinary kallikrein activities were significantly lower in SHR than in WKY rats already at 4 weeks of age. The difference, however, was gradually smaller and no difference of the urinary kallikrein activities were observed, when the systolic blood pressure reached a plateau at 10 weeks of age. Aprotinin infusion to SHR of 7 weeks of age for 3 days increased the blood pressure significantly over the natural increase (8). These results revealed that endogenous kallikrein-kinin system, particularly renal kallikrein-kinin system, plays the suppressive role in the early development of the hypertension.

This suppression may not be conducted through vasodilatation of kinin generated. The urinary kallikrein activity was increased under the DOCA-salt treatment and showed a peak at 10 weeks of age (3 weeks after the onset of the treatment) and the activities did not differ between mutant BN-Ka and normal BN-Ki rats. However, as shown in Figure 2, the urine volume and sodium excretion was increased by the DOCA-salt treatment only in normal BN-Ki rats and they were not increased in mutant BN-Ka rats, which were unable to generate urinary kinin. Urinary excretion of sodium and potassium was reduced in SHR, compared with those of WKY rats. These results may indicate that kinin generated in renal tubules modulates the absorption of electrolytes and/or water in the renal tubules. In fact, microperfusion of bradykinin into the cortical collecting tubules suppresses sodium reabsorption stimulated by mineral corticoids or antidiuretic hormone (9). However, in order for the renal effect of kinin to link to the reduction of the systolic blood pressure, an unknown factor may be necessary, since aprotinin treatment to WKY did not reduce the systolic blood pressure (8). In any case, it is evident that renal kallikrein takes essential role in prevention of the blood pressure increase in the early developmental stage of hypertension. Recently, it was reported that brief treatment of young SHR by an angiotensin converting enzyme inhibitor before the plateau of the blood pressure resulted in reduction of the pressure for long period (10). Thus, the importance of the developmental stage of the hypertension has been aware of. The missing link between the renal kallikrein and the blood pressure decrease should be clarified in further studies.

CONCLUSION

The kininogen-deficient BN-Ka rats increased the systolic blood pressure rapidly during the DOCA-salt treatment, compared with the gradual increase of the blood pressure in normal BN-Ki rats. Urinary excretion of kallikrein was less in SHR than in WKY. The kallikrein-kinin system, particularly renal kallikrein-kinin system, may take an essential role in suppression of the blood pressure increase in the developmental stage of the hypertension.

ACKNOWLEDGEMENT

The authors expressed sincere thanks to Ms. Michiko Takahara, Ms. Maki Saito, Mr. Osamu Yoshida and Ms. Harue Mihara for their skillful technical assistance and to Mr. Hiroshi Ishikawa for his excellent technique for experimental devices.

REFERENCES

1. Margolius HS, Geller R, Pisano JJ, Sjoerdsma A.   Altered urinary kallikrein excretion in human hypertension. Lancet 1971; 2: 1063-1065.
2. Margolius HS, Celler R, deJong W, Pisano JJ, Sjoerdsma A. Urinary kallikrein excretion in hypertension. Circ Res 1972; 30: 358-362.
3. Margolius HS, Horwitz D, Pisano JJ, Keiser HR. Urinary kallikrein excretion in normal man: Relationships to sodium intake and sodium-relating steroids.   Circ Res 1974; 35: 820-825.
4. Kaizu T, Margolius HS. Studies on rat renal cortical cell kallikrein: I. Separation and measurement.   Biochem Biophys Acta 1975; 411: 305-315.
5. Majima M, Katori M, Hanazuaka M, Mizogami S, Nakano T, Nakao Y, Mikami R, Uryu H, Okamura R, Mohsin SSJ, Oh-ishi S.   Suppression of rat deoxycorticosterone-salt hypertension by kallikrein-kinin system. Hypertension 1991; 17: 806-813.
6. Uchida Y, Majima M, Katori M. A method of determination of human plasma HMW and LMW kininogen levels by bradykinin enzyme immunoassay. Pharmacol Res Commun 1986; 18: 831-846.
7.Oh-ishi S, Katori M. Fluorometric assay for plasma prekallikrein using petidylmethyl-coumarinylamide as a substrate. Thromb Res 1979; 15: 127-134.
8. Mohsin SSJ, Majima M, Katori M, Sharma JN. Important suppressive roles of the kallikrein-kinin system during the developmental stage of hypertension in spontaneously hypertensive rats. Asia Pacific J Pharmacol 1992; 7: 1-10.
9. Tomita K, et al. Control of Na^+ and K^+ transport in the cortical collecting duct of rat. J Clin Invest 1986; 76: 132-.
10. Harrap SB, Van der Merwe WM, Griffin SA, Macpherson F, Lever AF. Brief angiotensin converting enzyme inhibitor treatment in young spontaneously hypertensive rats reduces blood pressure long-term. Hypertension 1990; 16: 603-614.

KALLIKREIN-KININ,ENKEPHALIN, RENIN-ALDOSTERONE AND CATECHOLAMINE SYSTEMS IN THE VANADATE (AS VANADYL)-INDUCED ARTERIAL HYPERTENSION

M. Carmignani[1,2], P. Preziosi[2], R. Del Carmine[2], G. Porcelli[3] and A.R. Volpe[3]

Dept. of Cell Biology and Physiology[1], University of L'Aquila, I-67010 Coppito (L'Aquila) and Inst. of Pharmacology[2] and Chemistry[3] (CNR Receptor Chemistry Center), Catholic University School of Medicine, I-00168 Rome, Italy

SUMMARY : Exposure to vanadate was found to induce arterial hypertension through effects on renin-angiotensin-aldosterone, renal peptidergic, and central and peripheral catecholaminergic systems. Vanadate increased, mainly in vascular myocells, both receptor-operated Ca^{2+} channel- and cyclic-AMP-dependent availability of Ca^{2+} for conctractile processes. Vanadate was selectively accumulated by tissues in the +4 oxidation state (vanadyl).

INTRODUCTION

Although vanadium (as VO_3^-, vanadate-metavanadate,V) is known to both enter certain cells and be reduced by glutathione to VO^{2+} (vanadyl, +4 oxidation state), it is yet to be confirmed whether cardiovascular effects of V depend on the redox state of the involved cells and are influenced by other ions such as Na^+, K^+, Mg^{2+} and Mn^{2+} [1,2,3]. It is also uncertain whether these effects of V are related to inhibition or activation of various enzymatic activities (like Na, K-ATPase and adenylate cyclase, respectively), insulin-like action and/or increased intracellular availability of Ca^{2+} for contractile processes in vascular and cardiac myocells (because of augmented influx and/or decreased efflux of this ion)[1,2,3].

V was augmented in the urine of patients with essential hypertension, increased cardiovascular reactivity to noradrenaline (NA) and angiotensin II and induced arterial hypertension in laboratory animals depending on experimental conditions; however, V was shown to inhibit renin secretion [2,3].

The purpose of this research was to investigate the possibility that long-term exposure to V represents a significant factor in inducing arterial hypertension. In particular, this study concerned autacoidal, neurogenic and endocrine mechanisms involved in cardiovascular regulation, homeostasis of Ca^{2+} in vascular and cardiac myocells as well as distribution and oxidation states of vanadium in various tissues.

MATERIALS AND METHODS

Male Sprague-Dawley rats (n=10 in each group) were given 0 (control), 10 or 40 µg/ml of V (as $NaVO_3$) in drinking water for seven months starting from weaning. Other rats of the same sex and strain (n=6 in each group) received 0 (control) or 100 µg/ml of V (as $NaVO_3$) in drinking water for the same period.

At the end of the treatment the activities of kallikrein[4], kininases I (K1) and II (K2), and enkephalinase (ENase)[5] were determined in the 24-hour urines, in which creatinine, total nitrogen, proteins and electrolytes were also determined.

Plasma renin activity (PRA) [6] and levels of plasma aldosterone [7] were evaluated in samples of arterial blood.

Aortic blood pressure (BP), heart rate (HR) and maximum rate of rise of the left ventricular isovolumetric pressure (dP/dt) were monitored polygraphically under thiopental anesthesia[3,8]. The following substances (μg/kg) were injected by i.v. route: bradykinin (BK;11.5), ile-ser-BK (TK;11.5), leu[5]- and met[5]-enkephalins (LE,ME;100), angiotensin I (1) and II (0.50), phenylephrine (10), histamine (10), serotonin (10), isoprenaline (ISO;0.25), papaverine (2000), dibutyryl-cyclic-AMP (dcAMP;5000), veratrine (VE;10), verapamil (VER;100) and hexamethonium (HEX;625,1250). In the rats treated with 100 μg/ml of V, these substances (but not TK,LE,ME and PAP) were administred before and after vagotomy; acetylcholine (2.50), tyramine (TYR;250), ouabain (OUA;20) and clonidine (CLO;10) were also assayed.

Rats were killed by decapitation, the brain was removed and the hypothalamus was rapidly dissected and frozen on dry ice. NA, adrenaline and dopamine were assayed in the hypothalamus, within ten days, using a modification of described isotopic radioenzymatic assays [9].

Samples of heart,kidney, brain, aorta, femur and liver from rats treated with 10 and 40 ppm of V were processed by electron paramagnetic resonance spectroscopy (EPR) for determining the oxidation state of vanadium inside the tissues; tissue samples from the rats exposed to 100 ppm of V were submitted to instrumental neutron activation analysis for determining the contents of vanadium [2,10].

Data were expressed as means ± S.E.M. and compared by analysis of variance. Only a p value less than 0.05 was considered to be significant.

RESULTS

In the rats exposed to 10 and 40 ppm of V, the activities of K1 and K2 were increased in a dose-related manner and those of kallikrein and ENase were also increased; BP, but not HR and dP/dt, was augmented by V, which potentiated selectively (mostly at the lower dose) the effects of BK, LE, ME, ISO and dcAMP on BP and dP/dt without changing cardiovascular responses (CR) to the other tested drugs (Tables 1,2).

PRA (μg/ml/h) was 1.0±0.3, 4.7±1.5 (p<0.05) and 4.1±2.0 (p<0.05), levels of plasma aldosterone (pg/ml) were 188±57, 554±160 (p<0.05) and 265±61, and levels of urinary K^+ (mEq/g creatinine) were 118±14, 169±18 (p<0.05) and 221±28 (p<0.05) in the rats exposed to 0, 10 and 40 ppm of V, respectively (n=5 in each group).

V (only at the dose of 40 ppm) increased the content of adrenaline in the hypothalamus and reduced that of dopamine (Table 3).

EPR showed that tissues contained vanadium only in the tetravalent oxidation state (vanadyl) (Fig. 1).

In the rats exposed to 100 ppm of V, BP and HR (but not dP/dt) and activities of kallikrein, K1 and K2 were increased; V reversed BP response   while increasing inotropic response to BK, potentiated CR to ISO and reduced those to VE, increased BP responses to dcAMP, TYR, VER, HEX and CLO, reduced the effects of vagotomy on

Table 1. Cardiovascular indices and enzymatic activities in the 24-hour urines of groups of rats exposed for seven months to 0 , 10 or 40 ppm of vanadate and for seven months to 0 or 100 ppm of vanadate in drinking water.

| | CONTROL | EXPOSED | | CONTROL | EXPOSED |
| | | 10 ppm | 40 ppm | | 100 ppm |
	(n=8)	(n=8)	(n=8)	(n=6)	(n=6)
Systolic blood pressure (mmHg)	107±4	131±4*	126±2*	122±3	144±4*
Diastolic blood pressure (mmHg)	85±3	105±3*	107±3*	95±3	155±5*
Heart rate (beats/min)	356±7	345±9	363±16	239±4	288±6
Kallikrein	8.4±1[a]	13.7±2.5[a]*	11.7±0.8[a]*	0.4±0.1[b]	0.9±0.1[b]*
Kininase I	32.0±4.2[c]	129.9±14.9[c]*	156.8±9.1[c]*	0.2±0.1[b]	6.7±2.6[b]*
Kininase II	1.8±0.3[c]	2.6±0.1[c]*	3.9±4.1[c]*	148.5±12.9	261.1±17.7[b]*
Enkephalinase	46.9±6.5[c]	97.2±11.3[c]*	73.8±11.3[c]*		

*Values are means±S.E.M. *Significantly different from the control (p<0.05)*

[a]*nM/mg of 24-hour urine creatinine*

[b]*mM/min/L of 24-hour urine creatinine*

[c]*$\mu M \times 10^{-3}$ of hydrolyzed substrate/mg of 24-hour urine creatinine*

BP and HR and opposed those of OUA and CLO on dP/dt without influencing CR to the other tested drugs (Tables 1,2).

Vanadium was accumulated (ng/g wet weight) mostly in the femur (7857±434) and in the kidney (6770±230 in the

Table 2. Cardiovascular responses to several drugs, given by i.v. route , in groups of rats exposed for seven months to 0, 10 or 40 ppm of vanadate and for seven months to 0 or 100 ppm of vanadate in drinking water.

| DRUG (μg/kg) | MEAN BLOOD PRESSURE (mmHg) | | | dP/dt (mmHg/sec^{-1} x 10^{-1}) | | |
(n=8 in each group)	CONTROL	10 ppm	40ppm	CONTROL	10 ppm	40ppm
Bradykinin (11.5)	-19±2	-43±3*	-35±3*	+ 218± 17	+294±18*	+372±22*
Leu[5]-enkephalin (100)	-10±2	-30±5*	-18±6	- 86± 1	-166±17*	-139± 9*
Met[5]-enkephalin (100)	-19±4	-43±2*	-19±3	- 102± 8	-182±13*	-171±16*
Isoprenaline (0.25)	-22±2	-37±2*	-29±5	+ 283± 17	+449±32*	+371±24*
Dibutyryl-cyclic AMP (500)	-17±4	-31±2*	-33±4*	+ 118± 8	+107±15	+171±12*

(n=6 in each group)	CONTROL	100 ppm		CONTROL	100 ppm	
Bradykinin (1)	- 9±2	+16±3		+1189±182	+1875±213	
Veratrine (10)	-16±5	- 6±1*		- 805± 13	- 160± 17*	
Tyramine (250)	+20±6	+37±5*		- 348± 27	- 374± 43	
VAGOTOMY	+18±3	+ 9±1*		+ 107± 71	+ 128± 21	
Ouabain (20)	+14±2	+18±4		+ 548± 32	+ 231± 68*	
Verapamil (100)	-15±4	-35±5*		- 428± 70	- 426± 68	
Hexamethonium (625)	-21±3	-39±3*		- 294± 27	- 257± 32	
Clonidine (10)[a]	+15±2	+29±3*		- 535± 58	- 342± 11*	

*Values are means±S.E.M.*Significantly different from the control (p<0.05)*

[a]*The initial hypertensive response (depending on activation of the peripheral α_2-adrenoreceptors) was followed by long-lasting hypotension.*

cortex and 3342±347 in the medulla); the contents of vanadium were 2710±397 (auricle) and 554±49 (ventricle) in the heart,870±51 in the aorta, 137±19 in the brain and 1533±

Table 3. Levels of catecholamines in the   hypothalamus   of rats exposed for seven months to 0, 10 or 40 ppm of vanadate in drinking water

| | CATECHOLAMINES (ng/mg protein) | | |
	NORADRENALINE	ADRENALINE	DOPAMINE
Control	12.65±1.32	2.22±0.18	1.56±0.07
Exposed (10 ppm)	10.14±0.70	1.62±0.47 (n=4)	1.46±0.30
Exposed (40 ppm)	11.96±0.58	3.94±0.63* (n=5)	0.87±0.10*

*Values are means ±S.E.M.*Significantly different from the control (p<0.05)*

71 in the liver; all corresponding tissues of the control rats had a vanadium content < 25 ng/wet weight (n=6 in each group).

In the kidney, V was accumulated in the cytoplasm of the proximal tubule cells; no morphological alterations were found in other tissues[3]. Levels of both K^+ and Na^+ in the 24-hour urines were in creased by V.

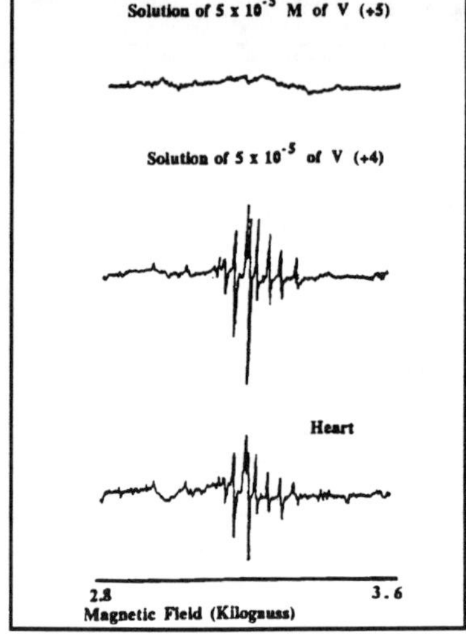

Fig. 1.
Electron paramagnetic resonance spectra showing that, in the heart of a rat exposed to 40 ppm of vanadate (see text), vanadium is present as vanadyl (+4 oxidation state).

DISCUSSION

V (following chronic exposure) was shown to induce arterial hypertension by acting on the renal kallikrein-kinin , enkephalin and renin-angiotensin-aldosterone systems (with increased CR to BK, LE and ME and potentiation of the BK_2-receptor-dependent positive inotropic response to BK). V increases CR to β_1- ,β_2- and α_2- adrenoreceptor agonists (also by effects on both cyclic-AMP-dependent availability of Ca^{2+} for contractile processes - mainly in vascular myocells - and influx of Ca^{2+} through the verapamil-sensible receptor-operated channels)[3,11]. Moreover, V reduces baro- afferent sensitivity and vagal activity, enhances central sympathetic tone (likely in relation to changed levels of adrenaline and dopamine in the hypothalamus) and stimulates the tyramine-sensible release of noradrenaline from the postganglionic sympathetic endings indipendently on nerve activity [2]. The observed effects of V on CR to OUA explain why the metal does not increase cardiac inotropism *in vivo*, as observed *in vitro* as consequence of inhibition of Na, K-ATPase activity[1].

Vanadium accumulates as vanadyl in aorta, heart (mostly in the auricles), kidney (mostly in the cortex) and brain. At these levels V activates, in the +4 oxidation state, the above mechanisms with the result of arterial hypertension related to increase of the total peripheral resistance and, at a higher level of exposure, also to moderate enhancement of cardiac chronotropism.

ACKNOWLEDGEMENTS: This work was supported by grants of Italian MURST to Marco Carmignani (40% and 60%: 1989 -1990 and 1990-1991), CNR nr. 88.01336.59 and Progetto Finalizzato Chimica Fine II.

REFERENCES

1. B. R. Nechay, L.B. Nanninga, P.S.E. Nechay, R.L. Post, J.J. Grantham, I.G. Macara, L.F. Kubena, T. Phillips,F. H. Nielsen. Fed. Proc. 1986;45:123-132.
2. M. Carmignani, E. Sabbioni, P. Boscolo, R. Di Pietra, G. Ripanti. In: 6th International Trace Element Symposium 1989 (Mo,V), Vol 1. M. Anke, W. Baumann, editors. Leipzig:Karl-Marx-University Press 1990:106-113.
3. M. Carmignani, P. Boscolo,A.R. Volpe, G. Togna, P. Preziosi. Arch. Toxicol. 1991;Suppl.14:124-127.
4. T. Morita, H. Kato, S. Iwanaga, K. Takeda, T. Kimura, S. Sakakibara. J.Biochem.1977;82:1495-1498.
5. A.R. Volpe, G. Porcelli, M. Di Iorio. J.Chromatogr.1989;414:427-431.

6.   A.E. Freedlender, T.L. Goodfriend. In: Methods of Hormone Radioimmunoassay. H. R.Behrman, editor. New York: Academic Press 1979:889-913.

7.   G.E. Abraham, A.R. Garza,F.S. Manlimos.In: Handbook of Radioimmunoassay. G.E. Abraham,editors.New York:M. Dekker inc.; 1977:591-618.

8.   P. Boscolo, M. Carmignani, G. Sacchettoni, F.O. Ranelletti, L. Artese,P. Preziosi. Toxicol. Lett.1988;41:129-137.

9.   J.M. Saavedra,R.Del Carmine,I. Iwai, N. Alexander.In: Radioimmunoassay of Drugs and Hormones in Cardiovascular Medicine. A. Albertini, M. Da Prada, B.A. Peskar, editors. Amsterdam: Elsevier/North-Holland Biomedical Press 1979:199-215.

10. E. Sabbioni,J.Edel, R. Di Pietra, E.Marafante, A. Springer, L. Ubertalli. Chemosphere 1984;13:88-93.

11. D. Marmé. Calcium and Cell Physiology. Berlin:Springer-Verlag 1987.

AAS 38/III
Recent Progress on Kinins
© 1992 Birkhäuser Verlag Basel

ACTION OF A NOVEL KININ PRECURSOR, MET-T-KININ-LEU, ON PROSTAGLANDIN I_2 AND BLOOD PRESSURE

H. Handa, K. Goto, T. Kanayasu[1], S. Murota[1], H. Izumi[2], S. Suzuki[3], K. Fujie[4] and W. Sakamoto[4]

Tonan Hospital, Sapporo, [1]School of Dentistry, Tokyo Medical and Dental University, Tokyo, [2]Dept. of Physiology, School of Dentistry, Tohoku University, Sendai, [3]College of Medical Technology, and [4]Dept. of Biochemistry, School of Dentistry, Hokkaido University, Sapporo, Japan

SUMMARY: The novel kinin precursor, Met-T-kinin-Leu, stimulated the release of prostaglandin I_2 from endothelial cells cultured using bovine carotid artery endothelial cells and minimum essential medium supplemented with 10% fetal calf serum.
However, it failed to stimulate the release of prostaglandin I_2 from the cells in the fetal calf serum-free medium conditions.
To examine the discrepancy of the release of prostaglandin I_2 from the cells by Met-T-kinin-Leu in the presence or absence of fetal calf serum, the products formed from Met-T-kinin-Leu by incubation of culture medium were analyzed by reverse-phase HPLC.
Regarding the blood pressure reaction of Met-T-kinin-Leu, it showed from one fifth to one hundredth of that compared with bradykinin in blood pressure reactivity of each species, such as rats, rabbits and cats.

INTRODUCTION

Kininogens are precursor proteins of vasoactive kinins involved in blood pressure regulation, cell proliferation, and vascular permeability (1).

In rat, three kinds of kininogens, namely high-molecular weight (HMW), low-molecular weight (LMW), and T-kininogens exist (2).

Recently, it was reported that T-kininogen increased in the plasma of rat after inflammation induced by Freund's adjuvant and carrageenin (3,4).

Previously, we reported that T-kinin and Met-T-kinin were released from T-kininogen by acid proteases of granulomatous tissues in rats with carrageenin-induced inflammation (5).

And also, we clarified that cathepsin D purified from the granulomatous tissues could not release T-kinin and Met-T-kinin from T-kininogen, but could release kinin precursors, T-kinin-Leu and Met-T-kinin-Leu (6).

However, it is not clear how or whether Met-T-kinin-Leu plays a physiological role in the inflammatory region.

In this paper, we report that Met-T-kinin-Leu stimulates the release of prostaglandin I_2 from endothelial cells cultured using bovine carotid endothelial cells and causes a fall of blood pressure in each species of rats, cats and rabbits.

MATERIALS AND METHODS

Isolation and culture of bovine endothelial cells

Bovine carotid endothelial cells were isolated by scraping the intimal surface with a surgical blade and suspended in Gibco's minimum essential medium (MEM) supplemented with 20% fetal calf serum according to the method described in our previous report (7) and maintained and subcultured in the medium supplemented with 10% fetal calf serum.

Release of prostaglandin I_2 from endothelial cells

The release of prostaglandin I_2 from cultured endothelial cells was determined by radioimmunoassay of 6-ketoprostaglandin $F_{1\alpha}$ (a hydrolyzed product of prostaglandin I_2). Namely, approximately 10^5 cells/well were cultured with 0.5 ml of medium supplemented with or without 10% fetal calf serum and Met-T-kinin-Leu for 1 h at 37°C in a 5% CO_2-incubator. The medium was then collected and 6-ketoprostaglandin $F_{1\alpha}$ was measured by radioimmunoassay with [^{125}I] - RIA KIT (du Pont de Nemours, Boston, U.S.A.).

<u>Cleavage of Met-T-kinin-Leu in MEM supplemented 10% fetal calf serum</u>

Met-T-kinin-Leu, 900 µg, and 36 ml MEM supplemented 10% fetal calf serum were incubated at 37°C. At intervals, samples of 2.5 ml were withdrawn, extracted with n-butanol, and subjected to a reverse-phase high performance liquid chromatography (HPLC) using a ODS-120T column (Toyo Soda Co., Japan), as described previously (6).

<u>Isolation of peptides generated from Met-T-kinin-Leu in MEM supplemented 10% fetal calf serum</u>

Met-T-kinin-Leu, 900 µg, and 36 ml of MEM supplemented 10% fetal calf serum were incubated at 37°C for 3 h. T h e peptides generated from Met-T-kinin-Leu were extracted and isolated by HPLC, as described above.

<u>Amino acid analysis</u>

Samples to be analyzed for amino acid compositions were hydrolyzed with 6 N HCl at 110°C for 24 h in an evacuated and sealed tubes. The hydrolysates were evaporated and analyzed in a Hitachi-835 amino acid analyzer.

<u>Blood pressure</u>

The measurement of blood pressure was carried out using rabbits (3-3.5 kg), cats (2-3 kg) and rats (350-400 g), as described previously (8).

RESULTS

<u>Effect of Met-T-kinin-Leu on prostaglandin I_2 production</u>

Cultured bovine vascular endothelial cells were incubated in MEM supplemented with 10% fetal calf serum and Met-T-kinin-Leu for 1 h. The amount of prostaglandin I_2 released into the medium during the incubation period was measured by the radioimmunoassay for 6-ketoprostaglandin $F_{1\alpha}$, which is a non-enzymatical metabolite of prostaglandin I_2. As shown in Fig. 1, Met-T-kinin-Leu significantly increased the release of prostaglandin I_2 in a dose dependent manner. The effect was observed at low concentration, 1 µg per ml of Met-T-kinin-Leu, and maximal release was elicited between 10 µg and 100 µg. However, it failed to stimulate the release of prostaglandin I_2 in the absence of fetal calf serum, as shown in Fig. 1. From these results it is suggested that Met-T-kinin-Leu indirectly stimulated the release of prostaglandin I_2 from the vascular endothelial cells.

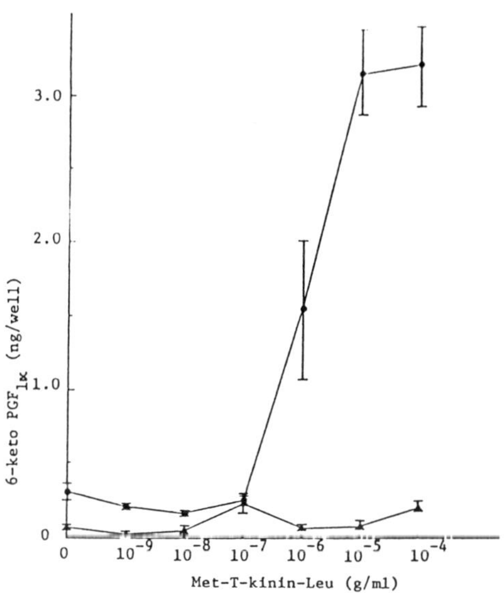

Fig. 1. Dose-response for stimulation of the release of prostaglandin I_2 from bovine carotid artery endothelial cells by Met-T-kinin-Leu in the presence or absence of 10% fetal calf serum.

    Cell monolayers (approximately 10^5 cells/well) were incubated at 37°C for 1 h with 0.5 ml MEM supplemented with 10% fetal calf serum (●) or without fetal calf serum (▲), in a concentration between 10^{-9} and 10^{-4} g/ml of Met-T-kinin-Leu. The medium was removed and centrifuged, and the supernatant was extracted. 6 ketoprostaglandin $F_{1\alpha}$ in the extracts was assayed serologically. Values are mean ± SD of triplicate.

Cleavage of Met-T-kinin-Leu in MEM in supplemented 10% fetal calf
serum

Met-T-kinin-Leu was incubated in MEM supplemented 10% fetal
calf serum at 37°C.

Samples were taken at intervals up to 180 min and subjected
to reverse-phase HPLC (ODS-120T, Toyo Soda, Japan).

As shown in Fig. 2, two peaks (fragments A and B),
increasing with the incubation time, newly appeared on
incubation for 180 min in MEM supplemented with 10% fetal calf
serum, but the peaks did not appear on incubation with MEM
alone.

In this incubation fragments A and B generated were 0.91 µg
and 2.4 µg respectively, whereas Met-T-kinin-Leu (54 µg) was
cleaved up to 27 µg.

From the analysis of HPLC and amino acid composition,
fragments A and B were estimated as Met-Ile-Ser-Arg-Pro-Pro-Gly-
Phe (octapeptide, Met^1-Phe^8) and Ser-Pro-Phe-Arg-Leu
(pentapeptide, Ser^9-Leu^{13}), respectively.

Effect of Met-T-kinin-Leu on blood pressure

Fig. 3 shows the effect of Met-T-kinin-Leu on the decrease
in systemic blood pressure (mm Hg) of each species, such as
rats, cats and rabbits.

As shown in Fig. 3, Met-T-kinin-Leu caused a decrease in
blood pressure, in dose-dependent amounts of intravenous
injection.

However, it showed from one fifth to one hundredth of that
compared with bradykinin in blood pressure reactivity of each
species.

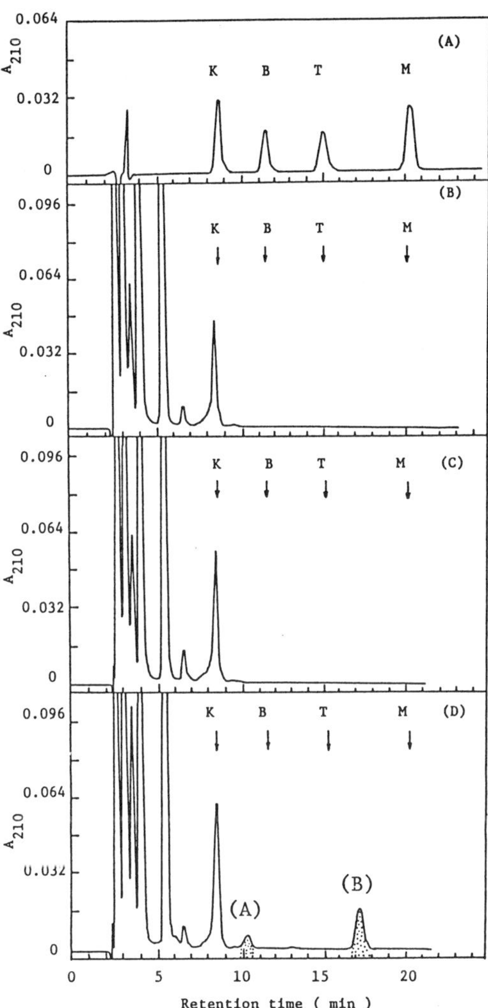

Fig. 2. Reverse-phase HPLC analysis of peptides released from Met-T-kinin-Leu in MEM supplemented with 10% fetal calf serum.

The reaction mixture was extracted with n-butanol and lyophilized, as described in the text. The extract was dissolved in 100 µl of distilled water, and aliquot of 10 µl was injected into the reverse-phase column (ODS-120T, 0.46 x 25 cm). The column was eluted isocratically with 20% acetonitrile in 0.05% trifluoroacetic acid at a flow rate of 1.0 ml/ml.

(A) 330 ng synthetic bradykinin, kallidin and T-kinin, and 500 ng Met-T-kinin-Leu; (B) the extracts from the reaction mixture of Met-T-kinin-Leu with MEM supplemented with 10% fetal calf serum; (C) the extracts from the reaction mixture of Met-T-kinin-Leu with MEM alone for 180 min incubation at 37°C; (D) the extracts from the reaction mixture of Met-T-kinin-Leu with MEM supplemented with 10% fetal calf serum for 180 min incubation at 37°C.

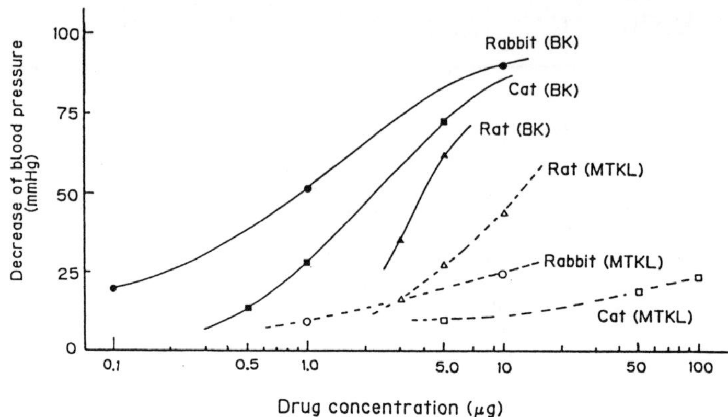

Fig. 3. Decrease of blood pressure of rats, cats and rabbits after i.v. administration of bradykinin and Met-T-kinin-Leu. BK, bradykinin; MTKL, Met-T-kinin-Leu.

DISCUSSION

Kinin, including bradykinin and kallidin, are well known to have potent actions mediating vasodilation, pain, smooth muscle contraction and blood coagulation (9).

Many of these actions are thought to result from the release of arachidonic acid and its metabolites.

On the other hand, Greenbaum and Okamoto reported that rat plasma contains T-kininogen, T-kinin precursor, besides HMW and LMW-kininogens, bradykinin and kallidin precursors (10).

Recently, we demonstrated that Met-T-kinin-Leu, a novel kinin precursor, was released from T-kininogen by cathepsin D of granulomatous tissues in rats with carrageenin-induced inflammation (6).

However, the physiological role of T-kininogen-kinin system remains unknown.

This study provides evidence for the stimulation of prostaglandin I_2 production by Met-T-kinin-Leu.

In cultured bovine carotid endothelial cells, Met-T-kinin-Leu stimulated the release of prostaglandin I_2 from the cells at a concentration of over 1 µg per ml, as shown in bradykinin (11).

However, the stimulatory effect was observed only in the presence of 10% fetal calf serum, but not in the absence.

Therefore, the release of prostaglandin by Met-T-kinin-Leu seems to occur after the peptide is metabolized. Alternatively, the release of prostaglandin may be stimulated by metabolites which are released from a biological inactive Met-T-kinin-Leu by protease of fetal calf serum.

In this study, the peptides generated from Met-T-kinin-Leu by the MEM supplemented with 10% fetal calf serum were identified as Met^1-Phe^8 and Ser^9-Leu^{13}, by the analysis of HPLC and amino acid composition.

However, it remains unknown whether these peptides stimulate prostaglandin I_2 production from endothelial cells.

Further investigations on these points are under way in our laboratory.

On the other hand, kinins are potent blood pressure lowering agents which develop their activity immediately via direct vasodilation and independent of sodium metabolism, the renin-angiotensin system, mineralocorticoids, prostaglandins, β-adrenoreceptors, and histamine receptors (12).

However, the action of Met-T-kinin-Leu on decrease in systemic blood pressure was less than that of bradykinin, in spite of showing similar capacity on the prostaglandin-release.

The discrepancy maybe due to the differences of receptor for bradykinin and Met-T-kinin-Leu and their susceptibility for endogenous protease in the vessel.

From these results, Met-T-kinin-Leu, which was released from T-kininogen by cathepsin D of granulomatous tissues in rats with carrageenin-induced inflammation (6), seems to play some physiological role in inflammatory responses via synthesis of prostaglandin E_2 and I_2, namely, vascular permeability, cell proliferation, and tissue metabolism.

In fact, Met-T-kinin-Leu stimulated prostaglandin E_2 production from macrophages, playing an important role in inflammatory reaction (unpublished data).

Recently Conklin et al. reported that bradykinin stimulates prostaglandin E_2-release from fibroblasts through activating phospholipase A_2, whereas it stimulates prostaglandin I_2-release from the endothelial cells through activating phosphatidylcholine-specific phospholipase C (13).

ACKNOWLEDGMENT

We wish to thank Dr. S. Nagasawa for his valuable suggestions, and Mrs. H. Matsumoto for amino acid analyses.

This research was supported in part by a grant from the Ministry of Education, Science and Culture of Japan.

REFERENCES

1. Carrentero OA and Scicli AG. In: The kallikrein-kinin system in health and disease. Fritz H, Schmidt I, and Dietze G, editors. Germany: Limbach-Verlag Braunschweig, 1989 : 63-78.

2. Greenbaum LM and Okamoto H. T-kinin and T-kininogen. Methods Enzymol. 1988; 163: 272-282.

3. Barlas A, Okamoto H, and Greenbaum LM. T-kininogen-the major plasma kininogen in rat adjuvant arthritis. Biochem Biophys Res Commun. 1985; 129; 280-286.

4. Sakamoto W, Yoshikawa K, Handa H, Uehara S, and Hirayama A. T-kininogen in rats with carrageenin-induced inflammation. Biochem Pharmacol. 1986; 35: 4283-4290.

5. Sakamoto W, Satoh F, Gotoh K, and Uehara S. T-kinin and Met-T-kinin are released from T-kininogen by an acid proteinase of granulomatous tissues in rats. FEBS Lett. 1987; 219; 437-440.

6. Sakamoto W, Satoh F, Nagasawa S, and Handa H. Identification of T-kinin-Leu released from T-kininogen by cathepsin D of granulomatous tissues in rats. Biochem Biophys Res Commun. 1988; 150; 1199-1206.

7. Morita I, Kanayasu T, and Murota S. Kallikrein stimulates prostacyclin production in bovine vascular endothelial cells. Biochim Biophs Acta. 1984; 792; 304-309.

8. Izumi H and Karita K. Vasodilator responses following intracranial stimulation of the trigeminal, facial and glossopharyngeal nerves in the cat gingiva. Brain Res. 1991; 560; 71-75.

9. Manganiello VC, Moss J, and Roscher R. Regulation of bradykinin action. Atemw-Lungenkrkh. 1988; 14; S57-S67.

10. Okamoto H and Greenbaum LM. Isolation and structure of T-kinin. Biochem Biophys Res Commun. 1983; 112; 701-708.

11. Crutchley DJ, Ryan JW, Ryan US, and Fischer GH. Bradykinin-induced release of prostacyclin and thromboxanes from bovine pulmonary artery endothelial cells. Biochem Biochim Biophys Acta. 1983; 751: 99-107.

12. Boenner G, Preis S, Schunk U, and Iwersen D. In: The kallikrein-kinin system in health and disease. Fritz H, Schmidt I, and Dietze G, editors, Germany: Limbach-Verlag Braunschweig, 1989: 79-96.

13. Conklin BR, Burch RM, Steranka LR, and Axelrod J. Distinct bradykinin receptors mediate stimulation of prostaglandin synthesis by endothelial cells and fibroblasts. J Pharmacol Exp Ther. 1988; 244: 646-649.

AAS 38/III
Recent Progress on Kinins
© 1992 Birkhäuser Verlag Basel

INFLUENCE OF A KININ ANTAGONIST ON ACUTE HYPOTENSIVE RESPONSES INDUCED BY BRADYKININ AND CAPTOPRIL IN SPONTANEOUSLY HYPERTENSIVE RATS

J.N. Sharma, J.M. Stewart*, S.S.J. Mohsin, M. Katori** and R. Vavrek*

Department of Pharmacology, School of Medical Sciences, Universiti Sains Malaysia, Kubang Kerian, Kelantan, Malaysia, *Department of Biochemistry, School of Medicine, University of Colorado Health Sciences Center, Denver, U.S.A., and **Department of Pharmacology, Kitasato University School of Medicine, Kitasato,1-15-1, Sagamihara, Kanagawa 228, Japan

SUMMARY: We have evaluated the effects of a B_2 receptor antagonist (B5630) of kinins on BK and captopril-induced acute hypotensive responses in anaesthetized SHR. Intravenous treatment of BK (1.0 µg) and captopril (0.3 mg/kg) caused significant ($p < 0.05$) fall in the SBP and DBP. Whereas BK caused greater fall in the SBP ($p<0.05$), DBP ($p < 0.01$) and duration of hypotension ($p<0.05$) when administered after captopril (Fig 1 and 2). All the hypotensive effects of BK and captopril were significantly antagonised ($p < 0.05$) in the presence of B5630 (2.0 mg/kg). Further, the duration of hypotensive responses of BK and captopril were blocked ($p < 0.05$) by B5630. The agonists and BK-antagonist did not cause significant ($p > 0.05$) alterations in HR during the entire investigation. These findings provide evidence to support the suggestion that B_2 receptor might be involved in the regulation of the hypotensive actions of BK and captopril. Kinins should also have valuable functions in the antihypertensive property of captopril-like drugs.

INTRODUCTION

The orally active angiotensin-converting enzyme (ACE) inhibitors are potent blood pressure (BP) lowering agents in human hypertensives, Dahl salt-sensitive hypertensive rats and spontaneously hypertensive rats (SHR) irrespective of the levels of plasma renin (1-5). Angiotensin converting enzyme inhibitors protect the biodegradation of kinins by inactivating kininase II. Kinins are powerful vasodepressor polypeptides that appear to be involved in the regulation of BP and physiopathology of hypertension (6,7). On the other hand, inhibition of the local and systemic angiotensin II formation may represent the major BP lowering mechanism of ACE inhibitors (8). Nevertheless, precise role of kinins in mediating and modulating the antihypertensive actions of ACE inhibitors in antihypertensive therapy remain uncertain.

Recently, a breakthrough was achieved by Stewart and colleague (9) who developed the first sequence of competitive bradykinin (BK) antagonist. Since then numerous BK-antagonists have been synthesized in an effort to increase the potency. In addition, BK-antagonists are able to block the activities of at least two classes of kinin receptors, mainly B_1 and B_2 (10).

We have, therefore, examined the influence of a new BK specific receptor antagonist B5630 (DArg-Arg-Pro-Hyp-Gly-Thi-Ser-DPhe-Thi-Arg.TFA) on acute hypotensive responses of captopril and BK in SHR.

MATERIALS AND METHODS

Bradykinin and pentobarbitone sodium were purchased from Chemopharm Malaysia. The kinin antagonist, B5630 or DArg-Arg-Pro-Hyp-Gly-Thi-Ser-DPhe-Thi-Arg.TFA (Thi=beta-(2-thienyl)-L-alanine; Hyp=L-4-hydroxyproline; TFA= trifluoroacetic acid salt), was synthesized as mentioned earlier (9). Captopril was a gift from Squibb (Princeton, New Jersey, USA). The kinin antagonist and BK were dissolved in freshly prepared Tyrode solution before use. Whereas captopril was dissolved in phosphate-buffered saline (10 mg/ml, pH 7.4). Fourteen week old male SHR (inbred at the School of Medical Sciences, Universiti Sains Malaysia), weighing 350-400 g, were used in this study. These animals were housed in a room with the constant temperature of $23^{\circ}C$. Food in the form of pellets of normal rat chow and drinking water were provided *ad libidum*.

Seven SHR were fasted overnight and given only water before commencement of the experiment. The animals were loosely restrained supine on the surgical table under pentobarbitone sodium (50 mg/kg ; i.p.) anaesthesia. Systolic blood pressure (SBP), diastolic blood pressure (DBP), heart rate (HR) and duration of hypotension were recorded via the left carotid arterial catheter using a Sthatham P23ID (USA) pressure transducer connected to a Grass model 7D polygraph recorder. The right jugular vein was cannulated with a polyethylene cannula (No. 7410, PE 50, Clay Adams, New Jersey, USA) for the administration of drugs. Following catheterization, the rats were allowed to stabilize for a period of 30 min before beginning the experimental procedure.

To determine the effectiveness of BK-antagonist, BK (1.0 µg), captopril (0.3 mg/kg) and BK (1.0 µg) were injected both prior to and after the administration of the BK-antagonist (2.0 mg/kg). The data are expressed as means±s.e mean. Statistical significance (p<0.05) was determined by Mann-Whitney U test.

RESULTS

As expected, administration of BK (1.0 µg) reduced (p<0.01) the SBP and DBP from 172.8±12.7 mmHg to 144.3±13.0 mmHg, and from 134.3±12.1 mmHg to 96.4±13.0 mmHg, respectively, in SHR. Intravenous injection of captopril (0.3 mg/kg) caused also a significant decrease (p<0.01) in both the SBP (from 172.8±12.7 mmHg to 126.0 ±7.1 mmHg) and DBP (from 134.3 ± 12.1 mmHg to 75.7 ±9.2 mmHg). These results are shown in figures 1 and 2. Following captopril, the magnitude of duration of hypotension induced by BK was significantly (p<0.05) enhanced (Figs. 3 and 4). In addition, there was futher reduction (p<0.05) in the SBP (144.3 ±13.0 mmHg to 114.3 ± 6.1 mmHg) and DBP (96.4 ±13.0 mmHg to 57.1 ± 6.4 mmHg) to the BK treatment after captopril (Figs. 1 and 2).

The BK-antagonist (B 5630 , 2.0 mg/kg) administration did not change the BP. However, BK (1.0 µg)- and captopril (0.3 mg/kg)-induced hypotensive responses (SBP and DBP) were markedly (p<0.05 and p<0.01) attenuated in the presence of B5630 (Fig. 1,2 and 3). It should be noted that the duration of hypotensive responses caused by BK and captopril were abolished (p<0.05) by B5630 (Figs. 3 and 4).

Figure 5 illustrates the effects of BK, captopril and B5630 on HR. These agents prodeced slight, but not significant (p>0.05) decrease in HR.

DISCUSSION

The present study demonstrates that intravenously administered BK and captopril to anaesthetized SHR can induce hypotension and that captopril enhances the

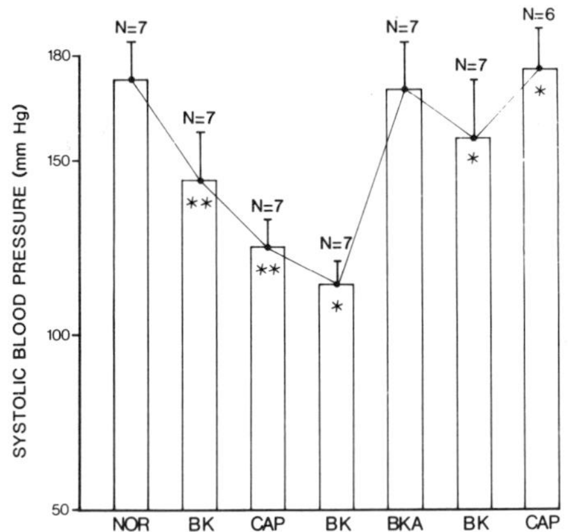

Figure 1. Changes in the systolic blood pressure of spontaneously hypertensive rats after intravenous administration of bradykinin (BK,1.0 mg), captopril (Cap 0.3 mg/kg) and BK (1.0 µg) in the presence and in the absence of BK antagonist (BKA, 2.0 mg/kg)). Baseline control column indicated as normal (Nor). Each column represents the mean value (vertical bars show s.e. mean). N represents number of rats. ** p<0.01 vs baseline (Nor); * p<0.05 vs cap treated rats before BKA; *<0.05 vs Bk and cap before BKA, respectively

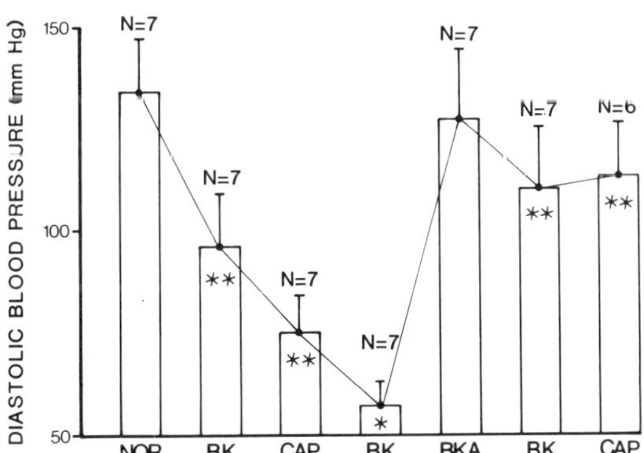

Figure 2. Changes in the diastolic blood pressure of spontaneously hypertensive rats after intravenous administration of bradykinin (BK,1.0 µg), captopril (Cap, 0.3 mg/kg) and BK (1.0 µg) in the absence and in the presence of BK antagonist (BKA, 2.0 mg/kg). Baseline control column indicated as normal (Nor). Each column represents the mean value (vertical bars show s.e.mean). N represents numbers of rats. ** p<0.01 vs baseline (Nor); *<0.05 vs cap treated rats before BKA; **<0.01 vs BK and cap before BKA, respectively.

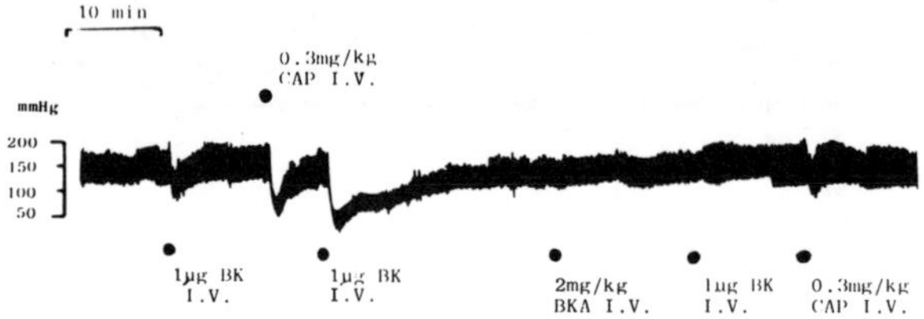

Figure 3. A typical tracing of bradykinin (BK, 1.0 µg) and BK (1.0 µg) following captopril (Cap, 0.3 mg/kg)-induced changes in the systolic and diastolic blood pressure, and duration of hypotension of spontaneously hypertensive rat in the absence and in the presence of bradykinin antagonist (BKA, 2.0 mg/kg) administered intravenously (i.v.).

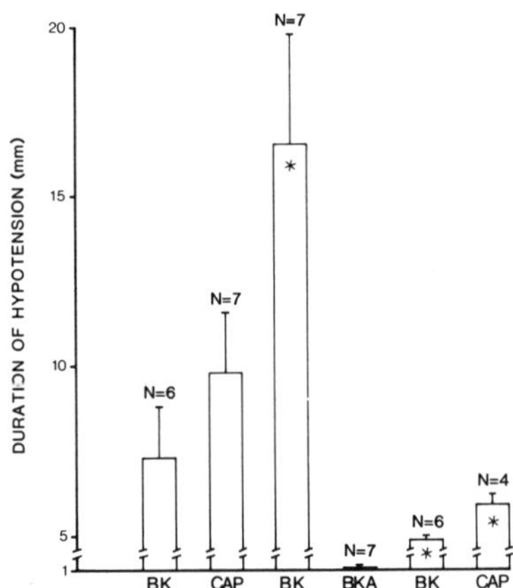

Figure 4. Effects of intravenous administration of bradykinin (BK, 1.0 µg), captopril (Cap, 0.3 mg/kg) and BK (1.0 µg) in the absence and in the presence of BK antagonist (BKA, 2.0 mg/kg) on duration of hypotension in spontaneously hypertensive rats. Each column represents the mean value (vertical bars show s.e. mean). N shows number of rats. * $p < 0.05$ vs BK and cap before BKA; * < 0.05 vs initial control values of BK and cap before BKA treatment.

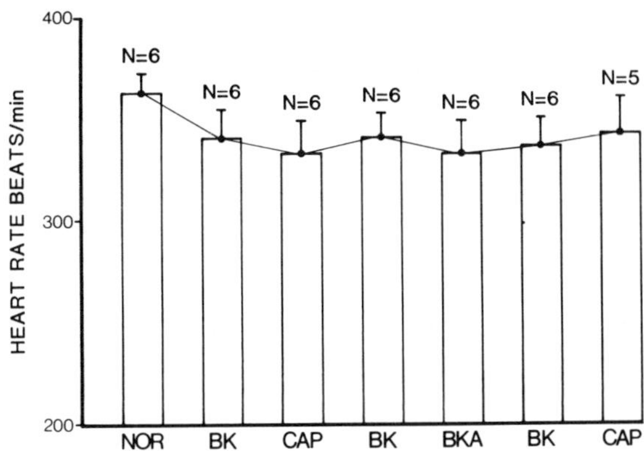

Figure 5. Effects of intravenous administration of bradykinin (BK, 1.0 μg), captopril (Cap, 0.3 mg/kg) and BK (1.0 μg) in the absence and in the presence of BK antagonist (BKA, 2.0 mg/kg) on heart rate in spontaneously hypertensive rats. Baseline control column indicated as normal (Nor). Each column represents the mean value (vertical bars show s.e. mean). N represents number of rats.

hypotensive effects of BK. These effects can be contributed to reduced inactivation of BK due to ACE (kininase II). The present results confirmed the observations made by other investigators (11-14).

The role of kinin in the antihypertensive action of captopril-like drugs is not clearly known. However, it is reported that BP lowering effect of ACE inhibitors in hypertensive subjects is primarily due to inhibition of angiotensin II release (8). It has also been proposed that the reduced plasma angiotensin II and raised plasma kinin levels might contribute to the antihypertensive action of ACE inhibitors in low renin class of hypertensive patients (8,15-17). Also the failure of captopril to lower BP in some hypertensive patients could reflect the blunted activity of the kallikrein-kinin system (18).Whether or not the systemic kinin is implicated in the BP lowering effects of ACE inhibitor remains, however, controversial. This lead us to investigate the effects of acute ACE inhibition with captopril on the SBP, DBP and duration of hypotension to a BK antagonist (B5630) in SHR.

Nowadays, several BK antagonists have been used as tools to define the biological functions of kinins with special reference to the mechanism(s) of action of ACE inhibitors. Our present results document that the BK-antagonist is not only an antagonist of BK-induced fall in the SBP and DBP, but is also antagonistic to captopril-produced decrease in the SBP and DBP (see Figs. 1,2 and 3). It is of interest to note that i.v. administration of BK-antagonist exerted neither a pressor or a depressor response (Fig. 3). This finding demonstrates that the antagonist under the present investigation is devoid of partial agonistic activity. Furthermore, our data agree with other investigators who have noted that the BK- antagonist (D-Arg-Arg-Pro-Hyp-Gly-Thi-Ser-DPhe-Thi-Arg-TFA) produced a significant decrease in hypotensive responses to exogenous BK in normotensive rats, even when the vasodepressor effect of BK was increased by ACE inhibitor, enalaprilat (19). In contrast, some investigators obtained variable results with various types of BK-antaginists and suggested that the antihypertensive action of ACE inhibitors is not due mainly to the kinin participation (14,20). These discrepancies can only be explained on the basis that other investigators may have utilised higher concentrations of BK and/or captopril, or low concentrations of BK-antagonist with variable potencies during their experimentations.

It is generally believed that the BK-induced hypotension is mediated by B_2 receptor, but B_1 receptor might also be involved under certain circumstances (21). In this regard, Sharma (22) has recently proposed that the hypotensive action of ACE inhibitors might be a reflection of the activation of kinin receptors. In our experiments, B5630 significantly abolished the hypotensive actions of captopril (see Fig. 3). Hence, it would be reasonable to suggest that BK-antagonist (B5630) is active against B_2 receptor in BP regulation, since hypotensive responses of BK are mediated mainly via B_2 receptors.

The accumulation of BK following administration of ACE inhibitors with subsequent release of endothelium-derived relaxing factor (EDRF), prostaglandin E_2 (PGE_2) and prostacyclin (PGI_2) could be additional mediators released in the process of antihypertensive effects of captopril-like drugs. Mullane and Moncada (23) indicated that after ACE inhibition, the renal production of PGI_2 can contribute significantly to the hypotensive effect of BK. It is interesting that PGI_2 showed a more powerful hypotensive action than PGE_2 in SHR (24)

Exogenously added BK-induced release of EDRF and PGI_2 from bovine aortic endothelial cells can be abolished by B_2 receptor antagonist (D-Arg$^\mathrm{o}$,Hyp^3,Thi^{5,8}, D-Ph^9)-BK (25). Kinins have been shown to stimulate the synthesis of vasodilator PGE_2 (26). Thus, the interactions between the BK and arachidonic acid products seem to be the most appropriate way of evaluating the pattern of

BP changes following the application of BK-antagonist. The depressor effect of BK can be prevented by PG synthetase inhibition (27). Several investigators observed that captopril is able to increase the synthesis of PGI_2 in hypertensive man and rats (28,29). This is also in accord with the finding that treatment of normotensive rats with BK antagonist inhibits vascular PGI_2 release stimulated by enalaprilat (30). This mechanism may contribute to the generalised vascular smooth muscle dilatation (31). It is conceivable, therefore, that captopril-like drugs may act via accumulating kinins, stimulating B_2 receptor, and causing release of PGE_2, PGI_2 and EDRF (Fig. 6) within the vascular smooth muscle in lowering high BP. Furthermore, ACE inhibitors are able to block the conversion of angiotensin I to angiotensin II. It would be of great value to examine these hypotheses extensively with the use of new BK specific receptor antagonists in various hypertensive models. These investigations may provide better knowledge regarding the role of kinins in BP regulation and the mode of action of ACE inhibitors.

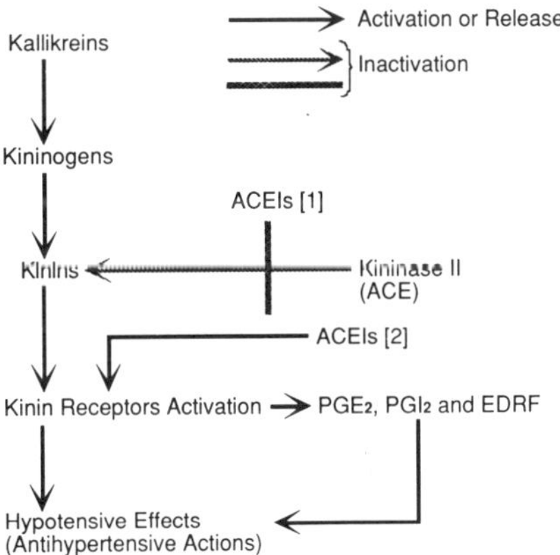

FIGURE 6. A hypothetical explanation of the antihypertensive actions of angiotensin converting enzyme inhibitors (ACEIs) mediated through the kallikrein-kinin system [1]. In addition, kinin may cause hypotension by activating kinin receptor (B_2), and by releasing prostaglandins E_2 (PGE_2), prostacyclin (PGI_2) and endothelium-derived relaxing factor (EDRF) from the endothelial cells [2].

CONCLUSION

The kinin receptor antagonist (B5630), when administered acutely by i.v. route at a dose 2.0 mg/kg, was found to block the hypotensive parameters of BK and captopril. In contrast, the agonists and BK-antagonist did not change HR significantly. It is possible to conclude that captopril might act as a kinin receptor agonist. Since endogenous BK would be raised by captopril, but hypotensive effects would be abolished after administration of a BK antagonist. The discovery of specific B_1 and B_2 receptor antagonists can offer novel approaches for examining the functions of kinins not only in human haemodynamics, but also in other physiopathological situations where kinins are thought to be important mediators and modulators.

ACKNOWLEDGEMENTS

This work was supported by short-term grants (112/0500/0200 and 122/0500/0210) from the Universiti Sains Malaysia. S.S.J. Mohsin is a RONPAKU fellow of the Japan Society for Promotion of Science.We thank Mr. A. Lokman, Mrs. Norhayati Md. Yassin and Miss Khor Lee Kean for technical assistance.

REFERENCES

1. Gavras H, Brunner HR, Turni GA, Kershaw GR, Tifft CP, Cuttelod S,Gavras I, Vukovich RA, McKinstry DN. Antihypertensive effect of the oral angiotensin converting enzyme inhibitor SQ 14225 in man. N Eng J Med 1978; 298: 991-995.

2. Oparil S, Horten R, Wilkins L, Irvin J, Hammett DK. Antihypertensive effect of enalapril in essential hypertension: role of prostacyclin. Am J Med Sci 1987; 294: 395-402.

3. Sharma JN, Fernandez PG, Kim BK, Triggle CR. Systolic blood pressure responses to enalapril maleate (MK421, an angiotensin converting enzyme inhibitor) and hydroclorothiazide in concious Dahl salt-sensitive and salt-resistant rats. Can J Physiol Pharmacol 1984; 62: 846-849.

4. Sharma JN, Fernandez PG, Kim BK, Idikio H, Triggle CR. Cardiac regression and blood pressure control in the Dahl rat treated with either enalapril maleate (MK 421, an angiotensin converting enzyme inhibitor) or hydroclorothiazide. J Hypertension 1983; 1: 251-256.

5. Richer C, Doussau M-P, Giudicell J-F. MK421 and prevention of genetic hypertension development in young spontaneously hypertensive rats. Eur J Pharmacol 1982; 79: 23-29.

6. Sharma JN. Kinin-forming system in the genesis of hypertension. Agents Actions 1984; 14: 200-205.

7. Sharma JN. Interrelationship between the kallikrein-kinin system and hypertension: a review. Gen Pharmacol 1988; 19: 177-187.

8. Bonner G. Haben die Kinin eine Bedeutung fur die antihypertensive Wirkung der ACE-Hemmer? Z Kardiol 1988; 77(Suppl.3): 23-27.

9. Vavrek RJ, Stewart JM. Competitive antagonists of bradykinin. Peptides 1985; 6: 161-164

10. Regoli D, Barabe J. Pharmacology of bradykinin and related kinins. Pharmacol Rev 1980; 32: 1-46.

11. Murthy VS, Waldron TL, Goldberg ME, Vollmer RR. Inhibition of angiotensin converting enzyme by SQ14225 in conscious rabbits. Eur J Pharmacol 1977; 46: 207-212.

12. Murthy VS, Waldron TL, Goldberg ME. The mechanism of bradykinin potentiation after inhibition of angiotensin converting enzyme by SQ14225 in conscious rabbits. Circ Res 1978; 43(Suppl.1): 41-45.

13. Waldron TL, Antonaccio MJ, Murthy VS. Reversal of bradykinin-induced reflex tachycardia to bradykardia by captoril : evidence for prostacyclin involvement. Eur J Pharmacol 1982; 79: 283-292.

14. Waeber B, Aubert J-F, Fluckiger J-P, Nussberger J, Vavrek RJ, Stewart JM, Brunner HR. Role of endogenous bradykinin in blood pressure control of conscious rats. Kidney Int 1988; 34(Suppl. 26) : S63-S68.

15. Sharma JN. Contribution of kinin system to the antihypertensive action of angiotensin converting enzyme inhibitors. Adv Exp Med Biol 1989 ; 247A: 197-205.

16. Iimura O, Shimamoto K. Role of kallikrein-kinin system in the hypotensive mechanisms of converting enzyme inhibitors in essential hypertension. J Cardiovas Pharmacol 1989 ; 13(Suppl 3) : S63-S66.

17. Schror K. Converting enzyme inhibitors and the interaction between kinins and eicosanoids. J Cardiovas Pharmacol 1990; 15(suppl 6): S60-S68.

18. Madeddu P, Oppes M, Rubattu S, Dessi-Fulgheri P, Gloriosa N, Soro A, Rappelli A. Role of renal kallikrein in modulating the antihypertensive effect of a single dose of captopril in normal- and low-renin antihypertensive effect of a single dose of captopril in normal- and low-renin essential hypertensives. J Hypertension 1987 ; 5: 645-648.

19. Carbonell LF, Carretero OA, Stewart JM, Scicli AG. Effect of a kinin antagonist on the acute antihypertensive activity of enalaprilat in severe hypertension. Hypertension 1988; 11: 239-243.

20. Waeber B, Niederberger M, Gavras H, Nussberger J, Brunner HR. Hemodynamic effects of a kinin antagonist. J Cardiovas Pharmacol 1990 ; 15(suppl 6): S78-S82.

21. Regoli D. Neurohumoral regulation of precapillary vessels : the kallikrein-kinin system. J Cardiovas Pharmacol 1984 ; 6(Suppl): S401-S412.

22. Sharma JN. Does kinin mediate the hypotensive action of angiotensin converting enzyme (ACE) inhibitors? Gen Pharmacol 1990; 21: 451-457.

23. Mullane KM, Moncada S. Prostacyclin mediates the potentiated hypotensive effect of bradykinin following captopril treatment. Eur J Pharmacol 1980 ; 66: 355-365.

24. Pace-Asciak CR, Carrara MC, Rangaraj G, Nicoalou KC. Prostaglandin I_2 has more potent hypotensive properties than prostaglandin E_2 in the normal and spontaneously hypertensive rats. Prostaglandins 1978; 15: 999-1010.

25. D'Orleans-Juste P, de Nucci G, Vane JR. Kinins act on B1 or B2 receptors to release conjointly endothelium-derivedrelaxing factor and prostacyclin from bovine aortic endotheliel cells. Br J Pharmacol 1989 ; 96: 920-926.

26. Conklin BR, Burch RM, Steranka LR, Axelrod J. Distinct bradykinin receptors mediate stimulation of prostaglandin synthesis by endothelial cells and fibroblasts. J Pharmacol Exp Ther 1988; 244: 646-649

27. Sharma JN, Fernandez PG, Triggle CR. The effects of indomethacin on the duration of the hypotensive action of bradykinin in Dahl salt-resistant rats : role of cyclooxygenase inhibition. Prostaglandins Leukotrienes Med 1984; 14: 131-135.

28. Vinci JM, Horowitz D, Zusman RM, Pisano JJ, Catt KJ, Keiser HR. The effect of converting enzyme inhibition with SQ 20881 on plasma and urinary kinins, prostaglandin E and angiotensin II in hypertensive man. Hypertension 1979; 1: 416-426.

29. Dusting R, Scherhag R, Landsberg G, Glanzer K, Kramer HJ. The converting enzyme inhibitor captopril stimulates prostacyclin synthesis by isolated rat aorta. Eur J Pharmacol 1983; 91: 501-504.

30. Beierwaltes WH, Carretero OA. Kinin antagonist reverses converting enzyme inhibitor-stimulated vascular prostaglandin I_2 synthesis. Hypertension 1989; 13: 754-758.

31. Sharma JN. The kallikrein-kinin system in hypertension. In: Renal Function, Hypertension and Kallikrein-Kinin System. Iimura O, Margolius HS, editors. Tokyo: University of Tokyo Press, 1988: 147-154.

AAS 38/III
Recent Progress on Kinins
© 1992 Birkhäuser Verlag Basel

ANGIOTENSIN CONVERTING ENZYMES FROM URINE OF TREATED AND UNTREATED ESSENTIAL MILD HYPERTENSIVE PATIENTS (EHP) WITH DIURETIC: PARTIAL PURIFICATION AND CHARACTERIZATION

Kaethy Bisan Alves[*], Dulce Elena Casarini[**], Rita Heloisa da Costa[*], Frida Liane Plavnik[*], Jorge Enrique Portela[**] and Odair Marson[**].

[*] Department of Biochemistry,Escola Paulista de Medicina, [**] Disciplina de Nefrologia, Escola Paulista de Medicina, Rua 3 de Maio, 100, Caixa Postal 20372, 04044, SP, Brasil.

SUMMARY: Urine of untreated EHP was eluted, on a ion-exchange chromatography, in two protein peaks with ACE activity , at 0.7 mS (B_I) and 1.25 mS (B_{II}), while urine of treated EHP, was eluted only in one peak with ACE activity (0.7 mS). B_I (Mr, 88 kDa) and B_{II} (Mr, 61 kDa) convert AI to AII, hydrolyze bradikinin, are inhibited by captopril, EDTA and metal ions.

INTRODUCTION

In the literature there are many papers showing that the urinary excretion of angiotensin converting enzyme (ACE) is altered in several kidney diseases (1,2,3). ACE is also involved with mechanisms related with blood pressure (sodium retention and hormone inactivation (4) and high levels of these enzymes are found in the urine of essential mild hypertensive patients (5). The present paper describes the partial, purification and some properties of angiotensin converting enzymes from urine of treated and untreated essential mild hypertensive patients with diuretic and a comparison with a control group.

MATERIAL AND METHODS

Three groups of patients were studied: A) patients with normal blood pressure (mean of 112.0/70.0 ±7.10/7.50), 5 men, 5 women, aging from 25 up to 40 years; B) mild hypertensive untreated patients, mean of blood pressure of 143.7/94.2 ±10.90/4.48, 1 male, 9 females, aging from 30 up to 50 years; C) mild hypertensive patients with blood pressure controled with chlortalidone for 2 months (50 mg/day), 10 women, aging from 30 to 50 years. Values of mean blood pressure were: 132.3/93.2 ± 4.61/8.97. All patients had normal renal function assessed by clearance of creatinine. They also showed normal chest X-Ray, renal angiography, blood sugar, serum potassium, EKG and plasma renin activity.

Angiotensin converting activity. a) The angiotensin converting activity was determined by the method of Santos et al. (6). Urine (300 ul) was incubated with 240 ul of assay solution containing 5 mM Hip-L-His-Leu (HHL) in 0.4 M sodium borate buffer, 0.9 M NaCl, pH 8.3, for 3 hours at 37°C. The reaction was stopped by addition of 1.2 ml of 0.28N NaOH. Then was added 100 ul of o-phthaldialdehyde (20 mg/ml in methanol), which was followed 10 minutes later by the addition of 200 ul of 3N HCl. The product His-Leu was quantified fluorometrically (365 nm excitation and 495 emission). Blanks were prepared by reversing the order of the addition of enzyme and NaOH. ACE activity was expressed as umol of His-Leu released per min/mg of protein. b) The convertion of Angiotensin I (AI) to angiotensin II (AII) was determined by bioassay using the isolated rat uterus (7). The uteri were collected from rats injected with diethyl-stilbestrol (100ug/100g) one day before the experiment.AI and AII (10ug) were incubated during 2 hours with enzyme samples (200ul) in 0.05 M Tris-HCl, pH 8.0 and 0.05 M NaCl in a final volume of 1.0 ml, at 37°C. The reaction was stopped by heating in a boiling water bath during 10 min.

Kininase activity. The isolated guinea pig ileum was used to monitor kininase activity. Bradykinin (5 nmol/ml) was incubated

with enzyme in 0.05 M Tris-HCl buffer, pH 8.0, at 37°C. Aliquots
(1 nmol) were removed at   0, 5, 10, 15 30 and 60 minutes of
incubation and diluted with Tyrode buffer to 1.0 ml. The
reaction was stopped mainteining the incubation mixture in
a boiling water bath during 10 min. Residual bradykinin
activity was determined by matching bioassay employing
the isolated guinea pig ileum bathed at 37°C in 10 ml
of tyrode buffer containing 1 mg/l atropine. Under these
conditions, one unit of kininase activity was defined as
the quantity of enzyme that hydrolyzes 1 nmol of bradykinin
per minute.

Protein determination. Absorbance at 280 nm in a 1.0 cm cuvette
was used to measure protein concentration in chromatographic
eluates.

Ion exchange chromatography. Each urine sample (180 mg of protein) after dyalize against 0.02 M sodium phosphate buffer,
pH 7.0, was chromatographed on a DEAE-cellulose D column,
equilibrated with   0.02 M sodium phosphate buffer, pH 7.0.
Elution was carried out with a linear gradient of 0.02 M to 0.5
M sodium phosphate buffer, pH 7.0. Angiotensin converting
activity was measured upon HHL as substrate.

Gel filtration. Bio-gel $A_{0.5m}$ (2 x 95 cm) and Sephadex G-50
(2 x 87 cm) columns were equilibrated and developed with 50 mM
Tris-HCl buffer, 150 mM NaCl, pH 8.0. The pools of active fractions obtained after gel fitration were lyophilized and dialyzed
against 50 mM Tris-HCl buffer, pH 8.o and submitted to an FPLC
on a Sepharose 12 (12 x 60 cm) column, equilibrated and developed with 50 mM Tris-HCl buffer, pH 8.0, containing 0.1 M NaCl.

Gel electrophoresis. SDS-polyacrylamide gel electrophoresis was
performed on 10% gel (17.0 x 14.5 x 0.2 cm) with non-reduced
samples, under the conditions described by Laemmli (8) and
stained with Coomassie Brilliant Blue.

Molecular weight. The molecular weight of the enzymes were
determined by SDS-polyacrylamide gel electrophoresis and gel
filtration chromatography (9) on a Bio-gel $A_{0.5m}$ column precalibrated with ovoalbumin (Mr,43 kDal), bovine serum albumine

(Mr, 67 kDa), catalase (Mr, 232 kDa) and ferritin (Mr, 440 kDa) or on Sephadex G-50 column precalibrated with ribonuclease (Mr, 13,7 kDal), soybean trypsin inhibitor (Mr, 21,5 kDa) and chymotrypsinogen (Mr, 25 kDal).

Inhibion studies. The inhibition studies of angiotensin converting activity were made using HHL as substrate. Four different inhibitors' concentrations (0.2, 0.4, 0.8 and 1.2 mM) were used with $CoCl_2$, $MnCl_2$, $CaCl_2$, $ZnCl_2$, EDTA and 0.01, 0.1, 0.3 and 0.5 mM with captopril. Inhibitors were preincubated during 30 min with the enzyme before the addition of the substrate.

RESULTS

Ion-exchange chromatographies of each urine from control group (A) and the group of patients treated with diuretic (C) were eluted in only one protein peak with angiotensin converting

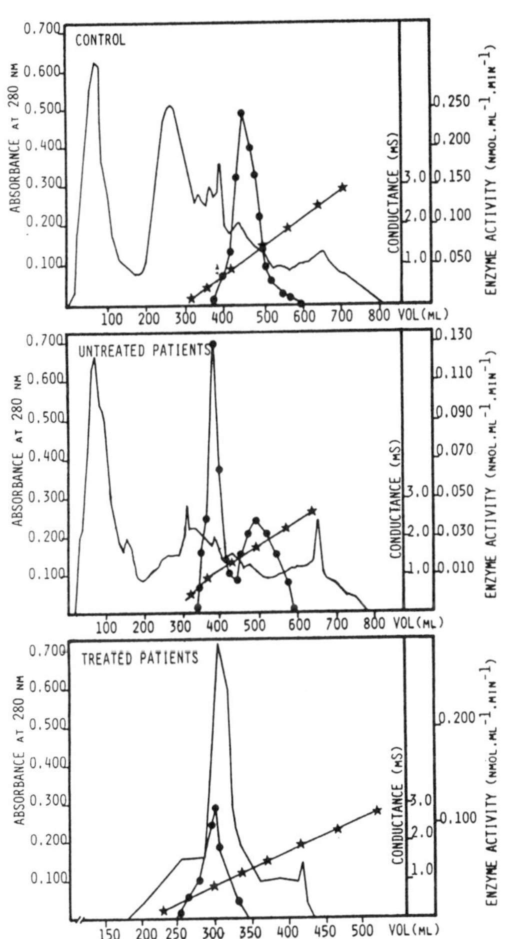

Fig. I - CHROMATOGRAPHY ON DEAE-CELLULOSE COLUMN. The chromatography was carried out as mentioned in Methods. (————), A_{280}/ml; (●——●), enzyme activity; (✶——✶), conductance.

TABLE 1.   ACE   ACTIVITY   UPON   HHL   AFTER   IONIC-EXCHANGE CHROMATOGRAPHY

ACE activity (nmols/min)	Control group A_I	EPH untreated		EPH treated C_I
		B_I	B_{II}	
mean	0.7	0.4	0.2	0.5
SD	0.34	0.14	0.05	0.23
SEM	0.07	0.03	0.01	0.05

SD, standard deviation; SEM, standard error of mean

activity at 0.7 mS (A_I and C_I respectively). The chromatographies of urine of untreated essential mild hypertensive group
(B) were eluted in two protein peaks with angiotensin converting
activities, at 0.7 mS (B_I) and 1.25 mS (B_{II}). Fig.I shows the
chromatogram of each group. The amount of angiotensin-converting
enzyme recovery after these chromatographies is shown in Table
1. Fractions , A_I B_I and C_I, with specific activity (SA, U/mg)
7.05, 0.08 and 2.0 respectively, were chromatographed on a
Bio-gel $A_{0.5m}$ column giving only one protein peak with ACE
activity upon HHL, ACE_{IA} (SA, 16.8) ACE_{IB} (SA, 5.2) and ACE_{IC}
(SA,10.2), respectively. Molecular weight of 88 kDal was found
for all these enzymes. Fraction B_{II} (SA,0.20) was chromatographed on a Sephadex G-50 column giving also only one protein
peak of ACE activity, ACE_{IIB} (SA, 0.6) with molecular weight of
61 kDal. All enzymes, ACE_{IA}, ACE_{IB}, ACE_{IC} and ACE_{IIB}, converted
angiotensin I to angiotensin II and hydrolyze bradykinin. The
effect of EDTA, captopril and the ions Ca^{2+}, Co^{2+}, Mn^{2+} and
Zn^{2+}, upon the angiotensin converting activity of the enzymes,
using HHL as substrate, are shown in   Table II.

DISCUSSION

Urine of each patient with essential mild hypertension (treated
and untreated with chlortalidone) and the control group were

submitted to an ion exchange chromatography. The chromatogram showed that only one protein peak with angiotensin converting activity was eluted at 0.7 mS, in control and treated groups. On the other hand, urine of untreated patients was eluted in two protein peaks with angiotensin converting activity: the first at the same condutivity as that obtained with urine of control and treated patients and the second at 1.25 mS.

TABLE 2. INHIBITION OF ACE ACTIVITY UPON HHL

INHIBITOR	INHIBITION (%)			
	CONTROL GROUP	EHP GROUP		EHP TREATED GROUP
	A_I	B_I	B_{II}	C_I
Ca^{2+} 4.0 mM	n.i.	n.i.	n.i.	n.i.
Co^{2+} 4.0 mM	100	100	100	100
Mn^{2+} 0.5 mM	100	100	100	100
Zn^{2+} 6,0 mM	100	100	100	100
EDTA 0.2 mM	100	100	100	100
Captopril 25 uM	100	100	100	100

n.i., no inhibition

The enzymes eluted at 0.7 mS had molecular weight of 88 kDa, hydrolyze bradykinin and convert angiotensin I to Angiotensin II, were not inhibited by ions Ca^{2+}, but were completely inhibited by EDTA, captopril, and ions Co^{2+}, Mn^{2+} and Zn^{2+}. The second enzyme found in the urine of untreated patients (eluted at 1.25 mS) had molecular weight of 61 kDa, converts AI to AII, hydrolyzes bradykinin, is totaly inhibited by EDTA, captopril, ions Co^{2+}, Mn^{2+}, and Zn^{2+} but not by ions Ca^{2+}. It seems that the enzymes eluted at 0.7 mS are the same for the three groups and are similar to the ACE previously described in human urine (10). Furthermore, our results show that a new form of active converting enzyme (Mr, 61 kDa) is also present in urine of untreated hypertensive patients. Comparing the results showed in Table I, we can observe that the total activity of angiotensin-converting enzyme was diminished in treated and untreated hypertensive patients. The amount of ACE activity in the treated patients' group is the sum of the two activities found in untreated group. Both enzymes found in the last group differ only in molecular weight. Further studies have to be done to determine whether this second enzyme is a fragment of the angiotensin converting enzyme normally found in urine or a new enzyme which appears with the hypertension and disappeares after treatment with diuretic.

ACKNOWLEDGEMENTS

This work was supported by FAPESP.

REFERENCES

1. Baggio B, Piccoli A, Favaro S, Antonello A, Bertaglia E
    Borsatti A. Urynary angiotensin-I-converting enzyme activity
    as a, marker of tubulo-interstitial involvement in kidney
    diseases. In: Biotechnology in Renal Replacement Therapy.
    Bonomini V, Scolari MP, Stefoni S, et al editors. Basel,
    Karger, 1989, 70: 208-212.

2. Maruhn D, Paar D, Bock KD. Lysosomal and brush border enzymes
    in urine of patients with renal artery stenosis and with
    essential hypertension. Clin. Biochem., 1979, 12: 228-230.

3. Kato I, Takada K, Nishimura K et al. Increased urinary
    excretion of angiotensin converting enzyme in patients with
    renal diseases. J.Clin.Chem. Clin. Biochem., 1982,20: 473-
    476.

4. Baggio B, Favaro S, Cantaro S, Bertazzo L, Funzio A and
    Borsatti A.Increased urine angiotensin converting enzyme in
    patients with upper urinary tract infection. Clinica Acta,
    1981; 211-218.

5. Bailie MD and Barbour JA. Effect of inhibition of peptidase
    activity on distribution of intrarenal blood flow.
    Am.J.Physiol., 228(3): 850-853.

6. Santos RAS, Krieger EM and Greene LJ. An improved flu-
    orimetric assay of rat serum and plasma converting
    enzyme. Hypertension, 1985, 7 (2), 244-252.

7. Borges DR, Limaos EA and Prado JL. Catabolism of vasoactive polypeptides by the perfused rat liver. Naunyn-Schmiedebergs Arch.of Pharmacol.,1976, 295: 33-40

8. Laemmli UK.Cleavage of structural proteins during the assembly of head of bacteriophage T4. Nature, 1970, 227: 680-685.

9. Andrews P Estimation of molecular weight of protein by gel filtration. Biochem J, 1964, 91: 222-233

10.Ryan J, Oza NB, Martin LC and Pena GA. Components of kallikrein-kinin system in urine. In: Kinin II, Biochemistry, Pathophysiology and Clinical Aspects, Fuji S, Mryia H and Suzuki T, editors. New York: Plenum Press. 1978, 10: 313-323.

AAS 38/III
Recent Progress on Kinins
© 1992 Birkhäuser Verlag Basel

COMPONENTS OF PLASMA AND TISSUE KALLIKREIN-KININ SYSTEM IN THE URINE OF PATIENTS WITH LATENT, NEPHROTIC AND HYPERTONIC FORMS OF GLOMERULONEPHRITIS

T.S.Paskhina, L.R.Polyantseva*, L.V.Platonova and G.O.Levina

Institute of Biological and Medical Chemistry, Pogodinskaya 10, 119832 Moscow, *Nephrological Laboratory of Moscow Sechenov Medical Academy

SUMMARY: Activities of main components of KKS were estimated in the urine of patients with latent, nephrotic and hypertonic forms of chronic glomerulonephritis (ChGN) and compared to those parameters in urine of healthy persons.The data obtained allow to make a conclusion concerning the pathogenetic and the compensatory role of plasma KKS in the nephrotic form of ChGN.

INTRODUCTION

Pathogenesis of glomerulonephritis - a renal disease of immunoinflammatory nature is influenced by the systems, responsible for microcirculation and enhanced vascular permeability, primarily on the kallikrein-kinin system (KKS).

Based on the notion that the urinary KKS components reflect, to a certain extent, the activity of this system in nephron structure and in kidney as a whole, we have carried out a detailed comparative study of distinct KKS components (tissue and plasma kallikreins, total kininase activity, kininases I and II, free kinins) in the urine of patients with 3 forms of chronic glomerulonephritis (ChGN) and in the urine of healthy individuals. The studies of such kind received no coverage in the literature.

MATERIALS AND METHODS

Examined were 44 ChGN patients (12 females and 32 males, aged 17 to 57 years); among them there were 13 patients with latent form (LGN), 18 with nephrotic syndrome (NS) and 13 with arterial hypertension form (AHF). The control group included 14 virtually healthy individuals.The study was done using DVal-Leu-Arg-pNA·2HCl (S-2266), DPro-Phe-Arg-pNA·2HCl (S-2302) and soyabean trypsin inhibitor (SBTI) (Serva, Germany), kallikrein inactivator (aprotinin) (Calbiochem, USA). DPhe-Phe-Arg-CH$_2$Cl was synthesized in our institute.

The activities of the KKS components were measured not in the native urine, but in the solution of the protein precipitates in 0.9% NaCl. Urinary protein acetone precipitates were obtained as in [1].

Kininogenase activity in urinary protein precipitates was determined by the amount of kinins liberated by urine kininogenases from heated human plasma as a kininogen source. Kinin content was determined using the isolated rat uterus horn with synthetic bradykinin (BK) (Reanal, Hungary) as standard [2]. Amidase activity of proteinases was estimated by the rate of hydrolysis of S-2266 and S-2302 as described in [3].

Total BK inactivating activity of urinary protein precipitates was measured by the rate of synthetic BK degradation [2]. Kininase I activity was determined by splitting of Bz-Gly-Lys [2]. Kininase II activity was determined fluorimetrically by splitting of Z-Phe-His-Leu [4].

Isolation of kinins from urine was carried out as in [5]. Separation and identification of kinins were performed on SP-Sephadex C-25 column [6].

RESULTS AND DISCUSSION

KALLIKREINS IN URINE. The studies done were based on the substrate-inhibitory analysis. Activities of plasma and tissue kallikreins were assayed by two enzymatic reactions: a highly specific kininogenase reaction and the amidase reaction (S-2266 or S-2302).

It was shown in model experiments with highly purified human urinary kallikrein and with human plasma kallikrein [7] that S-2266 has no selective specificity and is equally efficient for the measurement of both tissue and plasma kallikrein. Both enzyme hydrolyze S-2266 at the same rate with a V_{max} = 10 μmol/min·mg. The substrate S-2302 is also hydrolyzable by the two kallikreins but the rate of its hydrolysis by plasma kallikrein is 35-fold higher than that by tissue kallikrein (V_{max} = 70 μmol/min·mg and 2 μmol/min·mg, respectively).

Tables 1 and 2 summarize data on the kininogenase and amidase activities in donors' and patients' urine and show the effects of proteinase inhibitors on these activities.

Table 1. Kininogenase activity (M±m) in urine of healthy donors and patient with different nephrological syndromes

	Kininogenase activity		Inhibition by SBTI
	in 1 ml	in total daily volume	
	μg BK eq/hour		%
1. Donors (n=14)	0.70±0.14	800±150	0
2. LGN (n=13)	1.40±0.20	1860±260	7±3
3. NS (n=18)	2.90±0.30	3440±460	60±8
p_{2-1}	<0.01	<0.01	>0.05
p_{3-1}, p_{3-2}	<0.01	<0.01	<0.01

Donors urine is characterized by relatively low levels of the kininogenase and S-2266 amidase activities - an evidence of the presence of kallikrein (mostly tissue kallikrein because low S-2302 amidase levels point only to the traces of plasma kallikrein).

In LGN, the urine kininogenase, S-2266 and S-2302 amidase activities were elevated (2-, 4- and 30-fold, respectively); the elevation of the latter reflects the onset of plasma kallikrein penetration into the urine.

Sharp elevation of S-2266 and S-2302 amidase activities (15- and 10-fold, respectively) in NS patients as compared with LGN patients provides evidence of the already massive transport of plasma kallikrein and other trypsin-like plasma proteinases into the urine - as a result of enhanced renal filter permeability.

The amidase activities in the urine of patients with AHF were minimal and did not appreciably differ from those in the healthy donors' urine.

In all cases, enzymatic activities were inhibited by aprotinin ($8.7 \cdot 10^{-7}$ M) up to 90-100%. Data on inhibition of these activities by SBTI (10^{-5} M) and chlormetylketone ($5.5 \cdot 10^{-9}$ M) specify the contribution of plasma proteinases. A slight inhibition of proteinase activities in donors' urine is indicative of only trace amounts of these proteinases. As follows from the inhibitory assay data, in LGN and AHF patients, 10-14% of amidase activity is associated with the presence of plasma proteinases, including plasma kallikrein. In NS patients, the nearly 100% inhibition by SBTI of amidase activity is due to the trypsin-like plasma proteinases with plasma kallikrein accounting for 25% of this activity (chlormetylketone inhibition).

URINARY KININASES. The kininase activity in normal urine is determined by current data by the presence of kininase I (carboxypeptidase N), kininase II, and enkephalinase. From the Table 3 one can see that in LGN daily excreted BK-inactivating activity and activity of kininase II are enhanced 2.1- and 2.5-fold, respectively, whereas kininase I activity is 7 times higher than normal.

Abrupt elevation of all 3 forms of kininase activity was revealed upon assaying the urine of ChGN patients with NS: the BK-inactivating activity, as compared with that in the urine of LGN patients, increased 20-fold, the kininase I and II activities, 6.2- and 15-fold, respectively; as compared with those in the urine of healthy donors - 4-, 45- and 40-fold, respectively. These facts were interpreted as indirect evidence of a damage of definite sites and cells of the nephron in NS - a resultant effect of such diverse factors as an enhanced permeability of glomerules' capillary filter, dystrophy, and necrosis of vascular endothelium, dystrophy of the epithelium of the proximal compartment of the twisted tubule with exfoliation of brush border. In the urine of AHF patients, the kininase activity levels were the lowest (i.e. only 3-fold increased) as compared with those in other groups.

FREE URINARY KININS.It is known that free urinary kinins have renal origin . The levels of free kinins in urine characterize both the process of their formation in the kidney and the inactivation by kininases upon passage of kinins through the urine excreting system.

By current RIA data, healthy persons excrete 20-37 μg BK eq daily [8,9]. For the purpose of our investigation we have isolated kinins from daily urine of healthy persons and nephrotic patients using sorption method [5], with subsequent estimation of their daily excretion levels by the biotest.

Chromatography of the kinin fractions on SP-Sephadex C-25 column allowed us to separate the kinin mixture into individual components, identify them, and determine the content of each particular kinin by the biotest. The urine of all individuals tested (both healthy and affected) contains 3 kinins: BK, Lys-BK, and Met-Lys-BK. The kinin mixture of patients with AHF of ChGN was not subjected to separation.With healthy persons the value of kinin excretion in urine was found to be 6.6±0.9 μg BK eq (Table 4). The daily kinin excretion in ChGN patients with LGN,AHF and NS decreased constituting only 28, 48, and 42%, respectively, of those for healthy donors. The highest excretion of free kinins in the urine of the patients with AHF of ChGN correlated with the lowest kininase activity in these patients' urine. A lowering of the daily kinin excretion in the urine in renal diseases was observed by other authors as well; according to their data, ChGN patients excrete 2-5 μg BK eq daily [8] while in chronic renal deficiency the daily kinin excretion is only 0.9±0.4 μg BK eq [10]. Data on the qualitative composition of urine kinins are not available in the literature.

As known, the major urinary kinin is Lys-BK. Met-Lys-BK is formed by the action of uropepsin on urokininogen after urine excretion [5]; the formation of this kinin is facilitated by a specially prepared acid medium - favorable for kininase

Table 2. S-2266 and S-2302 amidase activities (M±m) in urine of healthy donors and patients with different forms of ChGN

	S-2266				S-2302			
	Amidase activity		Inhibition		Amidase activity		Inhibition	
	in 1 ml	in daily volume	SBTI	D-FFR-CH$_2$Cl	in 1 ml	in daily volume	SBTI	D-FFR-CH$_2$Cl
	nmol/min	μmol/min	%	%	nmol/min	μmol/min	%	%
Donors (n=14)	0.12±0.03	0.14±0.03	4.2±1.3	4.0±1.0	0.08±0.01	0.09±0.01	2.5±2.5	1.0±1.0
LGN (n=13)	0.46±0.09	0.64±0.13	14.5±0.1	7.0±1.5	1.2±0.2	1.6±0.3	78.0±11.0	9.0±1.0
NS (n=18)	7.0±2.2	8.0±3.0	91.0±4.0	26.0±6.0	47.0±8.0	57.0±9.0	88.0±12	27.0±5.0
AHF (n=13)	0.08±0.01	0.10±0.01	15.0±0.2	5.0±1.5	0.18±0.04	0.23±0.04	60.0±10	5.0±0.8

All p are significant excluding p$_{4-1}$

Table 3. Total BK-inactivating activity, kininase I and II activities (M±m) in urine of healthy donors and patients with different forms of ChGN

| | Total BK-inactivating activity | | Kininase I activity | | Kininase II activity | |
| | ng BK/min | | nmol Bz-GlyLys/min | | nmol Z-PheHisLeu/min | |
	in 1 ml	in daily volume	in 1 ml	in daily volume	in 1 ml	in daily volume
Donors (n=14)	1.7±0.2	2000±200	0.70±0.05	690±30	0.02±0.01	24.0±3.0
LGN (n=8)	4.0±0.8	4300±600	3.5±0.8	4900±900	0.04±0.01	57.0±7.0
NS (n=8)	71.0±17.0	81000±2800	28.0±5.0	31400±1500	0.70±0.10	860±110
AHF (n=13)	-	-	2.1±0.5	2800±100	-	-

All p are significant excluding p$_{4-2}$

inactivation - in the urine collector.

The kinin ratios of healthy persons and those of LGN and NS patients substantially differ (Table 4). Decrease levels of urinary kinins in LGN and NS are determined almost exclusively by the decrease in the absolute and relative Lys-BK content - this fact being in correlation with the lowered tissue kallikrein excretion in these pathologies. The stability of BK excretion in urine in spite of high urinary kininase activity (especially in NS patients) suggest the BK neogenesis in kidney and urine (against the background of ever increasing proteinuria) - at the expense of both plasma kallikrein and its substrate HMW kininogen.

Table 4.Urinary excretion of free kinins in healthy donors and patients with nephrological syndromes (μg BK eq/day)

| | Total kinin excretion | | Excretion of individual kinins | | | |
| | | | | BK | Lys-BK | Met-Lys-BK |
	n	M±m	n	M	M	M
Donors	16	6.6±0.9	5	0.53	4.62	1.45
LGN	13	1.7±0.3	4	0.32	0.85	0.53
NS	18	2.8±0.8	7	0.54	0.76	1.50
AHF	9	3.2±0.4	–	–	–	–

p_{2-1} <0.01
p_{3-1}, p_{4-1}, p_{4-2} <0.05
p_{3-2} >0.05

In spite of the known fact of the presence of [Hypro^3]Lys-BK in human urine we failed to find it in our species. It is highly likely that upon separation of kinins on the SP-Sephadex C-25 column, Lys-BK and [Hypro^3]-Lys-BK would not resolve. Whether or not this kinin is present in the daily urine of patients with 2 forms of ChGN is yet unclear.

CONCLUSION

Thus the substrate-inhibitor assay of enzymes together with the identification of different enzyme types, excreted in the urine of patients with different forms of ChGN allows the estimation of a particular contribution of the components of the two systems - the tissue (renal) and the plasma kallikrein-kinin systems - to their activity in urine.

In LGN, the excretion of tissue kallikrein is below normal, as indirectly evidenced by the lowest share of Lys-BK in the total amount of excreted kinins. Decreased excretion of tissue kallikrein in LGN may be due either to the destruction of the nephron's cellular structures responsible for the synthesis of this enzyme in kidney or to the binding of tissue kallikrein to proteinase inhibitors.

Low activity of kininases as well as the insignificant presence of plasma kallikrein in urine, both point only to the slightly damaged permeability of glomerular capillaries in patients with the hypertonic form of ChGN. At the same time the lowest excretion of tissue kallikrein in this group, as compared to all other groups of patients, is probably caused by its decreased synthesis in kidney as a result of a damage of the distal compartment of the nephron's tubule. The combination of these causes results in the below normal excretion of free kinins.

Maximal changes in the activity of all components in urine were revealed in patients with nephrotic form of ChGN. This form is characterized by a pronounced dysfunction of renal KKS: on the one hand, lowered tissue kallikrein levels; on the other, sharp elevation of kininase activity. Besides, it is with NS that the tissue KKS activity is supplemented by the substantial contribution of plasma KKS' component manifesting itself in a significant increase of plasma kallikrein content in urine, as well as in enhanced BK contribution and kininase I appearance.

These facts are in contradiction with the view of Cumming *et al.* [11] who consider total amidase (S-2266) activity in urine as being due to the kallikrein of renal origin. In contrast to this view, our data indicate a significant contribution of plasma

KKS' components to urinary kinin formation at the kidney level in NS. Also, these data allow to make a conclusion about the combination of the pathogenetic and the compensatory roles of plasma KKS in the nephrotic form of ChGN.

REFERENCES

1. Trautschold I, Werle E. Hoppe-Seyler's Z Physiol Chem 1961; 325:48-59.
2. Paskhina TS, Yegorova TP *et al*. In: Modern Methods in Biochemistry. Orekhovich VN, editor. Moscow, 1968:232-261.
3. Geiger R, Fritz H. Methods Enzym 1981; 80:466-493.
4. Yeliseyeva YuYe, Orekhovich VN, Pavlikhina LV. Vopr Med Khim 1976; 22:81-89.
5. Hial V, Keiser HR, Pisano JJ. Biochem Pharmacol 1976; 25:2499-2503.
6. Makevnina LG, Levina GO *et al*. Bioorgan Khim 1982; 8:1606-1614.
7. Nartikova VF, Yegorova TP *et al*. Biokhimia 1986; 51:463-471.
8. Abe K, Sato M *et* al. Kidney Int 1981; 19:869-880.
9. del Canizo-Gomez FJ, Duran F *et al*. Hormone Res 1984;20:143-149.
10. Weinberg MS, Azar P *et al*. Kidney Int 1985; 28:975-981.
11. Cumming AD, Robson JS. Nephron 1985; 39:206-210

THE RENAL KALLIKREIN-KININ SYSTEM AT THE PREHYPERTENSIVE STAGE OF HYPERTENSION

Kazuaki Shimamoto and Osamu Iimura

The Second Department of Internal Medicine, Sapporo Medical College,
S-1 W-16, Sapporo 060, Japan

SUMMARY: In essential hypertension, renal dopaminergic activity and PGE_2 was suppressed as was the renal kallikrein-kinin system. The suppression of the renal depressor systems may play an important role in the etiology of hypertension through sodium and body fluid retention. At the prehypertensive stage of essential hypertension, renal dopaminergic activity was suppressed, which induced the retention of body fluids and sodium. On the other hand, the renal prostaglandin E_2 was augmented, and it seemed to be a compensatory mechanism for sodium retention. We could not confirm the suppression of the renal kallikrein-kinin system at the prehypertensive stage of hypertension. Although all renal depressor systems are suppressed in essential hypertension, the suppression of the renal kallikrein-kinin system and PGE_2 seem to be very important in the etiology of hypertension. Further study is necessary to clarify when both systems are suppressed.

INTRODUCTION

In essential hypertension, the possibility exists that functional changes happen within the sympathetic nervous system, renin-angiotensin-aldosterone system, kallikrein-kinin system, prostaglandin (PG) system, atrial natriuretic peptide or other substances. These disturbances might provoke the abnormality of cardiovascular function and/or of water-sodium metabolism, which causes high blood pressure. In terms of the renal kallikrein-kinin system in essential hypertension, a clear reduction of urinary kallikrein excretion has been reported by Margolius et al (1) and other researchers (2-4).

In this manuscript, the pathophysiological significance of this system at the prehypertensive stage of essential hypertension and in the experimental hypertension of reduced renal mass hypertensive rats is discussed

RENAL KALLIKREIN-KININ, PROSTAGLANDIN AND DOPAMINE SYSTEMS IN ESSENTIAL HYPERTENSION

The renal kallikrein-kinin system is composed of several factors; prekallikrein, kallikrein, kininogen, kinin and kininases. The daily excretions of both kallikrein quantity and activity, total kallikrein, prekallikrein, and kinin were significantly lower in both the normal renin and low renin groups than in normal control (4,5). The daily excretions of total kininase and kininase activity without kininase II were significantly higher in both the normal renin and low renin groups, but kininase II was not (6). In comparing the normal renin group and low renin group (4,5), the urinary excretions of kallikrein activity and kinin were lower in the low renin group than in the normal renin group, and that of total kininase and kininase activity without kininase II were higher in the low renin group than in the normal renin group (6). It has been reported from our laboratory that neutral endopeptidase occupies a large part of the non-kininase II kininases and that urinary excretion of neutral endopeptidase is significantly higher in essential hypertensives, suggesting that neutral endopeptidase has a very important place in kinin metabolism (7,8).

On the other hand, in essential hypertension, renal dopaminergic activity and the PGE2 system were also clearly suppressed (9) as was the renal kallikrein-kinin system. There is a close relationship between the renal depressor systems composed of the dopamine, kallikrein-kinin and prostaglandin systems, and therefore the suppression of renal dopaminergic activity may be involved in the suppression of the renal kallikrein-kinin system (9). These data suggest that the suppressed renal depressor systems may play an important role in the etiology of essential hypertension through sodium and body fluid retention.

RENAL KALLIKREIN-KININ SYSTEM AT THE PREHYPERTENSIVE STAGE OF ESSENTIAL HYPERTENSAION

The question has been raised whether the suppression of the renal kallikrein-kinin system is the cause of hypertension or the result of hypertension. Zinner et al (10) and

Berry et al (11) reported that the suppression of renal kallikrein occurs at the prehypertensive stage of essential hypertension. However, their studies were done using out-patients. Therefore, 24 hours urine was not collected and sodium intake was not constant in their subjects. In studies done by our hospital,the renal kallikrein-kinin system in the prehypertensive stage of essential hypertension has been investigated using hospitalized subjects (12,13).

Ten normotensives with a family history (FH) of essential hypertension who had hypertensive parents or a hypertensive parent and at least one hypertensive grandparent of the other parent, and 10 normotensives without FH who had normotensive parents, were employed in the present study. Various parameters were determined and a dopamine infusion study was performed. All subjects were hospitalized and received a diet containing 120 mEq of Na and 75 mEq of K/day.

At the basal states of hemodynamics and renal water-sodium metabolism, no significant difference was found in mean arterial pressure(MAP), pulse rate(PR), urinary excretion of sodium (UNaV), creatinine clearance (Ccr) nor fractional excretion of sodium (FENa) between the two groups (12)(Table 1). After dopamine infusion, UNaV and FENa in both groups, and HR in FH(+) were significantly increased. The changes in PR,UV,UNaV and FENa were significantly higher in FH(+) than in FH(-) (11). Considering the mechanism of downward regulation in the renal dopamine receptor, these responses suggest that the dopaminergic system is suppressed in FH(+) at the prehypertensive stage. In fact, urinary excretion of dopamine was significantly lower in normal subjects with FH(+) than in those without FH (Table 1)(12).

In the basal state, we could not find any difference in urinary kallikrein-quantity, kallikrein-activity, kallikrein-specific activity nor kinin between the two groups (12,13). These findings were different from the early reports of Dadone et al (14), Zinner et al (10) and Berry et al (11), indicating a hereditary suppression of the renal kallikrein-kinin system. We cannot clearly explain this difference in evaluation of the renal kallikrein-kinin system at the prehypertensive stage, although the protocols of their studies were different from ours. Unexpectedly, urinary prostaglandin E$_2$ excretion was significantly higher in FH(+) than in FH(-)(12,13). These data suggest that an accelerated renal prostaglandin system may act as a compensatory mechanism for a latent disorder of renal sodium metabolism in the prehypertensive stage, and may be suppressed when accompanied by the progression of essential hypertension.

In FH(-), infused dopamine significantly augmented the urinary excretion of kinin and PG E$_2$, but did not significantly change urinary kallikrein-quantity , -activity nor - specific activity. On the other hand, in FH(+), the infusion of dopamine significantly augmented urinary kallikrein-activity, kallikrein-specific activity, kinin and PGE$_2$ with the exception being urinary kallikrein-quantity (12,13). Changes in urinary kallikrein-activity, -specific activity, kinin and PGE$_2$ were significantly higher or tended to be

Table 1
Basal level of variables in normotensive subjects with or without a family history of hypertension

		FH (-)	FH (+)
Body weight	(Kg)	67.7 ± 3.9	72.7 ± 6.2
Heart rate	(bpm)	58 ± 4	57 ± 3
MAP	(mmHg)	82.7 ± 2.8	77.3 ± 3.8
Ccr	(ml/min)	103.3 ± 6.5	106.5 ± 5.0
UNaV	(mEq/day)	100.6 ± 11.3	93.5 ± 11.3
pE	(pg/ml)	26.5 ± 2.1	32.6 ± 5.5
pNE	(pg/ml)	94.2 ± 9.4	118.7 ± 17.6
PRA	(ng/ml/min)	1.21 ± 0.30	1.31 ± 0.31
pAT-II	(pg/ml)	24.3 ± 12.3	12.9 ± 4.0
PAC	(pg/ml)	91.2 ± 14.8	97.2 ± 16.1
uDA	(ng/min)	288.0 ± 5.8	234.8 ± 18.2*
uKIN	(ng/min)	11.9 ± 1.9	15.1 ± 1.9
uKAL-A	(ng/min/min)	530.5 ± 92.0	494.6 ± 78.2
uKAL-Q	(ng/min)	121.1 ± 19.3	125.9 ± 20.0
uKAL-Sp		4.5 ± 0.4	3.9 ± 0.3

MAP:mean arterial pressure, Ccr:creatinine clearance, UNaV:urinary excretion of sodium, pE:plasma epinephrine, pNE:plasma norepinephrine, PRA:plasma renin activity, pAT-II:plasma angiotensin II, PAC:plasma aldosterone concentration, uDA:urinary dopamine, uKIN:urinary kinin, uKAL-A:urinary kallikrein activity, uKAL-Q:urinary kallikrein quantity, uKAL-Sp:urinary kallikrein specific activity
* p<0.05,

higher in FH(+) than FH(-), while the change in urinary kallikrein-quantity was not significantly different in the two groups. These responses of the kallikrein-kinin system and PGE2 also suggest that the dopaminergic system is suppressed in FH(+) at the prehypertensive stage. The stimulation of renal PGE2 synthesis by infused dopamine may be partly explained by the increase of kinin.

Thus, at the prehypertensive stage of essential hypertension, renal dopaminergic activity was suppressed, which induces the retention of body fluids and sodium. On the other hand, the renal PGE2 activity was augmented, and it seemed to be a compensatory mechanism for sodium retention. We could not confirm the suppression of renal kallikrein-kinin system at the prehypertensive stage of hypertension. Although all renal depressor systems are suppressed in essential hypertension, the suppression of the renal kallikrein-kinin system and PGE2 seems to be very important in the etiology of hypertension. Further study is necessary to clarify when both systems are suppressed.

RENAL KALLIKREIN-KININ SYSTEM IN PREHYPERTENSIVE STAGE OF REDUCED RENAL MASS HYPERTENSIVE RATS

In several hypotheses regarding the etiology of essential hypertension, the genetical disturbance of renal sodium handling seems to be the first step in the initiation of hypertension. Therefore, we used reduced renal mass hypertensive rats as kidney dysfunction models ,and the role of the renal kallikrein-kinin and prostaglandin systems on the etiology of hypertension was investigated in this experimental hypertensive animal model (15).

Male Sprauge-Dawly rats (150-200 g) were divided into 4 groups, control, 3/6 reduced renal mass (RRM), 4/6 RRM and 5/6 RRM hypertensive rats. In all rats, heminephrectomy or sham operations were done. One week after operation, additional subtotal nephrectomy or sham operations were performed. After two weeks as a control period, 1 % saline was administered to each rat. Then urine collection and blood pressure measurements were done every week during the next 4 weeks. Urinary kallikrein and PGE2 were also determined

UNaV increased significantly in each group after 1 % saline administration. In control, 3/6 and 4/6 RRM groups, UNaV decreased after 2 weeks of sodium administration, but a remarkable increase of UNaV was observed in 5/6 RRM rats during sodium loading. No significant change in blood pressure was found in the control,3/6 nor 4/6 RRM groups after the 1 % saline administration. However, in 5/6 RRM hypertensive rats, blood pressure increased significantly at one week after the 1 % saline administration.

In each group, urinary excretion of kallikrein was significantly lower in each group than in control at the basal state, in the order of the 3/6, 4/6, and 5/6 RRM rats with 5/6 RRM rats being the lowest. After 1 % saline administration, urinary kallikrein excretion did not show any significant changes , and the same tendency was maintained during the following 4 weeks. A negative correlation was observed between urinary kallikrein and blood pressure. Thus, the suppressed activity of renal kallikrein must have been very important in inducing the hypertension through water and sodium retention after the 1% saline administration.

On the other hand, urinary excretion of PGE2 was increased in each group at the basal state. After 1 % saline administration, urinary PGE2 increased in all groups. However, the increments of urinary PGE2 were largest in the 5/6 RRM group, followed by 4/6, and 3/6 RRM rats with control having the lowest increase. Significant positive correlations were observed in these relationships. These results suggest that renal PGE2 may act in a compensatory manner offsetting the water and sodium retention and the blood pressure elevation.

CONCLUSION

From these studies, we concluded that the renal kallikrein-kinin systems may contribute to the etiology of hypertension in these volume dependent hypertensive animals through body fluid and sodium retention by its suppression at the prehypertensive stage.

REFERENCES

1. Margolius HS, Geller RG, Pisano JJ et al., Altered urinary kallikrein excretion in human hypertension. Lancet 1971; 2:1063-65.

2. Seino M, Abe K, Ohtsuka Y et al., Urinary kallikrein excretion and sodium metabolism in hypertensive patients. Tohoku J Exp Med 1975; 116:359-67.

3. Levy SB, Lilley JJ, Frigon RP et al., Urinary kallikrein and plasma renin activity as determinants of renal blood flow. J Clin Invest 1977; 60:129-38.

4. Shimamoto K, Ura N, Nakao T et al., Role of the renal kallikrein-kinin system in sodium metabolism in normotensives and essential hypertensives. NZ Med J 1983; 96:905-7.

5. Iimura O, Shimamoto K, Ura N et al., Study on the renal kallikrein-kinin system in normal and low renin subgroups of essential hypertension. J Hypertens 1984; 2(Suppl 3):297-99.

6. Ura N, Shimamoto K, Tanaka S et al., Urinary excretions of kininase I and kininases II activities in essential hypertension: a sensitive and simple method for its kinin-destroying capacity. J Clin hypertens 1985; 1:15-22.

7. Ura N, Shimamoto K, Satoh S et al., Renal kininase I, kininase II and neutral endopeptidase 24.11 in hypertensive diseases. Hypertension: submitted.

8. Ogata H, Ura N, Shimamoto K et al., A sensitive method for differential determination of kininase I,II and neutral endopeptidase (NEP) in human urine. Adv Exp Med Biol 1989; 247B:356-62.

9. Iimura O, Shimamoto K , Ura N et al., The pathophysiological role of renal dopamine, kallikrein-kinin and prostaglandin systems in essential hypertension. Agents and Actions 1987; 22:247-56.

10. Zinner SH, Margolius HS, Rosner B et al., Familial agregation of urinary kallikrein concentration in childhood: relation to blood pressure, race and urinary electrolytes. Am J Epidemiol 1976; 104:124-32.

11. Berry TD, Hassted SJ, Hunt SC et al., A gene for high urinary kallikrein may protect against hypertension in Utah kindreds. Hypertension 1989; 13:3-8.

12. Iimura O, Shimamoto K, and Ura N. Dopaminergic activity and water-sodium handling in the kidneys of essential hypertensive subjects: Is renal dopaminergic activity suppressed at the prehypertensive stage ? J Cardiovasc Pharmacol 1990; 16(Suppl 7): S56-S58.

13. Shimamoto K, Ura N, Nishimura M et al., Renal kallikrein-kinin, prostaglandin and dopamine systems in young normotensive subjects with a family history of essential hypertension. Hypertens Res: submitted.

14. Dadone MM, Smith JBM, Anderton DL et al., Evidence for environmental familiarity of kallikrein excretion in Utah kindred. West J Med 1986; 144:559-63.

15. Nakagawa H, Shimamoto K, Nakagawa M et al., The role of endogenous digitalis-like substance and renal depressor systems in the pathogenesis of hypertension. Sapporo Med J: submitted.

AAS 38/III
Recent Progress on Kinins
© 1992 Birkhäuser Verlag Basel

LACK OF ORAL KALLIKREIN IN LOWERING SYSTEMIC BLOOD PRESSURE IN PRIMARY HYPERTENSION

G. Bönner[1], C. Toussaint[1], M. Claus[3], E. Fritschka[3],
T. Philipp[3], H. Vetter[2], H. Feltkamp[4]

[1]Department of Internal Medicine II, University of Cologne, Merheim Hospital,
Ostmerheimer Str. 200, DW 5000 Köln 91, F.R.G.
[2]Department of Internal Medicine/Policlinic, University of Bonn,
[3]Department of Internal Medicine/Nephrology, University of Essen,
[4]Division Clinical Investigations, Bayer AG, Wuppertal

Abstract

Primary hypertension is associated with a lack in renal kallikrein activity which might be one of the reasons for the blood pressure elevation. Some smaller and partially uncontrolled studies suggested that an oral substitution of glandular kallikrein lowers blood pressure by a kinin-mediated vasodilation and increased natriuresis. To test this hypothesis we treated in two studies over 100 patients with untreated mild to moderate primary hypertension (WHO I-II) for 5 resp. 12 weeks in a double blind randomized and placebo controlled manner with 1800 U glandular kallikrein orally. Blood pressure measurements were performed according to the two study designs after 3 and 5 resp. 8 and 12 weeks of treatment sphymomanometrically in the day time course. No significant changes in blood pressure by kallikrein treatment could be observed at any time. Neither renal kallikrein excretion, renin and ACE-activity nor blood glucose concentration in diabetics or non-diabetics was changed. Thus, we could undoubtedly demonstrate that oral applied glandular kallikrein has no effect on primary hypertension.

Introduction

Today we have a multitude of substances to lower blood pressure as drugs of first choice. Nevertheless some of these antihypertensive drugs develop side-effects which render such therapy not the best. In this regard natural human substances which lower blood pressure by physiological mechanisms could have some advantages as they are probably free of undesired effects in comparison to synthetic pharmaceutics. The kallikrein-kinin system is one of the most potent endogenous vasodilating principles lowering blood pressure (3). Exogenous kinins are peptides and are effective only if injected parenterally; they cannot be used for longterm oral therapy. On the other hand oral antihypertensive treatment with active glandular kallikrein, which will generate kinins in vivo, seems to be more promising because the enzyme can be resorbed in biological active form by the intestine after oral application (7, 12, 15, 23).

Until now kallikrein is used only in treatment of male infertility (22). However, recent investigations give a hint for some blood pressure lowering effect of orally applied kallikrein (9, 17, 18, 19, 27) where the kinin-mediated stimulation of the prostaglandin system especially the endothelial prostacyclin seems to play a role (16). In the previously published studies the results were not comparable in regard to blood pressure lowering effect as well as hormonal changes observed in urine. Therefore we studied again the effect of orally applied pig pancreatic kallikrein on blood pressure in primary hypertension.

Material and Methods

We performed two studies, both randomized, double blind and placebo controlled.

In the <u>first study</u> 60 patients of both sex (43m/17f, mean age 52 $\pm$ 3 years, mean weight 81 $\pm$ 2 kg, mean blood pressure 159/101 $\pm$ 3/1 mmHg) with untreated primary mild to moderate primary hypertension (WHO-stage I-II) were recruited from hospital. The systolic blood pressure had to be less than 200 mmHg. Serious concommitant diseases, renal or hepatic insufficiency as well as cardiac arrhythmias were excluded. The patients were observed for 24 hours under clinical conditions. The blood pressure was measured sphymomanometrically with a mercury manometer 11 times after a period of rest (5 min) in a sitting position (at 8, 9, 10, 11, 12, 14, 16, 18, 20, 24 and again at 8 o'clock). Blood pressure measurements were performed twice in an interval of 2 min, the mean value was used. All measurements were taken by the same investigator. Additionally, urinary sodium, potassium, creatinine and kallikrein excretion in 24-h-urine samples was determined on every day of investigation. Further parameters of liver and renal function, blood glucose, cholesterol, triglycerides, and hematologic parameters as well as an ECG were checked at the start and the end of the study. The patients took 1800 U highly purified pig pancreatic kallikrein (Bay d287) or placebo orally once a day for 3 weeks, some of them additional for another 2 weeks (this means 5 weeks altogether (see table 1)) in a double blind randomized manner. After 3 and 5 weeks all measurements of blood pressure and other investigations were repeated.

In the <u>second multicenter study</u> 40 out-patients (28m/12f, mean age 42 $\pm$ 2 years, mean weight 79 $\pm$ 3 kg, mean blood pressure 149/103 $\pm$ 3/1 mmHg) with the same including and excluding criterias as study I were recruited. After 2 weeks of wash-out all patients were treated with placebo for another 2 weeks. After that time patients were treated with 900 U kallikrein or placebo for 12 weeks in a double blind and randomized fashion. Blood pressure was determined 5 times in an interval of 2 min sphymomanometrically at the first day and after 8 and 12 weeks by automatical measurement (Tonoprint). The dose of each therapy was doubled after 8 weeks (kallikrein to 1800 U) if the blood pressure was not lowered sufficiently (diastolic blood pressure still over 90 mmHg). This happened to 17 patients in the verum group and to 18 patients in the placebo group. The following biochemical methods were used: The excretion of sodium and potassium in the 24-h-urine-samples were determined by a flame-photometer, the concentration of creatinine by the method of Jaffé, the concentration of glucose enzymatically by glucose-oxidase test, the activity of urine kallikrein by amidolytic assay (2) according to Amundsen (1), the concentration of ACE by radioassay (21) and plasma renin concentration by radioimmunoassay according to Hummerich (11). The statistical evaluation was performed by Mann-Whitney U-test (p<0,05). The given values are mean values $\pm$ standard error of mean (SEM).

Results

Study I

The mean day-time blood pressure before therapy was identical in the control and the kallikrein-group (Co 158/102 $\pm$ 3/1 mmHg; Kal 159/102 $\pm$ 3/1 mmHg). After 3 or 5 weeks treatment no significant change in mean day-time blood pressure was observed in both groups (see table 1). In one patient of each group study had to be stopped in the early treatment phase since blood pressure rose over study limit. Two patients of each group were non-compliant and did not come to the last study visit. Major side effects were reported neither in the control nor in the kallikrein group.

Table 1: Comparison of day-time mean blood pressure (study I) and of office blood pressure (study II) (systolic/diastolic) before and after therapy. $\bar{x} \pm$ SEM, number of patients.

		blood pressure (mmHg)		
		kallikrein		**placebo**
STUDY I				
3 weeks	before	$158\pm3/102\pm1$	27	$159\pm3/102\pm1$ 27
	after	$159\pm3/101\pm2$	27	$158\pm3/101\pm2$ 27
5 weeks	before	$159\pm6/102\pm3$	9	$159\pm4/102\pm2$ 15
	after	$156\pm7/99\pm4$	9	$157\pm5/103\pm3$ 15
STUDY II				
12 weeks	before	$153\pm3/101\pm1$	19	$149\pm3/103\pm1$ 21
	after	$153\pm3/99\pm2$	19	$144\pm3/98\pm2$ 21

Renal kallikrein excretion (Table 1), creatinine clearance (table 1) as well as sodium and potassium excretion and urinary volume remained unchanged throughout the study. Plasma renin and angiotensin I-converting enzyme activity was found unchanged too.

Blood glucose concentrations were 97 ± 3 mg/dl in the control (n = 24) and 98 ± 2 mg/dl (n = 24) in the kallikrein group before treatment and was not effected by the 5 week treatment (Co:95 $\pm$ 2 mg/dl; Kal: 96 $\pm$ 5 mg/dl). Even in the diabetic hypertensives (6 in each group) no changes in blood glucose levels were observed.

Study II
In study II the baseline blood pressure was similar to that measured in patients of study I (Co: 149/103 $\pm$ 3/1 mmHg; Kal: 153/101 $\pm$ 3/1 mmHg). Treatment with kallikrein even over 12 weeks had no effect on blood pressure too (table 1).

Kallikrein excretion and creatinine clearance remained stable throughout the whole treatment period, as already observed in study I (see table 2).

Table 2: Changes in kallikrein-excretion (3rd and 5th week related to creatinine-excretion: dUkal/Ucrea; 12th week as (24-h-excretion: dUkal/24h), creatinine-clearance (dcrea-clearence, ml/min) and day time blood pressure (dRR) by therapy. $\bar{x} \pm$ SEM

	kallikrein	placebo
dUkal/Ucrea and dU/24 h		
STUDY I		
3 weeks (U/g)	-0,1±0,18	-0,1±0,13
5 weeks (U/g)	-1,2±0,34	+0,2±0,14
STUDY II		
12 weeks (U/24 h)	-0,1±0,10	-0,1±0,00
dcrea-clearence (ml/min)		
STUDY I		
3 weeks	-0,4±0,2	-0,9±0,2
5 weeks	-1,3±0,2	±0,0±0,3
STUDY II		
12 weeks	-0,1±0,0	+0,1±0,1
dRR (systol/diastol)(mmHg)		
STUDY I		
3 weeks	-0,1±2/-0,5±1	-0,9±2/-0,1±1
5 weeks	-3,2±4/-2,4±1	-1,9±2/+1,1±2
STUDY II		
12 weeks	+0,0±3/-2,0±1	-5,0±3/-5,0±2

Discussion

As shown in three larger epidemiological studies (24, 26, 28) the kallikrein gen was suggested to have some protective effect against high blood pressure. Thus, renal kallikrein seems to play a role in so called saltsensitivity (the elevation of blood pressure after oral sodium intake) since even in normotensive subjetcs blood pressure reacts after sodium intake in a significant inverse manner to the activity of renal kallikrein (6). In established hypertension we find a markedly lowered activity of renal kallikrein which in addition can also be less stimulated in comparison to healthy persons with normal blood pressure (3, 10, 14, 19). As the renal kallikrein-kinin system develops a local natriuretic and diuretic activity and can act systemically as potent vasodilator a lack in the activity of this system might effect negatively the volume and sodium homoeostasis as well as the peripheral vascular resistance. Sodium and water retention and increased peripheral vascular resistance therefore will be the consequence of reduced renal kallikrein activity and might play an important role in the pathogenesis of arterial primary hypertension.

As a possible attempt of endogenous compensation an elevated level of kallikrein in saliva of hypertensiva patients was found by other investigators (8). Therefore experimental studies showed that orally applied kallikrein can actually be resorbed in the intestine in a biological active form (7, 20, 23) and a positive effect of orally applied kallikrein on male infertility can be observed (22). In regard to these results it was obvious to investigate whether the lack of endogenous renal kallikrein in hypertension could be compensated by exogenous kallikrein and whether by such therapy blood pressure of hypertensives could be normalized.

In both studies, which we carried out in a double blind randomized placebo controlled and uni- respective multicenter manner, kallikrein was applied in doses as it was used in other hypertension studies too. Nevertheless we did not observe any blood pressure lowering effect of the oral therapy in our patients with primary hypertension neither in occasional blood pressure nor in mean systolic or diastolic day-time blood pressure.

These findings are contradictory to four other studies (9, 17, 19, 27) which described a favourable effect of oral substitution of kallikrein on blood pressure. These studies are partly difficult to evaluate. For example two working groups describe a decrease in blood pressure by an oral therapy of kallikrein in a small number of hypertensives (17, 18). There were no control groups in both studies, therefore it could not be differentiated between the known placebo effect of repeated blood pressure measurements and a real therapeutical effect of the drug. A dose-dependent kallikrein-specific effect with three different doses of kallikrein could be shown in a single blind placebo controlled study (9). The rate of responders (90%) and the dimension of the blood pressure decrease (mean:21/17 mmHg) by the same dose of applied kallikrein, which did not show any response in our studies, are striking high and were much higher than the results of all other established antihypertensive drugs. This unbelievable high efficacy of oral kallikrein is contradictory to all other investigations with oral substitution of kallikrein and cannot be reproduced in our studies. Another placebo controlled study, the only older double blind study, describes a blood pressure lowering effect of oral kallikrein in primary hypertension, but this effect is less clear in comparison to the discussed single blind study above (19). An increase of renal prostaglandin E_2-excretion as biochemical effect of the kallikrein therapy was found, too. The initially described increase in excretion of renal kallikrein (18) had to be retracted by the same working group later on in a second evaluation of the study (19). Another problem of the later investigation could be seen in the small number of patients (n= 20) and the blood pressure measurements performed only as occasional blood pressure.

Our results based on a therapy with the highest doses of kallikrein and the greatest number of patients show that orally applied highly purified kallikrein of pig pancreas cannot sufficiently lower blood pressure in primary hypertension. Accordingly, the discussed effects of kallikrein on other parameters as glucose metabolism, activity of renin, creatinine clearance or renal kallikrein excretion itself are missed too in our investigations (13, 17, 25).

Kallikrein of pig pancreas is resorbed by small intestine in an active form (7, 12, 15, 23) and can be successfully used in treatment of male infertility (22). The in our studies observed negative effects on blood pressure exclude, however, the oral application of kallikrein in therapy of primary hypertension. But it still remains unknown, whether the kallikrein-mediated release of circulating kinins really achieve a sufficient concentration for a physiologic blood pressure reaction or if they are inactivated too fast by kininases I and II to develop a systemic effect. Therefore the oral therapy of kallikrein cannot be rejected on principle and it had still to be clarified whether the use of kallikrein in combination with an ACE-inhibitor, which can attenuate the inactivation of kinins in vivo (4), will develop some advantages for antihypertensive treatment.

Acknowledgement

We thank Mrs. K. Sommer and Mrs. R. Chrosch for their excellent and reliable technical assistance. We want to thank also Bayer AG (Dr. P. Huber) for preparation and supply of the highly purified pig pancreatic kallikrein.

References

1.   AMUNDSEN, E., J. PUETTER, P. FRIBERGER, M. KNOES, M. LARSBRATEN, G. CLAESON, Methods for determination of glandular kallikrein by means of a chromogenic tripeptide substrate. Adv. Exp. Med. Biol. 120 A, 83-96 (1979).

2.   BÖNNER, G., M. MARIN-GREZ, Measurement of kallikrein activity in urine of rats and man using a chromogenic tripeptide substrate. J. Clin. Chem. Biochem. 19, 165-8 (1981).

3.   BÖNNER, G., Kallikrein. In: D. Ganten, E. Ritz, Lehrbuch der Hypertonie. Schattauer Verlag, Stuttgart, 231-42 (1985).

4.   BÖNNER, G., Haben die Kinine eine Bedeutung für die antihypertensive Wirkung der ACE-Hemmer? Z. Kardiol. 77, Suppl. 3, 23-7 (1988).

5.   BÖNNER, G., S. PREIS, U. SCHUNK, C. TOUSSAINT, W. KAUFMANN, Hemodynamic effects of bradykinin on systemic and pulmonary circulation in healthy and hypertensive humans. J. Cardiovasc. Pharmacol. 15, Suppl. 6, S 46-56 (1990).

6.   BÖNNER, G., B. THIEVEN, C. TOUSSAINT, W. KAUFMANN, Urinkallikrein, eine genetische Determinante des Blutdrucks? Hochdruck 10, 42 (1990).

7.   FINK, E., T. DIETL, J. SIEFERT, H. FRITZ, Studies on the biological function of glandular kallikrein. Adv. Exp. Med. Biol. 120 B, 261-73 (1979).

8.   HEIDLAND, A., A. ROECKEL, G. SCHMIDT, Salivary kallikrein excretion in hypertension. Klin. Wschr. 57, 1047-52 (1979).

9.   HOFFMANN, J., Antihypertensive Wirkung oraler Therapie mit Kallikrein. Medizinische Welt 41, 193-7 (1990).

10.   HOLLAND, O. B., J. M. CHUB, H. BRAUNSTEIN, Urinary kallikrein excretion in essential and mineralocorticoid hypertension. J. Clin. Invest. 65, 347-56 (1980).

11.   HUMMERICH, W., D.K. KRAUSE, Improvement of renin determination in human plasma using a commonly renin standard in a radioimmunological method. Klin. Wschr. 53, 553-69 (1975).

12. LIEBOW, C., S. S. ROTHMANN, Enteropancreatic circulation of digestive enzymes. Science 189, 472-4 (1975).

13. MADEDDU, P., M. OPPES, A. SORO, P. DESSI'-FULGHERI, N. GLORIOSO, F. BANDIERA, F. MANUNTA, S. RUBATTU, A. RAPELLI, The effects of aprotinin, a glandular kallikrein inhibitor, on renin release and urinary sodium excretion in mild essential hypertensives. J. Hypertens. 5, 581-6 (1987).

14. MARGOLIUS, H. S., D. HORWITZ, J. J. PISANO, H. R. KEISER, Urinary kallikrein excretion in hypertensive man. Relationship to sodium intake and sodium-retaining steroids. Circ. Res. 35, 820-5 (1974).

15. MORIWAKI, C., K. MORIYA, K. YAMAGUCHI, K. KIZUKI, H. FUJIMORI, Intestinal absorption of pancreatic kallikrein and some aspects of its physiological role. In: Kininogenases. Kallikrein (G. L. Haberland, J. W. Rohen, T. Suzuki, eds.) S 57-66, Schattauer Verlag, Stuttgart, New York (1972).

16. MÜLLER, H. M., A. OVERLACK, K. O. STUMPE, R. KOLLOCH, C. RESSEL, Increased Prostaglandin E2 excretion with oral kallikrein treatment in hypertensive and normotensive objects. Adv. Exp. Med. Biol. 156, 1005-9 (1983).

17. OGAWA, K., T. ITO, M. BAN, M. MOCHIZUKI, T. SATAKE, Effects of orally administered glandular kallikrein on urinary kallikrein and prostaglandin excretion, plasma immunoreactive prostanoids and platelet aggregation in essential hypertension. Klin. Wschr. 63, 332-6 (1985).

18. OVERLACK, A., K. O. STUMPE, C. RESSEL, R. KOLLOCH, W. ZYWZOK, F. KRUECK, Decreased urinary kallikrein activity and elevated blood pressure normalized by orally applied kallikrein in essential hypertension. Klin. Wschr. 58, 37-42 (1980).

19. OVERLACK, A., K. O. STUMPE, R. KOLLOCH, C. RESSEL, F. KRUECK, Antihypertensive effect of orally administered glandular kallikrein in essential hypertension. Hypertension 3, Suppl. I, I18-I21 (1981).

20. OVERLACK, A., A. G. SCICLI, O. A. CARRETERO, Intestinal absorption of glandular kallikrein in the rat. Amer. J. Physiol. 244, G689-G694 (1983).

21. RYAN, J. W., A. CHUNG, C. AMMONS, M. L. CARLTON, A simple radioassay for angiotensin-converting enzyme. Biochem. J. 167, 501-4 (1977).

22.   SCHILL, W. B., Improvement of sperm motility in patients with asthenozoospermia by kallikrein treatment. Int. J. Fertil. 20, 61-3 (1975).

23.   MISKA, W., W. B. SCHILL, Resorptionsstudie mit Schweinepankreas-Kallikrein am Menschen. Arzneim. Forsch. 41, 1061-1064 (1991).

24.   SINAIKO, A. R., R. J. GLASSER, R. F. GILLUM, R. J. PRINEAS, Urinary kallikrein excretion in grade school children with high and low blood pressure. J. Pediatrics. 100, 938-40 (1982).

25.   WICKLMAYR, M., Evidence of an involvement of kinin liberation in the priming action of insulin on glucose uptake into skeletal muscle. FEBS Letters 98, 61 (1979).

26.   WILLIAMS, R. R., S. J.HASSTEDT, S. C. HUNT, L. L. WU, P. N. HOPKINS, T. D. BERRY, B. M. STULTS, G. K. BARLOW, H. KUIDA, Genetic traits related to hypertension and electrolyte metabolism. Hypertension 17 (1), Suppl. I, 69-73 (1990).

27.   YAMADA, K., K. ITO, Se YOSHIDA, M. WATANABE, K. YAMAMOTO, Y. IAMURA, Sh. YOSHIDA, Induction of endogenous renal kallikrein in salt-loaded hypertensive patients by hog pancreatic kallikrein administration. Savannah Kinin Congress (1984).

28.   ZINNER, S. H., H. S. MARGOLIUS, B. ROSNER, H. R. KEISER, E. H. KASS, Familial aggregation of urinary kallikrein concentration in childhood: relation to blood pressure, race and urinary electrolytes. Am. J. Epidemiol. 104, 124-32 (1976).

AAS 38/III
Recent Progress on Kinins
© 1992 Birkhäuser Verlag Basel

THE KALLIKREIN - KININ SYSTEM IN EARLY STATE OF DIABETES

S.B.Vila,   V.A.Peluffo,   C.Alvarado,   J.C.Cresto,   A.Zuccollo,
O.L.Catanzaro

Cátedra de Fisiologia, Facultad de Farmacia y Bioquimica;
Laboratorio de Péptidos Vasodepresores (CONICET). Unidad de
Endocrinologia, Hospital General de Niños Dr. Pedro Elizalde.
Buenos Aires . Argentina

SUMMARY: The kallikrein-kinin system was studied in 9 normals,
healthy subjects (6 men, 3 women, age range 1 to 14 years) and
15 diabetic patients (9 men, 6 women age range 2 to 14 years)
with an evolution of the disease between 1 to 14 years. Diabetic
patients with low microalbuminuria (6.62 ± 0.97 mg/24 h) show
increased total and pre-kallikrein respect to control (3 and 2
fold respectively). On the other hand patients with high
microalbuminuria (44.7 ± 13.2 mg/24 h) show a total and prekallikrein of more than 4 and 8 fold increased respectively,
compare with the control. According with these results we can
concluded: 1) The total kallikrein and pre-kallikrein is
increased in the diabetic state. 2) When microalbuminuria is
high, the total and pre-kallikrein correlates with those
increasing. 3) These changes could modified the renal
hemodinamic in diabetes.

INTRODUCTION

The renal Kallikrein-kinin system seems to participate in
the control of electrolytes, water excretion and renal vascular
resistant (1,2).

By the action of renal kallikreins on kininogen substrate it
generates a potent vasoactive kinins,which regulates the
hemodynamic and ion transporting processes.In kidney, kinins are
able to affect renal function directly or via prostaglandins
stimulation (3,4). As a matter of fact the osmotic diuretic
produced by a constant infusion of 30% glucose, did not alter
the GFR but increased the kallikrein excretion (5).

On the other hand, diabetic patients showed elevated urinary
kallikrein excretion compared to normal subjects (6).

Because the diabetic state is associated with proteinuria and
hemodinamics alterations, we decided study whether changes in

protein filtration of the diabetes are associated with changes in the urinary kallikrein-kinin system.

MATERIALS AND METHODS

Nine normals healthy subjects (6men,3 women), age range 1 to 14 years, and 15 diabetics subjects (9 men; 6 women, age range 2 to 14 years) were studied.At the time of this study the diabetic subject were divided in two different groups: A) 6 diabetic patients with a microalbuminuria of 44.7 ± 13.2 and B) 9 diabetics patients with a microalbuminury of 6.62 ± 0.97 mg/24h.

The water over weight method of uAlb (microalbuminuria) detection was determined according to Mogensen et al (7). Patients with 12 h fast and no insulin, received 470 ml of water per m^2 of b.s./h during 2.5hours.

Samples were collected for glucosuria (glucose oxidase) and GFR was calculated by the method of Schwartz et al (8). Samples for kallikrein and microalbuminuria were collected during 24h. Microalbuminuria was estimated by R.I.A. (9). Kallikrein was determined by its amidolytic activity on the chromogenic tripeptide sustrate. H-D-Val-Leu-Arg-pNA, S-2266 (Kabi-Diagnostica) as described by Amundsen et al (10) Pre-kallikrein was determined by Shimamoto et al.(11). Data are expressed as mean ± SEM. Student's test and Analysis of Variance were used for statistical analysis.

Probability p value ≤ .05 was considered statistically significant.

RESULTS

Table 1 shows different data in control and diabetic patients. When the diabetic patients were considered together (independently of microalbuminuria), urinary kallikrein excretion measured by their amidolytic activity was significantly different from normal subjects (T.Kall: D:316.3 ± 24.45; C:86.2 ± 12.21 μmol.min^{-1}/day, respecti-vely).

Diuresis and glycemia, also showed significantly differences compare to control.Measurements of GFR showed that in diabetic patients increased over 168% .

When total kallikrein was calculated by the amount of GFR, the increasing of urinary kallikrein in diabetic patients was associated with the altered GFR (168% increasing). In a second study diabetic patients were divided according to their microalbuminuria excretion.

Table 2 shows significantly differences between controls and diabetics patients with or without albuminuria.

In 6 patients (A) with high microalbuminuria excretion (44.7 ± 13.2), the total kallikrein and prekallikrein was: 405.9 ± 29.1 and 270.6 ± 22.1 (μmol.min^{-1}/day) respecti-vely.

In 9 patients (B) with low microalbuminuria (6.62 ± 0.97), the total kallikrein and prekallikrein was: 226.7 ± 19.8 and 64.2 ± 13.0, respectively. Normal control with a microalbuminuria of 5.53 ± 1.07 mg/24 h, show a total and prekallikrein of: 86.2 ± 12.3 and 31.4 ± 9.85 respectively. In both groups (A and B) of diabetic patients,total and prekallikrein were always associated with GFR alterations: G(A): 193 ± 14.2;G(B): 181 ± 12.7;Control: 116 ± 15.4 (ml.min^{-1}/1.73), but only in A group total and pre-kallikrein were associated to microalbuminuria.

Table 1: Volume, urinary kallikrein and GFR in diabetic and normal subjects (mean ± SEM).

	DIABETIC (15)		NORMAL (9)
Diuresis (ml/day)	2303.0 ± 397.00 **		825.00 ± 152.00
Glucose (mg%)	255.0 ± 109.00 **		72.00 ± 14.00
Total Kallikrein (μmol.min^{-1}/day)	316.3 ± 24.45 **		76.10 ± 25.00
Pre-Kallikrein (μmol.min^{-1}/day)	167.4 ± 18.00 **		23.00 ± 14.20
GFR (ml.min^{-1}/day)	187.0 ± 13.40 **		89.20 ± 11.00
T.Kall./GFR	1.7 ± 0.12 *		0.68 ± 0.28

** p $\leq$ 0.01 vs Normal
* p $\leq$ 0.05 vs Normal

Table 2: Urinary kallikrein excretion in diabetic with and without microalbuminuria.

	Group A (6)	Group B (9)	Normal (9)
Microalbuminuria (mg/24h)	44.70±13.2***	6.62± 0.97	5.53± 1.1
GFR(ml.min^{-1}/day)	193.00±14.2**	181.00±12.70**	116.20± 7.7
Total Kallikrein (μmol.min^{-1}/day)	405.90±29.1**	226.70±19.80**	86.20±12.2
Pre-Kallikrein (μmol.min^{-1}/day)	270.60±22.1**	64.20±13.00**	31.40± 9.6
T.Kall./GFR	2.24± 0.1**	1.17± 0.32**	0.73± 0.2
Pre.Kall/GFR	1.49± 0.1	0.33± 0.08*	0.26± 0.1

* p $\leq$ 0.05
** p $\leq$ 0.01
*** p $\leq$ 0.001

DISCUSSION

Diabetic nephropathy, defined as a progressive clinical condition leading to end-stage renal failure or premature mortality from cardiovascular disease, only affect a subset of about 35% of insulin-dependent diabetic patients (12).

Several studies have shown that microalbuminuria strongly predicts the development of diabetic nephropathy in IDDN patients (13).

Diabetic nephropathy is the single most importance cause of end-stage renal disease in the western world, accounting for > 25% of all endstage renal disease in adults (14).

The role of renal kallikrein,and the vasoactive kinin it can produce,in regulating renal blood flow,water and electrolyte balance and also involved in the mediation of insulin action on blood glucose uptake has been found in several studies (15,16).

The localization of kallikrein in connecting tubule cells adjacent to the afferent arteriole (17), and observations that kallikrein inhibitors and kinin antagonists affect renal function (18) suggest that this enzyme and vasoactive peptide may participate in regulating glomerular function,possibly modulating tubuloglomerular feedback (19).

In the present study we found that the increasing of urinary albumin in diabetic patient is follow by an excretion of total and pre-kallikrein, and that this excretion correlate with the GFR.

In diabetic patients where the amount of urinary albumin exceeded more than 8 fold, the total kallikrein and pre kallikrein is also higher.

An interesting observation is that in patients with urinary albumin excretion no significant from control,the total and pre kallikrein increased over 100%

One of the explanation because total and pre kallikrein increased in the diabetic group without microalbuminuria could be the osmotic diuresis.

Consequently the principal site of an osmotic diuretic in the nephron is at the proximal tubule.Thus the amount of water that reaches the distal tubular cells is increased at this site with a possible "wash out" effect on the kallikrein containing cells of the distal tubule,adjacent to the glomerular afferent arteriole (5).

Some others evidences suggest that osmotic diuresis transiently increases urinary kallikrein levels due to wash out (20).

On the other hand diabetic rats with marked osmotic diuresis show decreased renal and urinary kallikrein (21).

Anyway the present results with diabetic patients with or without microalbuminuria are consistent with the notion that the vasodilator activity of the renal kallikrein kinin system could be altered by the level of urinary albumin excreted. So, this is consistent with the observation that the urinary kallikrein excretion represents probably renal synthesis of the enzyme

(22). The mechanism of the transition from low to high levels of microalbuminuria is unknown, but may be a combination of hemodynamic abnormalities, with the metabolic derangement of synthesis of membrane glycosialoproteins and proteoglycans (23). Increased glomerular capillary pressure,together with some hyperfiltration states, may initiate glomerular damage and accelerate the progression of glomerular diseases(24).

In conclusion, it is of interest that the present findings confirm that the increasing of total and pre kallikrein in diabetic without microalbuminuria could precede the possible diabetic nephropathy.

This work was supported by CONICET.Argentina

REFERENCES

1. Scicli A.G., Carretero O.A. Kidney Int 1985 ; 29: 120-130.

2. Marin-Grez M. Biochem.Pharmacol 1982 ; 31: 3941-3947.

3. Tomita K., Pisano J.J., Knepper M.A. J. Clin.Invest 1980 ; 66: 757-762.

4. Mc.Giff J.C., Vio C.P. N.Z.Med.J. 1983 ; 96: 883-887.

5. Martinez Seeber A., Vila S.B., Catanzaro O.L. Arch.Intern.Phisiol.Biochim. 1985 ; 93: 83-88.

6. Mayfield R.K.,Margolius H.S.,Shimojo N.,Miller D.H.,Squires J.,Namm D.H. ; Hormone and Metabolic Res 1984 ; 15: 31-37

7. Mogensen C.E., Christensen C.K., Gundersen H.J.G. Diabetology 1980 ; 18: 453-457.

8. Schwartz G.J., Gauthier B. J.Pediatr 1985 ; 106: 522-526

9. Abdenur J.E., Eliceo M.T., Sires J.M., Cresto J.C. Medicina 1989 ; 49: 1-6.

10. Amundsen E., Putter J., Prieberger P., Know M., Larsbraten M., Gleason G. Adv. Exp. Med. Biol. 1979; 120: 83-95.

11. Shimamoto K., Chao J., Margolius H.S. Tohoku J. Exp. Med. 1982; 137: 269-274.

12. Kofoed-Enelvoldsen A., Borch-Johnsen K., Kreiner S.,
    Nerur J., D eckert T. Diabetes 1987; 36: 205-210.

13. Mogensen C.E., Christensen C.K. N.Engl.J.Med. 1984;
    311: 89-93.

14. Eggers P.N. N.Engl.J.Med 1988; 318: 223-229

15. Carretero O.A., Scicli A.G. Klin.Wochenschr 1978;56:I,
    113-125.

16. Dietze G. Mol.Cell.Endocrinol. 1982; 25: 127-149

17. Barajas L., Powers K., Carretero O., Scicli A.G.,
    Inagami T. Kidney Int 1986; 29: 965-970.

18. Beier Waltes W.N., Carretero O.A., Scicli A.G.
    Am.J.Phisiol 1988; 255:F408-414.

19. Schnermann J., Briggs J.P., Schubert G., Marin-Grez M.
    Am.J.Phisiol 1984; 247: F 912-918.

20. Bonner G., Marin-Grez M, Beek O., Deeg M., Gross F.
    Clin.Sci 1981; 61: 47-51.

21. Mayfield R.K., Margolius H.S., Bailey G.S., Miller
    D.H., Sens D.A., Squires J., Namm D.H. Diabetes 1985;
    34: 22-28.

22. Jaffa A.A., Miller D.H., Bailey G.S., Chao J.,
    Margolius H.S., Mayfield R.K. J.Clin.Invest. 1987; 80:
    1651-1659.

23. Van-Yu W., Wilson B., Panush-Cohen B.M. Diabetes 1987;
    36: 679-683.

24. Anderson S., Meyer T.W., Rennke H.G., Brenner B.M.
    Kidney Int. 1986; 30: 509-517

STUDIES ON COMPONENTS OF THE PLASMA KALLIKREIN SYSTEM AS A RISK FACTOR IN DIABETICS UNDERGOING CARDIOPULMONARY BYPASS

W. Heller, M.J. Gallimore and H.-E. Hoffmeister

Department of Thoracic, Heart and Cardiovascular Surgery, University of Tuebingen,
D-7400 Tuebingen, Germany

SUMMARY: Components of the FX11-kallikrein systems were determined in blood samples from non diabetics and diabetics undergoing cardiopulmonary bypass (CPB). FX11 and prekallikrein levels fell in both groups with the largest falls in the diabetic group. Kallikrein inhibition was also lower in the diabetic group. Alpha-2-macroglobulin levels were lower in the diabetic group before operation and were markedly lower throughout CPB. Kallikrein like activities were also lower in the diabetic group. Beta FX11a inhibition values were higher in the diabetic group and fell in both groups during CPB. From the results obtained we concluded that the lower levels of FX11, prekallikrein, kallikrein inhibition and alpha-2-macroglobulin in the diabetic patients during CPB reflect enhanced activation of the FX11-plasma kallikrein systems. Blood loss in the diabetic group was higher than the for non diabetic group.

INTRODUCTION

In previous studies (1) we have shown that the plasma kallikrein system is activated during cardiopulmonary bypass (CPB). We have also reported that the general proteinase inhibitor aprotinin when given in reasonably high doses reduces the activation of this system and has therepeutic benefits (2). In our studies we have used chromogenic peptide substrate assays to determine components of the FX11-plasma kallikrein systems and one of our major findings has been that kallikrein like activities (spontaneous amidolytic activity on a chromogenic substrate for plasma kallikrein) is markedly increased in CPB. Kallikrein like activity reflects plasma kallikrein bound to alpha-2-macroglobulin. The two major plasma inhibitors of plasma kallikrein determined in chromogenic substrate assays are C1-esterase inhibitor and alpha-2-macroglobulin.(3). Kallikrein complexed with C1-esterase inhibitor has no activity on natural or synthetic kallikrein substrates. Kallikrein complexed with alpha-2-macroglobulin attacks small chromogenic peptide substrates (4).

Very few studies on components of the FX11 plasma kallikrein systems have been performed on diabetics although it has been suggested that a hypercoaguable state exists in diabetics and the FX11-plasma kallikrein systems play a role in this. To our knowledge no studies on components of the FX11-plasma kallikrein systems have been reported on diabetics undergoing CPB and our findings comparing components of these systems in non diabetics and diabetics undergoing CPB are reported here.

MATERIALS AND METHODS

Two groups of patients were studied non diabetics (ND) and diabetics (D). All patients suffered from coronary artery disease and underwent aorto-coronary bypass surgery. Both groups of patients were treated with aprotinin 20,000 KIU aprotinin/kg body weight. Age, body weight and body surface were comparable in both groups. For the determination of the various plasma components blood samples were collected according to our previous sampling scheme (2). FX11,β FX11a inhibition (βFX11aI), prekallikrein (PKK), kallikrein like activity (KK), kallikrein inhibition (KKI), C1-esterase inhibitor (C1INH) and α2-macroglobulin (α2M) were determined by means of chromogenic peptide substrate assays using kits supplied by KabiDiagnostica, Munich, Germany, Channel Diagnostics, Deal, UK or Immuno, Vienna. The methods used were those recommended by the manufacturers. All values were corrected for haematocrit. Since heparin is known to influence the methods for KK and PKK heparin antagonisation with protamine sulphate was performed. Differences between groups were determined by Student's t-test.

RESULTS

FX11 levels were lower in the D group before and during CPB and were significantly lower ($p = < 0.05$) during CPB (Fig. 1). PKK levels and KKI were significantly lower ($p = < 0.01$) in the D group during CPB and remained lower after CPB (Figs. 2,3). KK activities were elevated in both groups during CPB with significantly ($p = < 0.01$) higher values in the ND group (Fig. 4). α2M levels were significantly lower ($p = < 001$) in the D group before CPB and remained lower throughout and after CPB (Fig. 5). No significant differences in C1INH and βFX11aI levels were found between the two groups although βFX11aI levels were slightly higher and C1INH levels were slightly lower in the D group during CPB.

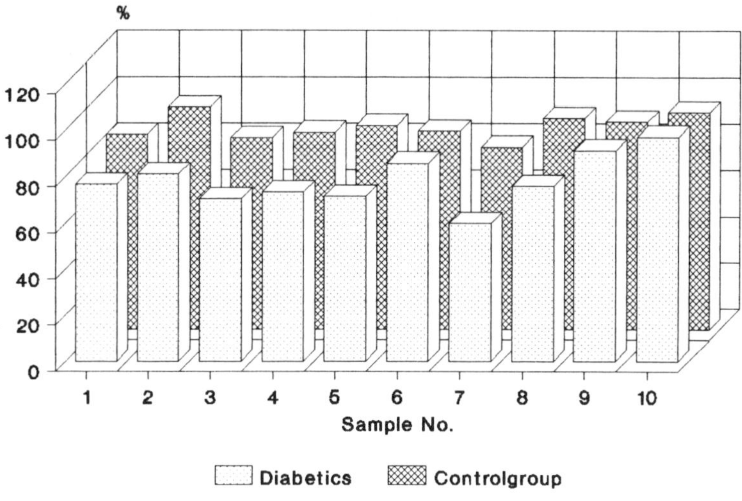

Figure 1. FX11 levels in diabetics and non diabetics undergoing cardiopulmonary bypass

DISCUSSION

In our previous studies in cardiopulmonary bypass we have observed that a pronounced rise in kallikrein like activity occurs early in bypass, accompanied by a fall in prekallikrein levels, reflecting activation of the FX11-plasma kallikrein systems In our present study this is seen in the ND group (Fig. 4). In the D group only a small rise in kallikrein like activity is seen and values remain significantly lower throughout CPB (Fig. 4). These results might suggest that less activation of the FX11-plasma kallikrein systems occurred in the D group of patients. However FX11 levels fell slightly in the D group (Fig. 1), prekallikrein levels in the D group fell to much lower levels than in the ND group (Fig. 2) and kallikrein inhibition was also lower in the D group (Fig. 3), indicating a more pronounced activation of these systems Kallikrein like activity reflects kallikrein bound to α2-macroglobulin and the D group had significantly lower levels of α2M than the ND group both before and during CPB (Fig. 5). This could account for the lower kallikrein like activity in the D group and suggests that the pattern of inhibition of the FX11-plasma kallikrein systems is different in diabetics undergoing CPB and that a more pronounced

Prekallikrein

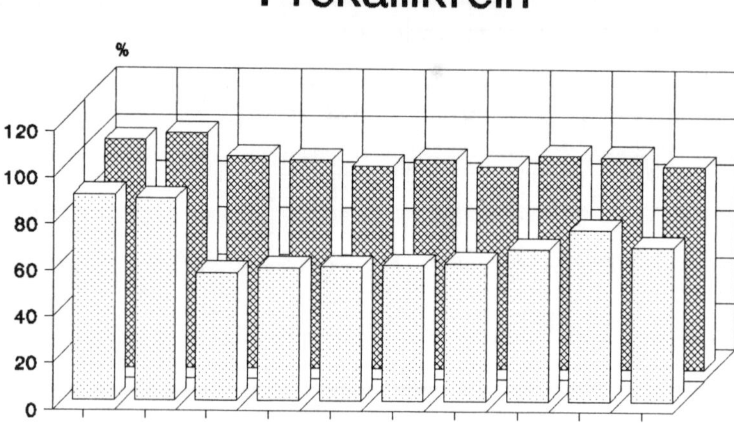

Figure 2. Prekallikrein levels in diabetics and non diabetics undergoing cardiopulmonary bypass

Kallikrein Inhibition

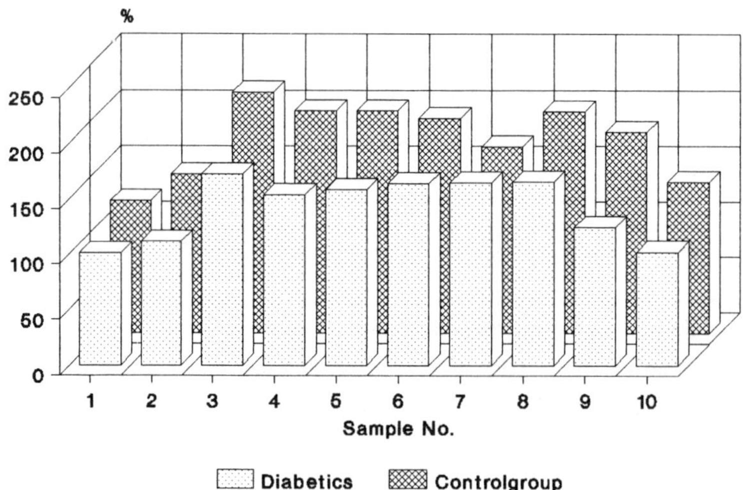

Figure 3. Kallikrein inhibition levels in diabetics and non diabetics undergoing cardiopulmonary bypass

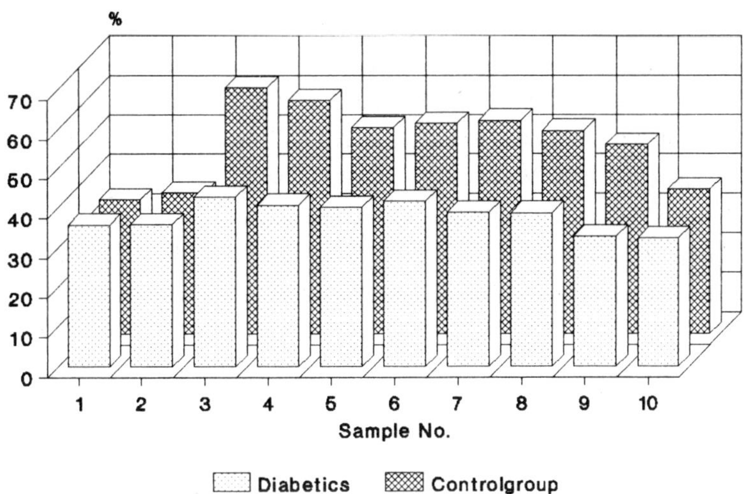

Figure 4. Kallikrein like activities in diabetics and non diabetics undergoing cardiopulmonary bypass

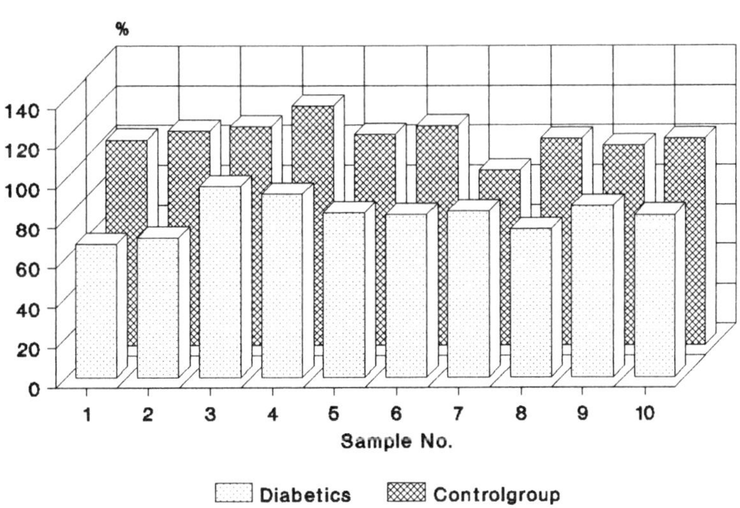

Figure 5. Alpha-2-macroglobulin levels in diabetics and non diabetics undergoing cardiopulmonary bypass

activation of these systems occurs in diabetics than in non diabetics during CPB. Our assay for βFX11a inhibition and the chromogenic substrate assay for C1-esterase inhibitor both measure the functional activity of C1-esterase inhibitor and no significant differences between the D and ND groups were seen in our study. We have postulated that because of cooling of the blood during CPB the major inhibitor of the FX11-plasma kallikrein systems (C1-esterase inhibitor) functions less effectively (2, 5) therefore α2M appears to play a more important role in CPB. We have also suggested that activation of the FX11-kallikrein systems during CPB leads to activation of the fibrinolytic system (2) and comparisoms between the blood loss between the D and ND groups where blood loss was apparently higher in the D group (data still undergoing evaluation) suggest increased fibrinolytic activity in the D group. Unfortunately we were unable to measure fibrinolytic parameters in the present study but we hope to do this and expand our observations on diabetics undergoing CPB in a future study.

REFERENCES

1. Fuhrer G, Gallimore MJ, Heller W, Hoffmeister H.-E. Studies on components of the plasma kallikrein-kinin system in patients undergoing cardiopulmonary bypass. In: KININS 1V Advances in Experimental Medicine and Biology. Greenbaum LW, Margolis HS, editors. New York 1986:198B:385-391.

2. Fuhrer G, Gallimore MJ, Heller W, Hoffmeister H.-E. Aprotinin in cardiopulmonary bypass-Effects on the Hageman factor (FX11)-kallikrein system and blood loss. Blood Coagulation and Fibrinolysis. In Press.

3. Gallimore MJ, Amundsen E, Larsbraaten M, Lyngaas K, Fareid E. Studies on plasma kallikrein using chromogenic peptide substrate assays. Thromb Res 1979; 16:695-703.

4. Amundsen E, Aasen AO, Gallimore MJ, Larsbraaten M Lyngaas K. The plasma kallikrein -kinin system in endotoxin shock in dogs. In: Chromogenic Peptide Substrates Chemistry and Clinical Usage. Scully MF, Kakkar VV, editors. Churchill Livingstone: Edinburgh 1979: 72-83.

5. Schapira M. Major inhibitors of the contact phase coagulation factors. Semin Throm Hemostas 1987; 13: 69-78.

Inflammation

STUDIES OF THE ACTIVATION AND INHIBITION OF THE PLASMA KALLIKREIN-KININ SYSTEM

Allen P. Kaplan, Sesha Reddigari, Tilo Brunnée,
Katsumi Nishikawa, Piotr Kuna, and Michael Silverberg

The State University of New York at Stony Brook
Division of Allergy, Rheumatology, and Clinical Immunology
Department of Medicine, Stony Brook, New York 11794

Introduction

The plasma cascade leading to the generation of bradykinin consists of three proteins, namely Hageman factor (factor XII), prekallikrein, and high molecular weight (HMW)-kininogen. All are part of the intrinsic coagulation cascade since each of these proteins plus coagulation factor XI forms part of the initiating contact activation complex. The cascade is initiated by binding of factor XII to certain negatively charged surfaces upon which an autoactivation reaction proceeds; small amounts of factor XIIa are generated[1]. Prekallikrein and factor XI are each substrates of factor XIIa and circulate in plasma as complexes with HMW-kininogen[2,3]. Separate complexes containing prekallikrein-HMW-kininogen or factor XI-HMW-kininogen bind to the surface. Factor XIIa converts prekallikrein to kallikrein and factor XI to factor XIa. The latter reaction then continues the intrinsic coagulation cascade while kallikrein digests HMW-kininogen to liberate bradykinin[4]. Kallikrein also converts factor XII to factor XIIa[5-8] at a rate that is 50-100 fold faster than the factor XII autoactivation reaction, so that the bulk of factor XIIa generated is kallikrein-dependent[9] and the rate of factor XI activation is also kallikrein-dependent. HMW-kininogen has a binding site for the surface[10] as well as a single site that binds either prekallikrein or factor XI[11,12] and it functions as a co-factor which markedly accelerates the rate of activation of prekallikrein and factor XI by surface-bound factor XIIa[6]. It also indirectly increases the rate of factor XII activation by kallikrein[13]. A schematic diagram summarizing these reactions is given in Figure 1.

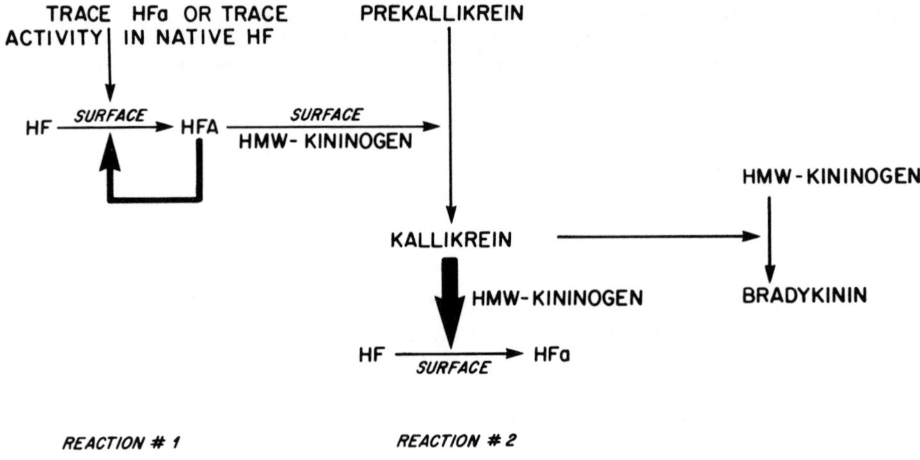

Fig. 1. Diagrammatic representation of the Intrinsic Coagulation-Kinin Pathway indicating those reactions that are inhibited by C1 INH.

There are three general areas of investigation that our laboratory has pursued that are further extensions of the aforementioned studies. These are: 1) The nature of the surface that catalyzes contact activation, 2) The interaction of proteins of the kallikrein system with vascular endothelial cells, 3) The activation of the kinin forming cascade in human disease and 4) Inhibition of kinin formation that might have therapeutic efficacy. This manuscript will briefly present recent data relevant to each of these areas.

Activation of Factor XII/Prekallikrein by Mucopolysaccharides

Initiation of the coagulation kinin cascade is routinely performed using negatively charged surfaces such as kaolin[13]. It appears that a particular charge-density and/or geometry is required to serve as an initiator. However, the "physiologic" surface that is involved in factor XII activation during tissue injury is unknown, but is suspected to be negatively charged (sulfated) proteoglycans and/or its constituent mucopolysaccharides. Dextran sulfate is a commonly used soluble factor XII activator[14,15] but it is a mucopolysaccharide that is not a component of normal tissues. In order to study naturally occurring proteoglycans or mucropoly-saccharides, we developed a method for the study of contact activation in which a mixture of prekallikrein and factor XII is allowed to autoactivate (and reciprocally activate) in the presence or absence of the substance to be studied. In Figure 2 we show a dose response activation of a factor XII-prekallikrein mixture using three concentrations of heparin as the macromolecular "surface". As the heparin concentration increases, the rate of activation is markedly accelerated. We hope to adapt this system to a variety of animal heparins, as well as heparan sulfate, dermatan sulfate, and chondroitin sulfates and then study intact proteoglycans consisting of hyaluronic acid, core protein, link protein, and mucopolysaccharide chains.

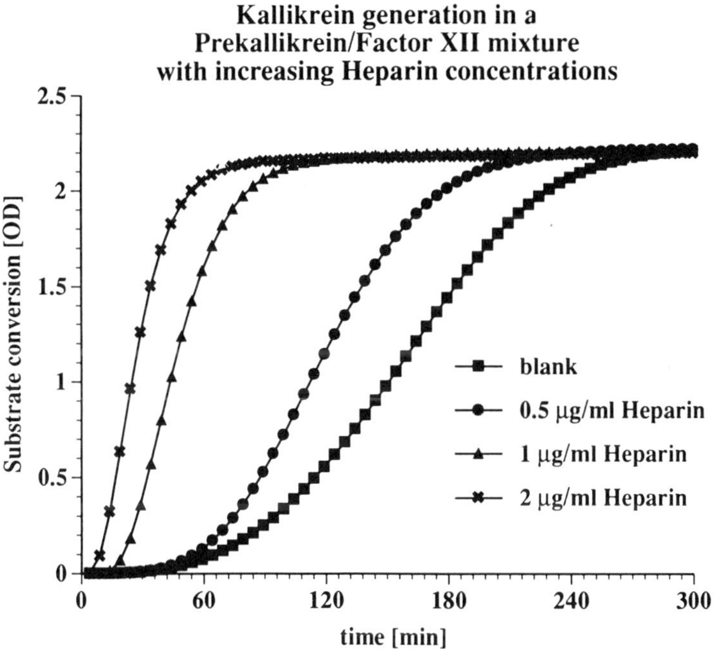

Fig. 2. Time course of activation of a Factor XII-Prekallikrein mixture by three concentrations of heparin compared to the buffer blank. Kallikrein formation was monitored by cleavage of pro-phe-arg-p-nitroanilide as monitored continuously at O.D. 405 nm by a recording spectrophotometer. The concentrations of Factor XII and prekallikrein were each 62.5 nM.

Interactions with Endothelial Cells

The primary functions of bradykinin involve interactions with vascular endothelial cells via B_2 receptors to cause vasodilatation and increased vascular permeability[16]. Since bradykinin is rapidly degraded within the circulation (particularly within the pulmonary vascular bed)[17], it would be advantageous to generate it in proximity to the endothelial cell surface. Earlier data suggested this possibility since HMW-kininogen was shown to bind to vascular endothelial cells and such binding appeared to require interaction with a specific kininogen receptor[18,19]. We sought to further characterize this receptor and incubated human or bovine umbilical vein endothelial cells with ^{3}H-HMW-kininogen and studied its binding in terms of dose-response, time course, and Scatchard analysis. We also attempted to further define the binding site on HMW-kininogen by displacement or competition for ^{3}H-HMW-kininogen binding by unlabeled kininogen, purified H or L chains, or LMW-kininogen.

In Figure 3 is shown the interaction of ^{3}H-HMW-kininogen with human vascular endothelial cells demonstrating saturable binding as the concentration of kininogen is increased and inhibition of binding by addition of a 50-fold molar excess of unlabeled HMW-kininogen or purified H chain. These data suggest that the binding site is specific and that a binding site resides on the heavy chain derived from cleaved, reduced, and alkylated kininogen. We have demonstrated that there are approximately 2×10^6 sites on endothelial cells with a calculated dissociation constant of 43 nM. Scatchard analysis reveals a single high affinity binding site. Yet, we have independent evidence that the light chain also can contribute to binding. It is not yet clear whether separate sites exist on the heavy and light chains with comparable affinity, or whether both are available in the native HMW-kininogen, or whether cleavage of kininogen affects the interaction and, perhaps, exposes an additional site. Evidence of heavy and light chain binding of HMW-kininogen to platelets has also been presented[19] and, like the endothelial cell-kininogen interaction, the interaction with platelets is zinc dependent[20,21]. We have also demonstrated that cell-bound HMW-kininogen can be cleaved by plasma kallikrein to release bradykinin at the same molar ratio that is active in solution. Our data suggest that vascular endothelium is normally saturated with HMW-kininogen, based on the aforementioned binding parameters, and bradykinin can be generated locally. Future studies will deal with interactions of the vascular endothelium with factor XII, prekallikrein, and factor XI.

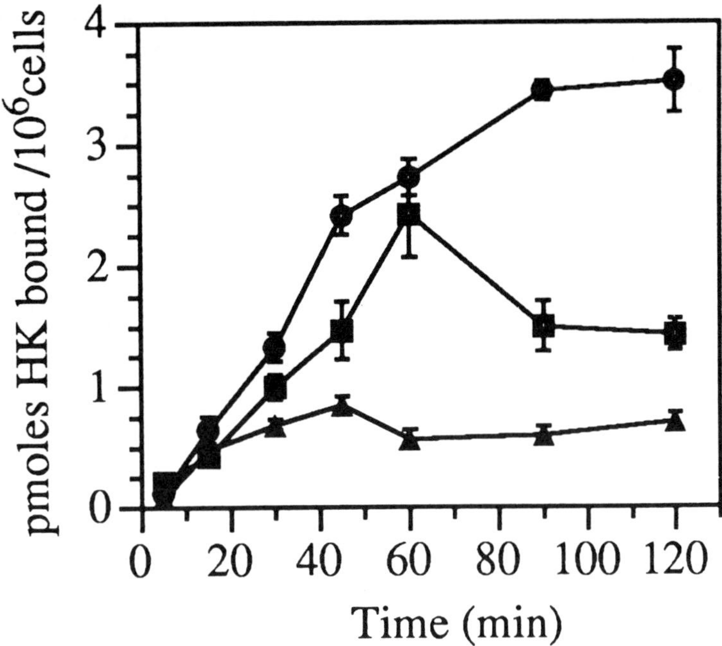

Fig. 3. Human umbilical vein endothelial cells were incubated with 1 nM ^{3}H-HMWK at 4° C and at indicated time points, cell-bound HMWK was determined. (●) = ^{3}H-HMWK; (▲) = ^{3}H-HMWK plus 50 nM unlabeled HMWK; (■) = ^{3}H-HMWK plus 50 nM purified heavy chain added at the 60 min time point. The unlabeled HMWK and the purified H chain inhibit ^{3}H-HMWK binding.

Assessment of the Kinin Forming System in Human Diseases

Assessment of the plasma kinin forming system in human disease is difficult because the concentration of activated enzymes is often low and the half-life of bradykinin in the circulation is brief. We have developed a series of assays that facilitate assessment of the entire cascade and allow distinction between activation of tissue kallikrein from activation of the plasma system. Since activated factor XII (factor XIIa or factor XIIf) binds to the C$\bar{1}$ INH, we have developed a double antibody ELISA which can be used to quantitate complexes of activated factor XII and C$\bar{1}$ INH in body fluids[22]. Kallikrein also binds to C$\bar{1}$ INH and α_2 macroglobulin

(α_2M), which together account for over 90% of complex formation between kallikrein and plasma inhibitors[23,24]. At 37°C, the ratio of kallikrein binding to $\overline{C1}$ INH and α_2M is approximately 65:35[25]. We have developed a double antibody ELISA for quantitation of plasma kallikrein - $\overline{C1}$ INH complexes[26]. Assays of kallikrein-α_2M complexes have also been developed[24], but cannot use the same methodology because kallikrein loses antigenicity upon binding α_2M. Monoclonal antibody to unique determinants in the complex can be used as a primary assay[27] or one can quantitate residual kallikrein activity within the complex [25] using anti-α_2M to isolate the complex.

We have also developed an assay for cleaved HMW-kininogen based upon an immunoblotting technique[28]. We can quantitate as little as 1% cleavage of HMW-kininogen in plasma while the double antibody ELISA's for enzyme-inhibitor complexes similarly detect 1-2% activation. If the concentration of HMW-kininogen is known, the absolute amount of cleaved HMW-kininogen can be calculated and the theoretical value for bradykinin released determined.

Cleavage of low molecular weight (LMW)-kininogen by tissue kallikrein is the other major route by which kinins are formed. Lys-bradykinin is the product and it is rapidly converted to bradykinin by a plasma aminopeptidase. Lys-bradykinin and bradykinin have similar biologic activities and are cross-reactive in the radioimmunoassay. They can be separated by HPLC[29] and this has been used to distinguish the source of the kinin. Any lys-bradykinin must be a result of the action of tissue kallikrein while bradykinin is formed either by activation of the plasma cascade or by conversion of lys-bradykinin. Tissue kallikrein is only very slowly inhibited in plasma or other biologic fluids, thus it is usually found free rather than complexed to an inhibitor. It can be quantitated by radioimmunoassay[30] and/or its activity determined using a specific synthetic substrate[31]. There is not, as yet, any specific assay for cleaved LMW-kininogen.

Activation in Specific Disorders

There are a variety of general mechanisms by which kinins are generated in disease states. There may be initiation of the plasma cascade by contact with negatively charged surfaces (? macromolecules) and this may occur at the site of local tissue damage where plasma or interstitial fluid makes contact with connective tissue. The liberation of kinins in allergic reactions may occur via such a mechanism. Secretion of tissue kallikrein from local mucous/serous glands lining the nose or bronchial tree would allow formation of lys-bradykinin. In Figure 4 is shown the formation of plasma kallikrein-C1 INH complexes during the late phase of a cutaneous allergic reaction which occurs at the same time there is cellular infiltration and local fibrin deposition.

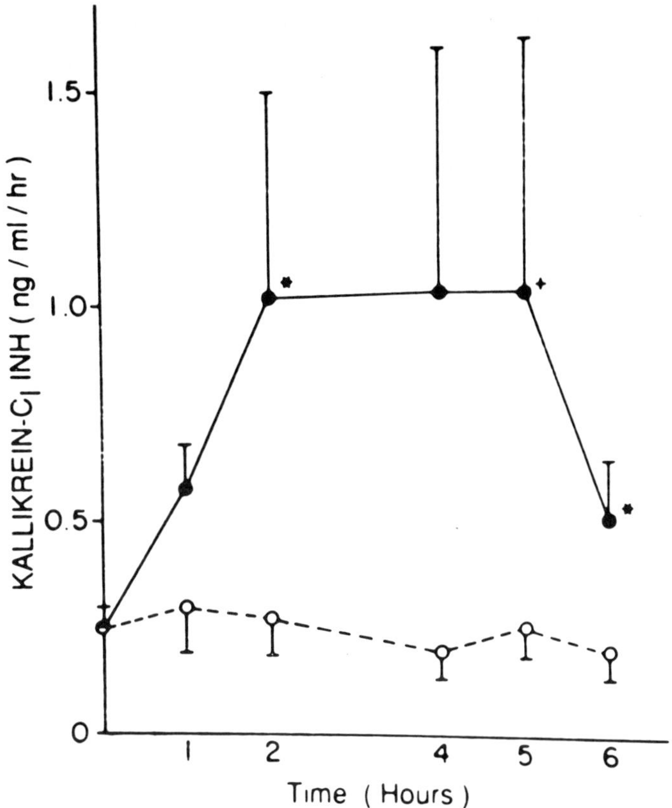

Fig. 4. Time course of formation kallikrein-C$\overline{1}$ INH complexes in a late phase cutaneous reaction to ragweed pollen. Suction blisters of 1 cm diameter were induced on the forearm, unroofed, and chambers containing antigen or buffer placed over the denuded skin. Chamber fluid was replaced every hour and the kinetics of kallikrein-C$\overline{1}$ INH complex formation determined. The antigen challenged site (•) is compared to the control (o) without antigen.

Other possible initiators of the plasma cascade in pathologic conditions are endotoxin[32], uric acid crystals, and pyrophosphate crystals[33,34]. Tissue kallikrein can be released (secreted) from the pancreas, kidney, salivary glands, brain, endocrine glands and serous/mucous glands, or secreted from cells such as the neutrophil [35]. Activation of the plasma cascade can also occur by liberation of any enzyme which can directly activate factor XII, prekallikrein, or

HMW-kininogen. These may be derived from activated or damaged cells or secreted by invasive organisms (bacteria).

Activation of the plasma system can also occur if there is insufficient inhibition, e.g., by congenital absence or dysfunction of C̄Ī INH (hereditary angioedema) or by dilution of body fluids so that the extent of inhibition is diminished. It appears that the plasma kinin-forming cascade is constantly turning over[36] and the extent is regulated by inhibitors. This may be due to the autoactivation of proteins such as factor XII. It is also known that the function of inhibitors is very dependent on concentration with rapid loss of inhibitory activity upon dilution. The effect of dilution on the enzyme cascade to generate bradykinin is small by comparison. Thus, dilution leads to activation[37].

Figure 5 demonstrates cleavage of HMW-kininogen in hereditary angioedema (HAE) plasma. A low dose of dextran sulfate, which leads to less than 1 percent HMW-kininogen cleavage in normal plasma leads to complete digestion of HMW-kininogen in 16 minutes in HAE plasma.

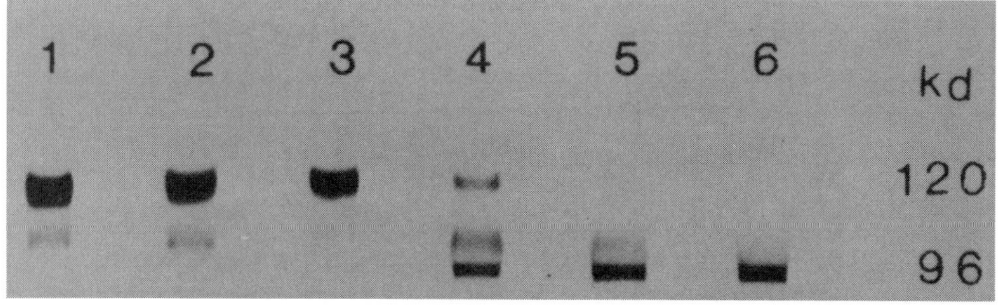

Fig. 5. Cleavage of HMW-kininogen in hereditary angioedema (HAE) plasma compared to normal plasma. Plasma was incubated with dextran sulfate (10 mg/ml) for varying times at 37° C. Lane 1 - normal plasma control. Lane 2 - normal plasma plus dextran sulfate in 16 min incubation. Lane 3 - HAE plasma control. Lane 4 - HAE plasma plus dextran sulfate - 4 min. Lane 5 - HAE plasma plus dextran sulfate - 8 min. Lane 6 - HAE plasma plus dextran sulfate - 16 min. Native HMW kininogen is seen at 120 Kd while cleavage products are seen at 105 Kd which converts to 96 Kd.

Inhibition of the Plasma Kinin Forming System

We have studied Neurotropin, a drug marketed in Japan, which is used as a "pain killer" and anti-inflammatory agent and is thought to act by inhibition of kinin formation and/or action. In Figure 6 is shown a dose-response inhibition of kallikrein-C̄Ī INH complex formation by Neurotropin when plasma is incubated with dextran sulfate. We also observed inhibition of kininogen cleavage and inhibition of plasma kallikrein formation which parallels the inhibition of kinin formation. Subsequent studies revealed that the major effect of Neurotropin is to inhibit interaction of HMW-kininogen with the surface by direct interaction with the kininogen heavy chain (rather than the surface). A lesser effect upon factor XII interaction with the surface was seen, plus a minor inhibition of the conversion of prekallikrein to kallikrein by surface-bound activated factor XII[38].

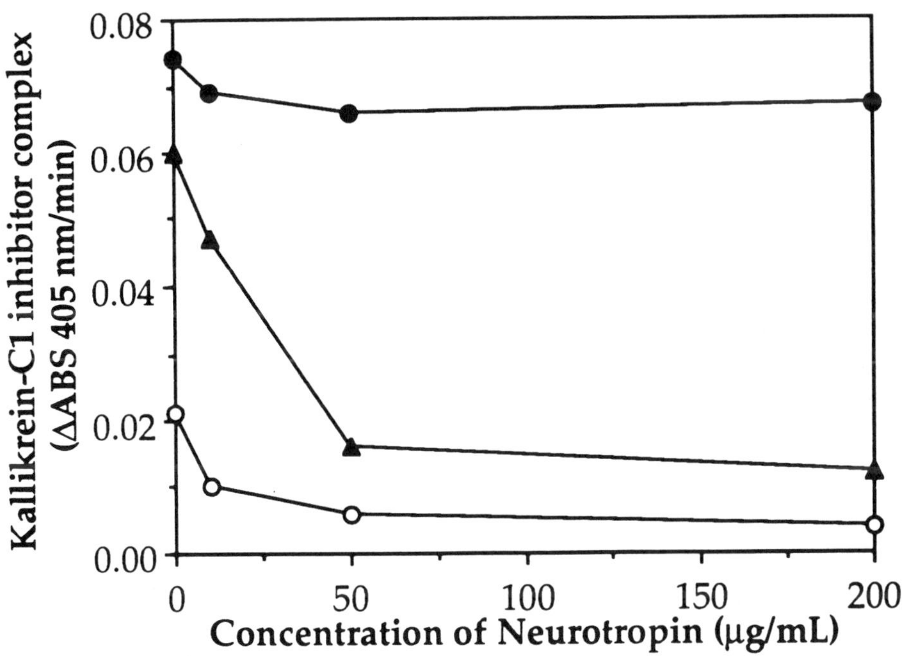

Fig. 6. Inhibition of kallikrein-C1 INH complex formation in dextran sulfate activated plasma by Neurotropin. Plasma was incubated with Neurotropin in the presence of 100 mg/ml (●) or 20 mg/ml (▲) dextran sulfate with 5mM 1,10 phenanthroline or 100 mg/ml (O) dextran sulfate without 1,10 phenanthroline. Each data point represents the mean of four experiments.

Concluding Comments

Considerable progress has been made in our understanding of the mechanism of kinin formation and our ability to assess each of these in human disease. Although kinin formation has been documented in many disorders, it is very difficult to assess the contribution of kinins (or direct actions of any of the proteins) to the symptoms or organ dysfunction seen. However, the discovery of kinin antagonists such as B_2 receptor ligands[39] and the use of inhibition of contact activation such as Neurotropin may allow assessment of the role of the kinin cascade(s) in human disease. An example is the observation that bradykinin is the only vasoactive peptide seen in nasal secretions in patients with the common cold[40]. Its formation may be due to liberation of tissue kallikrein and/or to contact of diluted plasma or interstitial fluid with exposed connective tissue. If a specific kinin antagonist could be administered which would ameliorate the symptoms of a cold, it would serve as evidence that its presence is of pathologic significance, and that a therapeutic maneuver that actually treats a cold is possible. As more potent and specific antagonists are developed, such studies will be performed and definitive answers obtained. It may also be possible to develop antagonists such as Neurotropin which act upon the earliest steps of contact activation. In fact, a combination of such a drug with a kinin receptor antagonist might yield synergistic inhibition of the plasma cascade, an approach that might be applicable to the treatment of acute attacks of swelling in patients with hereditary angioedema.

References

1. Silverberg, M., Dunn, J.T., Garen, L. and Kaplan, A.P. Autoactivation of human Hageman factor. J. Biol. Chem. 1980; 255:7281.

2. Mandle, R.J., Colman, R.W. and Kaplan, A.P. Identification of prekallikrein and HMW-kininogen as a complex in human plasma. Proc. Natl. Acad. Sci. USA. 1976; 73:4179.

3. Thompson, R.E., Mandle, R.J. and Kaplan, A.P. Association of factor XI and high molecular weight kininogen in human plasma. J. Clin. Invest. 1977; 60:1376.

4. Thompson, R.E., Mandle, R.J. and Kaplan, A.P. Characterization of human high molecular weight kininogen. Procoagulant activity associated with the light chain of kinin-free high molecular weight kininogen. J. Exp. Med. 1978; 147:488.

5. Cochrane, C.G., Revak, S.D. and Wuepper, K.D. Activation of Hageman factor in solid and fluid phases. J. Exp. Med. 1973; 138:1564.

6. Griffin, J.H. and Cochrane, C.G. Mechanisms for the involvement of high molecular weight kininogen in surface-dependent reactions of Hageman factor. Proc. Natl. Acad. Sci., USA. 1976; 73:2554.

7. Meier, H.L., Pierce, J.V., Colman, R.W. and Kaplan, A.P. Activation and function of human Hageman factor. The role of high molecular kininogen and prekallikrein. J. Clin. Invest. 1977; 60:18.

8. Dunn, J.T., Silverberg, M. and Kaplan, A.P. The cleavage and formation of activated Hageman factor by autodigestion and by kallikrein. J. Biol. Chem. 1982; 275:1779.

9. Tankersley, D.L. and Finlayson, J.S. Kinetics of activation and autoactivation of human factor XII. Biochemistry. 1984; 23:273.

10. Ikari, N., Sugo, T., Fujii, S., Kato, H. and Iwanaga, S. The role of bovine high molecular weight kininogen in contact-mediated activation of bovine factor XII: Interaction of HMW kininogen with kaolin and plasma prekallikrein. J. Biochem. 1981; 89:1699.

11. Tait, J. and Fujikawa, K. Identification of the binding site plasma prekallikrein in human high molecular weight kininogen. J. Biol. Chem. 1986; 261:15396.

12. Tait, J. and Fujikawa, K. Primary structure requirements for the binding of human high molecular weight kininogen to plasma prekallikrein and factor XI. J. Biol. Chem. 1987; 262:11651.

13. Silverberg, M., Nicoll, J.E. and Kaplan, A.P. The mechanism by which the light chain of cleaved HMW-kininogen augments the activation of prekallikrein, factor XI, and Hageman factor. Thromb. Res. 1980; 70:173.

14. Silverberg, M. and Diehl, S.V. The autoactivation of factor XII (Hageman factor) induced by low Mr heparin and dextran sulphate. Biochem. J. 1987; 248:715.

15. Silverberg, M. and Diehl, S.V. The activation of the contact system of human plasma by polysaccharide sulfates. Ann. NY Acad. Sci. 1987; 516:268.

16. Vavrek, R.J. and Stewart, J.M. Competitive antagonists of bradykinin. Peptides. 1985; 6:1610164.

17. Sheikh, I.A. and Kaplan, A.P. The mechanism of digestion of bradykinin and lysylbradykinin (kallidin) in human serum: The role of carboxy-peptidase, angiotensin converting enzyme, and determination of final degradation products. Biochem. Pharmacol. 1989; 38:993.

18. Van Iwaarden, F., deGroot, P.G. and Bouma, B.N. The binding of high molecular weight kininogen to cultured human endothelial cells. J. Biol. Chem. 1988; 263:4698.

19. Schmaier, A.H., Kuo, A., Lundberg, D., Murray, S. and Cines, D.B. The expression of high molecular weight kininogen on human umbilical vein endothelial cells. J. Biol. Chem. 1988; 263:16327.

20. Schmaier, A.H., Zuckerberg, A., Silverman, C., Kuchibhotla, J., Tuszynski, G.P. and Colman, R.W. High molecular weight kininogen: A secreted platelet protein. J. Clin. Invest. 1983; 71:1477.

21. Greengard, J. and Griffin, J.H. Receptor for high molecular kininogen on stimulated washed human platelets. Biochemistry. 1984; 23:6863.

22. Kaplan, A.P., Gruber, B. and Harpel, P.C. Assessment of Hageman factor activation in human plasma: Quantification of activated Hageman factor-C1 inactivator complexes by an enzyme linked differential antibody immunosorbent assay. Blood. 1985; 66:636.

23. Schapira, M., Scott, C.F. and Colman, R.W. Contribution of plasma protease inhibitors to the inactivation of kallikrein in plasma. J. Clin. Invest. 1982; 69:462.

24. van der Graaf, F., Koedam, J.A. and Bouma, B.N. Inactivation of kallikrein in human plasma. J. Clin. Invest. 1983; 71:149.

25. Harpel, P.C., Lewin, M.F. and Kaplan, A.P. Distribution of plasma kallikrein between C1 inactivator and α_2-macroglobulin-kallikrein complexes. J. Biol. Chem. 1985; 260:4257.

26. Lewin, M.F., Kaplan, A.P. and Harpel, P.C. Studies of C1 inactivator-plasma kallikrein complexes in purified systems and in plasma. Quantification by an enzyme-linked differential antibody immunosorbent assay. J. Biol. Chem. 1983; 238:6415.

27. de Agostini, A., Schapira, M., Wachtfogel, Y.T., Colman, R.W. and Carrel, S. Human plasma kallikrein and C1 inhibitor form a complex possessing an epitope that is not detectable on the parent molecules: Demonstration using a monoclonal antibody. Proc. Natl. Acad. Sci. USA. 1985; 82:5190.

28. Reddigari, S. and Kaplan, A.P. Quantification of human high molecular weight kininogen by immunoblotting with a monoclonal anti-light chain antibody. J. Immunol. Meth. 1989; 119:19.

29. Naclerio, R.M., Proud, D., Togias, A.G., Adkinson Jr., N.F., Meyers, D.A., Kagey-Sobotka, A., Plaut, M., Norman, P.S. and Lichtenstein, L.M. Inflammatory mediators in late antigen-induced rhinitis. N. Eng. J. Med. 1985; 313:65.

30. Shimamoto, K., Mayfield, R.K., Margolius, H.S., Chao, J., Stroud, W. and Kaplan, A.P. Immunoreactive tissue kallikrein in human serum. J. Lab. Clin. Med. 1984; 103:731.

31. Toki, N., Sumi, H. and Takasugi, S. Purification and characterization of kallikrein from plasma of patients with acute pancreatitis. Clin. Sci. 1981; 60:199.

32. Morrison, D.C. and Cochrane, C.G. Direct evidence for Hageman factor (factor XII) activation by bacterial lipopolysaccharides (endotoxins). J. Exp. Med. 1974; 140:787.

33. Kellermeyer, R.W. and Breckenridge, R.T. The inflammatory process in acute gouty arthritis, I. Activation of Hageman factor by sodium urate crystals. J. Lab. Clin. Med. 1965; 63:307.

34. Ginsberg, M., Jaques, B., Cochrane, C.G. and Griffin, J.H. Urate crystal dependent cleavage of Hageman factor in human plasma and synovial fluid. J. Lab. Clin. Med. 1980; 95:497.

35. Figueroa, C.D., MacIver, A.G., Dieppe, P., MacKenzie, J.C. and Bhoola, K.D. Presence of immunoreactive tissue kallikrein in human polymorphonuclear (PMN) leucocytes. In: Kinins V. Part B. Advances in Experimental Medicine and Biology, Vol. 247B. ed. Abe, K., Moriya, H., and Fujii, S. 1989, New York: Plenum Press. p. 207.

36. Weiss, R., Silverberg, M. and Kaplan, A.P. The effect of C1 inhibitor upon Hageman factor autoactivation. Blood. 1986; 68:239.

37. Kaplan, A.P. and Austen, K.F. A prealbumin activator of prekallikrein. J. Immunol. 1970; 105:802.

38. Nishikawa, K., Reddigari, S.R., Silverberg, M., Kuna, P.B., Yago, H., Nagaki, Y., Toyomaki, Y., Suehiro, S. and Kaplan, A.P. Effect of Neurotropin on the activation of the plasma kallikrein-kinin system. Biochem. Pharmacol. 1992; In press.

39. Stewart, J.M. and Vavrek, R.J. Bradykinin chemistry: Agonists and antagonists. In: Advances in Experimental Medicine and Biology. ed. H. Fritz, N. Back, G. Dietze, and G.L. Haberland. Vol. 156A, Kinins III - Part A. 1983, New York: Plenum Press. p. 585.

40. Naclerio, R.M., Proud, D., Lichtenstein, L.M., Kagey-Sobotka, A., Hendley, J.O., Sorrentino, J. and Gwaltney, J.M. Kinins are generated during experimental rhinovirus colds. J. Inf. Dis. 1988; 157:133.

HAGEMAN FACTOR DEPENDENT ACTIVATION AND ITS RELATIONSHIP TO LETHAL PSEUDOMONAS AERUGINOSA BURN WOUND INFECTIONS

Ian Alan Holder[1,3] and Alice N. Neely[2,3]

Departments of Surgery, Molecular Genetics, Biochemistry and Microbiology[1] Physiology and Biophysics[2], University of Cincinnati College of Medicine, and Shriners Burns Institute[3], 3229 Burnet Avenue, Cincinnati, Ohio 45229-3095

SUMMARY: Trauma causes increases in total protease load in the circulation of the traumatized host animal or patient. This increase is due, in part, to Hageman Factor activation followed by down-line cascade system activation. Concomitant with the activation of these systems is a reduced host capacity to resist infection. Infection superimposed on a traumatized host increases the total host protease load even more by additional activation of Hageman Factor and cascade systems.

INTRODUCTION

Total proteolytic activity in the circulation increases after humans or experimental animals are burned.[1] This increase in circulating protease load is more than can be attributed to proteases absorbed into the circulation from cells in the injured tissue. Further, thermally injured hosts become extremely susceptible to Pseudomonas aeruginosa infections[2,3] but only by strains that produce proteolytic enzymes.[3-5] Infections by these strains increase the total circulating protease level in the burned host even more.[1] These infections, if left untreated, are usually fatal[2,3] however, treatment using protease inhibitors enhances survival.[5,6,7]

Why host protease load increased above that which would have been expected from the tissue destruction that occurs after burns was unclear. One possibility was that burn associated host Hageman Factor activation caused activation of down-line protease generating cascade systems. Even if this were true, its relationship to the susceptibility of the burned host to lethal P. aeruginosa infection was unknown. Because of this, we were prompted to ask the following questions. Does Hageman Factor activation occur after burns and, if so, does this relate to predisposing the burned host to lethal P. aeruginosa infections?

METHODS

To answer these questions, we used the animal model outlined in Figure 1[8,9]. Male guinea pigs were anesthetized, injected intravenously with Evans blue dye (30mg/kg) and treated.

Animal Model

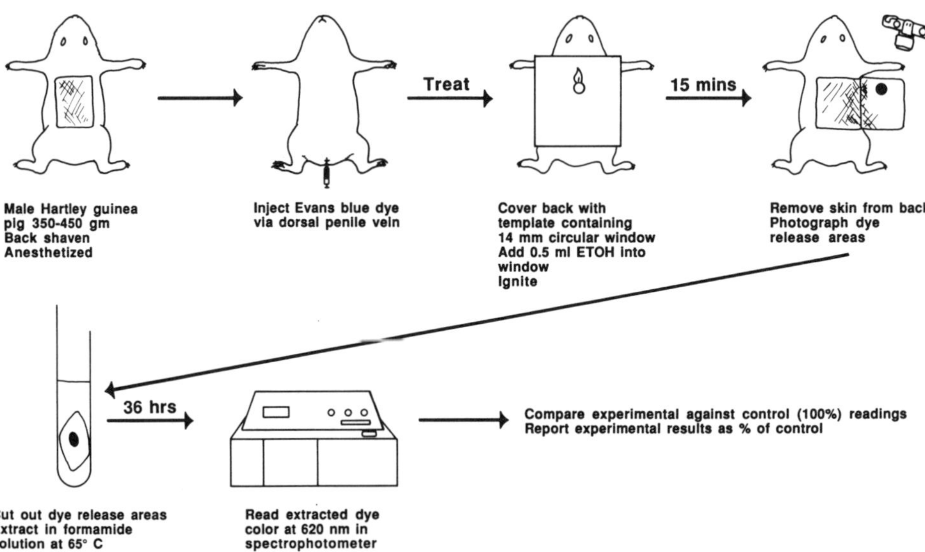

Figure 1. Animal model. Burn induced vascular permeability changes in the skin of guinea pigs, visualized as dye release lesions, were documented and sometimes quantified as outlined in this figure. In some instances injections of histamine, bradykinin and Pseudomonas eleastase were given and dye release lesions from these treatments were compared with those obtained by burning. These comparisons were made in both antihistamine pretreated and antihistamine non-treated guinea pigs. In other cases, substances which inhibit specific steps in the Hageman factor activated-kinin generating cascade were injected into sites on the skin of the guinea pig over which burns were made, subsequently.

Initially, a burn only was made on the guinea pigs' back. In some cases, subcutaneous injections of vascular permeability enhancing substances such as histamine and bradykinin were given to compare results with the thermal injury. In other cases, inhibitors of parts of the Hageman Factor- kinin generating sequence were injected and a thermal injury inflicted over the injection site.

When the burn was made, a flameproof template with a circular window was placed on the shaven back. Alcohol (0.5ml) was placed in the window and ignited. The flame was blown out after varying periods of time. Fifteen minutes later the animal was sacrificed, the skin of the back removed and the blue dye release lesions were photographed. In some cases, the dye release lesions were then cut out and extracted in warm formamide. The dye color extracted into the formamide was quantified by spectrophotometry at 620 nm. The results are presented as % dye relative to its control which is considered to be 100%.

RESULTS

Results, presented in Figure 2, show that thermal injury, and both histamine and bradykinin injected subcutaneously into the guinea pig skin causes dye to be released and visualized as a dark "lesion."

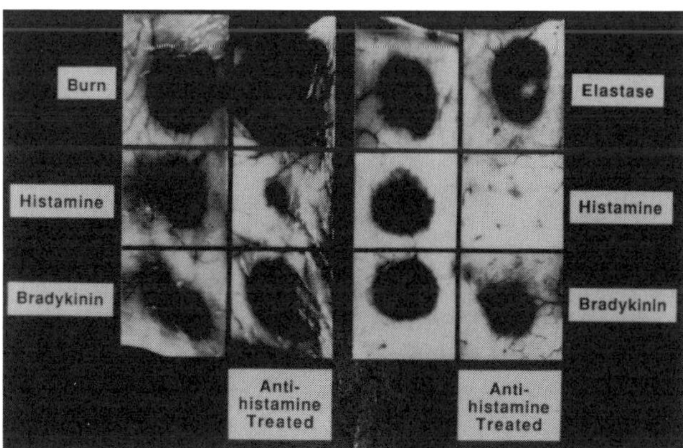

Figure 2. Comparison of burn, histamine, bradykinin and Pseudomonas elastase induced dye release skin lesions in non-antihitamine treated and antihistamine pretreated guinea pigs.

In animals preinjected with antihistamine, the histamine "lesion" is reduced whereas there is no effect in both the burn and bradykinin "lesion". This shows that the dye release associated with thermal injury is not histamine related but may be caused by bradykinin generation.[10] Likewise, when Pseudomonas elastase is injected into the back of a normal guinea pig, a dye lesion occurs.[11] This lesion is not affected by a prior injection with antihistamine.

As shown in Figure 3, bradykinin is generated in two steps after Hageman Factor is activated.[12] Each step can be inhibited by a specific inhibitor -- step 1 by corn trypsin inhibitor (CTI)[13] and step 2 by soy bean trypsin inhibitor (SBTI).[14] If bradykinin was generated from the burn activation of Hageman Factor, injection of each inhibitor into a site on the back over which a burn was placed would either reduce or eliminate the burn induced dye release lesion.

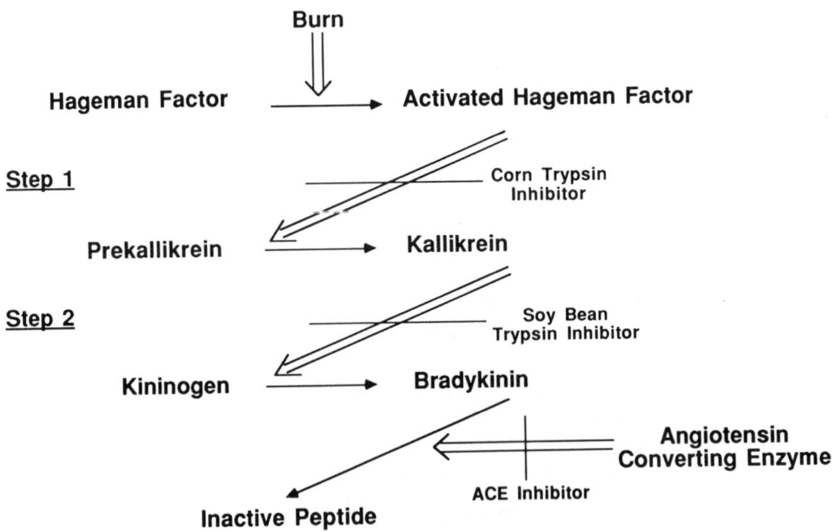

Figure 3. Steps in the burn induced Hageman factor actived kinin generating cascade showing points of specific inhibitor intervention.

Figure 4a presents results when burns are imposed over areas of the skin of the guinea pig which have been injected with either (CTI) or (SBTI). As would be expected, if bradykinin was generated through burn induced Hageman Factor activation, the dye release is reduced in treated areas. In addition, the response is inhibitor dose related.

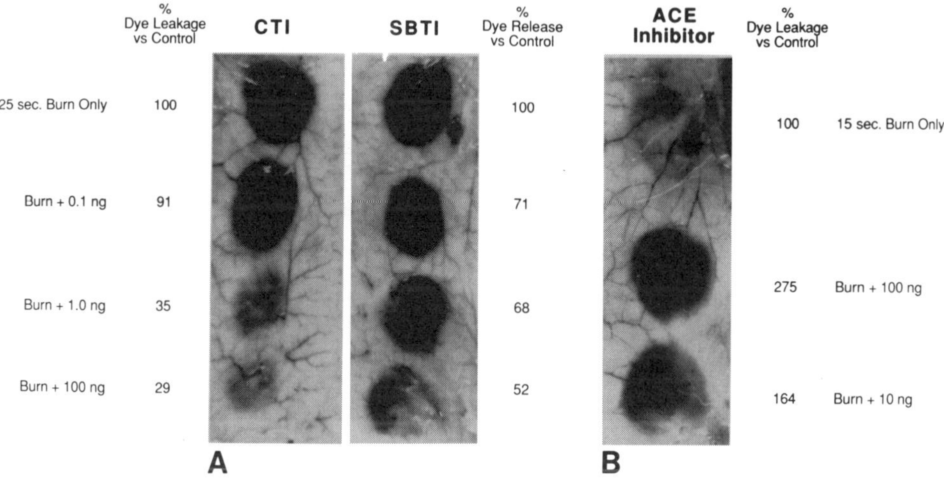

Figure 4. Effect of inhibitors of specific steps in the Hageman factor activated-kinin generating cascade on burn induced dye release. A) Using the optical density (620nm) of dye release extracted into formamide from a 25 sec burn as 100% dye release from 25 sec burns placed over guinea pig skin injected with either 0.1, 1.0 or 100 ng of corn trypsin inhibitor (CTI) or soy bean trypsin (SBTI) is reduced to 91, 35 and 29 or 71, 68 and 52% of control, respectively. B) Conversely, the optical density of dye release extracted into formamide taken as 100% from a 15 sec burn is increased to 275 and 164% when the burn is placed over skin sites injected with 100 or 10 ng of angiotensin converting enzyme inhibitor (ACE inhibitor), respectively.

When bradykinin is generated, in vivo, it is broken down by angiotensin converting enzyme (Figure 3).[14] If an angiotensin converting enzyme inhibitor (ACE inhibitor) is injected prior to burn, and bradykinin is generated from the injury, then the dye release lesion color should be increased due to continued accumulation of the bradykinin because the angiotensin converting enzyme is inhibited from breaking down the bradykinin into inactive peptides. The results presented in Figure 4b show that the dye release lesions are enhanced when ACE inhibitor is injected into the skin over which a burn is placed. The results are dose related, also.

Collectively, these results suggest that thermal injury causes generation of bradykinin and that it probably does this by activation of host Hageman Factor. Hageman Factor lies at the top of a complex series of interrelated protease activated and protease-generating cascade systems which include the kininogen-kinin, clotting, complement, fibrinolytic and arachidonic acid systems. A quite simplified view of these systems interactions is presented in Figure 5.

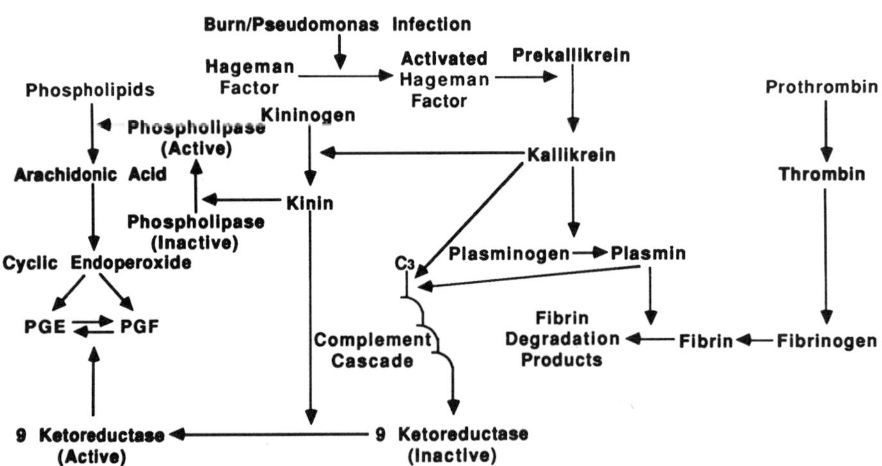

Figure 5. Simplified Schema of the interrelations among some protease actived cascade systems of human plasma.

We propose that protease generation from activation of down-line cascade systems occurs after the initial activation of Hageman Factor by the thermal injury. This would explain the increase in total circulating host proteolytic activity above that which might be expected from protease absorbed into host circulation from heat damaged cells only. Cascade activation, in addition to generating proteases which upset the homeostatic protease: protease inhibitor ratio in the host in favor of the proteases, has some immunoresistance consequences, as well.[15] Some of these are listed in Figure 6.

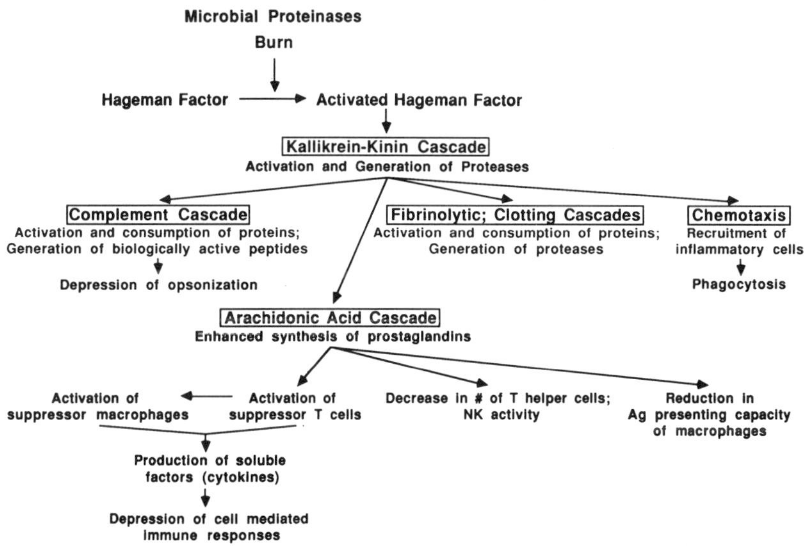

Figure 6. Some immunosuppressive consequences of activation of the Hageman factor dependent proteolytic cascade.

Thus, the burn reduces the capacity of the host to defend itself from infectious threat. It is in this host that infection with the protease producing P. aeruginosa infection becomes such a life threatening circumstance. We have seen that Pseudomonas elastase has the capacity to activate host Hageman Factor, also.[11] We believe that this additional activation with its attendant increase in both cascade related protease generation and additional burden on the host immunoresistance mechanisms reduces the host's capacity to fight the infection below a threshold from which the host can recover.

Our hypothetical view of these relationships are illustrated in Figure 7. A burn or P. aeruginosa infection alone generate increased host proteolytic load with reduced host immunoresistance and homeostasis. This reduction does not reach below a level from which the host cannot recover.

Figure 7. Relationship between circulating protease (P): protease inhibitor (PI) ratio, host resistance and survival from infection.

However, a burn followed by infection with an organism capable of activating Hageman Factor further, reduces the immunoresistance mechanisms of the host below a threshold from which it can recover, if left untreated. We thought that this is the case, when burned hosts are infected with protease producing P. aeruginosa. Treatments, with protease inhibitors for example, which presumably bring the host immunocompetence back above the survival threshold, enhance survival of these burned infected hosts.

Evidence which we have generated which support this hypothesis are as follows:

1.   Burned, but not unburned, animals are highly susceptible to P. aeruginosa infections;[2,3]

2.   Only protease producing P. aeruginosa strains are lethal in burned hosts.[3-5]

3.   Lethality of protease non-producing strains increased in burned hosts with exogenous protease supplementation.[4,5]

4.   Survival of burned protease producing P. aeruginosa infected animals improved by protease inhibitor treatment; inhibitors effective only against host proteases improve survival as well as inhibitors effective against pseudomonas proteases.[6,16,17]

Furthermore, we also have data which demonstrate that trauma other than thermal injury, eg, the surgical trauma of laparotomy, increases total circulating host protease load.[18] Superimposition of a Candida albicans yeast infection in the surgically manipulated animals increases the total circulating protease load even more. This is accompanied by significant mortality. Even though the surgical procedure or infection alone significantly increase host protease load above control, these events, by themselves, are not lethal. Only when the circulating level of protease is increased to the level observed when both trauma and infection are imposed, does significant mortality occur. Thus, we believe that similar events occur in a traumatized host, regardless of the type of trauma, when that host is subsequently infected by a microorganism capable of activating Hageman Factor.

Evidence in support of the universality of our hypothesis (Fig. 7) is as follows:

1.   Prior infection of mice with a protease producing but non-lethal P. aeruginosa strain increases susceptibility to lethal yeast (C. albicans) infection in burned, but not unburned mice.[19]   However, injection of thermolysin can substitute for the protease producing P. aeruginosa infection.[19]

2.   Kaminishi et al showed that C. albicans protease(s) activate Hageman Factor.[21]   We believe that it is because of this that a significantly lower challenge dose of a high protease producing C. albicans causes lethal infection in burned mice compared to challenge using a low protease producing strain.[20]

3.   Actual measurements of circulating proteolytic activity in burned mice infected with a protease producing C. albicans show protease load to be higher than in mice only burned or only infected.[20] Mortality correlates with protease load in this model, and treatment with protease inhibitors increases survival.[20]

4.   Other microbial proteases, as well as bacterial endotoxin, activate Hageman Factor.[22,23]

5.   Lower inocula of microorganisms of various genera (each having the capacity to activate Hageman Factor) cause lethal infection in burned and surgically traumatized mice compared to normal mice. (Neely, AN, Holder IA, Trauma, Sepsis and Increased Circulating Proteolytic Activity abstract PW1.1 Kinin'91 Munich Symposium).

In conclusion, data from our laboratory are consistent with the following theory. Trauma of many kinds inflicted on a host causes increases in total protease load in the circulation. This increase is due, in part, to trauma associated Hageman Factor activation followed by down-line cascade system activation. Attendant to this activation, is reduced capacity to resist infection. When infection by microorganisms, which have the capacity to activate Hageman Factor further, is superimposed on the traumatized host, the immunoresistance capacity of the host is reduced below a threshold from which the host can recover. Any treatment which has the potential to increase immunocompetence above this threshold will increase survival. Some possibilities are presented in Figure 8. Continued experimentation is underway to further investigate the validity of this theory.

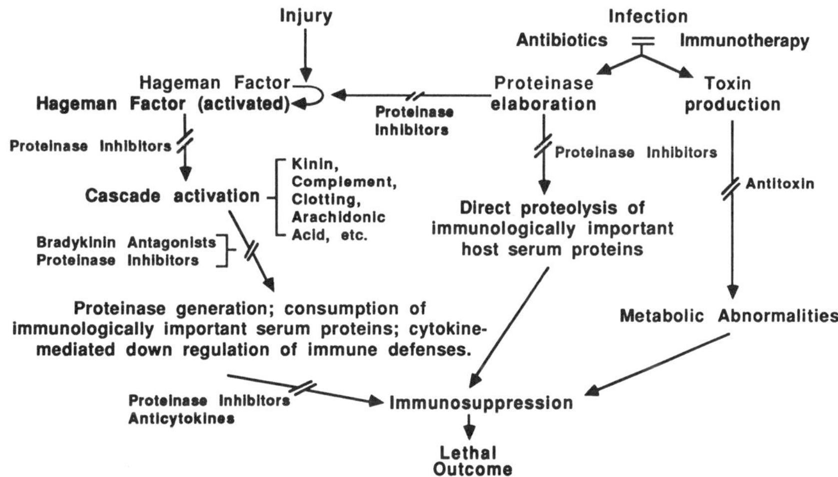

Figure 8. Traumatized host: microorganism interactions leading to lethal infection showing points at which conventional or experimental therapeutic intervention may increase host immunocompetence above the lethal threshold.

ACKNOWLEDGEMENT

These studies were supported, in part, from grants from the Shriners of North America.

REFERENCES

1.   Neely AN, Nathan P, Highsmith RF. Plasma proteolytic activity following burns. J Trauma1988; 28:362-367.

2.   Holder IA. Microbiology of the Burn Compromised Patient. In: Pathogenesis of Wound andBiomaterial-Associated Infections. Wadstrom T, Eliasson I, Holder I and Ljungh A (eds.). Springer-Verlag, London U.K., 1990: 9l-100.

3.   Stieritz DD, Holder IA. Experimental studies of the pathogenesis of infections due to Pseudomonas aeruginosa: description of a burned mouse model. J Infect Dis 1975; 131:688-691.

4.   Snell K, Holder IA, Leppla S, Saelinger DB. Role of exotoxin and protease as possible virulence factors in experimental infections with Pseudomonas aeruginosa. Infect Immun 1978; 19:839-845.

5.   Holder IA, Haidaris CG. Experimental studies of the pathogenesis of infections due to Pseudomonas aeruginosa: a extracellular protease and elastase as in vitro virulence factors. Can J Microbiol 1979;25:593-599.

6.   Holder IA. Experimental studies of the pathogenesis of infections due to Pseudomonas aeruginosa: effect of treatment with protease inhibitors. Rev Infect Dis 1983; 5:S914-S921.

7.   Holder IA, Wheeler R. Experimental studies of the pathogenesis of infections owing to Pseudomonas aerugionsa elastase, an IgG protease. Can J Microbiol 1984; 30:1118-1124.

8.   Kamata R, Yamatoto T, Matsumoto K, Maeda H. A Serratial protease causes vascular permeability reaction by activation of the Hageman factor-dependent pathway in guinea pigs. Infect Immun 1985; 48:747-753.

9.   Udaka K, Takeuchi Y, Movat HZ. A simple method for quantitation of enhanced vascular permeability. Proc Soc Exp Biol Med 1970;133:1304-1307.

10.   Holder IA, Neely AN. Hageman factor-dependent kinin activation in burn and its theoretical relationship to postburn immunosuppression syndrome and infection. J Burn Care Rehab 1990; 11:496-503.

11. Holder IA, Neely AN. Pseudomonas elastase acts as a virulence factor in burned hosts by Hageman factor-dependent activation of the host kinin cascade. Infect Immun 1989; 57:3345-3348.

12. Kaplan AP, Meier HL, Mandle R. The Hageman factor dependent pathways of coagulation, fibronolysis, and kinin-generation. Semin Thromb Hemostasis 1976; 3:1-26.

13. Homjima Y, Pierce JW, Pisano JJ. Hageman factor inhibitor in corn seeds: purification and characterization. Thromb Res 1980; 20: 149-162.

14. Matsumoto K, Yamamoto T, Kamata R, Maeda H. Pathogenesis of serratial infection:activation of the Hageman factor-prekallikrein cascade by serratial protease. J Biochem 1984; 96:739-749.

15. Bjornson AB, Bjornson HS. Theoretical relatioship among immunologic and hematologic sequelae of thermal injury. Rev Infect Dis 1984;6:704-714.

16. Neely AN, Holder IA. Experimental studies of the pathogenesis of infection due to Pseudomonas aeruginosa: unexpected results of treatment using Pseudomonas hyperimmune globulin plus minocycline and effects of minocycline on protease elaboration. Serodiag Immunother Infect Dis 1987; 1:193-200.

17. Holder IA, Neely AN. Combined host and specific anti-pseudomnonas directed therapy for Pseudomonas aeruginosa infections in burned mice: experimental results and theoretical considerations. J Burn Care Rehab 1989; 10:131-137.

18. Miller RG, Neely AN. Total serum protelytic activity in a murine model of post-operative Candida sepsis: prognostic implications and the effect of proteinase inhibition. J Surg Res, in press.

19. Neely AN, Law EJ, Holder IA. Increased susceptibility to lethal Candida infections in burned mice preinfected with Pseudomonas aeruginosa or pretreated with proteolytic enzymes. Infect Immun 1986; 52:200-204.

20. Neely AN, Holder IA. Effect of proteolytic activity on virulence of Candida albicans in burned mice. Infect Immun 1990; 58:1527-1531.

21. Kaminishi H, Tanaka M, Cho T, Maeda H, Hagihara Y. Activation of the plasma kallikrein-kinin system by Candida albicans proteinase. Infect Immun 1990; 58:2139-2143.

22.   Molla A, Yamamoto T, Akaike T, Miyoshi S, Maeda H.   Activation of Hageman factor and prekallikrein and generation of kinin by various microbial proteases.   J Biol Chem 1989;264:10589-10594.

23.   Morrison DC, Cochrane CG.   Direct evidence for Hageman factor (factor XII) activation by bacterial lipopolysaccharides (endotoxins).   J Exper Med 1974; 140:797-811.

INVOLVEMENT OF THE KININ-FORMING SYSTEM IN THE PHYSIO-PATHOLOGY OF RHEUMATOID INFLAMMATION

J.N. Sharma

Department of Pharmacology, School of Medical Sciences, University Sains Malaysia, Kubang Kerian 16150, Kelantan, Malaysia

SUMMARY: Kinins are potent mediators of rheumatoid inflammation. The components of the kinin-forming system are hyperactive in RA. Excessive release of kinins in the synovial fluid can produce oedema, pain and loss of functions due to activation of B_1 and B_2 receptors. These receptors could be stimulated via injury, trauma, coagulation pathways (Hageman factor and thrombin) and immune complexes. The activated B_1 and B_2 receptors might cause release of other powerful non-cytokines and cytokines mediators of inflammation, for example, PGE_2, PGI_2, LTs, histamine, PAF, IL-1 and TNF derived mainly from polymorphonuclear leukocytes, macrophages, endothelial cells and synovial tissue. These mediators are capable of inducing bone and cartilage damage, hypertrophic synovitis, vessels proliferation, inflammatory cells migration, and possibly angiogenesis in pannus formation. These pathological changes, however, are not yet defined in human model of chronic inflammation (RA). Hence , the role of kinin and its interacting inflammatory mediators would soon start to clarify the detailed questions they revealed in clinical and experimental models of chronic inflammatory joint diseases. Several B_1 and B_2 receptor antagonists are being synthesized in an attempt to study the molecular functions of kinins in inflammatory processes (RA, periodontitis and osteomyelitis), and they represent and important area for continued research in rheumatology. Future development of specific, potent and stable B_1 and B_2 receptor antagonists or combined B_1 and B_2 antagonists with y-IFN might serve as pharmacological basis of more effective rationally-based therapies for RA. This may lead to significant advances in our knowledge of the mechanisms and therapeutics of rheumatic diseases.

INTRODUCTION

Rheumatoid arthritis (RA) is a chronic inflammatory disease which primarily affects the joints with massive hyperplasia of the connective tissue, invasive destruction of the cartilage, capsule and supportive tissues, and bone erosions (1). Despite intensive research over the past four decades, the cause of RA is not yet known. In fact, the treatment of this disease is clearly based on relieving symptoms with the use of numerous antiinflammatory and analgesic agents. However, these drugs also pose the potential threat of severe adverse effects.

Hyperactivity of the kallikrein-kinin system (KKS) has long been implicated as an aetiological factor of inflamed rheumatoid joints (2,3) because kinins can cause inflammatory reactions, including vasodilatation, increased vascular permeability, leucotaxis and pain (4,5). In addition, kinins modulate the release of other potent inflammatory agents, such as prostaglandins (PGs), prostacyclin (PGI$_2$), leukotrienes (LTs), histamine, endothelial derived relexing factor (EDRF), platelet activating factor (PAF), tumor necrosis factor (TNF) and interleukin-1 (IL-1) in the process of chronic rheumatoid inflammation (6-8). In recent years, kinin-forming components have been included in the list of potential mediators of synovitis and bone resorption (9,10). Therefore, the purpose of this paper is to discuss the significant role of KKS in the pathogenic mechanisms of RA and thus to the development of new rationally-based therapies.

THE KININ GENERATING AND INACTIVATING FACTORS

Several components are involved in generating and metabolizing kinins in order to regulate their cellular and molecular activities. The kinins are pharmacologically active polypeptides, generated in the tissues and body fluids as a result of the enzymatic action of kallikreins on kininogens.

Kinins-forming, activating and inhibiting factors are summarised in figure 1. The kinin family includes bradykinin (BK, Arg-Pro-Pro-Gly-Phe-Ser-Pro-Phe-Arg), kallidin (Lys-BK,Lys-Arg-Pro-Pro-Gly-Phe-Ser-Pro-Phe-Arg) and methionyl-lysyl-BK (Met-Lys-BK, Mct-Lys-Arg-Pro-Pro-Gly-Phe-Ser-Pro-Phe-Arg). Lys-and Met-Lys Bk are converted into BK by aminopeptidases (11-13) present in plasma and urine. Kinins are normally present in BK receptors, mainly B_1 and B_2 (14) to produce biological actions. These peptides are rapidly (< 0.15 sec) inactivated by the circulating kininases (4).

Kininogens are multifunctional proteins derived mainly from α_2 globulin. In human, two forms of kininogens are high molecular weight kininogen (HMWK) and low molecular weight kininogens (LMWK) (15). They are synthesized in the liver and circulate in the plasma and other body fluids. These kininogens vary from one another in molecular weight, susceptibility to plasma and tissue kallikrein as well as in physiological activities (16). In addition T-kininogen has been discovered in the rat as an important acute phase reactant of inflammation (17,18). T-kininogen releases T-kinin (Ile-Ser BK) by the enzymatic action of T-kallikrein in rats (19).

The two natural kallikreins are plasma and tissue. Tissue kallikrein in found in various organs, including synovial tissue (20-22). These kinin-forming enzymes differ from each other in molecular weights, biological functions, physicochemical and immunological properties (23). The tissue kallikrein is synthesized in the cells as a precursor and converted into active from by the cleavage of an amino-terminal peptide (24). Active tissue kallikrein acts on LMWK to release Lys-BK. A kallikrein-like enzyme has been reported in basophils and mast cells (25). The plasma kallikrein is found in the circulation in an inactive form, called as prekallikrein or Fletcher factor (26). Inactive prekallikrein is activated to kallikrein by activated Hageman factor (factor XIIa), and remaining inactive factor XII to active factor XIIa by kallikrein through a positive feedback reaction (27). The plasma prekallikrein and HMWK are present together in a complex form (28). Whereas, factor XIIa and factor XI circulate with HMWK in bound form (29), therefore, factor XI is converted to XIa (30) for the participation in the intrinsic coagulation

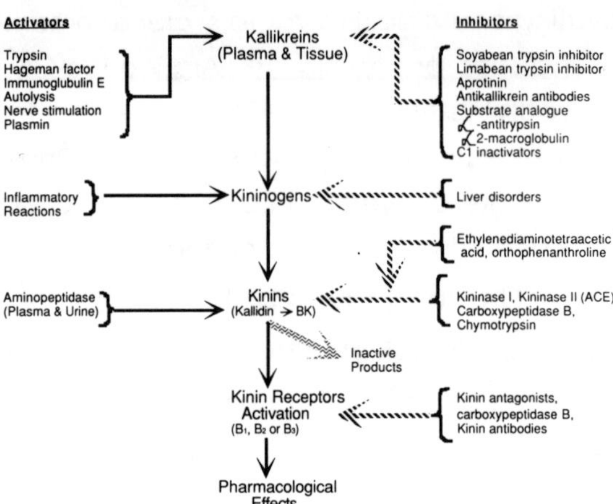

Figure 1. Kinin generating and inactivating system

cascade. During immunological reaction, the connective tissue proteoglycans and mas cells heparin might act as initiating surface for initial activation of Hageman factor (31,32). It seems, that the kinins may be generated in parallel with the formation of thrombin at the inflammatory sites, since inactive plasma kallikrein can be activated by coagulant Hageman factor (33).

The kininases, kinins inactivating enzymes, are generated in the plasma, endothelial cells and tissues to regulate the physiological functions of kinins. These are known as kininase I, kininase II (an angiotensin convering enzyme) and enkephalinase. In plasma, kninase I cleaves the C-terminal arginine to form des-Arg^9-BK (34), kininase II converts the octapeptides into a pentapeptide (Arg-Pro-Pro-Gly-Phe) and a tripeptide Ser-Pro-Phe) (34).

THE KININ COMPONENTS IN JOINT INFLAMMATION

Formation of kinin has been proposed as a pathogenic factor in the production of inflammatory symptoms in inflamed rheumatoid joints (2,3).

In fact, synovial fluids from patients with RA contain a kinin forming enzyme and its substrate (2,3,35-37). Free kinin levels are raised in the synovial fluids obtained from RA and gouty patients. In addition, high concentrations of Kallikrein-like activity have been demonstrated in the inflamed rheumatoid synovial (21,22). These kinin-generating components in inflamed joint may derived from plasma (37), since high concentrations of total plasma kininogen are found in the peripheral circulation (38-41). Kinins release can be activated in response to tissue injury to cause stimulation of sensary pain fibre and the release of other potent mediators of inflammation (4-7). As a result, vascular permeability and blood flow to the damaged parts are increased, and a full inflammatory response is obtained. The activation and release of kinin in the effusions from rheumatoid joints might be linked with the stimuli, such as interleukines and immune complexes to cause degranulation in the process of leukocyte tissue kallikrein release within the inflamed sites (42-44). The released tissue kallikrein could then generate kinin from kininogens, thereby providing inflammatory reactions. Interrelationship been the IL-1 and kinins is further substantiated by the fact that the IL-1 is produced by the inflamed synovium (45,46) and it can be detected in significant amounts in synovial fluids (47,48). However, formation of PGE_2, LTB_4 and PGI_2 in rheumatoid synovium may appear to be linked with the appearance of leucocytes in the synovial fluid (49,50). Thus, the kinin-activation might be mediated by hyperimmune activity and or instrinsic coagulation pathway action.

Recently, it has been discovered that kinins mediate osteoblasts formation and subsequent enhancement of bone resorption in vitro from cultured mouse calvarian bones (10). The bone resorption property of kinins has been atributed to high endogenous PGE_2 formation in bone tissue (10). Bradykinin is capable of stimulating bone resorption in vitro, which can be abolished by administration of several PG synthesis inhibitors (10). Hence, the kinins actions on bone resorption Is medlated through high concentrations of PG formation in bone tissue. This view is compatible with the findings that PGE_2 and PGI_2 are potent mediators for bone resorption in vitro (51). Kinins stimulate also PG formation in isolated osteoblast-like cells from mouse calvarial bone of neonatal mouse and in the cloned murine osteoblastic cell lineage MC3T3-EI (52), as well as in isolated human osteoblast-like cells (53, 9). It is indicated that osteoblasts are equipped with B_1 receptors mediating PGE_2 and PGI_2 formation and subsequent bone resorption (10). Whereas B_2 receptors are involved in mediating pain, increased vascular permeability, vasodilatation, oedema and PGE_2 release from the osteoblasts (1-3). It is possible therefore that in areas adjacent to inflammatory processes, bone tissue may become more sensitive to kinins due to availability of increased number of B_1 receptors in addition to receptors. In porcine cultured articular chondrocytes, kinins cause

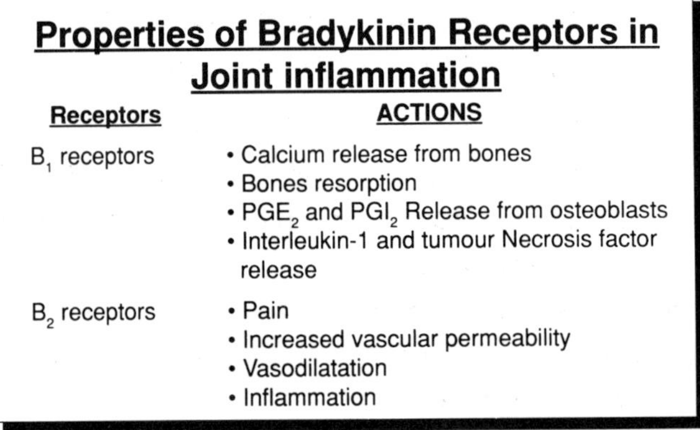

Figure 2. Possible actions of B_1 and B_2 kinin receptors in joint inflammatory disease.

increased inositol phosphate generation, intracellulal free calcium ions, and PGs synthesis via activating B$_2$ receptors (54). The properties of BK receptors in joint inflammatory disease are presented in figure 2. These findings suggest strongly that kinins could be considered as prime mediators in inducing significant pathological changes seen in arthritic joints. Although, further investigations on the functions of kinins in the contribution to bone and cartilage destruction in human arthritis require further investigations.

Kinins are known also to mediate the release of cytokines and non-cytokines mediators of inflammation.

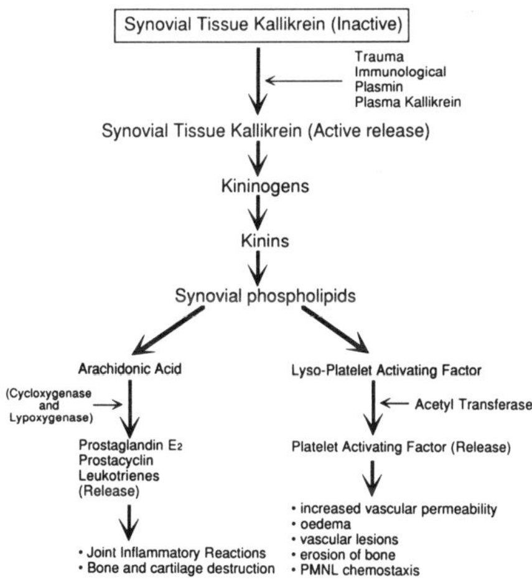

Figure 3.   A schematic presentation of the release of kinin and kinin-induced release of arachidonic acid products and platelet activating factor (non-cytokines mediators) from the synovial phospholipids in causing pathological changes in joint inflammation.

INTERACTIONS BETWEEN THE KININS AND NON-CYTOKINES INFLAMMATORY MEDIATORS

Non-cytokines inflammatory mediators, for example, PGE_2, PGI_2, LTs, histamine, PAF and EDRF are released by kinins (4-7, 12). High levels of extra cellular phospholipase A_2 have been detected from human synovial fluid (55). This might originate from macrophages granulocytes and or lymphocytes. Phospholipase A_2 is activated by kinins to release the arachidonic acid metabolites (PGs and LTs) as well as PAF in the process of rheumatoid inflammation (4,5,7). The endothelial cells and fibrobasts are capable of stimulating PG-synthesis mediated by kinins (56). The role of endothelium in chronic inflammatory synovitis has been discussed in detail by Ziff (59). Increased concentrations of PGE_2, LTB_4 and PAF are present in the synovial fluid obtained from arthritic patients and adjuvant-induced arthritis in rabbits (49,50,58). It is possible that PAF may be involved in attracting inflammatory cells into the arthritic joint to cause oedema, vascular lesions and erosion of bone (58,7,59). Figure 3 shows the kinins-induced release of PGE_2, PGI_2, LTs and PAF, and their relevance to inflamed rheumatoid joints.

INTERACTIONS BETWEEN THE KININS AND CYTOKINES INFLAMMATORY MEDIATORS

Kinins have significant stimulant effects on TNF and IL-1 release from macrophages (8), and most probably from endothelial cells. Bradykinin stimulated release of both (TNF and IL-1) cytokines from P388-D1 and RAW264.7 murine macrophages may be mediated by the activation of B_1 kinin receptors (8). Tumor necrosis factor and IL-1 can stimulate bone resorption, cartilage damage, hypertrophic synovitis, inflammatory cell migration, vessels proliferation and pannus formation (Fig.4). Both TNF

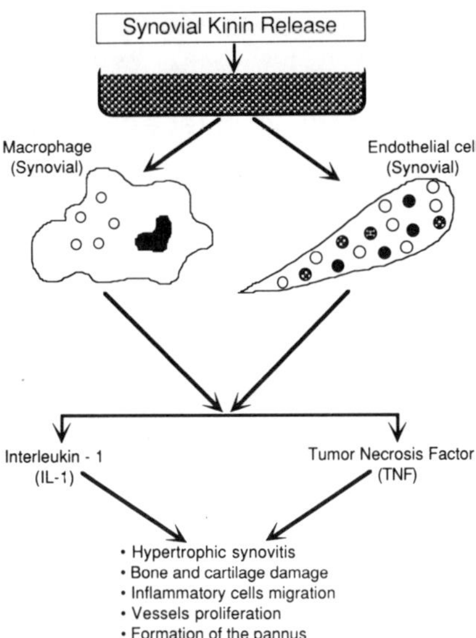

Figure 4. A hypothetical model showing kinin-induced release by cytokines inflammatory mediators (interleukin-1 and tumor necrosis factor) in the pathogenesis of rheumatoid inflammation.

and IL-1 could induce the release of at least three metaloproteinase, such as degrade gelatin, collagen and proteoglycan (60). These mediators might be released from mononuclear cells isolated from synovial tissue obtained from patients with RA and may contribute to the pathology of the inflamed joints by formation of the pannus (60). The roles of TAF and IL-1 in patho-physiology of RA have been extensively reported by Bonta et al. (61) and Miossec (62). Recently, inhibitory effects of y-interferon (y-INF) on BK-induced bone resorption and PGE_2 formation in cultured mouse calvarial bone have been demonstrated (63). Although, it has been shown that y-IFN can inhibit bone resorption by unknown mechanism unrelated to PGE_2 formation (63). Based on these findings, il is possible that the kinins might induce bone resorption by non-cyclooxygenase pathaway. The role of 5-lypooxygenase products release by kinins in causing bone resorption has yet to be investigated, since LTB_4 levels are raised in synovial fluid from patients with RA (50).

KININ ANTAGONISTS IN JOINT INFLAMATION

There are at least two subtypes of kinin receptors that are termed as B_1 and B_2 (14). These receptors have been pharmacologically recognized by using different kinin analogues with agonistic and antagonistic actions (14,64-66). It is interesting that the number of B_1 receptors increases during tissue injury by forming des-Arg^9 BK through the action of Kininase I (14). Activation of B_1 receptors causes bone resorption, calcium release from bone, PGE_2 and PGI_2 release from osteoblast (10). Stimulation of B_2 receptors mediated pain, vasodilatation, enhances the release of histamine and PGE_2 and bone resorption (67,10). These properties of B_1 and B_2 receptors are documented in figure 2.

In recent years, several BK antagonists have been synthesized (64-66) with major therapeutic goals to block the pathological conditions caused by either enhanced production or inadequate metabolism of kinins. A B_2 receptor antagonist, DArg Hyp^3 DPhe^7 BK, inhibits the development of carrageenan-induced oedema in rats (68). The B_1 receptors antagonist (des-Arg^9-Leu^8-BK) can inhibit des-Arg^9- BK-induced release of ^{45}Ca from prelabelled neonatal mouse calvarial bone in vitro (9,10). New highly potent BK antagonists include D-Arg (Hyp^3, D-Phe^7, Leu^8)-BK (66) and Hoe 140 (65). These antagonists are under experimental investigations to stablish the future prospects of their clinical utilities in rheymatology. It is suggested that the combinations of B_1 and B_2 antogonists with y-IFN may provide additive properties in acheiving greater anti-rheumatic therapeutic values, since y-IFN causes inhibition of abnormal biochemical changes taking place in joint inflammation (63). Figure 5 represents the possible mode of kinin release, and the functions of B_1 and B_2 receptors in RA.

CONCLUSION

Numerous in vitro and in vivo investigations conducted in the last few years strongly indicate that kinin system is hyperactive in RA. Kinins can directly as well as indirectly (by releasing non-cytokines and cytokines mediators of joint inflammatory disease) cause significant damage of rheumatoid joints. Although, extensive clinical and experimental investigations will have to be carried out to determine the exact functions of the kinin receptors with special emphasis in rheumatology. It is clear that the development of specific and effective new rationally-based B_1 and B_2 receptors antagonists may prove to be novel directions in pharmacotherapeutics of RA.

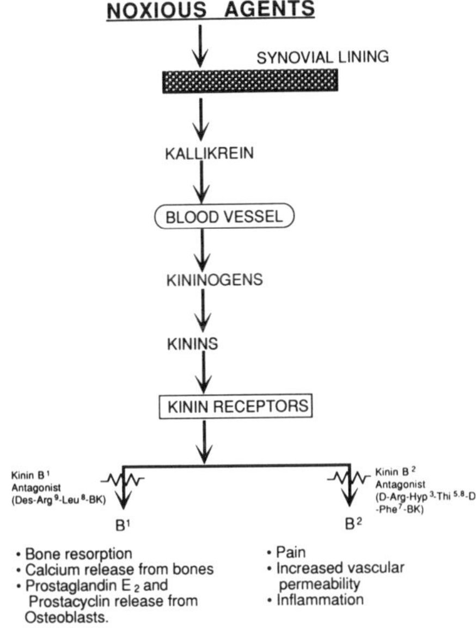

Figure 5. A schematic presentation of the possible mode of raised kinin-formation in the process of joint inflammatory disease. Kinin may cause pathogenic changes within the rheumatoid joints via stimulating B_1 and B_2 kinin receptors antagonists may serve as novel therapeutic agents for joint inflammatory diseases.

ACKNOWLEDGEMENTS

I would like to thank Prof. Dr. Mohd. Roslani A. Majid, Dean and Director for his support. This work has been funded by the U.S.M. short-term grants (122 500 0200 and 122 0500 0210). I appreciate Miss Halimah Mohd Noor for her secretarial assistance.

REFERENCES

1.   Dayer JM. Aspects of resorption and formation of connective tissue during chronic inflammation in rheumatoid arthritis. Eur J Rheumatoal Inflammation 1982; 5: 457-468.

2.   Melmon KL, Webster WE, Goldfinger SE, Seegmiller JE.   The presence of a kinin in inflammatory synovial effusions from arthritides of varing etiologies. Arthritis Rheumatism 1967; 10: 13-20.

3.   Keele CA, Eison V. Plasma kinin-formation in rheumatoid arthritis. Adv Exp Med Biol 1970; 8: 47-475.

4.   Sharma JN. The role of the kallikrein-kinin system in joint inflammatory disease. Pharmacol Res 1991; 23: 105-112.

5.   Sharma, JN. The role of kinin system in joint inflammatory disease. Eur J Rheumatol Inflammation 1991; 11: 30-37.

6.   Sharma JN. Mohsin SSJ. The role of chemical mediators in the pathogenisis of inflammation with emphasis on the kinin system. Exp Pathol 1990; 38: 73-96.

7.   Sharma JN. Pro-inflammatory actions of the platelet activating factor: relevance to rheumatoid arthritis. Exp Pathol 1991; 43: 47-50.

8. Tiffany CW, Burch RM. Bradykinin stimulates tumor necrosis factor and interleukin-1 release from macrophages. FEB 1989; 247: 189-192.

9. Lerner UH, Jones IL, Gustafson, GT. Bradykinin, a new potential mediator of inflammation-induced bone resorption. Arthritis Rheumatism 1987; 30: 530-540.

10. Ljunggren O, Lerner UH. Evidence for BK_1 bradykinin-receptor-mediated prostaglandin formation in osteoblasts and subsequent enhancement of bone resorption. Br J Pharmacol 1990; 101: 382-386.

11. Sharma JN. Interrelationship between the kallikrein-system and hypertension. a review. Gen Pharmacol 1988; 177-187.

12. Sharma JN. The kinin system and prostaglandins in the intestine. Pharmacol Toxicol 1988; 63: 310-316.

13. Sharma JN. Does kinin mediate the mod of action of angiotensin congerting enzyme (ACE) inhibitors. Gen Pharmacol 1990; 21: 451-457.

14. Regoli D, Barabe J. Pharmacology of bradykinin and related kinins. Pharmacol Rev 1980; 32: 1-42.

15. Nagayasu T, Nagasawa S. Studies of human kininogens. Isolation, characterization, and cleavage by plasma kallikrein of high molecular weight (HMW) kininogens. J. Biochem 1979; 85: 249-258.

16. Müller-Esterl W, Iwanaga S, Nakanishi S. Kininogens revisited. Trends Biochem Sci 1986; 11: 336-339.

17. Okamoto H, Greenbaum LM. Isolation and structure of T-kinin. Biochem Biophys Res Commun 1983; 112: 701-708.

18. Okamoto H, Greenbaum LM. Isolation and properties of two rat plasma T-kininogens. Adv Exp Med Biol 1986; 198A: 69-75.

19. Okamoto H, Greenbaum LM. Pharmacological properties of T-kinin. Biochem Pharmacol 1983; 32: 2637-2638.

20. Nustad K, Vaaje K, Pierce JV. Synthesis of kallikrein by rat kidney slices. Br J Pharmacol 1975; 53: 229-234.

21. Sharma JN, Zeitlin IJ, Deodhar SD, Buchanan WW. Detection of kallikrein-like activity in inflamed synovial tissue. Arch Int Pharmacodyn Ther 1983; 262: 279-286.

22. Al-Haboubi HA, Bennett D, Sharma JN, Thomas GR, Zeitlin IJ. A synovial amidase acting on tissue kallikrein-selective substrate in clinical and experimental arthritis. Adv Exp Med Biol 1986; 198B: 405-411.

23. Schachter M. kallikreins (kininogenases) - a group of serine proteases with bioregulatory actions. Pharmacol Rev 1980; 31: 1-17.

24. Takada Y, Skidgel RA, Erdor EG. Purification of human urinary prokallikrein, identification of the site of activation by the metalioproteinase thermolysin. Biochem J 1985; 232: 851-852.

25. Proud D, Togias A, Naclerio RM, Crush SB, Norman PS, Lichtenstein LM. Kinins are generated iun vivo following nasal airway challenge of allergic individuals with allergen. J Clin Invest 1983; 72: 1678-1685.

26. Weiss AS, Gallin JL, Kaplan AP. Fletcher factor deficiencyt: a diminished rate of Hageman factor activation caused by absence of prekallikrein with abnormalities of coagulation, fibrinolysis, chemotactic activity and kinin generation. J. Clin Invest 1974; 53: 622-633.

27. Cochrane CG, Revak SD, Wuepper KD. Activation of Hageman factor in solid and fluid phases. A critical role for kallikrein. J Exp Med 1973; 138: 1564-1583.

28. Griffin JH, Cochrane CG.   Mechanism for the involvement of high molecular weight kninogen in surface-dependent reactions of Hageman factor. Proc Natl Acad Sci 1976; 73: 2554-2558.

29. Mandle R, Coleman RW, Kaplan AP.   Identification of prekallikrein and HMW-Kininogen as a complex in human plasma.   Proc Natl Acad Sci 1976; 73: 4176-4183.

30. Thompson RE, Mandle R, Koplan AP.   Association of factor XI and high molecular weight kininogen in human plasma. J Clin Invest 1977; 60: 1376-1380.

31. Hojima Y, Cochrane CG, Wiggins RC, Austen KF, Stevens RL. In vitro activation of the contact (Hageman factor) system of plasma by heparin and controitin sulfate E. Blood 1984; 63: 1453-1459.

32. Silverberg M, Diehl S.   The autoactivation of factor XII (Hageman factor) induced by low-M heparin and dextran sulfate.   Biochem J 1987; 248: 715-720.

33. Meier HL, Pierce JV, Kaplan AP.   Activation and function of human Hageman factor-the role of high molecular weight kininogen and prekallikrein. J Clin Invest 1977; 60: 18-31.

34. Erdos EC. Some ole and some new ideas on kinin metabolism. J Cardiov Res 1990; 15 (Suppl. 6): S20-S24.

35. Eisen V. Formation and function of Kinin.   Rheumatology 1970; 3: 103-168.

36. Armstrong D, Jepson JB, Keele CA, Stewart JW. Pain-producing substance in human plasma. J. Physiol 1957; 135: 350-361.

37. Jasani MK, Katori M, Lewis GP, Intracellular enzymes and kinin enzymes in synovial fluid in joint diseases. Ann Rheum Dis 1969; 28: 497-511.

38. Sharma JN, Zeitlin IJ, Buchanan WW, Dick WC. The action of aspirin on plasma kininogen and other proteins in rheumatoid patients: relationship to disease acitivity. Clin Exp Pharmacol Physiol 1980; 7: 347-353.

39. Sharma JN, Zeitlin IJ, Brooks PM, Dick WC. A novel relationship between plasma kininogen and rheumatoid disease, Agents Actions 1976; 6: 148-153.

40. Zeitlin IJ, Sharma JN, Brooks PM, Dick WC. Raised plasma kininogen levels rheumatoid arthritis: response to therapy by non-steroidal anti-inflammatory drugs. Adv Exp Med Biol 1976; 70: 335-343.

41. Brooks PM, Dick WC, Sharma JN, Zeitlin IJ. Changes in plasma kininogen levels associated with rheumatoid acitivity. Br J Clin Pharmacol 1974; 1: 351P.

42. Figueroa CD, MacIver AG, Bhoola KD. Identification of a tissue kallikrein in human polymorphonuclear leucocytes. Br J Haematol 1989; 72: 321-328.

43. Figueroa CD, Bhool KD. Leucocyte tissue kallikrein: an acute phase signal for inflammation. In: The Kallikrein-Kinin System in Health and Disease. Fritz H, Schmidt I, Dietze G, editors. Germany: Limbach-Verlag Braun schweig, 1989 ; 311-320.

44. Epstein WV, Melmon KL, Tan M, Stoff J. Kinin generation caused by IgG-rheumatoid factor complex. J Clin Invest 1968; 47: 30-32.

45. Wood DD, Ihrie EJ. Hamerman D. Release of interleukin-1 from human synovial tissue in vitro. Arthritis Rheum 1985; 28: 853-862.

46. Danis VA, March LM, Nelson DS, Brooks PM. Interleukin-1 secretion by peripheral blood monocytes and synovial machrophages from patients with rheumatoid arthritis. J Rheumatol 1987; 14: 33-39.

47. Wood DD, Ihrie JE, Dinarello CA, Cohen PL. Isolation of an interleukin-1-like factor from human joint effusions. Arthritis Rehum 1983; 26: 975-983.

48. Nouri AME, Panayi GS, Goodman SM. Cytokines and the cyronic inflammation of rheumatic disease. I. The presence of interleukinin-1 in synovial fluid. Clin Exp Immunol 1984; 55: 295-302.

49. Higgs GA, Vane JR. Hart FD. Wojtulewski JA. Effects of anti-inflammatory drugs on prostaglandin in rheumatoid arthritis. In: Prostaglandin Synthetase inhibitors.Robinson HJ, Vane JR, editors. New York: Raven Press 1974; 165-173.

50. Davidson EM, Rae SA, Smith MJH. Leukotriene B_4 in synovial fluid. J Pharm Pharmacol 1982; 34: 410.

51. Harvey W, Bennett A. Prostaglandins and the mechanism of bone resorption. In: Prostaglandins inBone Resorption. Florida: CRC Press 1988: 43-46.

52. Lerner UH, Ransjo M, Ljunggren O. Bradykinin stimulates production of prostaglandin E_2 and prostacyclin in murine osteoblasts. Bone and Mineral 1989; 5: 139-154.

53. Ljunggron O, Rosonquist J, Ransjo M, Lerner UH. Bradykinin stimulated prostaglandin E_2 formation in isolated human osteoblast-like cells. Biosc. Res. 1990; 10: 121-128.

54. Benton HP, Jackson TR, Hanley MR. Identification of a novel inflammatory stimulant of chondrocytes. Biochem J 1989; 269: 861-867.

55. Hara S, Kudo I, Chang HW, Matsuta K, Miyamoto T, Inoue K. Purification and characterization of extracellular phospholipase A_2 from human synovial fluid in rheumatoid arthiritis J Biochem 1989; 105: 395-399.

56. Conklin BR, Burch RM, Steranka LR, Axelrod J. Distinct bradykinin receptors mediate stimulation of prostaglandin synthesis by endothelial cells and fibroblasts. J Pharmacol Exp Ther 1988; 244: 646-649.

57. Ziff M. Role of the endothelium in chronic inflammatory synovitis. Arthritis Rheum 1991; 34: 1345-1352.

58. Pettipher ER, Higgs GA, Henderson BN. PAF-acether in chronic arthiritis. Agents Actions 1987; 21: 98-103.

59. Archer CB, Page CP, Morley J, MacDonald DM. Accumulation of inflammatory cells in response to intracutaneous platelet activating factor (PAF-acether) in man. Br J Dermatol 1985; 112: 285-290.

60. Larrick JW, Kunkel SL. The role of tumor necrosis factor and interleukin-1 in immunoinflammatory response. Pharmaceutical Res 1988; 5: 129-139.

61. Bonta IL, Ben-Efraim S, Mozes T, Fieren MWJA. Tumour necrosis factor inflammation: relation to other mediators and to macrophage anti-tumour defence. Pharmacol Res 1991; 24: 115-130.

62. Miossec P. The role of interleukin-1 in the pathogenesis of rheumatoid arthritis. Clin Exp Rheumatol 1987; 5: 305-308.

63. Lerner UH, Ljunggren O, Ransjo M, Klaushofer K, Peterlik M. Inhibitory effects of y-interferonon bradykinin-induced bone resorption and prostaglandin formation in cultured mouse calvarial bones. Agents Actions 1991; 32: 305-311.

64. Vevrek RJ, Stewart JM. competitibve antagonists of bradykinin. Peptides 1985; 6: 161-164.

65. Hock FJ, With K. Albus U, Linz W, Gerhards HJ. Weimer G, Henke St, Breipohl G, Konig W, Knolle J, Scholkens BA. Hoe 140 a new potent and long acting bradykinin antogonist: in vitro studies. Br J Pharmacol 1991; 102: 769-773.

66. Regoli D, Rhaleb NE, Tousignant C, Rouissi N, Nantel F, Jukic d, Drapeau G. New highly potent bradykinin B_2 receptor antagonists. Agents Actions 1991; 34: 138-141.

67. Regoli D. Neurohumoral regulation of precapillary vessels: the kallikrein-kinin system. J Cardiovasc Pharmacol 1984; 6: S401-S412.

68. Burch RM, DeHaas C. A bradykinin antogonist inhibits carrageenan adema in rats. Nauyn-Schmiedeberg's Arch Pharmacol 1990; 342: 189-193.

AAS 38/III
Recent Progress on Kinins
© 1992 Birkhäuser Verlag Basel

MICROBIAL PROTEINASES AS AN UNIVERSAL TRIGGER OF KININ GENERATION IN MICROBIAL INFECTIONS

H. Maeda, K. Maruo, T. Akaike, H. Kaminishi and Y. Hagiwara

Department of Microbiology, Kumamoto University Medical School Kumamoto 860, Japan and Department of Oral Bacteriology, Fukuoka Dental College, Fukuoka 814, Japan

INTRODUCTION

Microbial proteinases induce multiple pathological, biochemical and clinical consequences as listed in Table 1. For instance, pain and edema at the sites of infection are the commonest symptoms and they are resulted from proteolytic action of microbial proteinases which trigger activation of kinin generating cascade (1). Through our recent research using serratial, pseudomonal, candidal and many other microbial proteinases, it became apparent that a key stone of above clinical symptoms was associated with kinin generation *in vivo* (1-7). Furthermore, by providing optimal environment for bacterial growth by degrading IgA, and other antibodies, complement systems etc (8-11), pathological events caused by these proteinases undergo multifactorially, and proteinases facilitate their propagation and spreading, and augment clinical symptoms (1,12,13). Here we describe detailed molecular mechanisms of kinin generation by microbial proteinases as well as house dust mite proteinase.

ACTIVATION OF BRADYKININ GENERATING CASCADE

We have initially found this cascade (Fig. 1) was highly enhanced when serratial 56K proteinase was injected into the skin resulting in a marked increase in vascular permeability (1-7). *In vitro* study using purified zymogens such as factor XII (*Hageman* factor) and prekallikrein showed all microbial proteinases activated one or more steps in the bradykinin generating cascade (Fig 2-4). One group of proteinases activates factor

Table 1.   Pathogenic mechanism of bacterial proteinases

1. Inactivation of plasma proteinase inhibitors and immunoglobulins.
2. Activation of bradykinin generating cascade: factor XII, PK, KNG.
3. Activation of clot forming cascade: factor XII, X, II, etc.
4. Inactivation of complement system and chemotaxis: C5a, etc.
5. Tissue degradation and spreading.
6. Cytotoxicity to cells.   Internalization via complex with α2M, and regeneration of proteinase activity in cells.
7. Activation of factor X and plasminogen, and activation of influenza virus infectivity (~ 100 fold)

XII (to XIIa) such as *Serratia* 56K; second group such as that of *Vibrio vulnificus* activates both factor XII and prekallikrein.   Further, all proteinases including acid proteinases of *Candida albicans* generated bradykinin directly from high molecular weight kininogen.   A few proteinases such as *Staphylococcus* V8 and *Streptomyces* proteinase affected only at this last step (1,3).

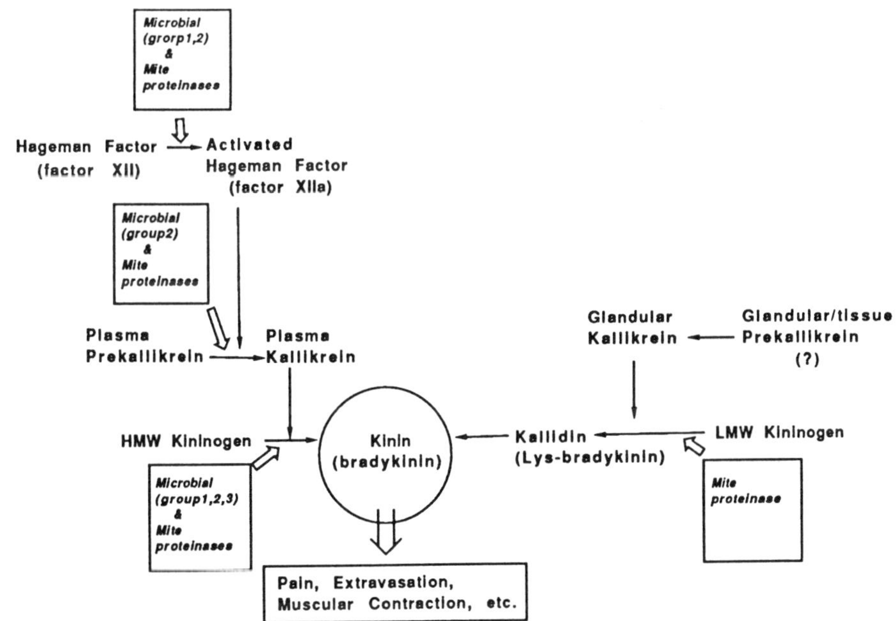

Figure 1.   A scheme of kinin generating cascade

These results demonstrated that microbial proteinases did generate bradykinin *in vivo*. Thus, it would be most reasonable to assign a cause of pain and edema to bradykinin. A possible involvement of histamine or other endogenous factors by these proteinases such as serratial 56K proteinase for the induced vascular permeability was denied in guinea pig experiments (2,6,7).

Candida albicans proteinase activate factor XII or generated kinin from high molecular weight kininogen even though its in vitro optimal pH is about 4 (acid proteinase). Although not shown in this article, *Serratia* 56K, *Pseudomonas* alkaline proteinase or *Candida* acid proteinase activate clotting cascade, ie., factor XII, XI, X and II even in the presence of whole human plasma.

Direct generation of bradykinin from high molecular weight kininogen is shown in Table 2. Three steps in kinin generating system (activations of both factor XII and prekallikrein, and direct kinin release) were examined using purified factor XII, prekallikrein and high molecular weight kininogen (all from guinea pig plasma), respectively, using appropriate synthetic peptide substrates. The results are shown in Fig. 2-4. The activated factor XII or kallikrein generation in normal plasma is examined and the results are shown in Fig. 4.

Table 2 Kinin generation from HMWKNG[a] from various proteinases

Proteinase	Kinin release (ng/ml)[b]	
	0.5 µg proteinase	1.0 µg proteinase
HMWKNG alone	<10	<10
Df-proteinase	240	912
Serratia proteinase	180	384
V8 proteinase	125	250
Streptomyces proteinase	96	224

[a] HMWKNG; high molecular weight kininogen
[b] Quantitated by enzyme immunoassay

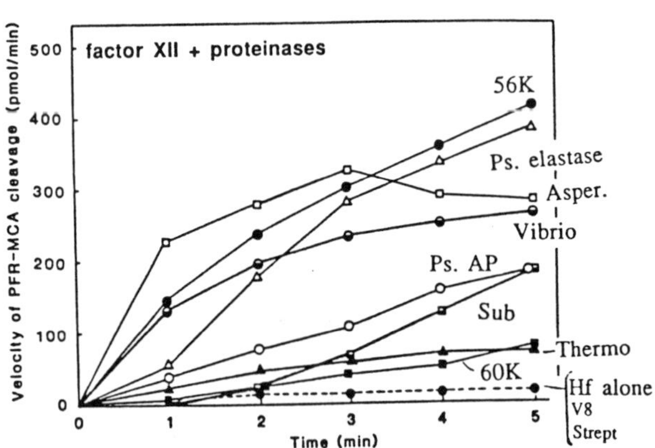

Figure 2. Activation of factor XII by microbial proteinase. See ref. 3 for details.

We have then tested a possible generation of lysyl-bradykinin (or kallidin) from low molecular weight kininogen by various microbial and house dust mite (*Dermatophagoides farinae*) proteinases (Fig. 1, Table 3) The mite proteinase we used here is serine type (14) and rapid generation of bradykinin is noted (15). It should be noted that the mite proteinase generated kinins from low and high molecular weight kininogens. All these results are consistent to the proposed scheme of Fig. 1.

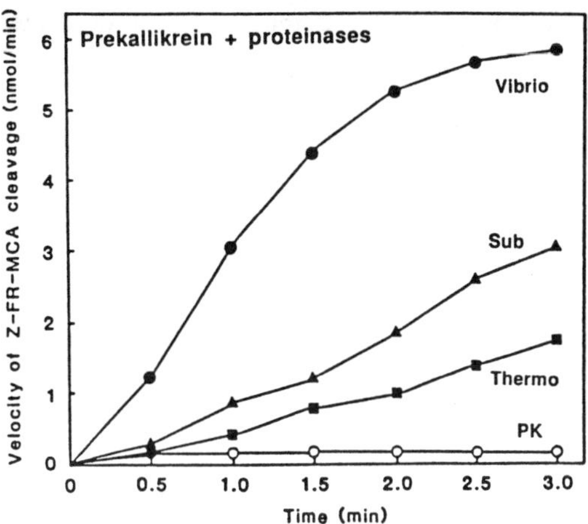

Figure 3. Activation of purified PK by microbial proteinases. See ref. 3 for details.

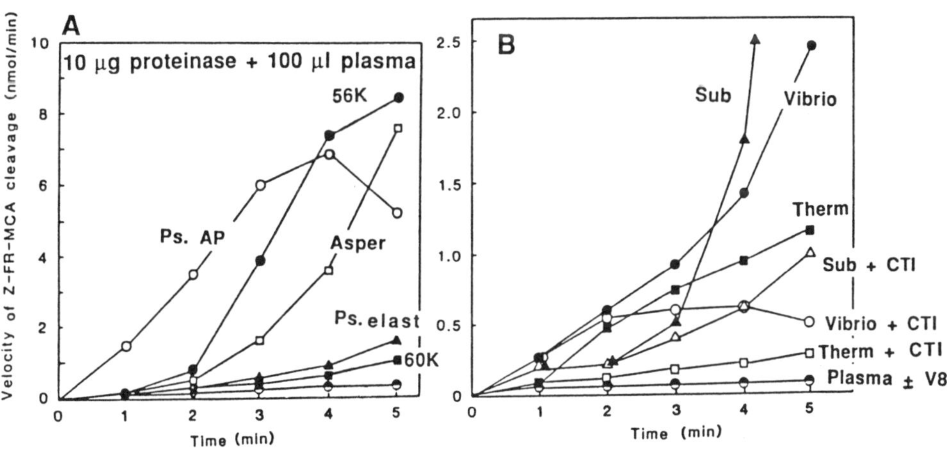

Figure 4. Generation of KK activity in guinea pig plasma by microbial proteinases and inhibition by SBTI or CTI (●——● all proteinases + STI or CTI). See ref. 3 for details.

Table 3. Activation of Hageman factor and prekallikrein and generation of bradykinin by various microbial and mite proteinases

Proteinase	Activation of		Generation of kinin from HMWKNG	Generation of kinin from LMWKNG	Inhibited by	
	Hf	PK			CTI	SBTI
Serratia 56KP	+	-	+	-	-	-
Serratia 60KP	+	-	+	n.d.	-	-
Serratia 73KP	+	-	+	n.d.	-	-
Ps. alkaline proteinase	+	-	+	-	-	-
Ps. elastase	+	-	+	-	-	-
Aspergillus proteinase	+	-	+	-	-	-
Candida proteinase	+	-	+	-	-	-
Vibrio proteinase	+	+	+	n.d.	-	-
Subtilisin	+	+	+	+	-	-
Thermolysin	+	+	+	-	-	-
Staphylococcus V8	-	-	+	-	+	+
Streptomyces	-	-	+	+	+	+
Df-proteinase	+	+	+	+	+	+

See ref. 1,3,7,15 for details.

DESTRUCTION OF VARIOUS PLASMA PROTEINASE INHIBITORS: A PREREQUISTE FOR PROTEINASE ACTION *IN VIVO*

First requiste for bacterial proteinases to exert pathological effects *in vivo* is that all relevant proteinase inhibitors must be inactivated to facilitate the action of thus activated proteinases. Among proteinase inhibitors, serine-proteinase inhibitors (serpins) play the most important

Table 4. Inactivation of human plasma proteinase-inhibitor by *Serratia* 56K proteinase

Incubation time with proteinase (h)	% inactivation of inhibitors: proteinase activity					
	Serratia proteinase/plasma inhibitor ratio (molar)					
	α_1-PI 1:200	C1-inhibitor 1:50	AT-III 1:25	α_2-AP 1:10	α_2-M 1:50	ovoM 1:50
0.5	80	3	10	7	43	37
1	100	5	35	14	48	33
2	100	15	90	31	62	28
4	n.d.	64	100	68	70	23
6	n.d.	100	100	90	80	21
25	n.d.	n.d.	n.d.	n.d.	90	8

See ref. 1 for details.

role in physiological and pathological conditions (for instance ref. 16). Destruction of α_1-proteinase inhibitor (α_1-PI) is, for instance, known to be involved in bronchial pathogenesis (17).

Pseudomonas, Serratia and other microbial proteinases can degrade serpins, and most preferablly near C-terminal end (18-20). We have tested *Serratia* proteinase against antithrombin III, α_1-PI, α_2-antiplasmin, C1-esterase inhibitor and α_2-macroglobulin. All these inhibitors were cleaved and inactivated within 2-6 hrs at enzyme/inhibitor (E/I, I being substrate) ratios of 1/10 to 1/200. Among these α_1-PI was most labile; i.e. at E/I ratio of 1/200, its antitryptic activity was lost 100 % within one hour (10) (Table 4).

CONCLUSION

All microbial proteinases are involved in activation of bradykinin generating cascade at one or more steps which takes place most efficiently utilizing high molecular weight kininogen. On the contrary, the house dust mite proteinase (a serine-type proteinase) activates the kinin generating cascade as well as utilizing the low molecular weight kininogen, which generates kallidin (lysyl-bradykinin). Kallidin will be rapidly converted to bradykinin. Most of these bacterial proteinases did inactivate endogenous serine proteinase inhibitors, thus pathological effects may be exerted in vivo. These results are consistent with the observation using human plasma.

REFFERENCES

1. Maeda H, Molla A. Pathogenic role of bacterial proteases. Clin Chim Acta 1989; 185: 357-368.
2. Matsumoto K, Yamamoto T, Kamata R, Maeda H. Pathogenesis of serratial infection: Activation of Hageman factor-prekallikrein cascade by serratial protease. J Biochem 1984; 976: 739-749.
3. Molla A, Yamamoto T, Akaike T, Miyoshi S, Maeda H. Activation of Hageman factor and prekallikrein and generation of kinin by various microbial proteinases. J Biol Chem 1989; 264: 10589-10594.
4. Molla A, Matsumura Y, Yamamoto T, Okamura R, Maeda H. Pathogenic capacity of proteases from *Serratia marcescens* and *Pseudomonas aeruginosa* and their suppression by chicken egg white ovomacroglobulin. Infect Immun 1987; 55: 2509-2517.

5. Kamata R, Yamamoto T, Matsumoto K, Maeda H. A serratial protease causes vascular permeability reaction by activation of the Hageman factor-dependent pathway in guinea pig. Infect Immun 1985; 48: 747-753

6. Matsumoto K, Yamamoto T, Kamata R, Maeda H. In: Kinin IV B. Greenbaum LM, Margolius HS, editors. New York: Plenum Publ Corp, 1986: 71-78.

7. Kaminishi H, Tanaka M, Cho T, Maeda H, Hagihara Y. Activation of the plasma kallikrein-kinin system by *Candida albicans* proteinase. Infect Immun 1990; 58: 2139-2143.

8. Oda T, Kojima Y, Akaike T, Molla A, Ijiri S, Maeda H. Inactivation of chemotactic activity of C5a by Serratial 56 kilodalton protease. Infect Immun 1990; 58: 1269-1272.

9. Molla A, Kagimoto T, Maeda H. Cleavage of immunoglobulin G (IgG) and IgA around the hinge region by proteases from *Serratia marcescens*. Infect Immun 1988; 56: 916-920.

10. Molla A, Akaike T, Maeda H: Inactivation of various proteinase inhibitors and the complement system in human plasma by the 56 killodalton proteinase from *Serratia marcescens*. Infect Immun 1989; 57: 1868-1871.

11. Molla A, Matsumoto K, Oyamada I, Maeda H. Degradation of protease inhibitors, immunoglobulins, and other serum proteins by *Serratia* protease and its toxicity to fibroblasts in culture. Infect Immun 1986; 53: 522-529.

12. Kamata R, Matsumoto K, Okamura R, Yamamoto T, Maeda H. The serratial 56K protease as a major pathogenic factor in serratial keratitis. Clinical and experimental study. Ophthalmol 1985; 92: 1452-1459.

13. Miyagawa S, Matsumoto K, Kamata R, Okamura R, Maeda H. Spreading of *Serratia marcescens* in experimental keratitis and growth suppression by chicken egg white ovomacroglobulin. Jpn J Ophthalmol 1992; (in press)

14 Takahashi K, Aoki T, Kohmoto S, Nishimura H, Kodera Y, Matsushima A, Inada Y. Activation of kallikrein-kinin system in human plasma with purified serine protease from *Dermatophagoides farinae*. Int Arch Allergy Appl Immunol 1990; 91: 80-85.

15. Maruo K, Akaike T, Matsumura Y, Kohmoto S, Inada Y, Ono T, Arao T, Maeda H. Triggering of the vascular permeability reaction by activation of the Hageman factor-prekallikrein system by house dust mite proteinase. Biochim Biophys Acta 1991; 1074: 62-68.

16. Carrell RW, Roswell DR. In: Proteinase Inhibitors. Barrett AJ, Salvesen G, editors. Amsterdam: Elsevier, 1986: 403-420.

17. Johnson DA, Carter-Hamm B, Dralle WM. Inactivation of human bronchial mucosal proteinase inhibitor. Am Rev Res Dis 1982; 126: 1070-1073.

18. Morihara K, Tsuzuki H, Oda K. Protease and elastase of *Pseudomonas aeruginosa*. Inactivation of α1-protease inhibitor. Infect Immun 1979; 24: 188-193.
19. Virca GD, Lyerly D, Kreger A, Travis J. Inactivation of human α1-proteinase inhibitor by a metalloproteinase from *Serratia marcescens*. Biochim Biophys Acta 1982; 704: 267-271.
20. Potempa J, Shieh BH, Travis J. Alpha-2-antiplasmin: a serpin with two separate but overlapping reactive sites. Science 1988; 241: 699-700.

AAS 38/III
Recent Progress on Kinins
© 1992 Birkhäuser Verlag Basel

LIVER AND SERUM LYSOSOMAL ENZYMES ACTIVITY DURING ZYMOSAN-INDUCED INFLAMMATION IN MICE

A.F. Safina[1], T.A. Korolenko[1], G.I. Mynkina[1], M.I. Dushkin[2] and G.A. Krasnoselskaya[1]

Institute of Physiology[1] and Institute of Therapy[2], Siberian Branch of the Academy of Medical Sciences of USSR, Novosibirsk

SUMMARY: Liver and serum lysosomal enzymes (acid glucosidases and cysteine proteinases) during zymosan-induced stimulation of MPS or MPR depression induced by $GdCl_3$ have been studied. Zymosan was used as a model for study of inflammation in vivo. The development of inflammation induced by zymosan was followed by increease activity of macrophage activation markers - ß-N-acetylglucosaminidase (NAGlu) and ß-N-acetylgalactosaminidase (NAGal) in liver and serum. There was enhance of liver cysteine proteinases activity. Similar, less prominent (partly) data were obtained during macrophage depression induced by $GdCl_3$.

INTRODUCTION

Zymosan has been used as a model for study of inflammation in vivo and in vitro (1). The stimulation of mononuclear phagocyte system (MPS) by zymosan is followed by the increse activity of some lysosomal hydrolases of macrophages (2, 3). These enzymes are able to activate different systems (blood clotting, fibrinolysis and complement cascades) and promote tissue injury (4, 5). The important role in this process belongs to lysosomal proteinases, closely related to inflammation.

Liver cysteine proteinases (CP) activity during zymosan-induced stimulation of MPS was studied. The results were compared with data obtained during depression of MPS function by gadolinium chloride ($GdCl_3$) (6). The activity of lysosomal glucosidases has been studied in liver and serum simultaneously. As a marker of macrophage activation we used ß-N-acetylglucosaminidase (NAGal) activity, increasing up to 3-5-fold in such cases (3).

METHODS

Male mice CBA weighing 16-20 g were used. Zymosan (Olaine, Latvia) was given i.v. as a single dose 100 mg/kg. $GdCl_3$ (kind gift of Prof. Hardonk M.J. and Prof. Bouma J.M.W., The Netherlands) was used in a dose of 20 μmoles per kg (6). Mice were sacrificed at 2 h, 1 day and 5 day after a single zymosan or $GdCl_3$ i.v. injections. The activity of glucosidases was determined according to Barrett (7), using MUF-derivated substrates (Fluka, Koch-Light and Imbimed, USSR). CP were determined with flourogenic peptidyl substrates: Z-Arg-Arg-NMec for cathepsin B, Z-Phe-Arg-NMec for cathepsin L (production of All-Union Institute of Molecular Biology, Koltosova, USSR) according to method described (8). Fluorescence was measured using Hitachi LTD model F-3000 spectrofluorimeter.

RESULTS AND DISCUSSION

Secretion of acid glucosidases in serum occurs both after stimulation of MPS by zymosan (Fig. 1) and during depression induced by $GdCl_3$ (Fig. 1). In zymosan-treated group enzyme secretion related to macrophage activation; in case of $GdCl_3$ the macrophage overloading by this compound and liver injury is possible. In the both experiments the most significant increase was noted for marker macrophage enzymes NAGlu and NAGal. Surprisingly, the most prominent raising of secretion was observed in case of $GdCl_3$ (2h, 1 day after) (Fig. 1). On the other hand in liver zymosan treatment was followed by the most significant (3-fold) increase of these markers of macrophage activation (Fig. 2). $GdCl_3$ administration (Fig. 2) resulted in raising of NAGal activity up to 170 % only. Activity of liver CP increased (Table 1) in 5th day after zymosan using, reaching 143 % for cathepsin B and 137 % (from the control values) for cathepsin L. There were no changes of CP studied in case of $GdCl_3$ injection (Table 1). Zymosan induced migration of peripheral monocytes in the liver, especially at 4-6th days after treatment. Significant increase of lysosomal hydrolases in liver and serum of zymosan-treated mice can be explained as synthesis of enzymes de novo (2). It's known also that macrophages enriched by cathepsin B and cathepsin H comparatively to other cell types (5, 8). $GdCl_3$ effect can be related to its uptake by Kupffer cells followed by releasing of lysosomal enzymes and liver injury. Less prominent increasing of CP (comparatively to acid glucosidases) was caused possibly by secretion of CP inhibitors occuring after stimulation of MPS. In conclusion it is necessary to note that symptoms of inflammation induced by zamosan included increase activity of acid glucosidases and cysteine proteinases in mice liver, serum secretion of acid glucosidases, but similar data were obtained during macrophage depression induced by $GdCl_3$.

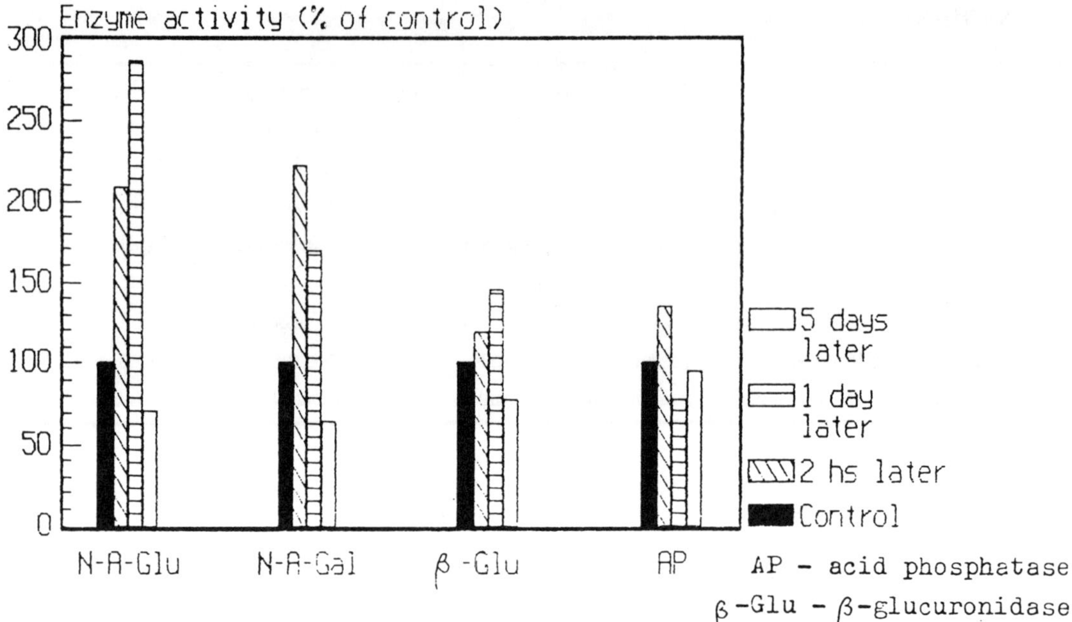

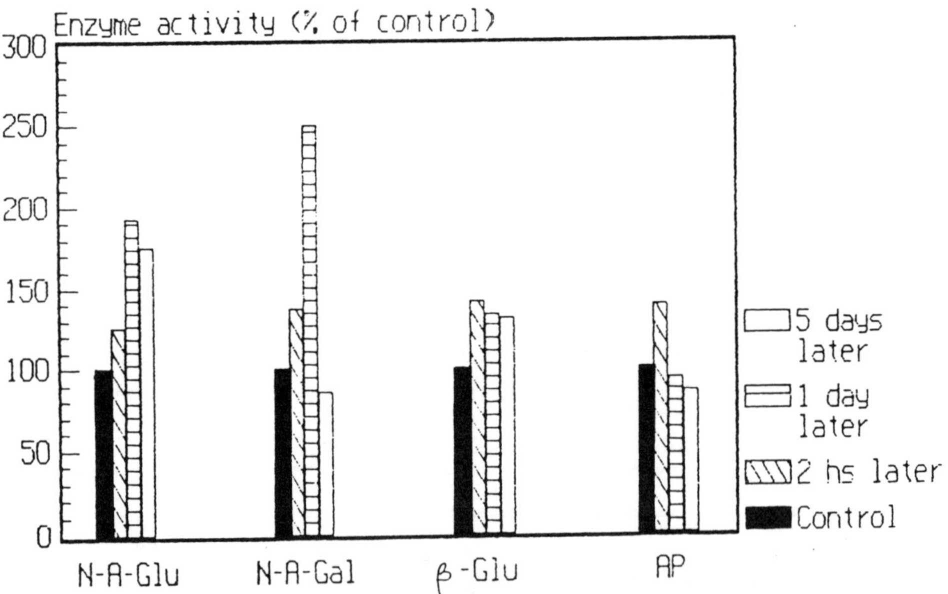

Figure 1. Serum secretion of lysosmal hydrolases after macrophages depression by GdCl$_3$ (upper panel) and after macrophages stimulation by zymosan (lower panel).

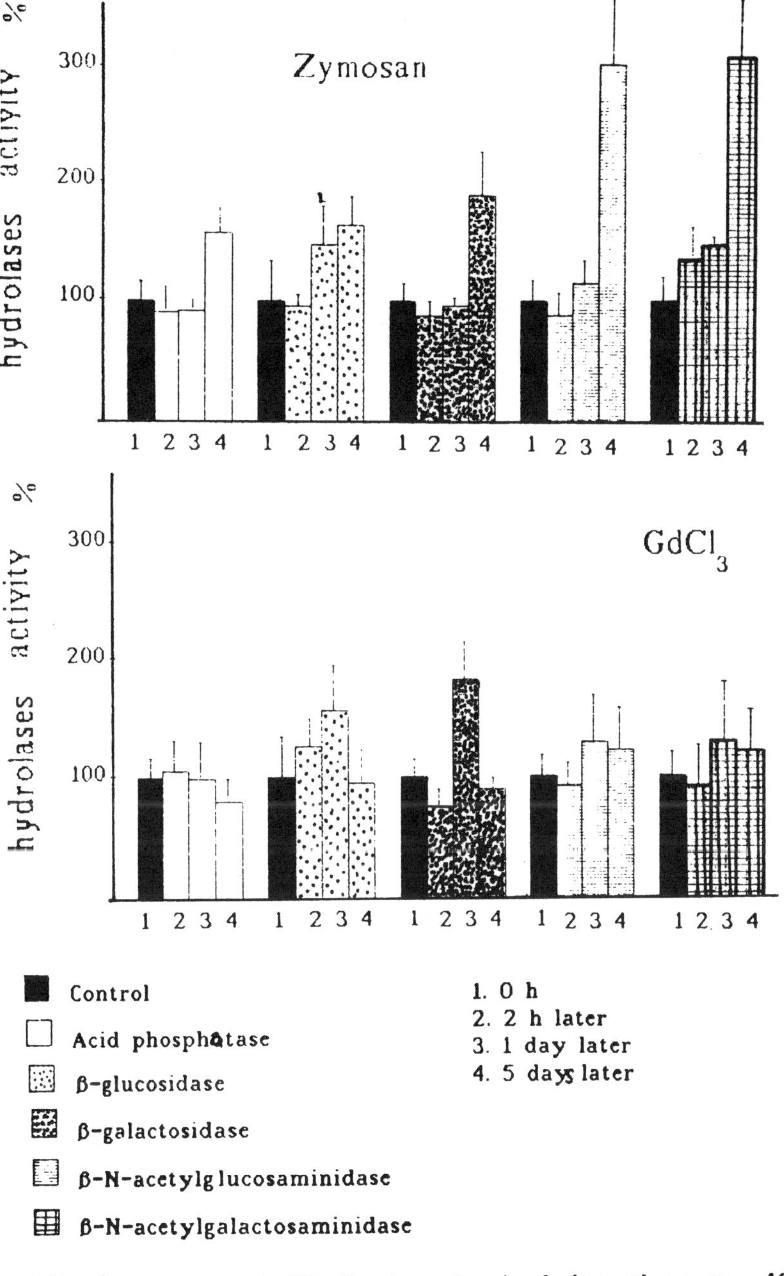

Figure 2. Effects of zymosan and GdCl$_3$ single administration to mice on liver acid hydrolases activity.

Table 1. Effect of zymosan and GdCl$_3$ administration to mice on liver cysteine proteinases activity

Group of animals, number	Cysteine proteinases, specific activity (nmol/min per mg of protein)	
	Cathepsin B	Cathepsin L
Control (5)	0.25 ± 0.006 (100 %)	0.79 ± 0.028 (100 %)
Zymosan, 2 h (6)	0.21 ± 0.008 (84 %)	0.78 ± 0.046 (99 %)
GdCl$_3$, 2 h (6)	0.22 ± 0.010 (88 %)	0.74 ± 0.056 (94 %)
Control (5)	0.19 ± 0.009 (100 %)	n.d.
Zymosan, 1 day (6)	0.17 ± 0.007 (91 %)	n.d.
GdCl$_3$, 1 day (6)	0.19 ± 0.010 (100 %)	n.d.
Control (5)	0.14 ± 0.007 (100 %)	0.51 ± 0.046 (100 %)
Zmyosan, 5 days (6)	0.20 ± 0.006 (143 %) p < 0.01	0.70 ± 0.027 (137 %) p < 0.01
GdCl$_3$, 5 days (6)	0.15 ± 0.005 (100 %)	0.56 ± 0.038 (110 %)

Zymosan was injected i.v. in a dose of 100 mg/kg, GdCl$_3$ was administered i.v. in a dose of 20 μmol/kg; control - mice with injection of the same volume of saline solution.

REFERENCES

1.   Tarayre JP, Delhon A, Aliaga M, Barbara M, Bruniquel F, Caillol V, Puech L, Consul N, Tisne-Versailles J. Pharmacological studies on zymosan inflammation in rats and mice. 1:zymosan-induced paw oedema in rats and mice. Pharmacological Research 1989; 21:375-395.

2.   Bouwens L, Wisse E. Proliferation, kinetics and fate of monocytes in rat liver during a zymosan-induced inflammation. J Leukocyte Biol 1985; 37:531-543.

3.   Warren L. Stimulated secretion of lysosomal enzymes by cells in culture. J Biol Chem 1989; 264:8835-8842.

4.   Jochum M, Duswald K-H, Neumann S, Witte I, Fritz H, Seemüller U. Proteinases and their inhibitors in inflammation: basic concepts and clinical implications. In: Proteinase inhibitors: medical and biological aspects, eds. Katunuma N et al. 1983; 85-95.

5.   Assfalg-Machleidt I, Jochum M, Joka T, Rothe G, Valet G, Zauner R, Scheuber H-P, Machleidt W. Cathepsin B-indicator for the release of lysosomal cysteine proteinases in severe trauma and inflammation. Biol Chem Hoppe-Seyler 1990; 371,Suppl:211-222.

6.   Bouma JMW, Smit MJ. Gadolinium chloride selectively blocks endocytosis by Kupffer cells. Kupffer Cell Foundation. Cells of Hepatic Sinusoid 1989; 2:132-133.

7.   Barrett AJ. Lysosomal enzymes. In: Lysosomes. A Laboratory Handbook, ed. by Dingle JT. North-Holland Publ Co, Amsterdam, London 1972; 46-136.

8.   Barrett AJ, Kirschke H. Cathepsin B, cathepsin H and cathepsin L. Methods Enzymol 1981; 80:535-561.

AAS 38/III
Recent Progress on Kinins
© 1992 Birkhäuser Verlag Basel

PLASMA EXUDATION BY ACTIVATED PLASMA KALLIKREIN AFTER

INTRAPERITONEAL INJECTION OF λ-CARRAGEENIN OR ENDOTOXIN IN RATS

K. Sugimoto*, M. Shindo*, T. Owada*, M. Majima**, M. Katori**

Department Traumatology & Critical Care Medicine*, and of Pharmacology**, Kitasato
University, School of Medicine, Kitasato 1-15-1 Sagamihara Kanagawa 228, JAPAN

SUMMARY: The experimental study was carried out to examine whether intraperitoneal
injection of carrageenin or endotoxin activates plasma kallikrein-kinin system to induce plasma
exudation in rats. Intraperitoneal injection of 2% λ-carrageenin induced plasma exudation in the
peritoneal cavity by activation of plasma prekallikrein. Intraperitoneal injection of endotoxin
(3mg/kg) also resulted in intraperitoneal plasma exudation, but plasma kallikrein-kinin system
did not seem to be involved.

INTRODUCTION

It has been believed that endotoxin generates a variety of biological effects, such as fever,
leucopenia, hypotension, shock, and death. Previous studies in several shock models have
suggested that bradykinin (BK) contributes to hypotension during shock [1,2,3,4]. But the
interaction between endotoxin and kallikrein-kinin (K-K) system still remains unclear and
controversial [5,6].

The aim of the present experiment is to determine whether intraperitoneal injection of endotoxin
activates plasma K-K system and induces plasma exudation in rats in comparison with that in
carrageenin induced peritonitis in rats.

MATERIALS AND METHODS

Peritonitis was induced by intraperitoneal injection of λ–carrageenin (2% saline solution, 1 ml)
or endotoxin (E.coli., 0127 B8., Difco. USA. : 3mg/kg; dissolved in sterile saline, 1 mg/kg)

in rats (male, S-D strain, SPF). The dye (pontamine sky blue: 50 mg/kg) was injected intravenously at 40 min, 160 min after intraperitoneal injection of λ–carrageenin or endotoxin. Rats were sacrified 20 min after intravenous injection of the dye and the peritoneal fluid was harvested for measurement of the intraperitoneal dye amount as a parameter of plasma protein exudation into the peritoneal cavity. The plasma kininogens (high molecular weight (HMW) kininogens and low molecular weight (LMW) kininogens) were measured by Uchida's method [7] and the degradation products of kinin (des-Phe^8-Arg^9-BK and [1-5] BK) in the peritoneal cavity and in the circulating blood were measured using an enzyme immunoassay. Under pentobarbital anesthesia, the right carotid artery was cannulated with a plastic tube (PE-50, Clay-Adams USA) for measurement of blood pressure with a pressure transducer (MPU-0.5 Nihonkoden, Tokyo, Japan). The concentrations of endotoxin in the arterial plasma were measured by chromogenic lymulus assay (Endospecy Kit. Seikagaku Kougyo, Tokyo, Japan) after pretreatment with perchloric acid.

All data are expressed as means±S.E.M. Statistical analysis of the results was made using a one-way analysis of variance and Student's t-test for paired or unpaired data. A probability level of less than 0.05 was considered as significant.

RESULTS

Endotoxin-induced peritonitis. Intraperitoneal injection of endotoxin resulted in the rapid increase in the level of endotoxin concentration in the arterial plasma at 20 min and in further increase until 60 min. The concentration was kept higher for longer than 420 min (Table-1).

Tab.1. ETX concentrations in arterial plasma (n=8)

Control	20 min.	60 min.	420 min.
0.98 ± 0.12	1.51 ± 0.05	7.10 ± 1.12	7.98 ± 2.10
(pg/ml)	(x10^{5}pg/ml)	(x10^{7}pg/ml)	

The mean arterial pressure was not changed at 20 min but was significantly reduced at 60 and 180 min after endotoxin injection compared with saline control (Table-2).

                                                K. Sugimoto et al.

Tab. 2. Mean arterial blood pressure changes after ETX injection (n=5)

	Control	60 min.	180 min.
ETX.(3mg/kg)	125 ± 6	111 ± 6[†]	100 ± 2[††]
Saline control	128 ± 10	125 ± 8	129 ± 11 (mmHg)

†: P<0.05, ††:P<0.01 by Student's t-test

The exudation rate of plasma into the peritoneal cavity, measured by dye amount exuded over a twenty min period, increased significantly at 60 and 180 min, compared with the saline control (Table-3).

Tab. 3. Plasma exudation rate after ETX injection (n=6)

	Control	60 min.	180 min.
ETX.(3mg/kg)	16±3	42±8[††]	156±25[††]
Saline control	15±5	16±4	16±3 (µg/20 min.)

††:P<0.01 by Student's t-test

As shown in Figure 1, the amount of des-Phe^8-Arg^9- BK in the peritoneal exudate tended to increase at 3 hrs after endotoxin injection, but that of [1-5] BK was not changed. The level of HMW kininogen in the peritoneal exudate was prone to be lower than that in plasma, but the difference was statistically not significant. The LMW kininogen level did not differ between plasma and peritoneal exudate, indicating that plasma prekallikrein was not clearly activated. In the circulating blood, the plasma levels of HMW and LMW kininogens as well as that of des-Phe^8-Arg^9-BK were not different 3 hrs after endotoxin injection.

Pretreatment of rats with a soy bean trypsin inhibitor (SBTI), which is known to inhibit plasma kallikrein and was injected 20 min before intraperitoneal injection of endotoxin, did not change the plasma exudation rate into the peritoneal cavity at 60 and 180 min after endotoxin (Table-4).

Tab. 4. Effect of the SBTI on the Plasma Exudation (n=6)

	Control	60 min.	180 min.
SBTI+ETX	15±5	44±10	162±22
ETX.(3mg/kg)	15±8	46±6	164±21 (µg/20 min.)

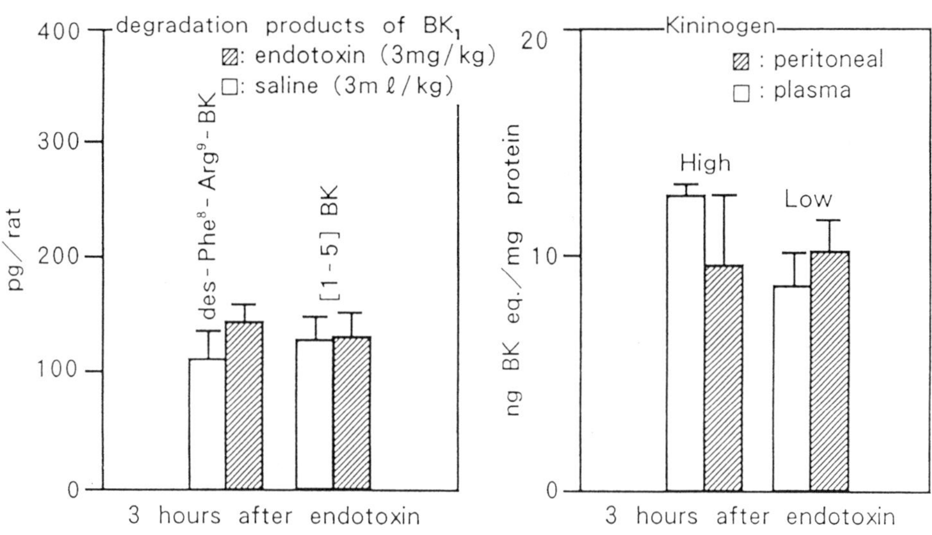

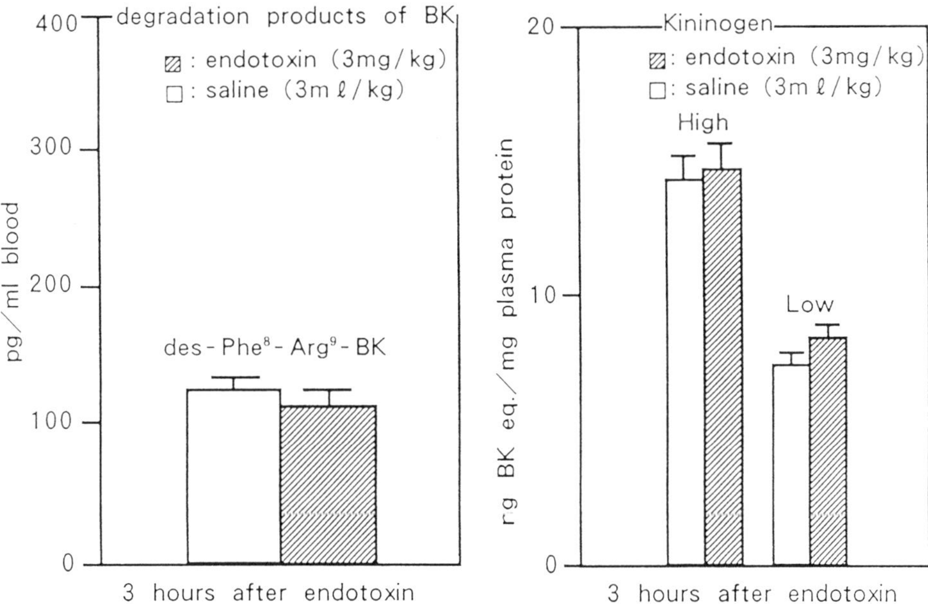

Figure 1: Changes in the levels of des-Phe^8-Arg^9-BK, [1-5]BK, and HMW and LMW kininogens in the peritoneal cavity (upper panel) and in the circulating blood (lower panel) in rats. The samples were collected 3 hrs after intraperitoneal injection of endotoxin (3mg/kg).

These results may indicate that the intraperitoneal injection of endotoxin did not cause the activation of plasma prekallikrein in the peritoneal cavity and the plasma exudation observed in the peritoneal cavity was attributable to mediators other than kinin.

Carrageenin-induced peritonitis. Intraperitoneal injection of λ-carrageenin also induced the plasma exudation. The exudation rates, measured by dye exudation over a 20 min period were increased at 60 and 180 min after the injection (Table-5).

Tab. 5. Plasma exudation rate after λ-carrageenin injection (n=6)

		Control	60min.	180 min.
Carrageenin	22±5	345±54[††]	198±32[††]	
Saline control	19±7	22±5	19±6	(µg/20 min.)

††: P<0.01 by Student's t-test

The levels of degradation products of bradykinin, des-Phe^8-Arg^9-BK and [1-5]BK, in the peritoneal exudate were increased markedly at 1 hr after carrageenin, compared with saline control (Figure 2). The HMW kininogen level in the peritoneal exudate was negligible in the peritoneal exudate, wheres the LMW kininogen level was kept at the same level as that in plasma. In contrast, The plasma levels of des-Phe^8-Arg^9-BK and HMW and LMW kininogens were not significantly changed after carrageenin injection (Figure-2).

Pretreatment of rats with SBTI inhibited the plasma exudation into the peritoneal cavity at 60 min by approximately 75% (Table-6).

Tab. 6. Effect of the SBTI on the Plasma Exudation after carrageenin

	Control	60 min.	
SBTI+Ca. (n=5)	23±6	78±9[††]	
Ca.(1mg/kg: n=5)	20±12	338±65	(µg/20 min.)

Ca : carrageenin ††:P<0.01 by Student's t-test

These results indicated that plasma prekallikrein was activated after the intraperitoneal injection of carrageenin and induced the plasma exudation in to the peritoneal cavity.

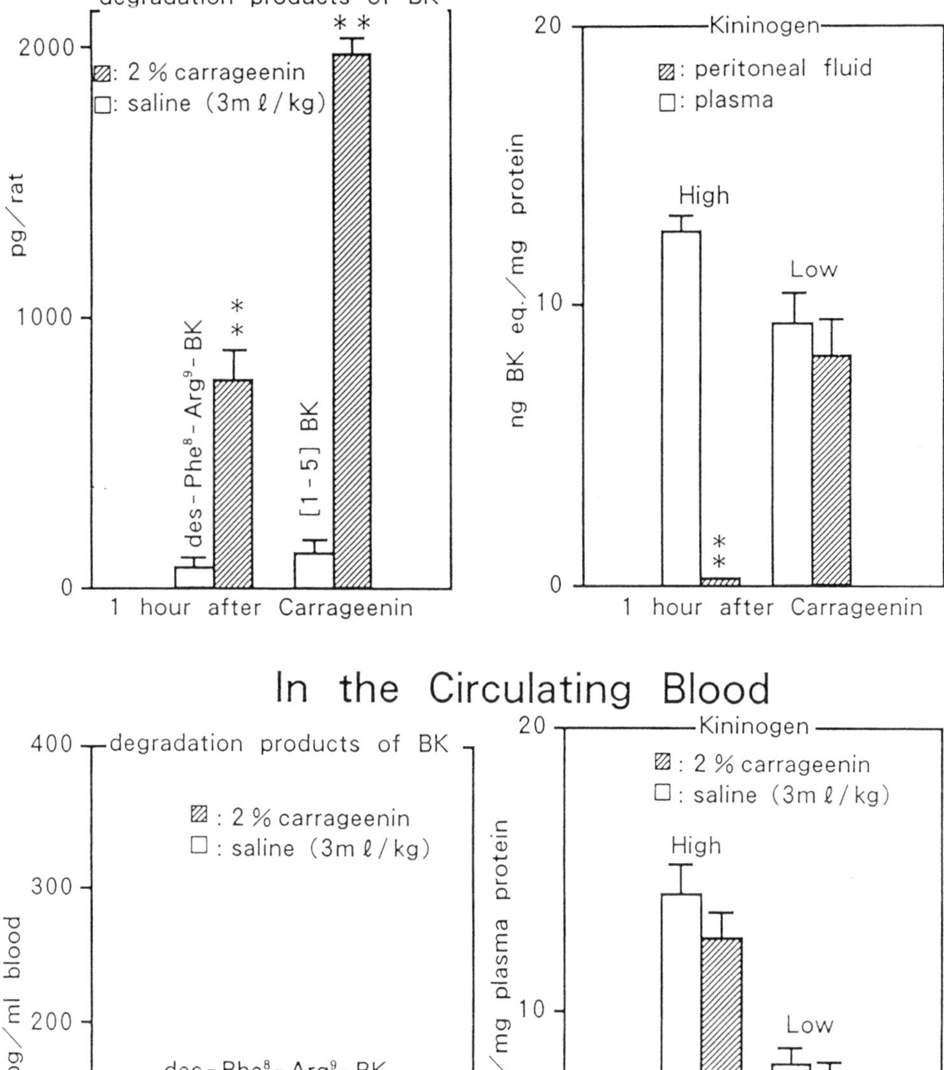

Figure 2 : Changes in the levels of degradation products of BK and kininogen in peritoneal cavity (upper panel) and in circulating blood (lower panel) 1 hr after intraperitoneal injection of carrageenin in rats. ** indicate p<0.01 by student's-t-test

DISCUSSION

We reported [8] that intrapleural injection of carrageenin in rats induces pleurisy by activation of plasma kallikrein-kinin system that and bradykinin is a major mediator of this pleurisy to cause the plasma exudation, since the level of HMW kininogen, but not of LMW kininogen, in the pleural exudate exhausted without change in the plasma levels of both kininogens [7]. Furthermore, the des-Phe^8-Arg^9-BK level is increased in the pleural exudate in the entire course of the pleurisy [8]. Thus, we interpreted that the plasma protein exuded into the pleural cavity contacts with carrageenin in this cavity and blood coagulation factor XII (Hageman factor) was activated. In fact, incubation of factor XII-deficient human plasma with carrageenin in vitro did not generate kinin, though human or rat plasma generated kinin in incubation with carrageenin.

In the present experiment, intraperitoneal injection of λ-carrageenin also resulted in the plasma exudation together with the markedly increased levels of degradation products of bradykinin, des-Phe^8-Arg^9-BK and [1-5]BK, in the peritoneal exudate. The increased levels of the degradation products of BK were accompanied with exhaustion of HMW kininogen without change in the LMW kininogen level. These results clearly indicate that carrageenin may contact with plasma factor XII in the peritoneal cavity, as in the pleural cavity, and activate factor XII and plasma prekallikrein. Hence, only HMW kininogen, which is the substrate for plasma kallikrein, was consumed without effect on the LMW kininogen level. Neither in pleurisy [7] nor in peritonitis, the plasma levels of kininogens were changed. The plasma exudation into the pleural cavity [7] as well as into the peritoneal cavity was mainly attributable to bradykinin generated, since the exudation was markedly suppressed by SBTI.

Endotoxin from E.coli is also a sulphated lipopoly-saccaride and can be expected to activate factor XII [9,10]. However, intraperitoneal injection of endotoxin into the peritoneal cavity in the present experiments did not cause significant reduction of HWM kininogen and clear increase in the levels of des-Phe^8-Arg^9-BK and [1-5]BK. This is further strengthened by the fact that SBTI did not modify the plasma exudation. The plasma exudation induced by endotoxin is attributed to prostaglandin generated in the peritoneal cavity (to be published).

Lack of the activation of plasma prekallikrein after intraperitoneal injection of E.coli endotoxin in rat may not be due to lack of the ability of this endotoxin to activate factor XII. Previously we reported [11] that intravenous injection of endotoxin from E.coli (3-30mg/kg) into the femoral vein on anesthetized rats resulted in dose dependent immediate hypotension. The plasma levels of plasma prekallikrein and HMW kininogen were significantly decreased together with transient increase in kinin and an active from of plasma kallikrein in blood one min after endotoxin injection. The des-Phe^8-Arg^9-BK was also increased in blood for longer period [11]. The hypotension by endotoxin was also suppressed by intravenous infusion of SBTI, indicating that bradykinin generated by endotoxin caused hypotension. Lack of the activation of plasma prekallikrein in the peritoneal cavity after intraperitoneal injection of endotoxin is not precisely known, but may be attributable to a weak activity of endotoxin from E.coli to activate factor XII [6] and/or to phagocytose the endotoxin by intraperitoneal macrophages and invaded leukocytes.

In the present experiments, endotoxin injected intraperitoneally entered the arterial blood even 20 min after the injection and the concentration was $1.51 \pm 0.05 \times 10^2$ ng/ml. Immediately, the arterial pressure however was not changed and the arterial pressure was reduced at 60 and 180 min after intraperitoneal injection of endotoxin. Lack of the immediate hypotension in the presence of endotoxin in the arterial blood may be due to the small quantity of endotoxin in the blood after intraperitoneal injection. The hypotension observed gradually may be reflection of generation of prostaglandin E_2 in the peritoneal cavity , since the hypotension as well as the plasma exudation in the peritoneal cavity were completely inhibited by indomethacin (reported separately).

CONCLUSION

Intraperitoneal injection of endotoxin induced the plasma exudation without activation of plasma prekallikrein, wheres that of λ-carrageenin activated plasma kallikrein probably through the activation of factor XII in the peritoneal cavity.

REFERENCES

1. Fal W. Kinin levels in experimental endotoxin shock. Arch Immunol Ther Exp 1969.17: 406-420,
2. Kimball HR, Melon KL, Wolff SM. Endotoxin-induced kinin production in man. Proc Soc Exp Biol Med 1971, 139: 1078-1082.
3. Nies AS, Forsyth RP, Williams HE et al. Contribution of kinins to endotoxin in unanesthetized Rhesus monkeys. Circ Res 1968, 22: 155-164.
4. O'Donnell TF, Clowes GHA. Kinin activation in the blood of patients with sepsis. Sur Gyn Ob. 1976 143: 539-545.
5. Janssen HF, Pugh JL, Lange DL. Badykinin does not contribute to hypotension in early canine endotoxemia. Circ Shock. 1987 23: 197-204.
6. Kaufmann J, Schapira M, Waeber J et al. Endotoxin-induced hypotension in rats is not mediated by prekallikrein activation. Circ Shock. 1987 23: 215-223.
7. Uchida Y, Tanaka K, Harada Y et al: Activation of plasma kallikrein-kinin system and its significant role in pleural fluid accumulation of rat carrageenin-induced pleurisy. Inflammation 1983 7: 121-129.
8. Majima M, Ueno A, Sunahara N et al: Measurement of des-Phe^8-Arg^9-BK by enzyme-immunoassay - a useful parameter of plasma kinin release. In kinin V part B 331-335. Adv Exp Med Biol New York ; Plenum Press, 1988,
9. Erdos EG, Miwa I: Effect of endotoxin shock on plasma kallikrein-kinin system of the rabbit, Fed Proc 1968 27: 92-95.
10. Morrison DC, Cochrane CG: Direct Evidence for Hageman factor (factorXII) activation by bacterial lipopoly-sacharides(endotoxins). J Exp Med 1974 140: 797-811.
11. Katori M, Majima M, Odoi-Adome R et al: Evidence for the involvement of plasma kallikrein-kinin system in the immediate hypotension produced by endotoxin in anaesthetized rats. Br J Pharamacol 1989 98:1383-1391.

THE ROLE OF α_1-PROTEINASE INHIBITOR IN SEMISOLUBLE GLUCAN INDUCED RESISTANCE TO HEMORRHAGE

E.I. Vereschagin, T.A. Korolenko, S.A. Arkhipov

Department of Cellular Biochemistry and Physiology,
Institute of Physiology, Siberian Branch of the Academy of Medical Sciences of USSR,
Novosibirsk, USSR

SUMMARY: The present study was undertaken to evaluate the mechanism of protective effect of semisoluble ß-1,3-carboxymethylglucan (CMG) during acute massive hemorrhage. CBA mice were injected i.v. with glucan (25 mg/kg) 24 h prior to hemorrhage (50 % of blood circulating volume). Survival data indicated that glucan increased survival compared with control. However, α_1-proteinase inhibitor activity in serum have been even decreased 24 h after glucan administration. Moreover enhancement of active oxygen form production and leakage of cytosol and lysosomal enzymes after CMG application were noted. Key words: hemorrhage, semisoluble glucan, α_1-proteinase inhibitor.

INTRODUCTION

A large amount of data indicate that there is a relationship between macrophages function and resistance to circulatory shock (1). Animals pretreated with macrophage stimulator - ß-1,3-glucan shows increased tolerance to several types of shock (2). It is known that liberated proteinases (elastase, cathepsins) may overcome inhibitory potential of their main antagonists, and significantly aggravate the shock (3). Antiproteinase inhibitors, first of all α_1-PI, α_2-macroglobulin, are known to be secreted by RES. Lipopolysaccharides increased more than 8-fold secretion of α_1-PI by monocytes and macrophages in vitro (4). Therefore, it is possible to suggest that protective effect of glucan to different types of shock is mediated, partly by α_1-PI enhancement. The present study was undertaken to study the effect of CMG during hemorrhage and to evaluate the role of α_1-PI in CMG-induced tolerance to hemorrhage.

MATERIALS AND METHODS

Experiments were carried out on male CBA mice (20-25 g body weight). Animals were anesthetized with ether and calculated blood volume was removed from retoorbital venous sinus during 2 min (about 50 % o blood circulating volume, or 3.5 % of body weight). Control animals underwent all similar procedures except hemorrhage. Semisoluble ß-1,3-D-carboxymethylglucan (CMG) (Chemical Institute, Bratislava, Czechoslovakia) used for RES stimulation was administered i.v. 24 h prior to hemorrhage in a dose of 25 mg/kg. The NBT test have been conducted on isolated peritoneal macrophages using nitroblue tetrasolium (Sigma, USA) (5). α_1-PI was assayed with method, based on interaction of α_1-PI with trypsin in the system with a low molecular substrate N-benzoyl-L-arginine (6). Lysosomal enzyme secretion was measured as increasing acid hydrolases level in serum. ß-galactosidase assays were performed spectrofluorimetrically using MUF-ß-D-galactropyranoside as a substrate (7). Acid phosphatase was determined with sodium ß-glycerophosphate as described (7). Activity of LDG and ALT in serum were estimated with standard test combination (Boehringer Mannheim, Germany). All data were statistically analyzed with the Student t-test.

RESULTS AND DISCUSSION

CMG pretreatment increased the survival rate in shocked mice comparatively to control. In group with CMG only one from twenty mice died after hemorrhage, meantime in control five animals from twenty have died. The percentage of isolated peritoneal macrophages, reduced NBT was amounted from 33 % in control to 53 % after CMG application. Increased production of oxygen active forms shows that CMG had potent stimulatory effect on macrophages. Activity of LDG and AcPh (but not ALT and ß-galactosidase) in serum enhances 24 h after CMG administration (Table 1). This effect of macrophage stimulators have been previously explained by leakage of cytosol and lysosomal enzymes due to changes of cell membrane (8). So stimulation of macrophages with CMG was accompanied with releasing of destructive potential of RES. These factors (active oxygen forms and acid hydrolases) could lead to tissue damage, especially during shock. CMG pretreatment significantly reduced cytosol and lysosomal enzymes leakage noted in hemorrhage (Table 1). This facts shows that preliminary administration of CMG protected the tissues from damage during hemorrhage. One of the possible mechanism of protection during shock might be the increasing of α_1-PI in blood after CMG application. Surprisingly it was shown that glucan i.v. administration even decreased the α_1-PI activity in serum. Compared with control

Table 1: Lysosomal and cytosol enzymes of serum during hemorrhage after CMG administration in mice

	LDG U/l	ALT U/l	AcPh nmoles Pi/min l	ß-Gal nmoles of MUF/min l
0 h prior to hemorrhage				
Control	296 ± 24	0.74 ± 0.03	4.4 ± 0.25	0.27 ± 0.05
CMG	921 ± 35 ***	0.7 ± 0.04	6.0 ± 0.4 *	0.23 ± 0.04
1 h after hemorrhage				
Control	643 ± 24 ***	1.62 ± 0.14 **	9.3 ± 1.0 ***	0.45 ± 0.03 ***
CMG	813 ± 86 ***	1.1 ± 0.04	6.9 ± 0.25 ***	0.24 ± 0.03

CMG was administered i.v. (25 mg/kg 24 h prior to hemorrhage).
* p < 0.05, ** p < 0.01, *** p < 0.001 compared to 0 h control.
N = 10 in each group.

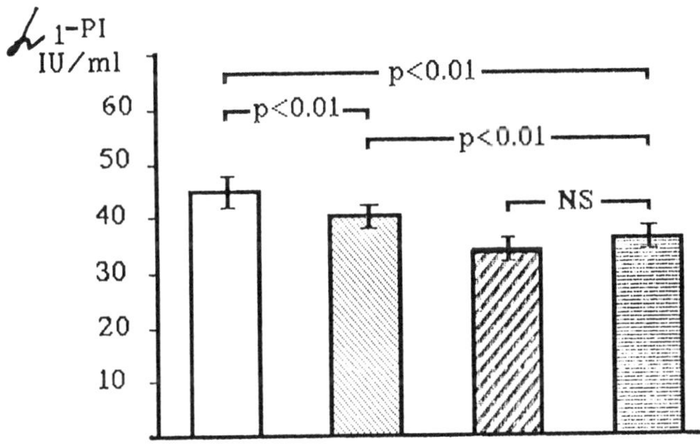

Figure 1. Effect of preliminary injection of ß-1,3-carboxymethylglucan (25 mg/kg) on α_1-proteinase inhibitor after hemorrhage in mice.
All values are mean ± SD. N = 10 in each group.

animals administration of CMG was associated with a decreasing of α_1-PI in blood of mice 24 h after injection (Fig. 1). Both in control and CMG groups α_1-PI have decreased after hemorrhage (1 h). There were no differences in α_1-PI between control and CMG groups. The possible mechanism of α_1-PI decrease could be depletion of proteinase inhibitory potential due to enhanced proteolytic enzymes secretion, and inactivation of α_1-PI by active forms of oxygen after glucan administration (9). The data obtained suggests that preliminary administered semisoluble CMG have protective effect during acute massive hemorrhage in despite of initial enhancement of active oxygen form production, and α_1-PI depletion after CMG administration.

REFERENCES

1.   Altura BM. Reticuloendothelial system function and histamine release in shock and trauma: relationship to microcirculation. Klin Wochenschr 1982; 60:882-890.

2.   Di Luzio NR. Immunopharmacology of glucan: Broad spectrum enhancer of host defense mechanisms. Trend Pharmacol Sci 1983; 4:344-347.

3.   Jochum M, Witte J, Duswald KN, Inthorn D, Welter H, Fritz H. Pathobiochemistry of sepsis: role of proteinases, proteinase inhibitors and oxidizing agents. Behring Inst Mitt 1989; 79:121-130.

4.   Barbey-Morel C, Pierce JA, Campbel EI, Perlmutter DN. Lipopolysaccharide modulates the expression of α_1-proteinase inhibitor and other serine proteinase inhibitor in human monocytes and macrophages. J Exper Med 1987; 166:1041-1048.

5.   Herscowitz HB, Holden HT, Bellanti JA. In: Manual of Macrophage Methodology. Collection, Characterization and Function. Immunol Ser. New York & Basel: Marcel Dekker, 1981: 190-198.

6.   Van Wees J, Tegtmeyer FK, Otte J, Wood WG, Brun J. Proteinase-antiproteinase imbalance in meningitis: determination of α_1-proteinase inhibitor (α_1-PI), elastase-α_1PI complex and elastase inhibition capacity in cerebrospinal fluid. Klin Wochenschr 1990; 68:1054-1058.

7.   Barrett AJ. Lysosomes: A Laboratory Handbook. Amsterdam, 1972: 46-135.

8.   Decker K. Biologically active products of stimulated liver macrophages. Eur J Biochem 1990; 193:245-261.

9.   Travis J, Guzdek A, Potempa J, Watorek W. Serpins: Structure and mechanism of action. J Biol Chem 1990; 371:3-11.

DETERMINATION OF TISSUE KALLIKREIN AND α_1-ANTITRYPSIN-TISSUE KALLIKREIN COMPLEXES IN SYNOVIAL FLUID OF PATIENTS WITH RHEUMATOID, OSTEO AND PSORIATIC ARTHRITIS

M.M. Rahman, Karen Worthy, C.J. Elson[1], P.A. Dieppe[2] and K.D. Bhoola

Departments of Pharmacology, Pathology[1] and Rheumatology[2], School of Medical Sciences, University of Bristol, Bristol BS8 1TD

SUMMARY: An enzyme–linked immunosorbant assay was developed to determine tissue kallikrein and α_1–antitrypsin–tissue kallikrein complexes in pooled synovial fluid of patients with rheumatoid, osteo and psoriatic arthritis. Even though basal values could be determined, the addition of synovial fluid shifted the standard curves for both tissue kallikrein and α_1–antitrypsin–tissue kallikrein complex to the right, because of the presence of a novel inhibitor.

INTRODUCTION

An important feature of inflamed joints is the accumulation of neutrophils. Tissue kallikrein has been localised by immunocytochemistry in neutrophils present in synovial membranes and fluid of patients with RA (1). Tissue kallikrein becomes a constituent of synovial fluid (2) either by transudation from plasma and or by release from degranulating neutrophils. The enzymic activity of the kallikreins is constrained by endogenous inhibitors; tissue kallikrein by α_1–antitrypsin, and plasma kallikrein by α_2–macroglobulin and C1 inhibitor. Although increased levels of these inhibitors have been observed in the synovial fluid of inflamed joints, an increased capacity may not be available for the inhibition of the kallikreins because of the presence of many other proteases. The purpose of this study was to characterise and determine precisely the levels of tissue kallikrein (TK) and α_1–antitrypsin–tissue kallikrein (α_1–AT–TK) complexes in the synovial fluid (SF) of patients with rheumatoid (RA), osteo (OA) and psoriatic (PA) arthritis.

METHODS

Synovial fluid, aspirated from the knee joints of patients with rheumatoid, osteo and psoriatic arthritis, was collected by Clinical Colleagues at Rheumatology Clinics in the Bristol Royal

Infirmary, and pooled human saliva, urine and serum was provided by the Haematology Department, Southmead Hospital, Bristol.

The fluid was centrifuged at 5,000 RPM for 7 min. The cells were removed, and the fluid aliquoted and stored at $-20^{\circ}C$. Just prior to each experiment, fluids were incubated with hyaluronidase (22.5 IU/ml for 30 min at $37^{\circ}C$). The rheumatoid factor was removed from RA synovial fluid by incubating it for 30 min at room temperature with heat aggregated rabbit IgG coupled to CnBr Sepharose (Pharmacia, Milton Keynes U.K.).

Sandwich immunosorbant assays were developed to measure both TK and α_1–AT–TK complexes, using 'Immulon 1' (3) microtitre plates.

(i) For the TK ELISA, microtitre plates were coated with anti–TK rabbit IgG 5 μg/ml (a polyclonal antibody, raised in our laboratory), which was diluted in coating buffer (pH 9.6; 0.015 M Na_2CO_3 and 0.035 M $NaHCO_3$). After washing, the unoccupied spaces were blocked by incubating for 30 min at room temperature (RT) with 5% low fat bovine milk dissolved in 0.02 M Tris–NaCl buffer (pH 7.2). Next, a TK standard concentration curve was prepared, using purified human urinary kallikrein at concentrations that ranged from 10 ng/ml to 0.79 ng/ml, as well as test samples. After application of the second antibody, 10 μg/ml of Goat anti–human urinary kallikrein (Protogen AG, Switzerland), a further incubation was performed with 1:250 of anti-goat IgG alkaline phosphatase conjugate (Sigma Chemicals, U.K.). The final incubation involved application of the substrate (disodium para–nitrophenyl phosphate, 1mg/ml) for alkaline phosphatase. The change in the chromophore in each well of the plate was read in an ELISA reader, and the result plotted on semilog graph paper. All the incubation steps lasted 1 h unless stated otherwise, and in between each step the microtitre plate was washed 3 times (for 10 sec periods) with phosphate buffer saline (PBS) containing Tween 20 (pH 7.4; 137 mM of NaCl, 2.7 mM of KCl, 10 mM of phosphate buffer and 0.5% Tween 20).

(ii) α_1–AT–TK complexes were prepared by incubating, 25μl of 0.8 μg/ml of lyophilised human urinary kallikrein (Green Cross Corporation, Osaka, Japan) with 25 μl of 100 μg/ml of α_1–antitrypsin (Sigma Chemicals, U.K.) at $37^{\circ}C$ for 24 h. Both reagents were diluted in 0.2 M Tris–HCl buffer (pH 8.2).

For the α_1–AT–TK complex ELISA, the microtitre plate was coated with the polyclonal, the anti–TK rabbit IgG, diluted in coating buffer at a concentration of 5 μg/ml. Next, the microtitre plate was incubated for 30 min at RT, with 5% low fat bovine milk in 0.02 M Tris–NaCl buffer (pH 7.2) to block the unoccupied sites. Subsequently, the plate was incubated with α_1–AT–TK complexes to generate a standard curve with concentrations that ranged from 0.39

μg/ml to 50 μg/ml, as well as test samples. Then sheep antihuman α_1–antitrypsin conjugated with alkaline phosphatase (Serotec, England) was added in a dilution of 1:250. Finally, the well content was incubated with the substrate (disodium para–nitrophenyl phosphate, 1mg/ml) for alkaline phosphatase for 20 h at RT to develope the chromophore. The values were determined, using an ELISA reader, and the values plotted on semilog graph paper. For each step, unless otherwise stated, the plate was incubated for 1 h at 37ºC, and was washed 3 times (for 10 sec periods) between steps with PBS–Tween 20 buffer (pH 7.4). A more sensitive α_1–AT–TK complex ELISA was developed using avidin–biotin conjugates. After applying the samples to the microtitre plate, it was first incubated with 8 μg/ml of sheep anti–human α_1–antitrypsin (Serotec, England), then with 1:2000 anti sheep IgG (whole molecule) biotin congugate (Sigma Chemicals, U.K.), and thereafter that with 1:250 dilution of extravadin alkaline phosphatase (Sigma Chemicals, USA). Finally, the wells were incubated with the substrate (disodium para–nitrophenyl phosphate, 1mg/ml) for alkaline phosphatase for 20 h at RT. All the steps were performed as already described. The dilution buffer used in all the steps for both TK and α_1–AT–TK ELISA was 5% low fat bovine milk in 0.02 M Tris–NaCl buffer (pH 7.2), unless otherwise stated.

RESULTS

The TK standard curve gave a lower limit estimate of 0.45 ng/ml. In the case of the TK–$\alpha_1$antitrypsin complex assay, it was necessary to develop the more sensitive avidin–biotin system that gave a lower limit value of 1.5 ng/ml (Figure 1). After obtaining standard curves for both TK and α_1–AT–TK, dilution curves were performed for each biological fluid (rheumatoid, osteo and psoriatic synovial fluids (SF), serum, saliva and urine) sample. Dilution curves were generated, commencing from the parent sample to 1:128. Measurable amounts of both TK and α_1–AT–TK were observed when dilution curves for synovial fluids (RA, OA and PA), serum and urine were constructed (Figures 2 and 3). In order to examine the question whether the levels obtained in the dilution curve represented true values, the standard curves were spiked with various amounts of each biological fluid. After spiking each standard curve with SF (Figures 4 and 5) and serum (Figures 6 and 7), the curves showed a marked shift to the right. The reduction was not so pronounced when a 1:10 dilution of SF and serum was used. When the TK standard curve was spiked with saliva (1:1000) and urine (1:10 and 1:50), there was a marked shift of the curve to the left.

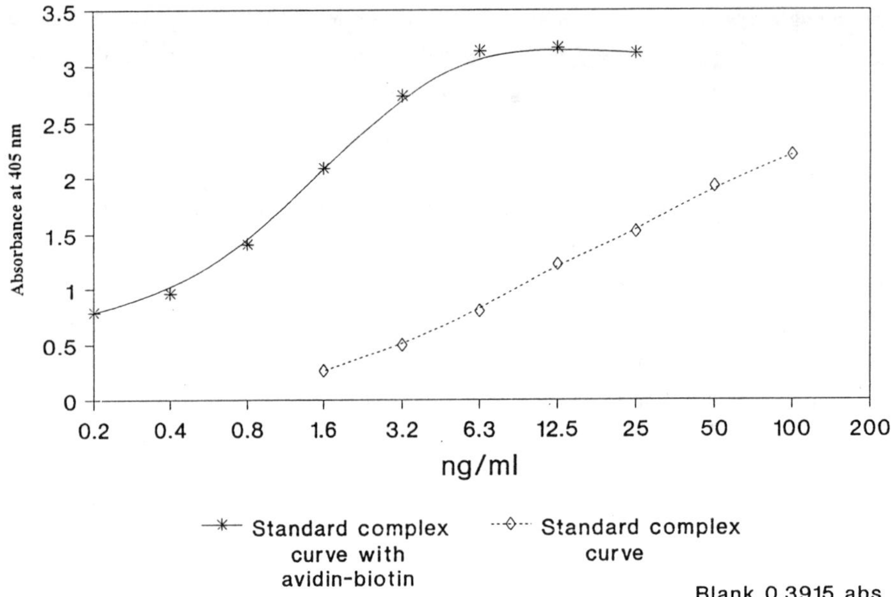

Figure 1. Concentration–dependent standard curves for α_1–antitrypsin–tissue kallikrein complexes performed with and without the avidin–biotin system.

DISCUSSION

Since TK and α_1–AT–TK complexes were detected in RA, OA and PA synovial fluids and in pooled serum, an attempt was made here, to quantify the precise values of both free and complexed tissue kallikrein that occur in biological fluids. At first each ELISA was standardised. Following standardisation the minimum detectable limit for TK equalled 0.45 ng/ml, and that for α_1–AT–TK complex was 1.5 ng/ml with the more sensitive avidin–biotin system. As measurable amounts of TK were found in urine and saliva, the TK standard curve was spiked with saliva (1:1000) and urine (1:10 and 1:50). On spiking there was a marked shift of the TK standard curve to the left. In contrast, and because α_1–AT–TK could not be detected in both urine and saliva, no attempt was made to spike the complex standard curve with these fluids.

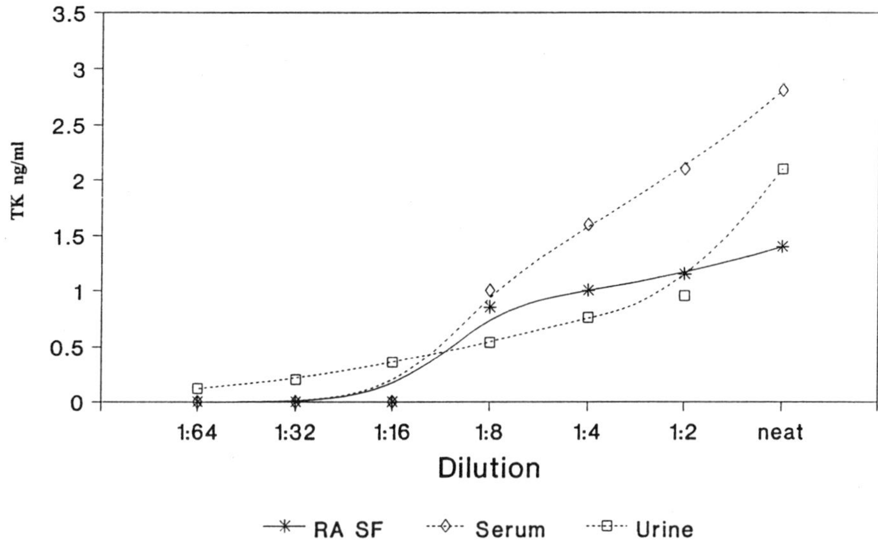

Figure 2. Dilution curves of tissue kallikrein in synovial fluid (RA), human serum and human urine.

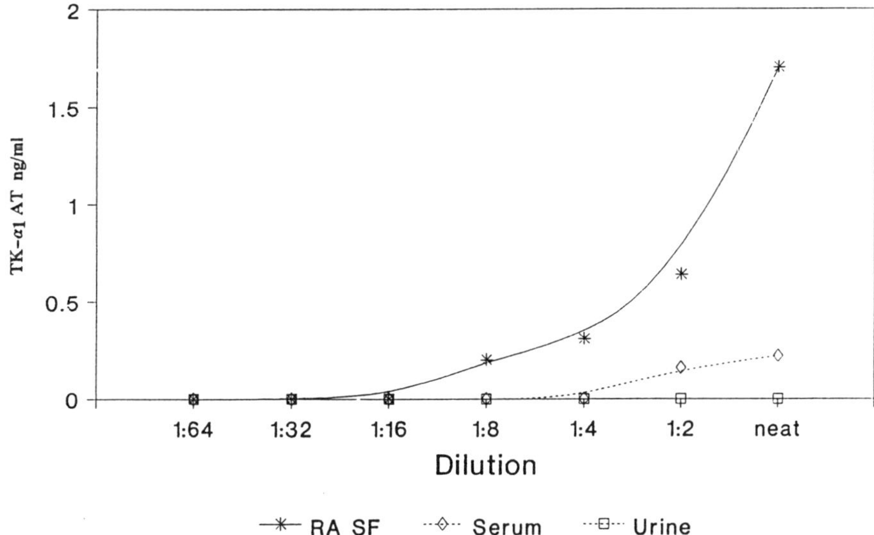

Figure 3. Dilution curves of α_1–antitrypsin–tissue kallikrein complexes in synovial fluid (RA), human serum and human urine.

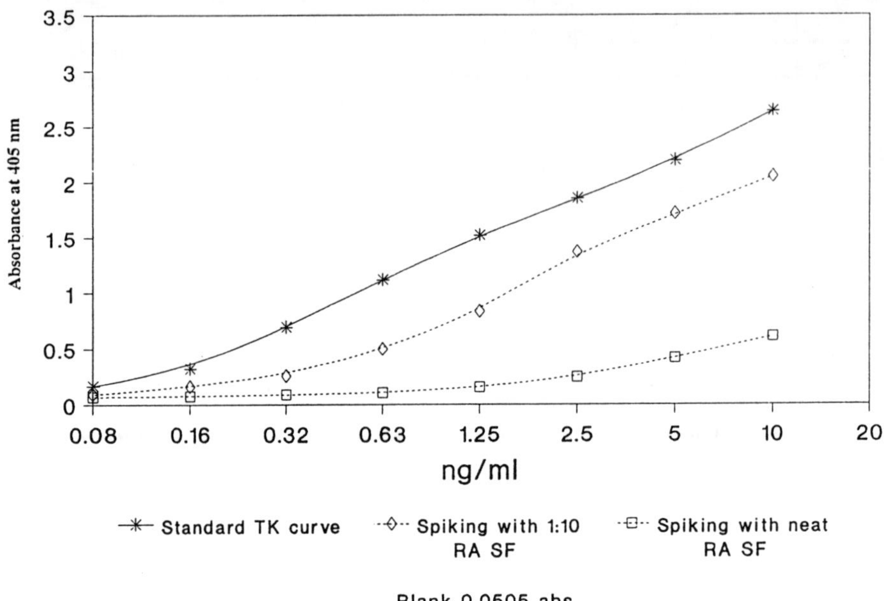

Figure 4. Concentration–dependent curves of tissue kallikrein spiked with RA synovial fluid.

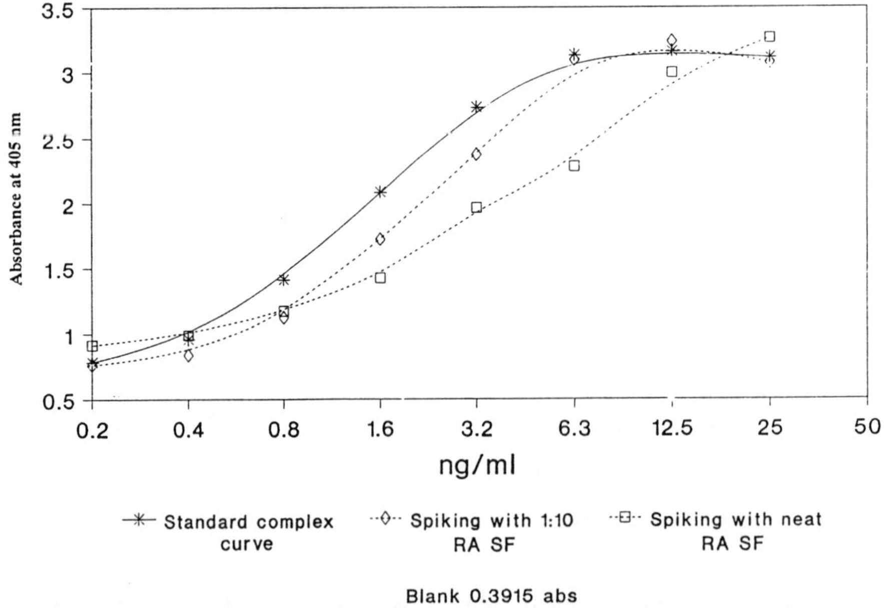

Figure 5. Concentration–dependent curves of α_1–antitrypsin–tissue kallikrein complexes spiked with RA synovial fluid.

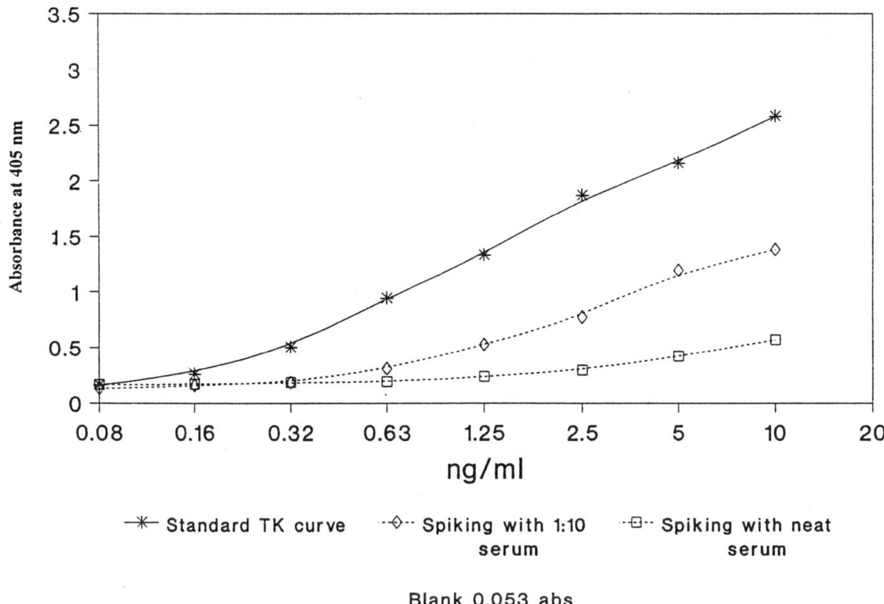

Figure 6. Concentration–dependent curves of tissue kallikrein spiked with human serum.

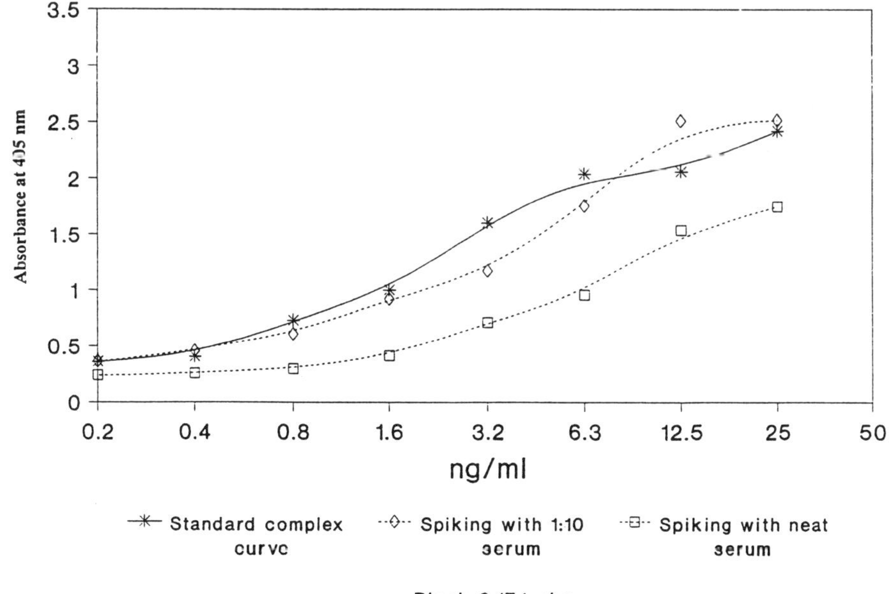

Figure 7. Concentration–dependent curves of α_1–antitrypsin–tissue kallikrein complexes spiked with human serum.

Experiments, to ascertain whether the synovial fluid and serum dilution curves represented true values, revealed that there was a marked flattening and a shift to the right when both the standard curves were spiked with these fluids. One approach to dissociate the possible effects of an inhibitor and/or interfering substance was to dilute the various synovial fluids as well as serum. Although the reduction and shift of the standard curves to the right was less marked, no additional recovery was observed by this procedure. Therefore, we believe that the difficulty in obtaining precise values for TK and α_1–AT–TK may be due to interference by some component of the synovial fluids and serum with the binding of TK to its antibody, or to the formation of complexes with a novel endogenous inhibitor other than α_1–antitrypsin.

ACKNOWLEDGMENTS

Dr. M.M. Rahman is a WHO Fellow. We thank the Medical Research Council, U.K. and WHO for research grants.

REFERENCES

1. Figueroa CD, MacIver AG & Bhoola KD. Identification of a tissue kallikrein in human polymorphoneculear lecucytes. Br J Haematol 1989, 72:321–328.

2. Selwyn B, Figueroa CD, Fink E, Swan A, Dieppe PA, Bhoola KD. A tissue kallikrein in synovial fluid of patients with rheumatoid arthritis. Ann Rheum Dis 1989, 48:128–133.

3. George E. Kenny, Carol L. Dunsmoor. Principal Problems and Strategies in use of Antigenic Mixture for ELISA. Journal of Clinical Microbiology 1983, 17:655–665.

COMPARISON OF PLASMA PREKALLIKREIN LEVELS IN HUMAN SKIN BLISTER AND CHAMBER FLUIDS FROM VOLUNTEER SUBJECTS AND PSORIATIC PATIENTS

N.K. Thomas, G. Marshman[1], K. Worthy, J. Burton[1] and K.D. Bhoola

Departments of Pharmacology and Dermatology[1], School of Medical Sciences, University of Bristol, Bristol BS8 1TD.

SUMMARY: Suction blisters were successfully raised on lesional, perilesional and non–lesional skin of psoriatic patients. Similar blisters were created on the skin of volunteer subjects. A specific method has been established for the measurement of plasma prekallikrein, and validated for blister fluid, skin chamber fluid, plasma and synovial fluid. Combined levels of plasma prekallikrein and its active form were compared in blister fluids obtained from psoriatic patients and the volunteers. Significantly higher values for plasma prekallikrein were found in the non-lesional blister fluids of psoriatic patients when compared to the volunteer subjects.

INTRODUCTION

Psoriasis is an inflammatory skin disease characterized by itch, inflammation and an increased turnover in epithelial keratinocytes. The presence of various components of the kallikrein–kinin system in the plasma and skin of psoriatic patients have been suggested in previous reports. An important feature of psoriasis is the infiltration of neutrophils into the epidermis. Recent studies show that neutrophils in sections of psoriatic scale and epidermis stained positively for tissue kallikrein (TK) (1). Both enzymes, plasma and tissue kallikrein, have been demonstrated in homogenized, psoriatic epidermis (2). Preliminary, conflicting results have been reported by Wolf et al (3) on the presence of plasma prekallikrein (prePK) and plasma kallikrein (PK) in plasma of patients with chronic plaque type psoriasis. In addition to these findings, elevated levels of kininogen have been reported in the plasma of 33 psoriatic patients when compared to values in volunteers (4).

Previous methods for the measurement of PK levels in biological samples have lacked specificity. Therefore, in the present experiments we set out to validate the determination of prePK and PK in a range of biological fluids with preliminary measurements being made in skin blister fluid collected from the psoriatic patients and the volunteer subjects (non–psoriatic controls). For the first time methods of sample collection from skin have been compared, and blisters raised on plaques of psoriatic skin.

MATERIALS AND METHODS

Patients and sample collection: Patients with a diagnosis of chronic plaque type psoriasis who had not received any treatment two weeks prior to the study were recruited from the Dermatology Department, Bristol Royal Infirmary, Bristol. An equal number of volunteer subjects (non-psoriatic controls) were also recruited. Exudates were collected from the skin of psoriatic patients and the volunteers using the suction blister and skin chamber techniques. Both techniques were performed on lesional, perilesional and non-lesional sites. Blisters were induced on the upper thighs of psoriatic patients and normal volunteers via suction, according to the method of Kiistala and Mustakallio (5). After 2–3 h at a negative pressure of 280–300 mm Hg, the fluid was removed from the uniform blisters using a fine needle syringe. 500μl to 2½ ml of blister fluid (BF) was obtained from each patient. The samples were immediately frozen in liquid nitrogen then stored at –20°C until required. The skin chamber technique relies on transudation from the skin (6). A plastic chamber was secured onto a patch of Sellotape–stripped forearm which had been stripped for approximately 10 min until the glistening layer was revealed. The glistening was produced by granules in the stratum granulosum once the stratum corneum had been removed. After securing the plastic chamber on the forearm, 5 ml of sterile phosphate buffered saline (PBS: 137 mmol sodium chloride, 2.7 mmol potassium chloride, 10 mmol phosphate buffer, pH 7.4; Sigma Chemicals, U.K.) was applied as a wash for 5 min. A further 5 ml was introduced and left in place for 30 min then removed and spun at 1200 rpm for 10 min at room temperature (RT). The supernatant was equally aliquoted, all samples frozen in liquid nitrogen and then stored at –20°C until required. Before measurements were made, the skin chamber fluid from both psoriatic patients and volunteers underwent a concentration step. This was performed using Millipore 10K nominal, molecular weight cut off filters (Millipore, U.K.). 2 ml of fluid was spun at 5000 rpm in a fixed angle centrifuge for 1 h at RT; the residue remaining on the filter was reconstituted in 200 μl microassay buffer (0.05 M Tris–HCl containing 0.1% poly ethylene glycol, pH 7.9).

Synovial fluids were obtained from the knee joints of patients with rheumatoid arthritis attending Rheumatology Clinics at Bristol Royal Infirmary. Samples were centrifuged (10 min, 1200 rpm, 2°C), and the pooled supernatants aliquoted and stored at –20°C. Pooled plasma was obtained from the Haematology Department, Southmead Hospital, Bristol.

Determination of PK values: Measurement of functionally activable prePK was adapted and improved from a method developed by de la Cadena et al (7). The assay is a colorimetric one that is performed on a 96 well microtitre plate (Immulon) and uses the selective, synthetic chromophore D–Pro–Phe–ArgPNA (S–2302, Kabivitrum, Stockholm, Sweden). Samples (blister fluid, skin chamber fluid, synovial fluid or plasma) were acidified before measurements were made in order to destroy inhibitors, thereby preventing inhibition of active PK when formed. The assay permits the measurement of both active enzyme and total enzyme which is determined after activating the samples.

Assay procedure: One volume (50 μl) of 0.167 M HCl was added to an equal volume of sample and incubated for 25 min at RT; One volume of both acidification buffer (0.1 M phosphate, 0.15 M sodium chloride and 0.001 M EDTA) and 0.167 M NaOH was added. Dilutions were performed in microassay buffer and 50 μl of sample added to the appropriate wells. An equal volume of activator mixture (see below) was then added to allow measurement of total PK. The plate was incubated for 15 min at 37°C. The activator mixture contained, 5 μg/ml Activated Hageman Factor (Protogen AG, Switzerland); 10^{-4} M Ellagic acid (Sigma Chemicals, U.K.) ; 1 μg/ml phospholipid (L–α–phosphatidylethanolamine, Sigma Chemicals, U.K.) and 25 μg/ml kininogen (a partially purified mixture of H– and L– kininogens prepared in our laboratory). Finally, one volume of substrate S–2302 was added, the plate incubated for 30 min and the absorbance read at 405 nm.

RESULTS

For the validation and optimization of the prePK and PK assay, plasma and synovial fluid pools at 1:32 dilutions were used. All incubations were performed at 37°C and the acidification step performed for 25 min using 0.167 M HCl. Time–courses for activation of prePK were performed using 0.025 mg/ml Dextran Sulphate (Sigma Chemicals, U.K.). Samples were incubated with activator for a variety of time points using acidified and non–acidified samples. For non–acidified samples there was an initial increase in activity which then decreased as time went on as the inhibitors bound the active PK which had been formed. For acidified samples the activity increased and remained high since the acid present prevented inhibitors binding active PK (Figure 1 a, b). The optimum time for activation was found to be 15 min. A range of methods of activation were compared for prePK in both the purified preparation and biological fluids. The optimum concentration of the various components was found to be: 5 μg/ml Hageman factor; 1 μg/ml phospholipid; 10^{-4} M ellagic acid; 25 μg/ml kininogen.

Specificity of the assay was determined by using anti–PK antibody (Protogen AG, Switzerland) linked to CnBr Sepharose. The Sepharose–linked anti–PK antibody was used to remove the proenzyme and its active form from the biological samples, prior to activation of prePK and the subsequent measurement in the assay.

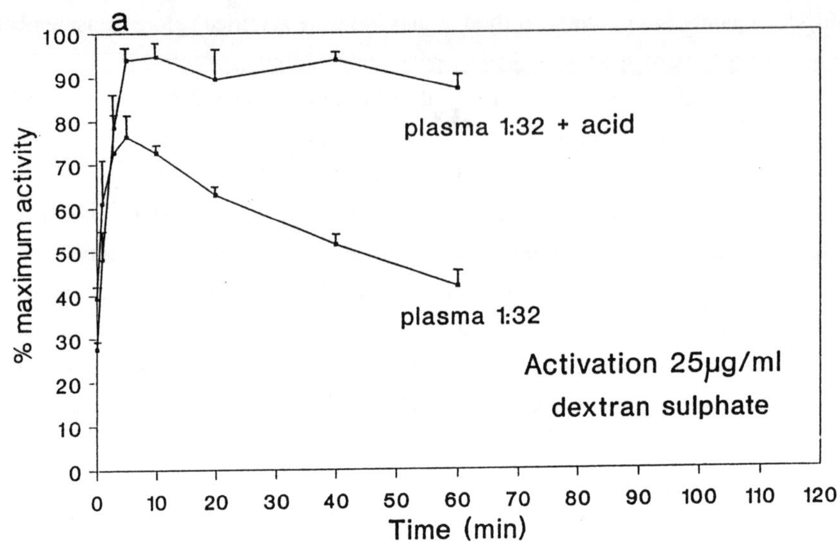

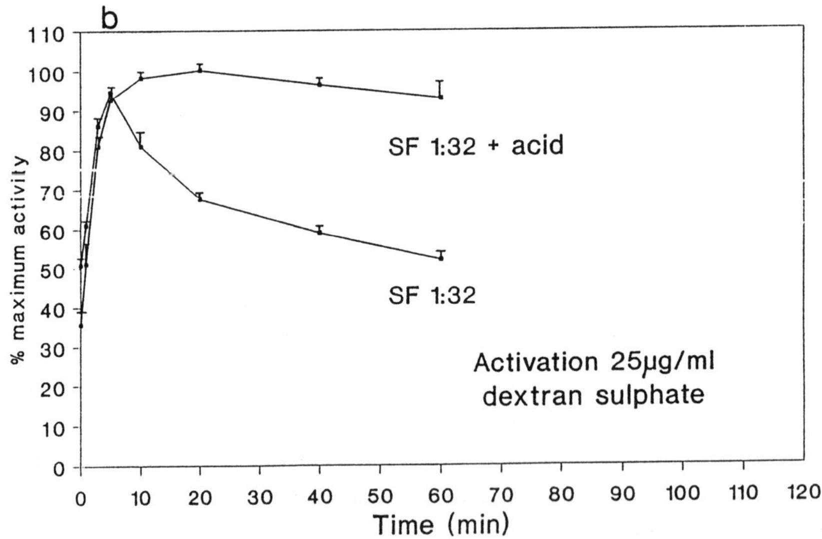

Figure 1. Activation of plasma prekallikrein in **a)** plasma **b)** synovial fluid (SF).

Measurement of total PK values in prePK (Fletcher factor) deficient plasma (Sigma Chemicals, U. K.) confirmed the efficiency with which total PK had been removed from the biological samples by the Sepharose–linked anti–PK antibody. The removal of prePK and PK is shown in figure 2 a , b.

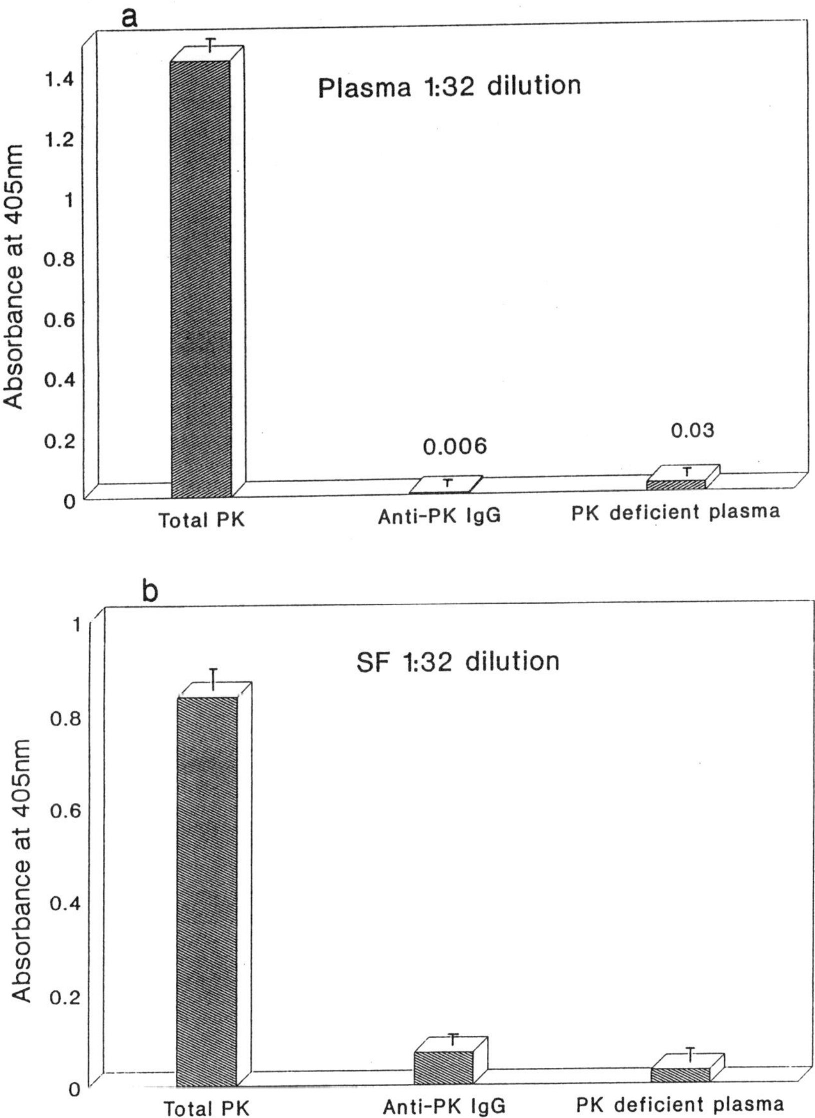

Figure 2. Confirmation of the presence of total PK in a) plasma b) synovial fluid. SF= synovial fluid; Anti–PK IgG= anti plasma kallikrein IgG linked to Sepharose

Blister fluid from the non–lesional site of psoriatic skin (2.768 ± 0.29 μg PK / mg protein) contained significantly more total PK (Fischer test, p <0.05) than from the skin of normal volunteers (2.124 ± 0.196 μg PK / mg protein) (Figure 3). However, neither prePK nor PK was detected in the skin chamber fluid collected from psoriatic patients or the volunteer subjects.

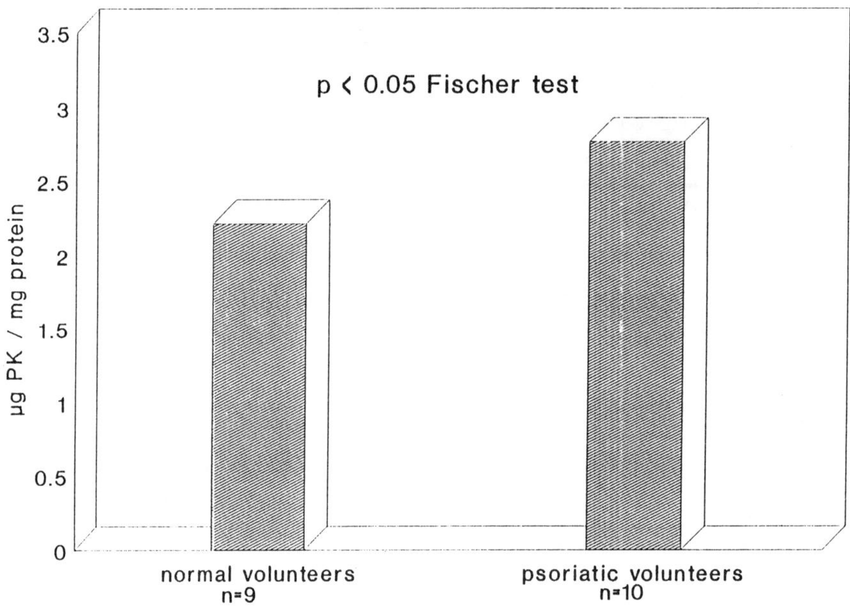

Figure 3. Plasma prekallikrein values in blister fluid collected from psoriatic patients and volunteer subjects (non–psoriatic, normal volunteers).

DISCUSSION

In this study we have compared different methods of sample collection from the skin of psoriatic patients and volunteers. The skin chamber technique was first reported in 1979 by Norris and co–workers. However, although this method is less invasive than the suction blister technique, unfortunately there has been little or no information on plasma or tissue kallikrein levels. Clearly, the skin chamber technique has not been useful for the aims of our study. The suction blister technique is relatively non–invasive, and can be performed on the plaques of psoriasis as well as on the surrounding regions. This report is one of the first, concerned with the creation of suction blisters on skin plaques of psoriatic patients, and generate information about the role of kallikreins and kinins in this disease.

Accurate and specific measurement of prePK and PK has been established in blister fluid and other biological fluids. The assay was specific for PK, fast, reliable and could be performed with great accuracy, therefore minimizing intra- and inter-assay variation. The acidification step prevented endogenous inhibitors binding to the activated enzyme; acid treatment did not damage the PK. The assay provided information regarding active and inactive PK with the activator mixture providing all the essential components to allow the activation of prePK to take place in both purified prePK and in the biological samples.

A significantly greater amount of total PK was present in non-lesional blister fluid of psoriatic patients than in the blister fluid of volunteer subjects. This would imply that the increase in total enzyme in the psoriatic blister fluid is a reflection of an increased transudation of the enzyme into the skin. This would then lead to greater formation of bradykinin which may be involved in many of the inflammatory changes that occur in the skin of psoriatic patients namely, increase in vascular permeability, arteriolar dilatation, promotion of cell proliferation and release of other inflammatory mediators such as prostaglandins and cytokines. Many of these actions are important in psoriasis; the scaly nature of psoriatic scale is caused by an increased epidermal cell proliferation; the heat and redness of psoriasis is due to epidermal blood vessel proliferation.

Measurements of the other components of the kallikrein-kinin system need to be performed to provide a clearer picture of its role in inflammatory skin disease. If an overactive kallikrein-kinin was found to be important in psoriasis, the use of kinin antagonists as therapeutic agents should be possible, and thereby provide an end to much suffering.

ACKNOWLEDGEMENTS

We thank the Medical Research Council and the Psoriasis Association for financial support.

REFERENCES

1. Poblete MT, Reynolds NJ, Figueroa CD, Burton JL, Müller-Esterl W, Bhoola KD. Tissue kallikrein and kininogen in human sweat glands and psoriatic skin. Br J Dermatol 1991; 124:236-241.

2. Hibino T, Isaki S, Kimura H. Partial purification of tissue kallikreins in psoriatic epidermis. J Invest Dermatol 1988; 90:505-510.

3. Wolf R, Machety I, Feuerman EJ, Creter D. The kallikrein-kinin pathway in psoriasis. Preliminary observations. Biomed 1989; 35:77-78.

4. Winkelmann RK. Total plasma kininogen in psoriasis and atopic dermatitis. Acta Derm Venereol (Stckh) 1984; 64:261–263.

5. Kiistala U, Mustakallio KK. In–vivo separation of epidermis by production of suction blisters. Lancet 1964; 1:1444.

6. Norris DA, Lipman SH, Weston WL. Human monocyte chemotaxis: A quantitative in vivo technique. J Invest Dermatol 1979; 72:81–84.

7. De la Cadena RA, Scott CF and Colman RW. Evaluation of a microassay for human plasma prekallikrein. J Lab Clin Med 1987; 109:601–607.

CHANGES IN POLYMORPHONUCLEAR NEUTROPHIL-ELASTASE IN PANCREATITIS

S. Uehara, K. Gotoh, H. Handa, K. Honjo, A. Hirayama and W. Sakamoto*

Department of Internal Medicine, Tonan Hospital N1 W 6, Chuo-ku, Sapporo 060, Japan
Department of Biochemistry*, School of Dentistry, Hokkaido University N 15 W5, Kita-ku, Sapporo 060, Japan

SUMMARY: In cases of acute and chronic pancreatitis, we measured the amount of polymorphonuclear neutrophil (PMN)-elastase. There was a significantly larger increase in PMN-elastase in patients with pancreatitis than normal adults. Especially, there was a particularly notable increase in amount of PMN-elastase in patients with severe pancreatitis. Furthermore, the peak of PMN-elastase increase throughout the course of the pancreatitis was seen to be 1-2 days after peak increase in pancreatic enzymes. In the experiment in which pancreatic juice and pig pancreatic kallikrein were added to granulocytes in vitro, we recognized a gradual release of PMN-elastase. From these data, we suggested that timely measurements of PMN-elastase are useful to marker of monitoring clinical changes in severe pancreatitis.

INTRODUCTION

Granulocytes released proteases which are 2 types by endotoxin, antigen-antibody complex, complement, and radical oxygens. The neutral protease plays a more important role under physiological condition.[1] This elastase was shown to possess elastinolytic and collagenolytic properties, and its role in mediating tissus damage received even wider attention.

It may play a role in vascular injury. Furthermore, this protease has recently received attention as a fibrinolytic factor of the non-plasmin system. It has been reported in clinical medicine that this PMN-elastase was known to increase in severe cases of respiratory diseases including acute respiratory distress syndrome(ARDS), sepsis, leukemia, and DIC.[1],[2]

The amount of PMN-elastase has been measured in cases of pancreatitis, and its clinical importance and its relationship, particularly with severe pancreatitis, have been reported.[3]

When we measured the amount of PMN-elastase in cases of pancreatitis, we confirmed its increase so that, based upon classification according to the severity of the conditions, we were able to examine its usefulness in monitoring clinical changes. In addition, we measured other pancreatic enzymes and examined the correlation between them and PMN-elastase.

MATERIALS AND METHODS

The cases in this study included : 20 cases representing a normal group ; 17 cases of acute pancreatitis, including 6 severe, 8 moderate, and 3 mild cases ; and 27 cases of chronic pancreatitis, including 1 severe, 11 moderate, and 15 mild cases.

The diagnosis of pancreatitis based on typical clinical symptoms, examination of raised various pancreatic enzymes, ultrasonic analysis, computerized tomography, and ERCP. In accordance with the report by Ranson et al.,[4] all patients were classified into three groups : severe, moderate, and mild.

1) The measurement of granulocytic elastase was determined in plasma in complex with α_1-antitrypsin (α_1-AT) by ELISA (Merck, Darmstadt, FRG)[5] and Substrate (S-2238 ; Kabi, Stockholm, Sweden).

2) Serum trypsin was examined by RIA (Behing, Ltd. Marburg, FRG), Serum elastase-I by RIA (Dainabott, Ltd. Tokyo, Japan) and phospholipase-A by RIA (Shionogi, Ltd. Osaka, Japan) .

3) Serum, urine amylase, serum lipase and leukocyte count was examined by the routine method.

4) The experiment of released elastase from granulocytes in vitro was determined using human granulocytes to which pancreatic juice and pig pancreatic kallikrein (Sanwa Chemical, Ltd. Tokyo, Japan) were added.

RESULTS

<u>PMN-elastase levels in pancreatitis patients</u> : As shown in Table 1, there was a significantly large increase in PMN-elastase in patients with acute or chronic pancreatitis than in normal adults.

Table 1. Plasma PMN-elastase levels in pancreatitis

normal subjects	20	101.3 ± 40.0
acute pancreatitis	12	627.1 ± 370.0**
chronic pancreatitis	27	409.3 ± 266.5**

mean $\pm$ SD(ng / ml)   ** $P < 0.01$ for comparison to normal subjects

Furthermore, the values of PMN-elastase according to the severity of patients are shown in Fig. 1.

While many severe cases showed values of 400 ng/ml or higher, the majority of moderate and mild cases showed values between 200-500 ng/ml.

In severe cases of pancreatitis, the peak PMN-elastase increase was seen to occur 1 to 2 days after the peak increase in pancreatic enzymes.

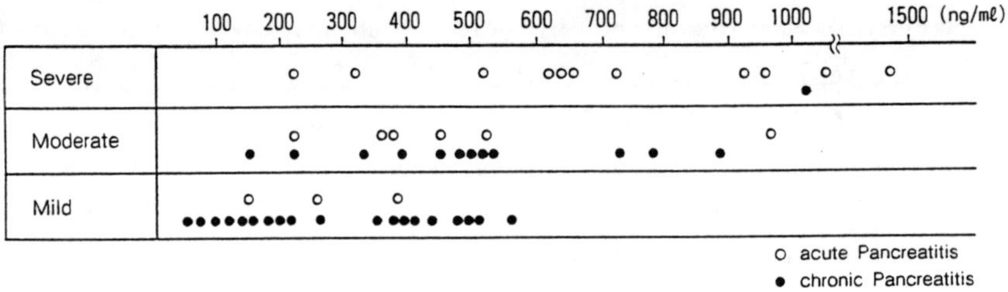

Figure 1. Plasma PMN-elastase levels related to clinical stage in pancreatitis

Correlation between some parameters : As shown in Table 2, there was no significant correlation between amylase and leukocyte counts on the one hand, and PMN-elastase on the other.

Table 2. Comparison of correlation coefficient between PMN-elastase, leukocyte counts and pancreatic enzymes in pancreatitis

	S-Amylase	u-Amylase	Trypsin	Elastase-1	PL-A$_2$	WBC
PMN-elastase	0.107 (n=62)	0.184 (n=62)	−0.03 (n=49)	0.256 (n=49)	0.024 (n=54)	0.222 (n=62)

Addition of pancreatic juice and pig pancreatic kallikrein to granulocytes in vitro : As shown in Fig.2, the release of PMN-elastase was seen to occur approximately 4-6 hours after the addition of pancreatic juice and pig pancreas kallikrein to granulocytes.

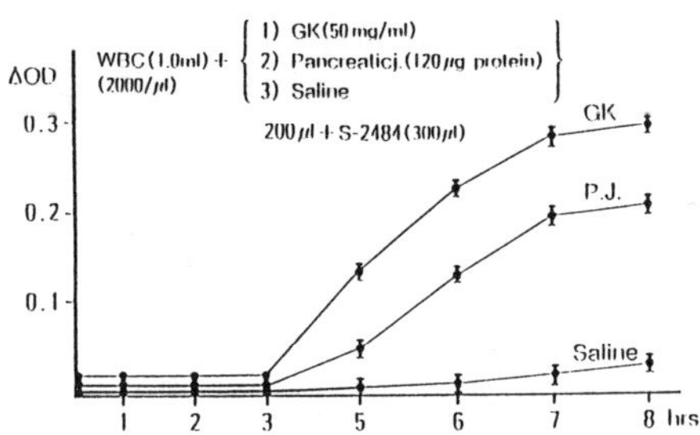

Figure 2. Addition of pancreatic juice and pig pancreatic kallikrein to granulocytes in vitro

DISCUSSION

Multiple Organ Failure (MOF) occurs in severe pancreatitis, and the death rate in these cases is about 30 % in Japan.

Complications accompanying MOF of severe pancreatitis include respiratory and circulatory failure, renal failure, hemorrhage, and sepsis. Various factors have been investigated as the mediator of such MOF.

In the case of pancreatitis in particular, the occurrence of autodigestion by pancreatic enzymes was considered the important factor in MOF. However, since it has been determined in recent years that there is no correlation between the symptoms observed during the course of severe pancreatitis and the fluctuation of pancreatic enzymes, we have to look for other mediators as the cause of MOF.

PMN-elastase is not necessarily an indicator of pancreatitis, even though it is known to increase in severe inflammation.

However, there has been a increasing number of reports recognizing an increase in PMN-elastase in pancreatitis, particularly in severe cases.

Our results showed that there was a significantly larger increase in the amount of PMN-elastase in patients with acute and chronic pancreatitis than in normal persons.

There was a particularly notable increase in the amount of PMN-elastase in patients with severe pancreatitis. Furthermore, the peak PMN-elastase increase throughout the course of the pancreatitis was seen to be 1-2 days after the peak increase in pancreatic enzymes. Our results seem to be in accordance with the reports by Büchler[6] and Gross.[7] Gross and others pointed out that, in severe cases of acute pancreatitis, the level of PMN-elastase was high above 400 μg/l in the early stage of the pancreatitis, and that, in cases in which the patients survived, PMN-elastase gradually decreased throughout the course of the pancreatitis while in cases in which patients died, it began to increase about 2 days after the development of this disease and continued at a high level. However, they concluded that the reason for such an increase in PMN-elastase was unclear.

On the other hand, the state of shock which always accompanies severe pancreatitis leads us to suspect that there is a relationship between PMN-elastase activation and the k-k system. This relationship is suspected because PMN-elastase is activated by the complement C5a or plasma kallikrein.[8,9]

The k-k system is also activated in acute or chronic pancreatitis. Due to this activation, high and low molecular weight kininogen are activated. Next, plasma kallikrein is thought to be activated.[9]

In the experiment in which pancreatic juice and pig pancreatic kallikrein were added to granulocytes, we recognized a gradual release of PMN-elastase. This result should explain the clinical observation mentioned earlier.

The release of a large quantity of trypsin and pancreatic kallikrein into the blood stream clearly indicates a serious failure of the pancreas. It seems that, as a result, the coagulation factors of the intrinsic pathway from the k-k system is activated as well.

Moreover, if endotoxemia develops due to bacterial infection, it may lead to the activation of the coagulation factors of the extrinsic pathway, through the activation of the tissue factor, which is under the influence of cytokinine, resulting in an increase in thrombin production. This should result in the development of DIC, or, in other words, thrombosis formation in each remote vital organs, and their subsequent failure.

Therefore, it is thought that an increase in PMN-elastase will result in the elastinolysis and fibrinolysis of the non-plasmin system. This will lead to the development of hemorrhages throughout the systemic body due to disorder of connective tissue, mostly in' the blood vessels.

REFERENCE

1) Janoff, A. ; Elastase in tissue injury, Ann. Rev. Med. 1985. 36 ; 207-216.

2) Struelens, M., Dilville, J., Luypaert, P., and Wybran, J. ; Granulocyte elastase compared to c-reactive protein for early diagnosis of septicemia in critically ill patients. Eur. J. Clin. Microbiol, Infect. Dis. 1988. 7 ; 193-195.

3) Büchler, M., Malfertheiner, P., Uhl, W., and Begar, H. G. ; Leukocyte elastase in human acute pancreatitis. Digestion, 1987. 38 (9) ; 9.

4) Ranson, J. H. C., and Pastermack, B. S., ; Statistical methods for guantifying the severity of clinical acute pancreatitis. J. Surg, Res. 1977. 22 ; 79-91.

5) Fritz, J., Schiessler, H., and Geiger, R. ; Nationally occurring low molecular weights inhibitors of neutral proteases from PMN-granulocytes and kallikreins. Agents Actions. 1978. 8 ; 57-64.

6) Buchler, M., Malfertheiner, P.; Schoetensack, C., Uhl, W., Scherbaum, W., and Berger, H.G.; Wertigkeit biochemicher und bildgebender Verfahren fur Diagnose und Prognose der akuten Pankreatitis. Z. Gastroenterol, 1986. 24 ; 100-109.

7) Gross, V., Scholmerich, J., Leser, H.G., Salm, R., Lansen, M., Ruckauer, K., Schoffel, U., Lay, L., Heinisch, A., Farthmann, E.H., and Gerok, W.; Granulocyte elastase in assessment of severity of acute pancreatitis. Digestive. Diseases and Sciences, 1990. 35 ; 97-105.

8) Idell, S., Kucich, U., Fein, A., Kuppers, F., James, H.L., Walsch, P.N., Weinbaum, G., Colman, R., and Cohen, A.B.; Neutrophil elastase-releasing factors in bronchoalveolar lavage from patients with adults respiratory distress syndrome. Am. Rev. Respir. Dis. 1985. 132 ; 1098-1105.

9) Uehara, S., Kyosuke Honjo, Satoshi Furukawa, Akio Hirayama, and Wataru Sakamoto ; Role of the kallikrein-kinin system in human pancreatitis. In Kinin V Part B, (Ed. Keishi Abe, Hiroshi Moriya, and Setsuro Fujii) Plenum Publishing Corporation, New York. 1989. 643-648.

CP-0127, A NOVEL POTENT BRADYKININ ANTAGONIST, INCREASES SURVIVAL IN RAT AND RABBIT MODELS OF ENDOTOXIN SHOCK

E.T. Whalley, J.A. Solomon, D.M. Modafferi, K.A. Bonham, and J.C. Cheronis

Cortech, Inc., 6840 North Broadway, Denver, Colorado 80206

SUMMARY: The bradykinin antagonist dimer CP-0127 was found to be a potent and selective inhibitor of the depressor response to bradykinin in the anaesthetized rat and rabbit. When given as a single dose s.c. (3.6 µmol/kg), the depressor response to bradykinin was blocked for the duration of the experiment (4 hours). In anaesthetized control rats, LPS from *E. coli* produced a profound and immediate hypotensive response, while in rats infused with CP-0127, the response to LPS was almost totally reversed. In addition, CP-0127 given as a single subcutaneous dose (3.6 µmol/kg) to rats 1 hour before LPS challenge produced a 93% survival rate, compared to 14% in control animals. Finally, a survival rate of 86% was achieved in rabbits infused with CP-0127 at 0.36 nmol/kg/min i.v., compared to 45.5% in saline-infused control animals given LPS (500 µg/kg i.v.). The results of these experiments provide evidence for a significant role for the kallikrein-kinin system in these models of endotoxic shock, and indicate the therapeutic potential of a bradykinin antagonist such as CP-0127 for treating this disorder in man.

INTRODUCTION

Plasma kinins have been implicated in the pathogenesis of septic shock in man and several studies have clearly demonstrated that the fatal outcome in patients with severe septicemia is accompanied by activation of the plasma kallikrein-kinin system [1]. The kallikrein-kinin system is one of the first systems to be activated by endotoxin in experimental studies [2,3,4], as well as in clinical sepsis [5]. In the rat, immunoreactive kinins are elevated up to 50-fold 2 h after endotoxin administration [6]. The early peripheral vascular changes that accompany endotoxic shock or sepsis are mimicked by bradykinin and previous studies have demonstrated the ability of bradykinin antagonists to partially attenuate the immediate hypotensive response to endotoxin in the rat [6,7].

Based on these data, the potential for use of a bradykinin antagonist in septic shock seems clear. Studies using NPC-567, a first generation bradykinin antagonist, have demonstrated a reduced mortality in a rat model of endotoxic shock [6]. These studies, described below, evaluate the effect of a novel potent bradykinin antagonist, CP-0127 [8], on blood pressure responses to bradykinin in the anaesthetized rat and rabbit, on lipopolysaccharide (LPS)-induced hypotension in the rat and on survival in lethal endotoxin shock in both species.

MATERIALS AND METHODS

<u>Blood Pressure Studies:</u> Male Sprague-Dawley rats (300-350 g) and New Zealand White rabbits (2-2.5 kg) were anaesthetized with pentobarbitone sodium (50 mg/kg i.p.) and blood pressure recorded from an indwelling femoral artery cannula connected to a Statham pressure transducer and displayed on a 4 channel Grass polygraph recorder. Drugs were administered via a catheter placed in the femoral vein (rat and rabbit) or the carotid artery (rat). Dose response curves were constructed to bradykinin in the absence and the presence of an infusion of CP-0127 at different doses. The ED_{50} (i.e., the dose of CP-0127 reducing the maximum response to bradykinin by 50%) was calculated. In the rat, the effect on blood pressure of LPS from *E. coli* (Serotype 0127:B8; Sigma) at a dose of 15 mg/kg i.v. in the absence and presence of CP-0127 infused at 18 nmol/kg/min i.v. was studied. Finally, the effect of CP-0127 at 3.6 µmol/kg s.c. on bradykinin-induced (i.a. via the carotid artery) hypotension was evaluated.

<u>Survival Studies:</u> Rabbits were implanted with Alzet 7-day mini-osmotic pumps to infuse saline or CP-0127 continuously into the jugular vein at a dose of 0.36 nmol/kg/min. Twenty-four hours later, LPS from *E. coli* at 500 ug/kg was given i.v.. The animals were observed continuously and the time to death was recorded. A separate group of rabbits received CP-0127 alone at 0.36 nmol/kg/min i.v. for 7 days. On day 4 and day 7, selected animals were prepared for recording blood pressure (as above) and the selectivity of CP-0127 assessed.

Rats were pre-implanted with an indwelling jugular catheter which was exteriorized at the back of the neck. Twenty-four hours later, the animals were injected s.c. with either saline or CP-0127 at 3.6 µmol/kg, followed 1 h later with an i.v. bolus injection of LPS at 15 mg/kg and the time to death was recorded.

<u>Statistical Analysis:</u>   Statistical analysis of the effect of CP-0127 on the duration of inhibition of the depressor response to bradykinin in rats was performed using a Student t-test.   The survival data was analysed using a two sample proportion test (SOLO Statistical package, BMDP Statistical Software, Inc., Los Angeles, CA, U.S.A.).

RESULTS

<u>Blood Pressure:</u>   From dose response curves to bradykinin in the absence and presence of CP-0127, the ED_{50}s for CP-0127 against bradykinin in the rabbit and rat were found to be 0.06 and 0.9 nmol/kg/min, respectively.   CP-0127 was selective in that blood pressure responses to acetylcholine, substance P, angiotensin I, angiotensin II and noradrenaline were unaffected in both species.   In the rabbits prepared with 7 day mini-osmotic pumps delivering CP-0127 at 0.36 nmol/kg/min i.v., an equivalent degree of blockade of the bradykinin responses to that seen on day 0 was observed on day 4 and day 7, the effect being selective in that CP-0127 did not antagonize responses to the other pressor or depressor substances listed above.

In control rats, LPS at 15 mg/kg i.v. produced an immediate decrease in blood pressure, followed by a recovery to normal and then a second fall. In this study, two of the control animals died shortly after administration of the LPS, while the third rat had severe respiratory distress.   CP-0127 infused at 18 nmol/kg/min i.v. significantly attenuated the hypotensive response to LPS (Fig 1).   In addition, no significant respiratory distress was seen in any of the CP-0127 trested animals.   Finally, in rats pretreated with 3.6 µmol/kg CP-0127 s.c., the response to bradykinin (0.01 nmol i.a.) was almost totally blocked within 15 min and for the duration of the experiment (4 h) compared to controls (Fig 2).   This effect was selective since responses to the other above mentioned pressor and depressor agents were completely unaffected (data not shown).

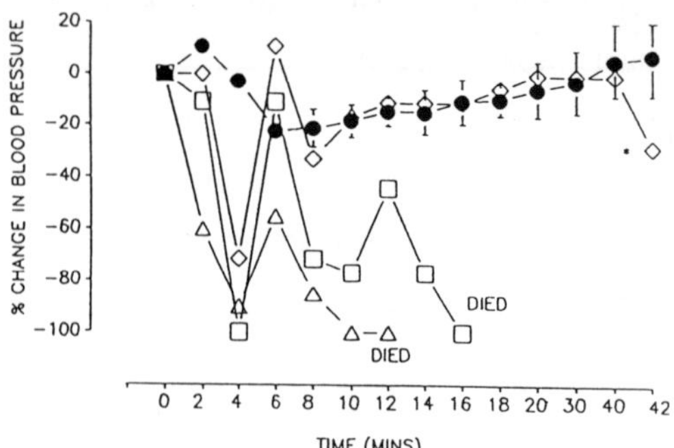

Figure 1. Blood pressure responses to LPS from E.Coli, 15 mg/kg i.v. in rats infused intravenously with saline (open symbols) or CP-0127 at (•) 18 nmol/kg/min. The saline animals are plotted individually for clarity. Vertical bars represent standard errors of the mean of n=3. (*) Saline treated animal suffering severe respiratory distress, causing the experiment to be terminated.

Figure 2. Blood pressure response to bradykinin (0.01 nmol i.a.) 15 min before and after a subcutaneous injection of saline (solid bars) or CP-0127 at 3.6 µmol/kg/min (open bars). Vertical bars represent standard errors of the means of n=5 (CP-0127) and n=4 (controls). * P< 0.05.

<u>Survival Studies:</u> The survival data for the rat and rabbit studies are shown in Figures 3a and 3b, respectively.   After 72 h, 45.5% of the saline control rabbits survived compared to 86% (P< 0.05) in the group infused with CP-0127. There were no deaths

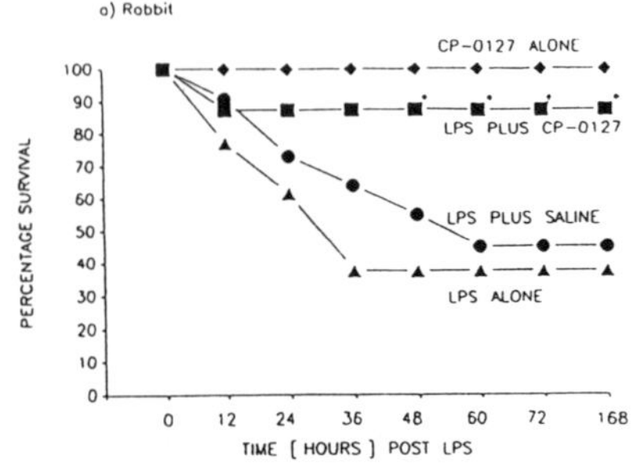

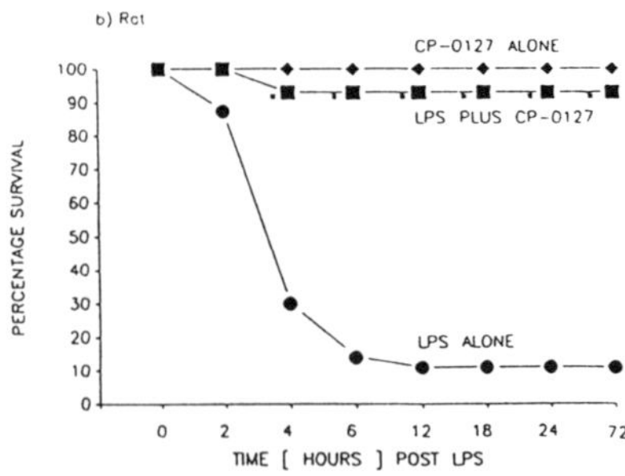

Figure 3.The effect of CP-0127 on survival rates in lethal endotoxic shock in:
(a) rabbits: (▲) LPS alone, n = 13; (•) LPS + saline infusion, n = 11; (■) LPS + CP-0127, 0.36 nmol/kg/min i.v., n = 14; (♦) CP-0127 alone, 0.36 nmol/kg/min, n = 6. *P < 0.05 compared to LPS + saline group.
(b) rats: (•) LPS + saline, n = 19; (■) LPS + CP-0127, 3.6 µmol/kg/min s.c., n = 14; (♦) CP-0127 alone, 3.6 µmol/kg/min s.c., n = 6. * P < 0.05 compared to LPS + saline group.

in the group of rabbits infused with CP-0127 alone. The CP-0127 treated rats were found to have 93% survival compared to 11% in the saline control group 24 h after administering the LPS (P < 0.001), and again there was 100% survival in the group of rats given CP-0127 alone. There were no further deaths up to 168 h for the rabbit and 72 h for the rat, after which the experiments were terminated.

DISCUSSION

CP-0127 was found to be a potent selective antagonist of bradykinin-induced hypotension in the rat and the rabbit, even after 7 days of continuous intravenous infusion (rabbit). On the basis of these observations and previous *in vitro* studies [8], we evaluated the effect of the compound on the immediate severe hypotensive response, which is seen following an intravenous injection of endotoxin from *E. coli* in the rat [7], an effect thought to involve immediate activation of the kallikrein-kinin system in this species [2]. Profound hypotension in response to LPS was also seen in the present study and this response was almost totally blocked by an infusion of the bradykinin antagonist CP-0127. This is in contrast to previous studies [6,7], where only partial attenuation was observed. This previous finding may reflect the low potency and partial agonist activity of the antagonists used in these studies [7]. It is noteworthy in the present study that two of the control LPS-treated animals died shortly after receiving LPS, and the third had severe respiratory distress. This was in contrast to the CP-0127 treated rats, which appeared normal with respect to respiratory function. These data clearly suggest a significant role for bradykinin in the early events in this model and are consistent with observations in other species, including man [2-5].

The efficacy of CP-0127 in increasing survival in two animal models of endotoxic shock has been demonstrated. CP-0127 given to rats as a single dose subcutaneously one hour before LPS resulted in 93% survival compared to 11% in the controls. In rabbits, a significantly increased survival was also seen in CP-0127 treated animals when compared to controls. Similarly, the bradykinin antagonist NPC-17731 has recently been demonstrated to delay the onset of hypotension and reduce and delay mortality in response to lethal doses of endotoxin in rats and mice [9]. These data are consistent pathogenesis of septic shock, including interleukin-1 alpha (IL-1) [10], tumour necrosis factor alpha (TNF) [11] and endothelium-derived relaxing factor or nitric oxide (NO) [12]. The relative roles and interactions of these and other mediators is complex; however, a major pivotal role for bradykinin can be suggested. Not only is the kallikrein-kinin system one of the first systems to be activated [2-5], but this activation appears to be sustained throughout the course of septic shock [6,13], unlike IL-1 and TNF, which appear to be generated, peak and decline over a specific time period [14-16]. In addition, bradykinin is a potent stimulant for the production and release of NO, IL-1 and TNF [17,18]. Moreover, IL-1, for example, can feed back, upregulate and sensitize bradykinin receptors [19,20]. Finally, the bradykinin antagonist NPC-17731 has been shown to be effective in reducing the LPS-induced circulating plasma concentrations of TNF and 6-keto-PGF1 alpha [9]. In conclusion, we have clearly demonstrated that CP-0127: (a) is a potent selective inhibitor of bradykinin *in vivo*, (b) almost totally blocks the severe hypotensive response to endotoxin in the rat, and (c) significantly increases survival in two animal models of endotoxic shock. These data suggest that bradykinin receptor antagonists such as CP-0127 may be useful therapeutic approaches for the treatment of septic shock in man.

REFERENCES

1.   Aasen AO, Smith-Erichsen N, Amundsen E. Plasma kallikrein-kinin system in septicemia. Arch Surg 1983; 118:343-345.

2.   Katori M, Majima M, Odoi-Adome R, Sunahara N, Uchida Y. Evidence for the involvement of a plasma kallikrein-kinin system in the immediate hypotension produced by endotoxin in anaesthetized rats. Br J Pharmacol 1989; 98:1383-1391.

3.   Nies AS, Forsyth RP, Williams HE, Melmon KL. Contributions of kinins to endotoxin shock in the unanesthetized rhesus monkey. Circ Res 1968; 22:155-164.

4.   Al-Kaisi N, Parratt JR, Siddiqui HH and Zeitlin IJ. Feline endotoxin shock: effects of methylprednisolone on kininogen depletion on the pulmonary circulation and on survival. Br J Pharmacol 1977; 60:471-476.

5.   Kimball HR, Melmon KL, Wolff S. Endotoxin induced kinin production in man. Proc Soc Exp Biol Med 1972; 139:1078-1082.

6.   Wilson DD, De Garavilla L, Kuhn W, Togo J, Burch R, Steranka LR. D-Arg-[Hyp^3-D-Phe^7]-Bradykinin, a bradykinin antagonist, reduces mortality in a rat model of endotoxic shock. Circ Shock 1989; 27:93-101.

7.   Weipert J, Hoffman H, Siebeck M, Whalley ET. Attenuation of arterial blood pressure fall in endotoxin shock in the rat using competitive bradykinin antagonist Lys-Lys-[Hyp^2, Thi^{5,8}, D-Phe^7]-Bk(B4148). Br J Pharmacol 1988; 94:282-284.

8.   Cheronis JC, Whalley ET, Ngyen KT, Eubanks SR, Allen LG, Duggan MJ, Loy SD, Bonham KA, Blodgett JB. A new class of bradykinin antagonists: synthesis and *in vitro* activity of bissuccinimidoalkane peptide dimers. Accepted for publication in J Med Chem, November 1991.

9.   Noronha-Blob L, Prosser JC, Lone VC, Sullivan JP, Kyle DJ, Martin JA, Burch RC. NPC-17731 delays the onset of hypotension and reduces mortality in response to lethal doses of endotoxin in rats and mice. Proc KININ '91 Munich, 1991; 365.

10.   Ohlsson K, Björk P, Bergenfeldt M, Hageman R, Thompson RC. Interleukin-1 receptor antagonist reduces mortality from endotoxin shock. Nature 1990; 348:550-552.

11.   Offner F, Philippe J, Vogelaers D, Colardyn F, Baele G, Baudrihaye M, Vermeulen A, Leroux-Roels G. Serum tumor necrosis factor levels in patients with infectious disease and septic shock. J Lab Clin Med 1990; 116:1:100-105.

12.   August M, Guillon J-M, Delaflotte S, Etiemble E, Chabrier P-E, Braquet P. Endothelium independent protective effect of N^G-monomethyl-L-arginine on endotoxin-induced alterations of vascular reactivity. In Life Sciences, Volume 48. Elmsford, New York: Pergamon Press, 189-193.

13.   DeLa Cadena RA, Colman RW. Structure and functions of human kininogens. TIPS 1991; 12:272-275.

14.   Wakabayashi G, Gelfand J, Burke JF, Thompson RC, Dinarello CA. A specific receptor antagonist for interleukin 1 prevents *Escherichia coli*-induced shock in rabbits. FASEB J 1991; 5:338-343.

15.   Michie R, Michie MB, Manogue KR, Spriggs DR, Revhaug A, O'Dwyer S, Dinarello CA, Cerami A, Wolff SM, Wilmore DW. Detection of circulating tumor necrosis factor after endotoxin administration. New Engl J Med 1988; 318:1481-1486.

16.   Engelberts I, Von Asmuth EJU, Van der Linden CJ, Buurman WA. The interrelation between TNF, IL-6, and PAF secretion induced by LPS in an *in vivo* and *in vitro* murine model. Lymph Cyt Res 1991; 10:2:127-131.

17.   D'Orleans-Juste P, De Nucci G, Vane JR. Kinins act on B_1 or B_2 receptors to release conjointly endothelium-derived relaxing factor and prostacyclin from bovine aortic endothelial cells. Br J Pharmacol 1989; 96:920-926.

18.   Burch RM, Connor JR, Tiffany CW. The kallikrein-kininogen-kinin system in chronic inflammation. Agents and Actions 1989; 27:3/4:258-260.

19.   DeBlois D, Bouthillier J, Marceau F. Effect of glucocorticoids, monokines and growth factors on the spontaneously developing responses of the rabbit isolated aorta to des-Arg^9-bradykinin. Br J Pharmacol 1988; 93:969-977.

20.   DeBlois D, Bouthillier J, Marceau F. Pulse exposure to protein synthesis inhibitors enhances vascular responses to des-Arg^9-bradykinin: possible role of interleukin-1. Br J Pharmacol 1991; 103:1057-1066.

EFFECTS OF PROTEASE INHIBITORS IN EXPERIMENTAL SEPTIC SHOCK

Siebeck M, Spannagl M, Hoffmann H, Fink E

Departments of Surgery, Medicine, Clinical Chemistry and Clinical Biochemistry
University of Munich, Klinikum Innenstadt, Munich, Germany

SUMMARY: Endotoxin shock and contact system activation were used to study effects of hirudin, antithrombin III, eglin C, aprotinin, and [Arg^{15}]-aprotinin in anesthetized pigs. Alterations in the systemic and pulmonary circulation were in part prevented by administration of the inhibitors.

INTRODUCTION

The proteolytic cascade systems of plasma are activated both in patients with septic shock and in animals with endotoxin-induced shock. In severe inflammation, digestive proteases are released from inflammatory cells such as phagocytes. In order to study potential therapeutic applications of various protease inhibitors we investigated their effect in experimental shock, induced by bacterial lipopolysaccharide (LPS) or dextran sulfate (DXS), an agent which activates the contact system of blood coagulation

MATERIALS AND METHODS

We used miniature pigs (body weight ≈ 20 kg) under general anesthesia and controlled ventilation. All animal procedures were approved by the Regierung von Oberbayern and complied with the Tierschutzgesetz in der Fassung vom 18. 8. 1986. We infused LPS from *S. abortus equi* for 6 h or for 8 h, or DXS in a dose of 5 mg/kg in 1 h or saline. We used the following proteinase inhibitors, each in randomized, controlled trials: a purified human antithrombin III-heparin complex (AT III-heparin) or recombinant hirudin for thrombin; recombinant eglin C for PMN elastase; bovine aprotinin or recombinant [Arg^{15}]-aprotinin for plasma kallikrein. Case numbers were mentioned in the legends. For full experimental details, the reader is referred to the original descriptions (**1, 2, 3, 4**). [Arg^{15}]-aprotinin was given intravenously as a bolus before and as a continuous infusion during stimulation with dextran sulfate. The study design and plasma levels thus achieved are given in table 1.

Table 1. Dextran sulfate (DXS) and [Arg15]-aprotinin: Doses and plasma concentrations

Group	n	Activation with	[Arg15]-aprotinin Bolus Injection (mg)	Continuous Infusion (mg)	Plasma Level at 1 h (μM)
NaCl Control	3	NaCl	0	0	0
A15A Control	2	NaCl	120	240	3.8
DXS + 360 mg	5	DXS	120	240	1.4
DXS + 180 mg	5	DXS	60	120	0.8
DXS + 90 mg	5	DXS	30	60	0.7
DXS + 45 mg	5	DXS	15	30	0.2
DXS Control	5	DXS	0	0	0

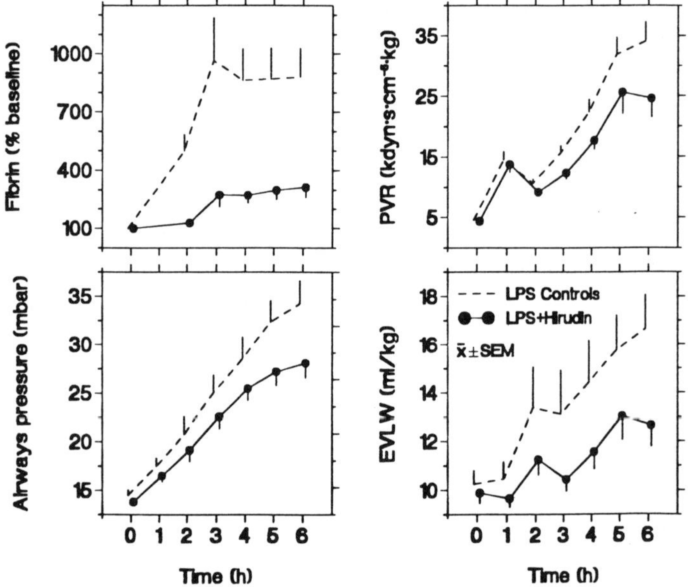

Figure 1. Effect of hirudin in LPS shock. Upper left: Soluble fibrin concentration in plasma. Upper right: Pulmonary vascular resistance (PVR). Lower left: Peak airways opening pressure. Lower right: Extravascular lung water (EVLW). All four parameters were lower in hirudin-treated animals (n = 18) than in control animals (n = 18).

The total intravascular protein content was calculated as the product of plasma volume as assessed by Evans' Blue dye dilution times total protein concentration in plasma (Biuret reaction), corrected for body weight. The difference between measurements obtained at baseline and after 4 h represents the loss of intravascular protein. Cardiac output and extravascular lung water were measured by single indicator thermodilution (Model 9310, Edwards), the pulmonary vascular pressure with a Swan-Ganz catheter, and the systemic blood pressure with a catheter in the abdominal aorta.

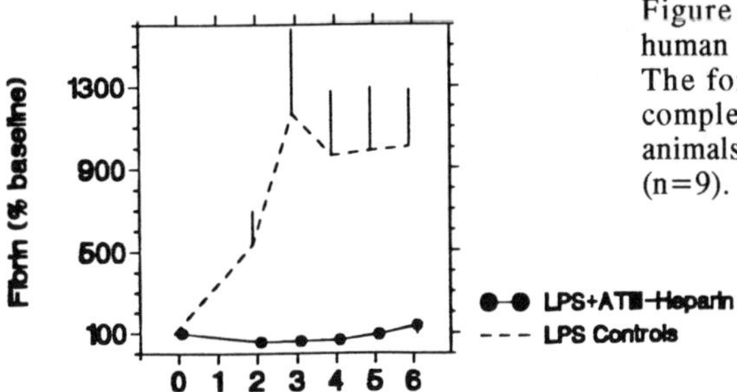

Figure 2. Effect of a purified complex between human antithrombin III and heparin in LPS shock. The formation of monomeric fibrin was almost completely blocked in the AT III-heparin-treated animals (n=8) as compared to the control animals (n=9).

Kinin-containing kininogen was determined on the basis of the amount of kinin releasable in plasma samples by incubation with trypsin. C3a in plasma was measured by radioimmunoassay. In the hirudin and AT III-heparin studies, soluble fibrin was detected using a functional assay (5). In the aprotinin study, fibrin monomer was detected in plasma by an enzyme immunoassay using a monoclonal antibody to the N-terminal α-chain of fibrin.

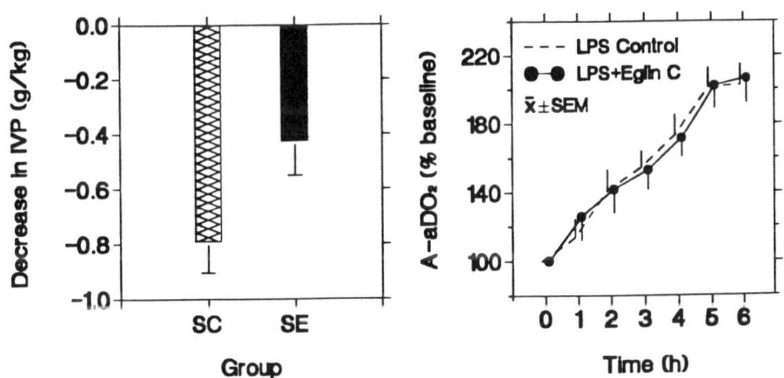

Figure 3. Effect of eglin C in LPS shock. Left: Decrease in intravascular protein content (IVP). Eglin-treated animals (n=18) had substantially less decrease in IVP than control animals (n=18). Right: Alveolar-arterial O2 pressure gradient (A-aDO2). There was no difference in A-aDO2 between controls and eglin-treated animals.

RESULTS

Hirudin (0.1 μM plasma concentration) reduced fibrin formation and ameliorated LPS-induced lung dysfunction (cf. fig. 1). AT III-heparin (120% of normal AT III activity in plasma) blocked fibrin formation (cf. fig. 2), but did not improve respiratory dysfunction. Eglin C (2 μM) reduced the loss of protein from the intravascular space, but did not affect respiratory dysfunction (cf. fig. 3).

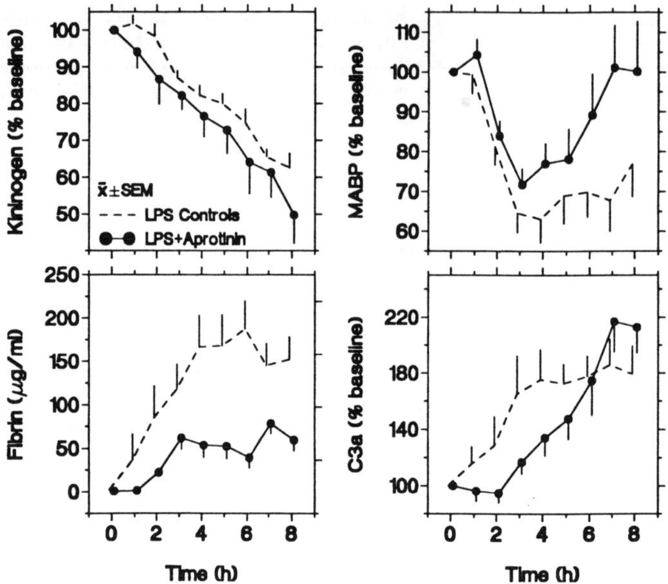

Figure 4. Effect of aprotinin in LPS shock. Upper left: Kinin-containing total kininogen. There was no difference in turn-over of kininogen between aprotinin-treated animals (n=10) and control animals (n=10). Upper right: Mean arterial blood pressure (MABP). Aprotinin-treated animals had higher blood pressure than control animals. Lower left: Monomeric fibrin concentration in plasma. Lower right: Complement fragment 3a (C3a) concentration in plasma. Aprotinin-treated animals had substantially lower fibrin levels and, at least initially, lower C3a levels than control animals.

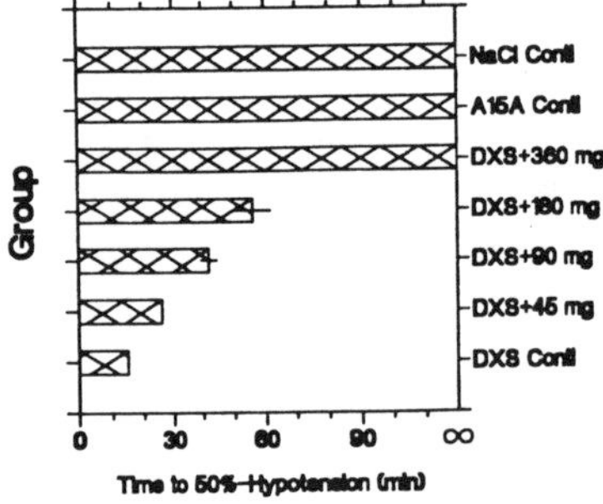

Figure 5. Effect of [Arg15]-aprotinin in dextran sulfate (DXS)-induced hypotension. Time from onset of DXS infusion to 50% arterial hypotension was progressively delayed with increased doses of [Arg15]-aprotinin; hypotension was completely abolished in the highest dose administered.

Aprotinin (10 μM) reduced LPS-induced hypotension, fibrin formation and C3a formation, but not the decrease in plasma kininogen levels (cf. fig. 4). [Arg15]-aprotinin (1.4 μM) blocked DXS-induced arterial hypotension (cf. fig. 5) and kininogen breakdown.

DISCUSSION

Hirudin had the anticipated effect on fibrinogen degradation and fibrin formation. Furthermore, it reduced some of the effects of LPS on lung function. In contrast, antithrombin III in complex with heparin had no effect on respiratory dysfunction, although it was even better at inhibiting fibrinogen degradation and completely blocked fibrin formation. Studies with heparin alone in a similar model revealed analogous results (6). Although these studies were not designed specifically to compare the effectiveness of hirudin and antithrombin III, we assume that hirudin in contrast to antithrombin III was able to block some of the cellular effects of thrombin (mediated by formation of thromboxane (7) or endothelin (8) or directly (9)), and this is a possible reason why hirudin ameliorated LPS-induced lung dysfunction and reduced the increase in pulmonary vascular resistance.

Eglin C was found only to have an effect in the systemic circulation where it reduced protein extravasation. Similar effects have been observed in two other independent experiments (10, 11). Three possible explanations exist for this finding. PMN elastase can influence the metabolism of arachidonic acid (12). Chymase, another enzyme that is inhibited by eglin C, is involved in mast cell degranulation, and it is possible that the observed effects of eglin C have nothing to do with PMN elastase, but rather with inhibition of mast cell degranulation (13). The third possibility is an action mediated *via* cytokines. Human alpha-1 antitrypsin (14) and, similarly, eglin C (A. Wendel, Konstanz, FRG, personal communication) are able to block a final proteolytic step in the generation of bioactive tumor necrosis factor α. The absence of an effect of eglin C on respiratory dysfunction suggests that the endotoxin-induced lung damage in our model was produced by other mechanisms than the release of PMN elastase.

Aprotinin produced effects that suggest that the contact system of plasma was activated in LPS shock. Hypotension and vasodilation are typical effects of bradykinin. Complement activation is associated with activation of the contact system. The observed effect on the C3a plasma levels however might also have been produced by plasmin inhibition, and the aprotinin plasma levels in our experiment were far higher than what is necessary to achieve plasmin inhibition. Significant was the finding that fibrin formation, assessed by an immunologic method, was reduced -- a fact that underscores the importance of this pathway for coagulation activation in LPS shock.

[Arg15]-aprotinin is an aprotinin homologue with excellent inhibition of plasma kallikrein (15). In our *in vivo* model using infusion of dextran sulfate it blocked at lower concentrations than aprotinin (16). It should be useful to test whether it has additional protective effects in LPS shock.

CONCLUSION

We have shown that different proteases are involved in different ways in the events during endotoxin shock, and that the effect of proteinase inhibition largely depends on the enzyme being inhibited and on the relative role of the enzyme in this very complex sequence of events. The impact of enzymes on organ dysfunction in shock, however, may depend on their ability to stimulate cells, in addition to their cleaving of substrates.

ACKNOWLEDGEMENT

This work was supported by the Deutsche Forschungsgemeinschaft, the Bundesinnenministerium and the Bundesministerium für Forschung und Technologie. Compounds were generously made available by C. Galanos, Freiburg, Bayer AG, Wuppertal, Plantorgan KG, Bad Zwischenahn, Ciba-Geigy AG, Basel, Switzerland, and Immuno AG, Vienna, Austria. Substantial experimental contributions were made by M. Weis, C. Keser, J. Bichler, J. Kohl, J. Weipert, P. Kroworsch, P. Scheuber, S. Moravec, M. Jochum, A. Braune, A. Oettl, and G. Godec. The authors wish to thank L. Schweiberer, W. Schramm, and H. Fritz for generous support and valuable discussion.

REFERENCES

1.   Hoffmann H, Siebeck M, Spannagl M, Weis M, Geiger R, Jochum M, Fritz H (1990). Effect of recombinant hirudin, a specific inhibitor of thrombin, on endotoxin-induced intravascular coagulation and acute lung injury in pigs. Am Rev Respir Dis 142:782-788.

2.   Spannagl M, Hoffmann H, Siebeck M, Weipert J, Schwarz HP, Schramm W (1991). A purified antithrombin III - heparin complex as a potent inhibitor of thrombin in porcine endotoxin shock. Thromb Res 61:1-10.

3.   Siebeck M, Hoffmann H, Weipert J, Fritz H (1992). Effect of the elastase inhibitor eglin C in porcine endotoxin shock. Circ Shock 36:3, in press.

4.   Siebeck M, Fink E, Weipert J, Jochum M, Fritz H, Spannagl M, Kroworsch P, Shimamoto K, Schweiberer L. Inhibition of plasma kallikrein with aprotinin in porcine endotoxin shock. J. Trauma, submitted.

5.   Wiman B, Rånby M (1986). Determination of soluble fibrin in plasma by a rapid and quantitative spectrophotometric assay. Thromb Haemostas 55:189-193.

6.   Griffin MP, Gore DC, Zwischenberger JB, Lobe TE, Hall M, Traber DL, Herndon DN. Does heparin improve survival in experimental porcine Gram-negative septic shock? Circ Shock 1990, 31:343-349.

7.   Seeger W, Neuhof H, Hall J, Roka L (1988). Pulmonary vasoconstrictor response to soluble fibrin in isolated lungs: possible role of thromboxane generation. Circ Res 62:651-659.

8. Boulanger CM, Lüscher TF (1991). Hirudin and nitrates inhibit the thrombin-induced release of endothelin from the intact porcine aorta. Circ Res 68:1768-1772.

9. Sonne O (1988). The specific binding of thrombin to human polymorphonuclear leukocytes. Scand J Clin Lab Invest 48:831-838.

10. Hock CE, Lefer AM (1985). Beneficial effects of a neutral protease inhibitor in traumatic shock. Pharmacol Res Comm 17: 217-226.

11. Siebeck M, Hoffmann H, Jochum M, Fritz H (1989). Inhibition of proteinases with recombinant eglin C during experimental *Escherichia coli* septicemia in the pig. Eur Surg Res 21:11-17.

12. LeRoy EC, Ager A, Gordon JL (1984). Effects of neutrophil elastase and other proteases on porcine aortic endothelial prostaglandin I_2 production, adenine nucleotide release, and responses to vasoactive agents. J Clin Invest 74:1003-1010.

13. Fink E, Nettelbeck R, Fritz H (1986). Inhibiton of mast cell chymase by eglin c and antileukoprotease (HUSI-I). Biol Chem Hoppe-Seyler 367: 567-571.

14. Niehörster M, Tiegs G, Schade UF, Wendel A (1990). In vivo evidence for protease-catalysed mechanism providing bioactive tumor necrosis factor alpha. Biochem Pharmacol 40:1601-1603.

15. Auerswald EA, Hörlein D, Reinhardt G, Schröder W, Schnabel E (1988). Expression, isolation and characterization of recombinant [Arg^{15},Glu^{52}]aprotinin. Biol Chem Hoppe-Seyler 369 (Suppl):27-35.

16. Hoffmann H, Siebeck M, Thetter O, Jochum M, Fritz H (1989). Aprotinin concentrations effective for the inhibition of tissue kallikrein and plasma kallikrein *in vitro* and *in vivo*. Adv Exp Med Biol 247B:35-42.

AAS 38/III
Recent Progress on Kinins
© 1992 Birkhäuser Verlag Basel

THE BRADYKININ ANTAGONIST HOE 140 INHIBITS CARRAGEENAN- AND THERMICALLY INDUCED PAW OEDEMA IN RATS

K.J. Wirth, H.G. Alpermann, R. Satoh[*] and M. Inazu[*]

(HOECHST AG, H 821, D-6230 Frankfurt / M. 80; FRG [*]Hoechst Japan Limited Kawagoe)

SUMMARY: The new and highly potent B_2 bradykinin (BK) antagonist Hoe 140 was tested for its ability to inhibit oedema of rat paws induced by scalding and carrageenan. The data show that Hoe 140 inhibits scalding and carrageenan oedema for more than four and six hours, respectively. Based on its potency against actions of enodogenously generated kinins Hoe 140 is appropritate to investigate the role of kinins in human inflammatory diseases.

INTRODUCTION

Kinins are potent proinflammatory peptides due their ability to increase microvascular permeability and blood flow and to stimulate phospholipase A_2 (1). Via activation of sensory nerve endings bradykinin (BK) has potent algesic effects. Pain is a cardinal symptom of inflammation. Based on their manifold actions commonly associated with inflammation the kinins are implicated as important mediators in many diseases associated with inflammation and pain (2). Since a variety of inflammatory mediators exist the relative contribution of a mediator to the symptomatology of a certain disease can only be evaluated when potent and selective inhibitors are available. The same applies to experimentally induced inflammatory conditions. In the present study, the bradykinin antagonist Hoe 140 was used to determine the role of the kinins in carrageenan induced inflammation and scalding. Hoe 140 (D-Arg[Hyp^3,Thi^5,D-Tic^7,Oic^8]BK) (1,2,3) is a highly potent and stable B_2 bradykinin antagonist recently described. Considerable inhibition of such inflammatory conditions by a selective BK antagonist would indicate contribution of BK to the inflammatory process. On the other hand, high potency and long duration of action of an antagonist would allow to conclude that the compound under investigation is appropriate to evalute the therapeutic usefulness of kinin antagonism in human disease.

MATERIALS AND METHODS

Carrageenan induced paw oedema: 0.1 ml of a 0.5 % solution was injected into the left hind paw of male Sprague-Dawley rats (n=5/group) weighing approximately 120 g. Rats received 0.1 or 1 mg/kg Hoe 140 either intravenously or subcutaneously immediately (i.v.) or 15 min (s.c.) before carrageenan injection

Thermic oedema: Oedema was induced in 9 weeks old male Sprague-Dawley rats (n=6/group) by immersing paws of the right and left hindlimb into 55°C hot water for 10 sec under anaesthesia, and immediately thereafter, rats received intravenously 0.1 or 1 mg/kg Hoe 140. Paw volume was measured by water displacement before and at regular intervalls after injection of carrageenan or hot water immersion.

RESULTS AND DISCUSSION

Hoe 140 markedly reduced the carrageenan induced rat paw oedema following both intravenous and subcutaneous administration. A maximum inhibitory effect of about 90% was achieved at a dose of 0.1 mg/kg i.v. two hours after carrageenan, and after 4 hours inhibition was still 48% (Fig.1.). The effect of the 1 mg/kg s.c. dose exceeded the observation period and still showed 42% inhibition after 6 hours. This model has been shown to be highly predictive of antiinflammatory drug activity in humans. The fact that Hoe 140 is effective upon s.c administration delivers a further argument in favour of metabolic stability. First generation antagonists failed to show efficacy after systemic administration due to metabolic instability, and their biological half lives were in the range of few minutes only whereas the biological half life of Hoe 140 was estimated to several hours (3). Nevertheless, the first generation antagonists were found to be effective against carrageenan when injected locally into rat paws (6).

At 1 mg/kg of Hoe 140, scalding oedema was markedly inhibited two and four hours after induction by about 76 % and 59 %, respectively whereas the effect of 0.1 mg/kg was not significant (Fig.1.) Since Hoe 140 is specific for B_2 bradykinin receptors our data allow to conclude that kinins are important mediators in carrageenan- and scalding induced inflammation in rats. The finding that Hoe 140 seems to be less potent against scalding than against carrageenan induced swelling might reflect a different relative contribution of the kinins to oedema formation of these models. In scalding, other mediators such as histamine or cyclooxygenase products might be more important than in carrageenan induced oedema, and physical damage of tissue by scalding can not be excluded. Taken together, it has been shown that Hoe 140 effectively antagonizes the actions of endogenously generated kinins for several hours.

In the last years, evidence has accumulated that kinins play a role in inflammatory joint diseases. Increased BK levels are found in the plasma of patients with rheumatoid arthritis (6), and kallikrein levels are increased in the joints of patients with rheumatoid arthritis (7). Although these observations suggest that kinins may be important mediators in human inflammatory diseases, definitive proof of this concept can only be achieved by the demonstration that selective blockade of these peptides leads to an improvement of symptoms. The availability of the potent kinin receptor antagonist Hoe 140 enables to address this issue and to determine the usefulness of bradykinin antagonism as a novel therapeutic approach in selected disease states such as inflammatory joint disease and burns.

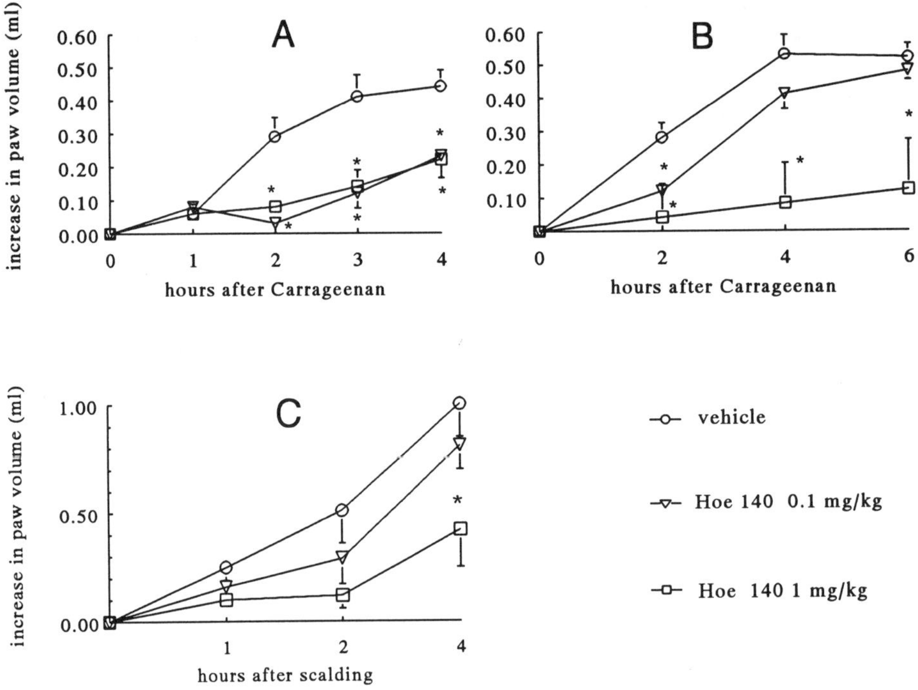

Figure 1. Effect of Hoe 140 on carrageenan and scalding induced rat paw oedema in rats. Panel A shows the effect of i.v. doses of Hoe 140 against carrageenan, panel B the effect of s.c. doses. Panel C depicts the inhibitory effect of an i.v. dose on scalding induced oedema. Data points show means ± S.E.M.. p < 0.05.

REFERENCES

1.   Marccau F, Lussicr A., Rcgoli D, Giroud JP. Pharmacology of kinins: Their relevance to tissue injury and inflammation. Gen. Pharmacol 1983; 14: 209-229.

2.   Taylor JE, Defeudis FV, Moreau JP. Bradykinin-antagonists: Therapeutic perspectives. Drug Develop Res 1989; 16: 1-11.

3.   Wirth K, Hock FJ, Albus U, Linz W, Alpermann HG, Anagnostopulos H, Henke St, Breipohl G, König W, Knolle J, Schölkens BA. Hoe 140 a new potent and long acting bradykinin-antagonist: in vivo studies. Br J Pharmacol 1991; 102: 774-777.

4.   Lembeck F, Griesbacher T, Eckhardt M, Henke S, Breipohl G, Knolle J. New, long-acting, potent bradykinin antagonists. Br J Pharmacol 1991; 102: 297-304.

5.   Hock FJ, Wirth K, Albus U, Linz W, Gerhards G Wiemer, Henke St, Breipohl G, König W, Knolle J, Schölkens BA. Hoe 140 a new potent and long acting bradykinin- antagonist: in vitro studies. 1991; Br J Pharmacol 102: 769-773.

6.   Costello AH, Hargreaves KM. Suppression of carrageenan-induced hyperalgesia, hyperthermia and edema by a bradykinin antagonist. Eur J Pharmacol 1989; 171: 259-263.

7.   Selwyn BM, Figourerou CD, Fink E, Swan A, Dieppe PA, Bhoola KD. A tissue kallikrein in the synovial fluid of patients with rheumatoid arthritis. Ann Rheum Dis 1989; 48: 128-133.

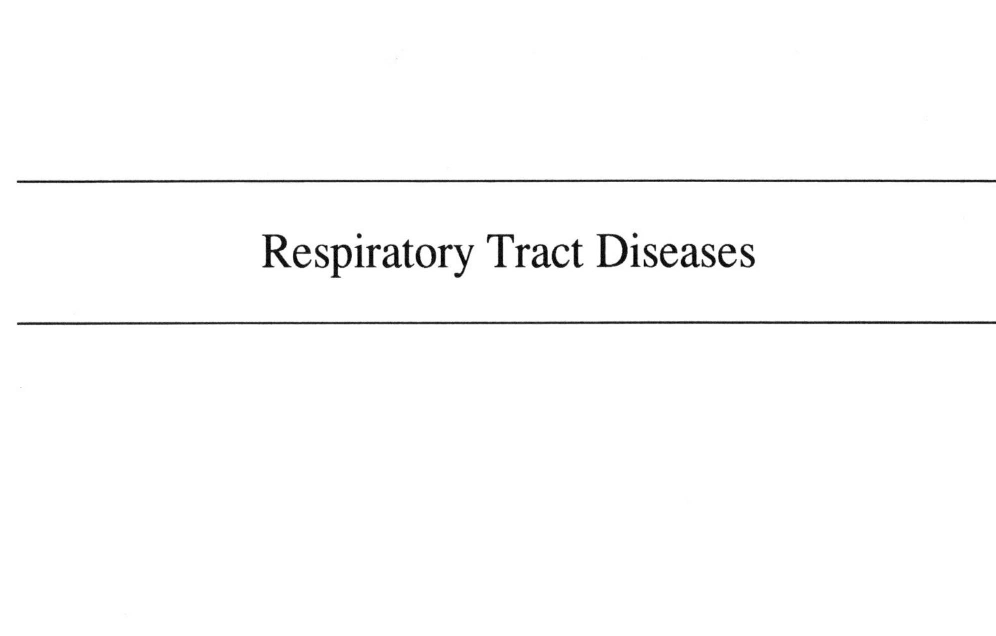

Respiratory Tract Diseases

AAS 38/III
Recent Progress on Kinins
© 1992 Birkhäuser Verlag Basel

EFFECT OF BRADYKININ ON AIRWAY FUNCTION

Peter J. Barnes

Department of Thoracic Medicine, National Heart and Lung Institute, Dovehouse St, London SW3 6LY, UK

SUMMARY: Bradykinin (BK) has several effects on airway function which may be relevant in obstructive airways disease. These effects are mediated via B_2-receptors. BK is a potent bronchoconstrictor in animals and humans *in vivo*. Bradykinin contracts airway smooth muscle, is a potent bronchial vasodilator, increases microvascular leakage, stimulates epithelial cells to release bronchodilators and stimulates mucus secretion. Perhaps its most important action is the activation of sensory nerves in airways, leading to reflex bronchoconstriction, coughing and neurogenic inflammation through the release of neuropeptides from sensory nerves.

INTRODUCTION

Many mediators have been implicated in the pathophysiology of asthma (1) and the role of individual mediators will only become apparent with the use of potent and selective antagonists or synthesis inhibitors, which are now becoming available for many mediators. BK has long been considered to be a mediator involved in asthma since the first demonstration of bronchoconstriction after inhaled BK in asthmatic patients (2). BK has many properties which are relevant to asthma and other airway diseases and the recent development of potent antagonists may soon reveal its role in the complex pathophysiology of asthma. There is evidence for kinin activity in BAL fluid of asthmatic patients (3) and it is likely that BK is formed from plasma which has exuded from the inflamed airways, by the action of plasma and tissue and inflammatory cell derived kallikreins. BK is therefore likely to be formed within the lumen of the airway and to have access to superficially localised structures in the airway.

BRADYKININ RECEPTORS

The effects on the airways which are mediated via specific surface receptors. The effects of BK on airways are mediated via B_2-receptors, and there is no evidence for functional B_1-receptors. A B_3-receptor has also been described in airway smooth muscle of sheep (4,5), but there are some doubts about its existence, since it has been defined with weak antagonists. The distribution of B_2-receptors has been mapped out in human and guinea pig lung by autoradiography using [^{3}H]BK (6). There is a high density of binding sites in bronchial and pulmonary vessels, particularly on endothelial cells. Epithelial cells, airway smooth muscle (particularly in peripheral airways), submucosal glands and nerves are also labelled, indicating that BK may have diverse effects on airway function. A particularly high density of labelling is observed in the lamina propria immediately beneath the epithelium; it is not clear what cellular structures are labelled but a very similar pattern of labelling has been observed in other epithelialised structures (7).

AIRWAY SMOOTH MUSCLE

Inhaled BK is a potent bronchoconstrictor in asthmatic patients, but has no effect even in high concentration in normal individuals (8,9), suggesting an increased responsiveness of airway smooth muscle as for other spasmogens. *In vitro* BK is only a weak constrictor of human airways, suggesting that its potent bronchoconstrictor effect in asthmatic patients is mediated indirectly. In guinea pig airways *in vitro* BK has weak and variable effects which are influenced by the presence of airway epithelium and by the activity of local degrading enzymes. BK causes relaxation of intact guinea pig airways *in vitro*, but constricts airways if epithelium is removed mechanically (10,11). The effect of epithelium removal is likely to be due to release of prostaglandin E_2 from epithelial cells by bradykinin and due to the activity of the enzyme neutral endopeptidase 24.11 (NEP) which is localised in high density to epithelial cells. A combination of indomethacin and phosphoramidon (which inhibits NEP) mimics the effect of epithelial removal. Epithelial shedding is commonly observed in asthmatic airways and this could be a factor contributing to the increased bronchoconstrictor effect of BK in asthma. The bronchoconstrictor effect of BK in ferrets *in vitro* and in guinea pig *in vivo* is enhanced by the inhibition of both NEP by phosphoramidon and of angiotensin converting enzyme by captopril (12,13).

Intravenous BK causes intense bronchoconstriction in guinea pig *in vivo* which is markedly inhibited by indomethacin, suggesting that a bronchoconstrictor cyclo-oxygenase product (probably thromboxane) largely mediates this effect (14). The bronchoconstrictor response to BK instilled into the airways is not reduced by indomethacin, however, suggesting a different mechanism of bronchoconstriction after airway delivery of the mediator (14). This is more likely to mimic the situation in airway inflammation when BK would be formed from plasma precursors exuded into the airway lumen. Similarly in human subjects inhibition of cyclo-oxygenase by aspirin has no effect on the bronchoconstrictor effect of inhaled BK. The bronchoconstrictor response to both intravenous and inhaled BK is mediated via a B_2-receptor, since the B_2-receptor antagonists NPC 349 and HOE 140 inhibit the bronchoconstrictor response, whereas a B_1-selective antagonist is ineffective (15-17).

NEURAL EFFECTS

Perhaps the most important property of BK is its ability to activate C-fibre nociceptive sensory nerve endings. BK is the mediator of inflammatory pain (18), and in the airways this may be manifest as cough and tightness of the chest which are commonly observed after inhalation of BK in patients with asthma (9). BK stimulates bronchial C-fibres in dogs (19). In guinea pigs the bronchoconstrictor response to instilled BK is reduced by atropine and by capsaicin pretreatment to deplete neuropeptides from sensory nerves, indicating that both a cholinergic reflex and release of neuropeptides from sensory nerves are involved (14). Indeed a combination of atropine and capsaicin pretreatment largely abolishes the bronchoconstrictor response to instilled BK, but has little effect on the bronchoconstrictor response to intravenous BK (which is largely inhibited by indomethacin) (14). BK also releases tachykinins from perfused guinea pig lung (20) and enhances the bronchoconstrictor response to electrical field stimulation (mediated by release of endogenous tachykinins) in guinea pig bronchi *in vitro* (21). The effect of BK on airway sensory nerves is blocked by the B_2-antagonist HOE 140. Single fibre recordings from sensory nerves of guinea pig airways indicate that BK is a potent activator of C-fibres, and that this is a direct action since it is not blocked by cyclo-oxygenase inhibition (22). BK has no direct effect on the release of neurotransmitters from airway cholinergic nerves (21).

In asthmatic patients the bronchoconstrictor response to BK is also reduced by anticholinergic pretreatment, indicating that cholinergic mechanisms are involved (9) and by pre-treatment with sodium cromoglycate and nedocromil sodium, indicating the likely involvement of C-fibre activation (23). Thus BK may be an important mediator of cough and chest discomfort in asthma. BK has also been implicated in ACE inhibitor induced cough which is seen in about 10% of patients on chronic therapy (24). ACE inhibitor cough is reduced by cyclo-oxygenase inhibitors, suggesting that prostaglandins (such as PGE_2 or $PGF_{2\alpha}$ may be involved. Endogenous BK may stimulate the release of these prostaglandins in the larynx and trachea leading to cough, although it is not clear why only a proportion of patients are affected. ACE inhibitors do not worsen asthma, presumably because other enzymes, such as NEP, are more important in degrading BK in airways. Indeed a potent ACE inhibitor has no effect on the bronchoconstrictor response to inhaled BK in asthmatic patients, indicating that ACE is not of critical importance in degrading intralumenal BK in human airways (25).

VASCULAR EFFECTS

BK is a potent inducer of airway microvascular leak and causes a prolonged leakage at all airway levels, which is partly mediated via the release of platelet activating factor, since a PAF antagonist markedly inhibits the prolonged leak (26). The immediate leakage response to BK is partly mediated via the release of neuropeptides (probably substance P) from airway sensory nerves. The effect of BK on leakage is mediated via B_2-receptors which are localised to endothelial cells on post-capillary venules since B_2-antagonists inhibit the leakage response (16,17). The microvascular leakage induced by BK is enhanced by inhibition of both NEP and ACE (27).

BK is a potent vasodilator of bronchial vessels and causes an increase in airway blood flow (28,29). This is consistent with the high density of BK receptors on bronchial vessels (6), and suggests that a major effect of BK in asthma may be hyperaemia of the airways.

SECRETORY EFFECTS

BK stimulates airway mucus secretion from canine and feline airways *in vitro* (30,31), presumably indicating a direct effect of BK on submucosal glands. This is consistent with the demonstration of B_2-receptors on these glands by autoradiographic mapping. BK also stimulates the release of mucus glycoproteins from human nasal mucosa *in vitro* (31). BK also stimulates ion transport in airway epithelial cells, which is mediated via the release of prostaglandins (32).

ROLE IN AIRWAY DISEASE

The role of BK in asthma and airway diseases such as chronic bronchitis is still not clear. The development of potent and stable B_2-receptor antagonists offers the possibility of clarifying its role in airway disease in the near future (33,34). Perhaps the most relevant property of BK is its ability to activate sensitised afferent nerves; BK may be the most important mediator contributing to cough and tightness of the chest, which are such common symptoms of asthma.

REFERENCES

1.   Barnes PJ, Chung KF, Page CP. Inflammatory mediators and asthma. Pharmacol Rev 1988; 40:49-84.

2.   Herxheimer H, Streseman E. The effect of bradykinin aerosol in guinea-pigs and in man. J Physiol 1961; 158:38P.

3.   Christiansen SC, Proud D, Cochrane CG. Detection of tissue kallikrein in the bronchoalveolar lavage fluid of asthmatic patients. J Clin Invest 1987; 79:188-197.

4.   Farmer SG, Burch RM, Meeker SA, Wilkins DE. Evidence for a pulmonary B_3-bradykinin receptor. Mol Pharmacol 1989; 36:1-8.

5.   Farmer SG, Ensor JE, Burch RM. Evidence that cultured airway smooth muscle cells contain bradykinin B_2 and B_3 receptors. Am J Resp Cell Mol Biol 1991; 4:273-277.

6.   Mak JCW, Barnes PJ. Autoradiographic visualization of bradykinin receptors in human and guinea pig lung. Eur J Pharmacol 1991; 194:37-44.

7.   Manning DC, Snyder SH. Bradykinin receptors localized by quantitative autoradiography in kidney, ureter and bladder. Am J Physiol 1989; 256:F909-918.

8.   Simonsson BG, Skoogh BE, Bergh NP, Anderson R, Svedmyr N. *In vivo* and *in vitro* effect of bradykinin on bronchial motor tone in normal subjects and in patients with airway obstruction. Respiration 1973; 30:378-388.

9.   Fuller RW, Dixon CMS, Cuss FMC, Barnes PJ. Bradykinin-induced bronchoconstriction in man: mode of action. Am Rev Respir Dis 1987; 135:176-180.

10.	Frossard N, Stretton CD, Barnes PJ. Modulation of bradykinin responses in airway smooth muscle by epithelial enzymes. Agents Actions 1990; 31:204-209.

11.	Bramley AM, Samhoun MN, Piper PJ. The role of epithelium in modulating the responses of guinea-pig trachea induced by bradykinin *in vitro*. Br J Pharmacol 1990; 99:762-766.

12.	Dusser DJ, Nadel JA, Sekizawa K, Graf PD, Borson DB. Neutral endopeptidase and angiotensin converting enzyme inhibitors potentiate kinin-induced contraction of ferret trachea. J Pharmacol Exp Ther 1988; 244:531-536.

13.	Ichinose M, Barnes PJ. The effect of peptidase inhibitors on bradykinin-induced bronchoconstriction in guinea-pigs *in vivo*. Br J Pharmacol 1990; 101:77-80.

14.	Ichinose M, Belvisi MG, Barnes PJ. Bradykinin-induced bronchoconstriction in guinea-pig *in vivo*: role of neural mechanisms. J Pharmacol Exp Ther 1990; 253:1207-1212.

15.	Jin LS, Seeds E, Page CP, Schachter M. Inhibition of bradykinin-induced bronchoconstriction in the guinea pig by synthetic B_2 receptor antagonist. Br J Pharmacol 1988; 97:598-602.

16.	Ichinose M, Barnes PJ. Bradykinin-induced airway microvascular leakage and bronchoconstriction are mediated via a bradykinin B_2-receptor. Am Rev Respir Dis 1990; 142:1104-1107.

17.	Sakamoto T, Elwood W, Barnes PJ, Chung KF. Effect of HOE 140, a new bradykinin antagonist, on bradykinin and platelet-activating factor-induced bronchoconstriction and airway microvascular leakage in guinea pig. Eur J Pharmacol 1992; in press.

18.	Steranka LR, Manning DL, de Haas C, *et al*. Bradykinin as a pain mediator: receptors are localized to sensory neurons and antagonists have analgesic effects. Proc Natl Acad Sci USA 1988; 85:3245-3249.

19.	Kaufman MP, Coleridge HM, Coleridge JCG, Baker DG. Bradykinin stimulates afferent vagal C-fibres in intrapulmonary airways of dogs. J Appl Physiol 1980; 48:511-517.

20.	Saria A, Martling CR, Yan Z, Theodorsson-Norheim E, Gamse R, Lundberg JM. Release of multiple tachykinins from capsaicin-sensitive nerves in the lung by bradykinin, histamine, dimethylphenylpiperainium, and vagal nerve stimulation. Am Rev Respir Dis 1988; 137:1330-1335.

21.	Miura M, Belvisi MG, Barnes PJ. Modulation of excitatory non-adrenergic non-cholinergic constrictor responses in guinea pig bronchi by bradykinin. Br J Pharmacol 1991; 104:132P.

22.	Fox A, Dray A, Barnes PJ. Single fibre recordings from guinea pig trachea. J Physiol 1992; in press.

23.	Dixon CMS, Barnes PJ. Bradykinin induced bronchoconstriction: inhibition by nedocromil sodium and sodium cromoglycate. Br J Clin Pharmacol 1989; 270:8310-8360.

24.	Fuller RW. Cough associated with angiotensin converting enzyme inhibitors. J Hum Hypert 1989; 3:159-161.

25.	Dixon CMS, Fuller RW, Barnes PJ. The effect of an angiotensin converting enzyme inhibitor, ramipril, on bronchial responses to inhaled histamine and bradykinin in asthmatic subjects. Br J Clin Pharmacol 1987; 23:91-93.

26.	Rogers DF, Dijk S, Barnes PJ. Bradykinin-induced plasma exudation in guinea pig airways: involvement of platelet activating factor. Br J Pharmacol 1990; 101:739-745.

27.	Lotvall JO, Tokuyama K, Lofdahl C-G, Ullman A, Barnes PJ, Chung KF. Peptidase modulation of noncholinergic vagal bronchoconstriction and airway microvascular leakage. J Appl Physiol 1991; 70:2730-2735.

28.	Laitinen LA, Laitinen A, Widdicombe JG. Effects of inflammatory and other meditors on airway vascular beds. Am Rev Respir Dis 1987; 135:S67-S70.

29.	Parsons GH, Nichol GM, Barnes PJ, Chung KF. Peptide mediator effects on bronchial blood velocity and lung resistance in conscious sheep. J Appl Physiol 1992; in press.

30.	Baker AP, Hillegass LM, Holden DA, Smith WJ. Effect of kallidin, substance P, and other basic polypeptides on the production of respiratory mcromolecules. Am Rev Respir Dis 1977; 115:811-817.

31.   Baraniuk JN, Lundgren JD, Goff J, *et al*. Bradykinin receptor distribution in human nasal
      mucosa, and analysis of in vitro secretory responses in vitro and in vivo. Am Rev Respir Dis
      1990; 141:706-714.

32.   Leikhauf GD, Ueki IF, Nadel JA, Widdicombe JH. Bradykinin stimulates chloride secretion
      and prostaglandin E_2 release by canine tracheal epithelium. Am J Physiol 1985;
      248:F48-55.

33.   Hock FJ, Wirth K, Albus U, et al. HOE 140 a new potent and long acting
      bradykinin-antagonist: *in vitro* studies. Br J Pharmacol 1991; 102:769-773.

34.   Wirth K, Hock FJ, Albus U, et al. HOE 140, a new potent and long acting bradykinin
      antagonist: *in vivo* studies. Br J Pharmacol 1991; 102:774-777.

AAS 38/III
Recent Progress on Kinins
© 1992 Birkhäuser Verlag Basel

THE POTENTIAL ROLE OF BRADYKININ ANTAGONISTS IN THE TREATMENT OF ASTHMA

William M. Abraham, Ph.D.

Mount Sinai Medical Center, Division of Pulmonary Disease, Miami Beach, Florida 33140

SUMMARY:

The potential role of bradykinin and the use of bradykinin antagonists in the treatment of allergic airway disease (asthma) was studied by determining the effects of the bradykinin B_2 receptor antagonist, NPC-567 (D-Arg-[Hyp^3,D-Phe^7]-bradykinin) on antigen-induced early and late bronchial responses, airway inflammation, mediator release and airway hyperresponsiveness in the sheep model of allergic asthma. In the control trial, antigen challenge produced an early bronchoconstrictor response that was associated with an increase in immunoreactive-kinins in bronchoalveolar lavage fluid (BAL), and a late bronchial response that was associated with increased concentrations of leukotriene [LT] B_4 and LTC_4 and inflammatory cells (neutrophils and eosinophils) in the BAL. Aerosol NPC-567 given before, during and 4 h after antigen challenge had no protective effect on the early bronchoconstrictor response, but significantly inhibited the late bronchial response. This inhibition was correlated with a reduction in the inflammatory lipid mediators and inflammatory cells in BAL. These results suggest that antigen-induced kinin generation may be important for controlling the release of arachidonic acid metabolites from inflammatory cells that contribute to the development of late responses and airway hyperresponsiveness.

INTRODUCTION

Recent evidence indicates that asthma is an inflammatory disease of the airways and therefore, experimental studies both in human subjects and in animal models have been directed at understanding the association between the airway inflammation and the airway dysfunction of asthma. This dysfunction is expressed as airway obstruction, airway hyperresponsiveness to a variety of stimuli, and a slowing of mucociliary clearance. One means of studying the interrelationships among these processes in the laboratory is to use controlled challenges with specific antigen. These studies can be designed such that the observed changes in pulmonary function can be correlated with the airway inflammatory response as assessed by bronchoalveolar lavage or bronchial biopsy. Using these techniques, it has become apparent that allergen challenge can elicit two types of airway responses: a single acute bronchial response or an acute bronchial response which is followed 6-8 h later by a late response (1,2). Each pattern of airway response (i.e. the acute response only, or both an acute and late response) is associated with a different inflammatory response, with the late response being characterized by the presence of eosinophils (3-5). Late bronchial responses have been of special interest because their development has been associated with characteristics of chronic asthma (1,2) including refractoriness to beta adrenergic agents, a dependence on glucocorticosteroids, a heightened inflammatory response and a prolonged and increased airway hyperresponsiveness (1,2,6-11). The increased airway responsiveness renders the airways more sensitive to subsequent exposures to allergen, viral infections, exercise, cold air or other noxious airborne stimuli (e.g. pollutants) and so is thought to result in a worsening of symptoms (cough, bronchoconstriction, mucus accumulation).

Bradykinin has received increased attention as a putative mediator of asthma because asthmatic subjects are hyperresponsive to inhaled bradykinin (12-14) and recent studies have demonstrated the presence of immunoreactive kinins and kininogenase activity in bronchoalveolar (BAL) fluid obtained from asthmatic subjects following allergen provocation (15). Increased immunoreactive kinins have also been found in nasal washings from patients with allergic rhinitis undergoing nasal allergen challenge (15). The exact role bradykinin plays in the pathophysiology of allergic airway disease is not well understood, but the recent availability of peptide analog antagonists of bradykinin (16) provide useful tools to begin to clarify these mechanisms.

We assessed the usefulness of one of these bradykinin antagonists, D-Arg-[Hyp^3,D-Phe^7]-bradykinin (NPC-567) (17), in modifying bradykinin and antigen-induced bronchial obstruction, bronchial hyperresponsiveness and airway inflammation in a sheep model of asthma (18,19).

MATERIALS AND METHODS

The characteristics of the animal model, the specific techniques used to assess changes in pulmonary function in response to inhaled agonists or <u>Ascaris suum</u> antigen and the procedures used to assess airway hyperresponsiveness have been published in detail (18,20-22).

RESULTS

Before assessing the potential role of NPC-567 in allergen-induced airway responses, it was important to show that the actions of bradykinin in sheep airways were similar to those described in human airways. We found bradykinin induced cough in sheep (23) and inhaled bradykinin caused a short-lived (~20 min) dose dependent bronchoconstriction (21,22) which had a similar pharmacological profile to that described in man (22). Furthermore, the sheep allergic to <u>Ascaris suum</u> antigen were approximately 6 times more sensitive to the constrictor effects of inhaled bradykinin than were their non-allergic counterparts (Abraham, unpublished observations). Pretreatment with inhaled NPC-567 blocked bradykinin-induced bronchoconstriction in a dose-dependent fashion (Fig. 1) (21,22) indicating the presence of functional bradykinin receptors in sheep airways as has been suggested from ligand binding studies (24).

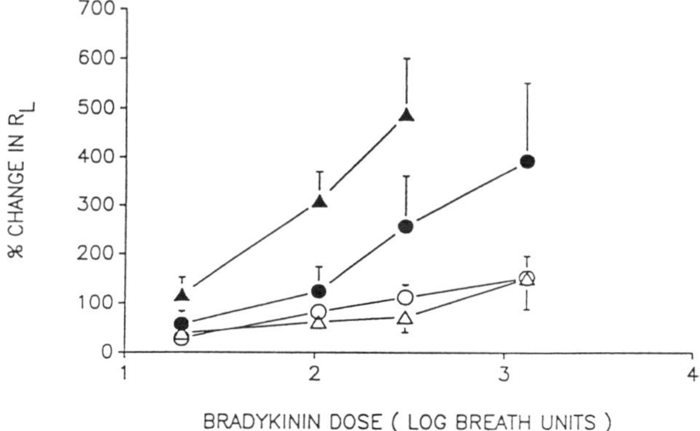

Figure 1: Dose-response curve to inhaled bradykinin in allergic sheep after placebo (▲), and after 20 breaths of 2 mg/ml (●), 5 mg/ml (△) and 10 mg.ml (o) NPC-567. Each point is the mean ±SE mean of six observations. Where not shown, SE lies within the symbol. From reference (22).

Sodium metabisulfite (MBS) is a food and wine preservative that causes bronchoconstriction in asthmatic subjects and some animal models. In sheep, inhaled MBS caused a bronchoconstriction, which had a pharmacology similar to that seen with bradykinin. Further, studies indicated that MBS challenge was associated with increased levels of immunoreactive kinins (25) in bronchoalveolar lavage and therefore it was not surprising that NPC-567 blocked the MBS-induced bronchoconstriction (Fig. 2). These findings suggest that bradykinin may mediate some of the adverse airway effects produced by irritant stimuli in asthmatics.

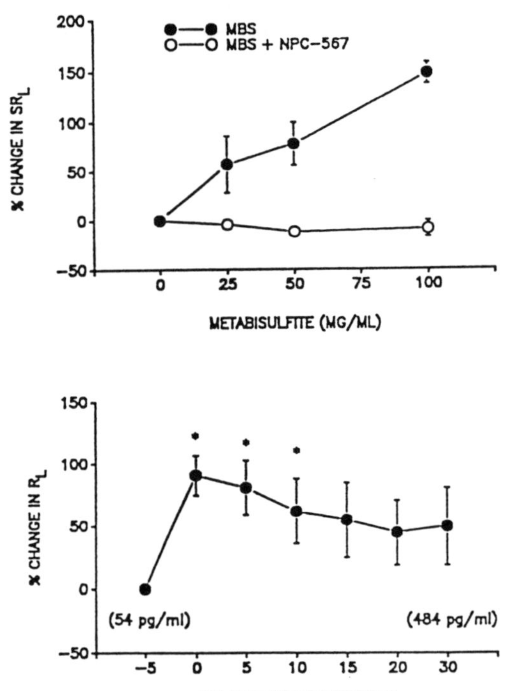

Figure 2: TOP - Effect of BK B$_2$-receptor antagonist NPC-567 on MBS-induced bronchoconstriction. NPC-567 (20 breaths of 5 mg/ml solution) completely inhibited responses to MBS. Values are means ± SE for 6 sheep. **BOTTOM** - Time course of MBS-induced bronchoconstriction. RL was measured after a single concentration (30 breaths) of 100 mg/ml MBS. Response to MBS was immediate (0-5 min) and short-lived, with RL returning toward baseline by 30 min. Values are means ± SE for 6 sheep. Values in parentheses are mean immunoreactive kinin concentrations in bronchoalveolar lavage. *P <0.05 vs. baseline. From reference (25).

We then used NPC-567 to determine if bradykinin contributed to antigen-induced airway responses. In our initial studies, we found that NPC-567 had no effect on the immediate bronchoconstrictor response to inhaled antigen, but did block the cell influx and airway hyperresponsiveness to inhaled carbachol observed after (~2 h) antigen challenge (21).

Because airway hyporesponsiveness and the recruitment of inflammatory cells are important determinants of the late bronchial response, these results suggested to us that bradykinin may be involved in the secondary inflammatory events that contribute to the late asthmatic response. Such a hypothesis is consistent with ability of bradykinin to cause the release of inflammatory lipid mediators which may mediate the late response (26-28). For these studies, the frequency of treatment and the dose of NPC-567 were increased (29) from the single pretreatment protocol in order to increase the bioavailability of the peptide. As was seen in the initial study, NPC-567 had no protective effect on the immediate bronchoconstrictor response (early response) but did protect against the late bronchial response (Fig. 3).

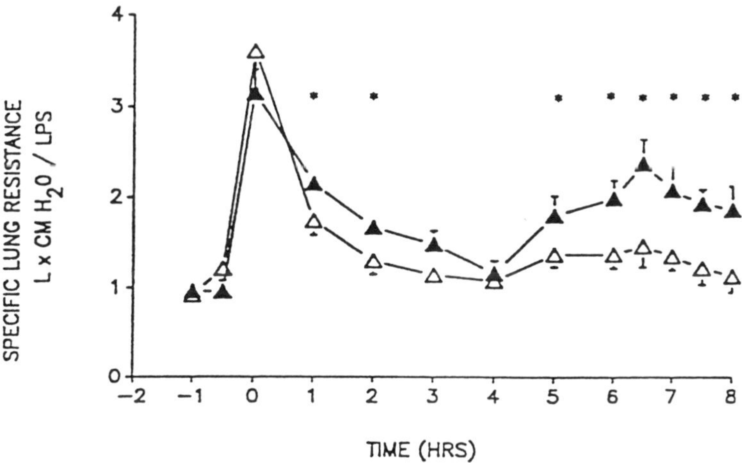

Figure 3:   Effect of NPC-567 (20 breaths 10 mg/ml 30 min before antigen challenge, 400 breaths 2 mg/ml during challenge and 20 breaths 10 mg/ml 4 h after challenge) on antigen-induced early and late responses in allergic sheep. NPC-567 (△) had no effect on the acute bronchoconstrictor response represented as an increase in specific lung resistance ($\uparrow SR_L$) but caused a significant inhibition of the late response. The lack of inhibition in the acute response and the blockade of the late response correlated with mediator measurements obtained from BAL (see Fig. 4). From reference 28.

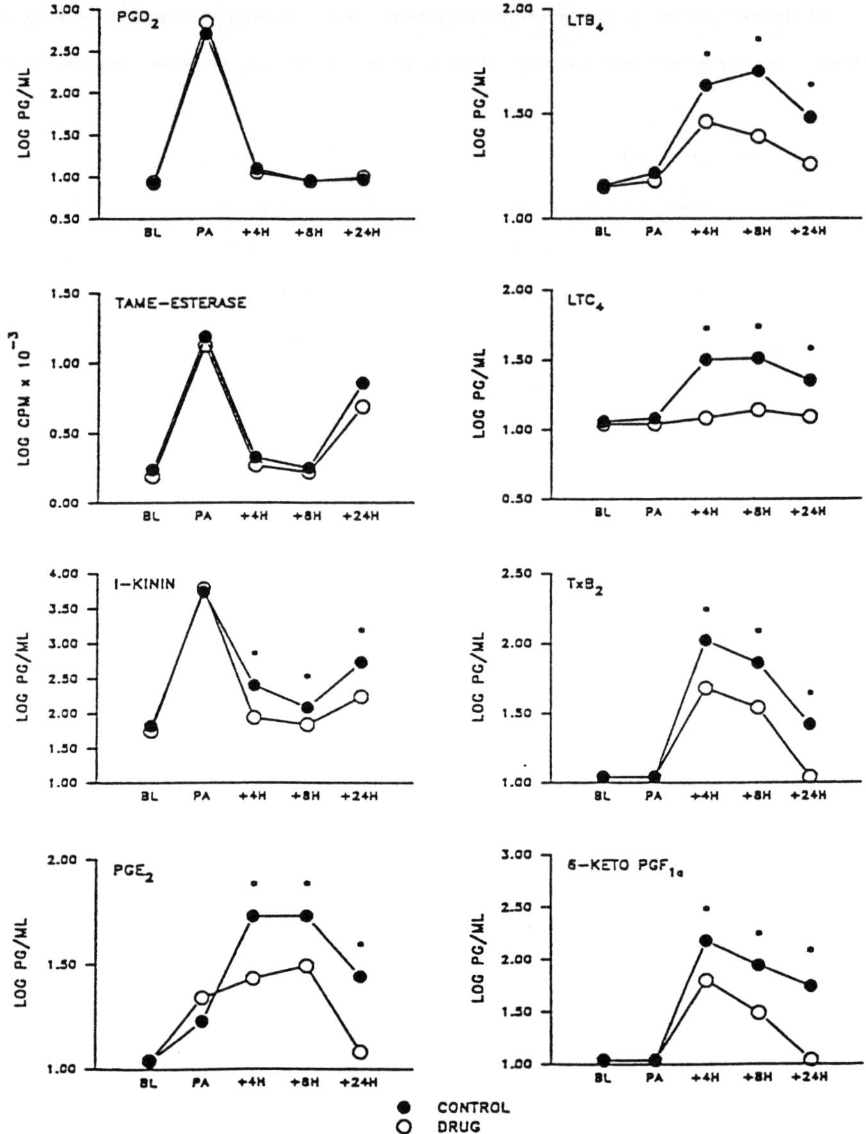

Figure 4: Mediator concentrations in BAL before and after antigen challenge in allergic sheep with and without NPC-567 (10 mg/ml) treatment. Concentrations are plotted as the log of the absolute value for the control trial (●) and for the NPC-567 trial (o). Values are mean for 6 sheep. * P<0.05 vs NPC-567 trial. Baseline (BL), post ascaris (PA). From reference 26.

These changes in airway function were associated with concomitant changes in inflammatory mediators in BAL. In the control trial, the early response was associated with significant increases in the concentrations of immunoreactive (i)-kinins, TAME-esterase, prostaglandin (PG) D_2 and PGE_2 (Fig. 4), and the late response was associated with significant increases in the concentrations of PGE_2, 6-keto PGF_1, thromboxane (Tx) B_2, leukotriene (LT)B_4 and LTC_4. I-kinins were also significantly elevated during the late response, but the concentration was less than that found during the early response. The late bronchial response and the late increase in inflammatory mediators was associated with a significant increase in the total percentage of granulocytes (neutrophils and eosinophils) in the BAL at 8 h post challenge (Fig. 5). When the animals were treated with NPC-567, there was no protection against the increased concentrations of i-kinins, TAME-esterase, PGD_2, and PGE_2 found immediately after challenge (consistent with the lack of effect in the airway response (Fig. 3), but during the late response there was a significant reduction in the secondary increase in the level of i-kinins, LTB_4, LTC_4, TxB_2 and the PGs (Fig. 4). The blockade of the late response and the reduction in the mediator concentrations observed in the NPC-567 treated sheep were concomitant with a reduction in the total percentage of granulocytes in the BAL (Fig. 5). It is important to note that NPC-567 reduced the release of all mediators and not just the spasmogenic leukotrienes (i.e. LTC_4, LTD_4) that have been proposed to mediate the late bronchial response in this model (26,28,30). Thus, while LT antagonists and NPC-567 may provide similar protection against the late constrictor response in allergic sheep, the mechanisms of action are probably different. The LT antagonists prevent the action of a specific contractile mediator released from inflammatory cells whereas the bradykinin antagonist attenuates overall mediator release by reducing the inflammatory cell influx. These findings extend the previous observations using NPC-567 in acute responders and lend support to the argument that bradykinin plays a contributory role in post antigen-induced inflammation and subsequent mediator release, the pathophysiologic consequence of which leads to the development of hyperresponsive airways.

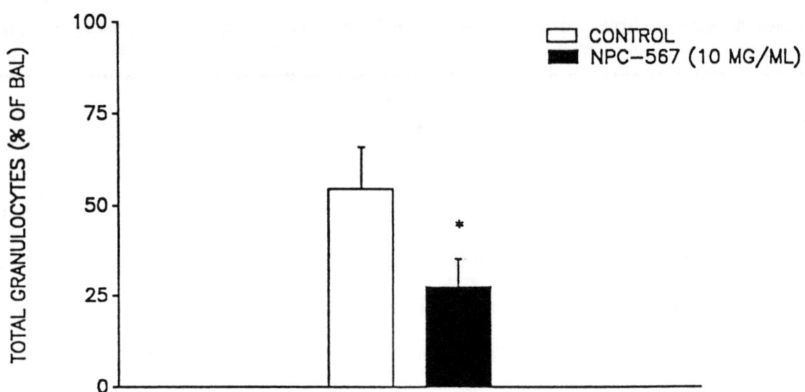

Figure 5: Effect of NPC-567 (10 mg/ml) on inflammatory cell influx in allergic sheep. NPC-567 significantly inhibited the granulocyte (expressed as percentage of neutrophils and eosinophils in BAL) infiltrate into sheep airways 4-8 h after antigen challenge. This inhibition was associated with a reduction in the concentration of inflammatory mediators in the BAL (Fig. 4) and a reduction in the late response (Fig. 3). *P <0.05 vs. control. From reference (31).

CONCLUSION

In summary, experiments using a well characterized animal model of allergen-induced early and late responses that were designed to determine if the inflammatory mediator bradykinin plays a role in these pathophysiologic responses have been described. Although it appears that bradykinin contributes little, if any, to the immediate bronchial response to inhaled antigen, the findings provide substantive evidence for the release of kinins and their modulatory role in allergen-induced inflammation and mediator release from these inflammatory cells that results in late responses and airway hyperresponsiveness. If similar findings are obtained in man, bradykinin antagonists may prove to be potentially useful in the therapy of inflammatory diseases of the airways, such as asthma.

REFERENCES

1. Pepys J, Hutchcroft BJ. Bronchial provocation tests in the etiologic diagnosis and analysis of asthma. Am Rev Respir Dis 1975; 112:829-859.

2.   O'Byrne PM, Dolovich J, Hargreave FE. State of Art: Late asthmatic responses. Am Rev Respir Dis 1987; 136:740-751.

3.   de Monchy JGR, Kauffman HF, Venge P, et al. Bronchoalveolar eosinophilia during allergen-induced late asthmatic reactions. Am Rev Respir Dis 1985; 131:373-376.

4.   Rossi GA, Crimi E, Lantero S, et al. Late-phase asthmatic reaction to inhaled allergen is associated with early recruitment of eosinophils in the airways. Am Rev Respir Dis 1991; 144:379-383.

5.   Abraham WM, Sielczak MW, Wanner A, et al. Cellular markers of inflammation in the airways of allergic sheep with and without allergen-induced late responses. Am Rev Respir Dis 1988; 138:1565-1571.

6.   Cockcroft DW, Murdock KY. Comparative effects of inhaled salbutamol, sodium cromoglycate, and beclomethasone dipropionate on allergen-induced early asthmatic responses, late asthmatic responses, and increased bronchial responsiveness to histamine. J Allergy Clin Immunol 1987; 79:734-740.

7.   Abraham WM, Perruchoud AP, Wanner A, et al. Bronchoalveolar lavage fluid from allergic sheep with allergen-induced late responses contains increased inflammatory cells and slow reacting substance of anaphylaxis (SRS-A). Am Rev Respir Dis 1986; 133:A240.

8.   Proud D, Togias A, Naclerio RM, Crush SA, Norman PS, Lichtenstein LM. Kinins are generated in vivo following nasal airway challenge of allergic individuals with allergen. J Clin Invest 1983; 72:1678-1685.

9.   Lanes S, Stevenson JS, Codias E, et al. Indomethacin and FPL-57231 inhibit antigen-induced airway hyperresponsiveness in sheep. J Appl Physiol 1986; 61:864-872.

10.   Wanner A, Ahmed T, Abraham WM. Drug actions on mediators. Drug therapy for asthma. In: Jenne JW, Murphy S, eds, Lung Biology and Health and Disease. New York: Marcel Dekker, Inc., 1987:413-461.

11.   Cortes A, Ahmed A, Sielczak MW, Abraham WM. Reversal of antigen-induced airway hyperresponsiveness by a novel anti-allergic compound TYB-2285. Am Rev Respir Dis 1992; 145:A726.

12.   Dixon CMS, Barnes PJ. Bradykinin-induced bronchoconstriction: inhibition by nedocromil sodium and sodium cromoglycate. Br J Clin Pharmacol 1989; 27:831-836.

13.   Fuller RW, Dixon CS, Cuss FMC, Barnes PJ. Bradykinin-induced bronchoconstriction in humans: Mode of action. Am Rev Respir Dis 1987; 135:176-180.

14.   Simonsson BG, Skoogh B-E, Bergh NP, Andersson R, Svedmyr N. In vivo and in vitro effect of bradykinin on bronchial motor tone in normal subjects and patients with airways obstruction. Respiration 1973; 30:378-388.

15.   Christiansen SC, Proud D, Cochrane CG. Detection of tissue kallikrein in the bronchoalveolar lavage fluid of asthmatic subjects. J Clin Invest 1987; 79:188-197.

16.   Steranka LR, Farmer SG, Burch RM. Antagonists of B_2 bradykinin receptors. FASEB J 1989; 3:2019-2025.

17.   Steranka LR, Manning JC, Dehaas CJ, et al. Bradykinin as a pain mediator: receptors are localized in sensory neurons, and antagonists have analgesic actions. Proc Natl Acad Sci USA 1988; 850:3245-3249.

18.   Abraham WM, Delehunt JC, Yerger L, Marchette B. Characterization of a late phase pulmonary response following antigen challenge in allergic sheep. Am Rev Respir Dis 1983; 128:839-844.

19.   Abraham WM. Pharmacology of allergen-induced early and late airway responses and antigen-induced airway hyperresponsiveness in allergic sheep. Pul Pharmacol 1989; 2:33-40.

20.   Wanner A, Reinhart M. Respiratory mechanics in conscious sheep. J Appl Physiol 1978; 44:479-482.

21.   Soler M, Sielczak MW, Abraham WM. A bradykinin-antagonist blocks antigen-induced airway hyperresponsiveness and inflammation in sheep. Pul Pharmacol 1990; 3:9-15.

22.   Abraham WM, Ahmed A, Cortes A, et al. Airway effects of inhaled bradykinin, substance P, and neurokinin A in sheep. J Allergy Clin Immunol 1991; 87:557-564.

23.   Cortes A, Ahmed A, Ahmed T, Abraham WM. Citric acid and bradykinin produce cough in allergic sheep. FASEB J 1990; 4:A614.

24.   Farmer SG, Burch RM, Meeker SM, Wilkins DE. Evidence for a pulmonary B3 bradykinin receptor. Mol Pharmacol 1989; 36:1-8.

25.   Mansour E, Ahmed A, Cortes A, Caplan J, Burch RM, Abraham WM. Mechanisms of metabisulfite-induced bronchoconstriction. Evidence for bradykinin B_2 receptor stimulation. J Appl Physiol 1992; 72:1831-1837.

26.   Abraham WM. The importance of lipoxygenase products of arachidonic acid in allergen-induced late responses. Am Rev Respir Dis 1987; 135:S49-S53.

27.   Abraham WM. The role of leukotrienes in allergen-induced late responses in allergic sheep. Annals of the New York Academy of Sciences: Biology of the Leukotrienes 1988; 524:260-270.

28.   Delehunt JC, Perruchoud AP, Yerger L, Marchette B, Stevenson JS, Abraham WM. The role of SRS-A in the late bronchial response following antigen challenge in allergic sheep. Am Rev Respir Dis 1984; 130:748-754.

29.   Abraham WM, Burch RM, Farmer SG, Sielczak MW, Ahmed A, Cortes A. A bradykinin antagonist modifies allergen-induced mediator release and late bronchial responses in sheep. Am Rev Respir Dis 1991; 143:787-796.

30.   Tagari P, Abraham WM, McGolrick J, et al. Increased leukotriene E_4 excretion during antigen-induced bronchoconstriction in allergic sheep. J Appl Physiol 1990; 68:1321-1327.

31.   Abraham William M. Bradykinin Antagonists in a Sheep Model of Allergic Asthma. In: Burch Ronald M, eds, Bradykinin Antagonists; Basic and Clinical Research. New York, New York: Marcel Dekker, Inc., 1991:261-276.

AAS 38/III
Recent Progress on Kinins
© 1992 Birkhäuser Verlag Basel

KININ RECEPTORS ON ASTHMATIC AIRWAYS: FUNCTIONAL SUBTYPING

Riccardo Polosa

Medicine 1, Level D, Centre Block, Southampton General Hospital,
Tremona Road, Southampton, SO9 4XY, England

SUMMARY

Bradykinin, kallidin and [desArg^9]-bradykinin are oligopeptides that may contribute as mediators in the pathogenesis of bronchial asthma by interacting with specific cell surface receptors. The structure-activity relationship of kinins within the lower airways has been investigated in 16 asthmatic subjects. The findings of the present study suggest a complex action of kinins on asthmatic airways probably involving more than one receptor subtype.

INTRODUCTION

Kinins are potent vasoactive peptides formed *de novo* in body fluids and tissues during inflammation. They are derived from the alfa-2-globulins, high and low molecular weight kininogens, through proteolytic cleavage by tissue and plasma kallikreins (1). In the upper airways, kinin generation occurs, and correlate with symptoms, both in the early (2) and late (3) responses to allergen challenge. Kinins have also been measured in the lower airways of asthmatic subjects both spontaneously (4) and following allergen challenge of the bronchial mucosa (5,6). These findings together with their known abilities to increase vascular permeability (7,8) and to provoke bronchoconstriction when administered by inhalation in asthmatic subjects (9), support the view that these peptides may be important mediators of the pathogenesis of airway obstruction in bronchial asthma.

Bradykinin is a nonapeptide generated as cleavage product from the action of kallikreins on high molecular weight kininogen (HMWK). An additional substrate for the kallikreins is low molecular weight kininogen (LMWK) cleavage of which generates lys-bradykinin (kallidin). Once generated both kinins may undergo enzymatic cleavage of the N-terminal arginine residue forming [desArg^9]-bradykinin and [desArg^{10}]-kallidin (1). Two main kinin receptor subtypes have been described, designated B1 and B2, on the basis of studies of differing agonist potencies in separate tissue preparations (1). In the cat ileal preparations (10) and in isolated canine tracheal strips (11) both bradykinin and kallidin induce responses through an action on the B2 receptors whilst the B1 agonist [desArg^9]-bradykinin is without effect. On rabbit aorta the situation is reversed, with [desArg^9]-bradykinin inducing contraction whereas bradykinin and kallidin are inactive (1).

We have shown that both bradykinin and kallidin, but not [desArg^9]-bradykinin, are potent bronchoconstrictor stimuli when administered by inhalation in asthmatic subjects (9). Because bradykinin and kallidin are agonists of B2 receptors and [desArg^9]-bradykinin is an agonist for B1 receptors, these *in vivo* structure activity studies suggest that this potent bronchoconstrictor action may result from a specific pharmacological effect compatible with the stimulation of B2 receptors. Confirmation of a common receptor for bradykinin and kallidin could also be obtained if cross-tachyphylaxis between the two would be shown. To characterise further the receptor mediating the bronchoconstrictor response to inhaled kinins we have carried out crossed bronchoprovocation tests with bradykinin, kallidin and [desArg^9]-bradykinin in a group of six asthmatic subjects.

MATERIALS & METHODS

A total of sixteen asthmatic subjects who were all non-smokers participated in the study. All subjects were atopic, as defined by positive prick skin tests (>2 mm weal response) to two or more of five common aeroallergens. Their baseline FEV1 (forced expiratory volume in one second) was >70% of their predicted values and none were receiving oral corticosteroids or theophylline within the preceding 3 weeks. Inhaled bronchodilators were discontinued for at least 8 h prior each visit to the laboratory, although subjects were allowed to continue inhaled corticosteroids as usual.

Airway calibre was recorded as FEV1 by means of a dry wedge spirometer. On each study day, bradykinin triacetic acid, kallidin acetic acid and [desArg^9]-bradykinin were freshly prepared in 10% ethanol in 0.9% sodium chloride to produce stock solutions of 8 mg/ml. Each stock solution was then diluted with its respective diluent to produce a concentration range of 0.0037 - 4 mg/ml (0.0035 - 3.77 mmol/l, 0.0031 -

3.37 mmol/l, 0.0041 - 4.43 mmol/l) for bradykinin, kallidin and [desArg^9]-bradykinin respectively. The solutions were administered as aerosols generated from a starting volume of 3 ml in a disposable Inspiron Mini-nebuliser driven by compressed air at 8 l/min. Under these conditions the nebuliser had an output of 0.48 ml/min and generated an aerosol with a mass median particle diameter of 4.7 μm (12). Subjects inhaled the aerosolized solutions in five breaths from end-tidal volume to full inspiratory capacity via a mouthpiece as described by Chai et al. (13).

In the first phase of the study, 10 subjects attended the laboratory on three separate occasions at least 72 h apart to undertake concentration-response studies with inhaled bradykinin, kallidin and [desArg^9]-bradykinin to derive their PC20FEV1 values as described in details in a previously reported study (9). In brief, increasing fourfold concentrations of kinin were inhaled at approximately 5 min intervals until FEV1 had fallen by >20% of the post-diluent value and the corresponding PC20FEV1 values derived.

In the second phase of the study, a further 6 asthmatic subjects attended the laboratory on three occasions separated by at least 5 days during which two consecutive concentration-response studies with inhaled bradykinin, kallidin and [desArg^9]-bradykinin were undertaken in a double blind randomized manner. During this phase an initial bronchial provocation test with bradykinin or kallidin was performed until FEV1 fell to > 20% of post-diluent baseline value. The initial inhalation test with [desArg^9]-bradykinin was carried out until the maximal concentration of agonist had been administered. The airways were then allowed to recover spontaneously until FEV1 had returned to within 5% of their post-diluent baseline value where appropriate. On achieving this, after approximately 35-60 min, a second kinin bronchoprovocation was undertaken with the same agonist until FEV1 fell > 20% of the original post diluent value, or the highest concentrations of each kinin had been administered. Again once the FEV1 after the second inhalation test had returned to within 5% of the post diluent baseline value, a third concentration-response study was undertaken this time with kallidin on the sequential bradykinin-bradykinin day, and with bradykinin on the kallidin-kallidin and [desArg^9]-bradykinin- [desArg^9]-bradykinin days. When possible the PC20 values for each of the kinin challenges were derived.

The final phase of the study was carried out to determine the specificity of the three kinins on subsequent non-specific contractile stimuli. The 6 asthmatic subjects attended the laboratory on three further visits, at least 5 days apart, to undertake a single blinded concentration-response study with inhaled histamine after receiving two consecutive inhalation challenges with bradykinin, kallidin and [desArg^9]-bradykinin administered in an identical fashion to that described in the second phase.

Figures refer to the mean ±SEM unless otherwise stated and the p<0.05 level of significance was accepted. Baseline FEV1 values prior to bronchial challenges were compared between study days by analysis of variance (ANOVA). The lowest FEV1 values recorded with each concentration of the inhaled agonists was used for analysis. From these data, concentration-response curves were constructed by plotting the percentage change in FEV1 from the post-diluent baseline value against the cumulative concentration of the agonist administered on a logarithmic scale and the concentration of agonist required to produce a 20% fall in FEV1 from the post-diluent baseline value (PC20FEV1) determined by interpolation. The repeatability of the kinin challenge procedure was determined according the method described by Altman and Bland (14). Values of PC20 bradykinin, kallidin and histamine following consecutive kinin challenges were logarithmically transformed to normalise their distribution and compared by the Student's t test for paired data. Any relationship between the airway responses to histamine and the kinins tested was examined by least-squares linear regression analysis of the logarithmically transformed values and a relative potency derived in molar terms.

RESULTS

There was no significant difference in baseline values of FEV1 between and within any of the study days. The challenge procedure with bradykinin and kallidin in this group of patients was found to be repeatable, with a C.R. of 1.7 and 1.6 doubling dilutions respectively. These findings were consistent with the repeatability data obtained in previous studies from our laboratory with bradykinin bronchoprovocation tests (15,16).

In phase 1 inhaled bradykinin and kallidin, but not [desArg^9]-bradykinin produced concentration-related falls in FEV1 in all the subjects studied. The geometric mean (range) of PC20 values obtained were 0.03 (0.01 - 0.3) and 0.08 (0.01 - 0.66) mg/ml for bradykinin and kallidin respectively. A significant correlation was obtained between PC20 values for bradykinin and kallidin (r= 0.87; p< 0.01) with bradykinin in molar terms being approximately 3 times more potent than kallidin.

In phase 2, we confirmed that in a further 6 asthmatic subjects, inhalation of bradykinin, kallidin but not [desArg^9]-bradykinin elicited concentration-related falls in FEV1. For the group as a whole, the geometric mean PC20 values were 0.07 (0.01 - 1.59) mg/ml and 0.22 (0.01 - 4.62) mg/ml for bradykinin and kallidin respectively. A significant correlation was obtained between the PC20 values for bradykinin and kallidin (r = 0.81; p <0.05). When inhaled up to a cumulative concentration of 10.62 mg/ml, [desArg^9]-bradykinin failed to produce any significant fall in FEV1 from diluent baseline in any of the asthmatic subjects studied.

Following recovery from the first concentration-response challenge with bradykinin or kallidin, the airways showed a substantially reduced response to a second challenge with the same agonist. Thus the geometric mean PC20 values for bradykinin and kallidin during the first challenge of 0.07 and 0.22 mg/ml increased after the second challenge 6.4- and 3.2-fold to 0.45 (0.07 - 10.62)(p<0.01) and to 0.70 (0.03 - 10.62)(p<0.01) mg/ml respectively. Consecutive inhalation tests with [desArg^9]-bradykinin showed no apparent change from the lack of a response in all the subjects studied.

Once the airways had recovered from the second bradykinin challenge, provocation with kallidin also revealed a reduced response to this agonist, the geometric mean PC20 value increasing 4-fold from 0.22 at baseline to 0.87 (0.03 - 10.62) mg/ml post-bradykinin (p<0.01)(Figure 1).

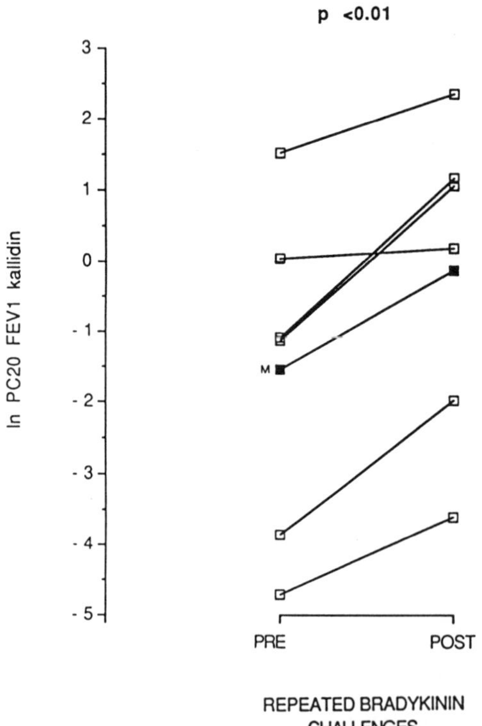

Fig.1 PC$_{20}$FEV$_1$ kallidin values obtained on baseline condition (PRE) and after having maximally induced bradykinin tachyphylaxis (POST). Closed squares represent geometric mean values (M).

Similarly, once the airways had recovered from the second kallidin challenge, provocation with bradykinin also showed a reduced response, the geometric mean PC20 value increasing 10-fold from 0.07 to 0.71 (0.02 - 10.62) mg/ml (p<0.01)(Figure 2).

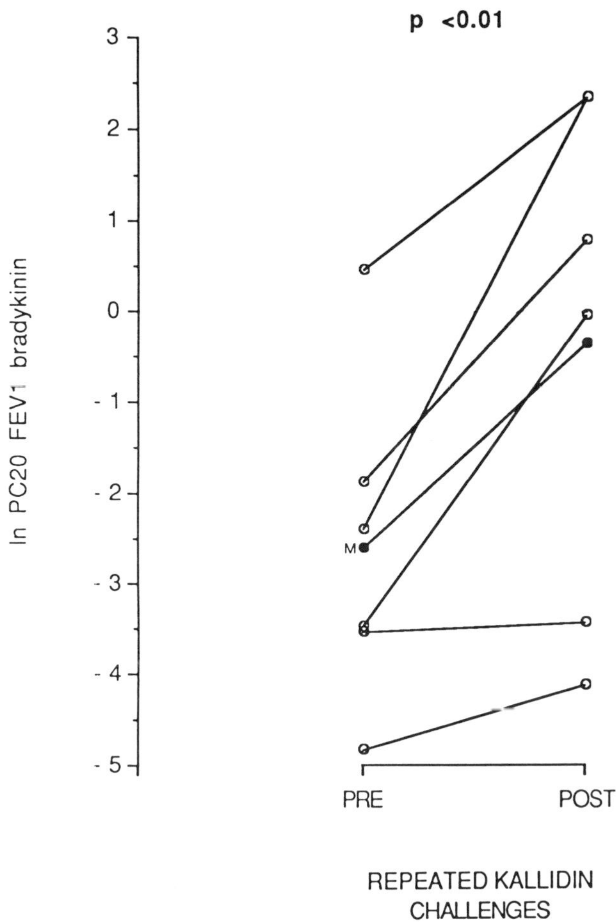

Fig.2 $PC_{20}FEV_1$ bradykinin values obtained on baseline condition (PRE) and after having maximally induced kallidin tachyphylaxis (POST). Closed circles represent geometric mean values (M).

Surprisingly, repeated exposure of the airways with the inactive B1 agonist, [desArg[9]]-bradykinin, reduced the response to inhaled bradykinin in 5 out of 6 subjects and the geometric mean PC20 value increased 6-fold from 0.07 to 0.39 (0.01 - 1.80) mg/ml (p<0.01)(Figure 3).

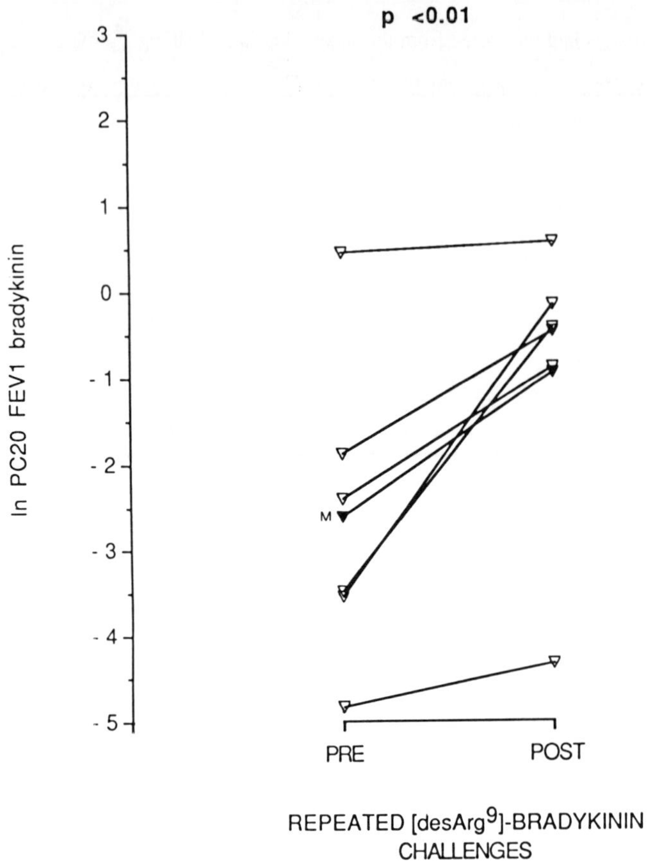

Fig.3 $PC_{20}FEV_1$ bradykinin values obtained on baseline condition (PRE) and after repeated [desArg^9]-bradykinin inhalations (POST). Closed triangles represent geometric mean values (M).

In phase 3 of the study, repeated challenge of the airways with bradykinin and kallidin produced an almost identical (8.5- and 4.2-fold respectively) loss of responsiveness to that observed with consecutive challenges with the same agonists in phase 2. However in the presence of reduced kinin responsiveness the airway response to a subsequent inhalation with histamine remained unchanged, the geometric mean PC20 value of 0.79 mg/ml not being significantly different from that of 0.77 (0.25 - 1.57) mg/ml and of 0.89 (0.29 - 1.88) mg/ml obtained on the bradykinin and kallidin study day respectively. The PC20 value for histamine obtained after the second exposure of the airways with [desArg^9]-bradykinin of 0.75 (0.30 - 2.51) mg/ml was not significantly different from the PC20 histamine values obtained when the airways developed maximal tachyphylaxis to bradykinin or kallidin.

DISCUSSION

The results of the present study confirm our previous finding that bradykinin and kallidin but not [desArg^9]-bradykinin cause concentration-related bronchoconstriction in asthmatic subjects (9) and that sequential bronchial provocation with bradykinin results in a repeatable loss of responsiveness to this peptide (16). In the same subjects, tachyphylaxis to kallidin inhalation has been demonstrated and found to be repeatable. We have extended these observations by showing cross-tachyphylaxis between the different kinins tested. Of particular interest is the demonstration that repeated bronchial challenge with [desArg^9]-bradykinin, a selective agonist for B1-receptor systems, produces a significant reduction in the bronchoconstrictor response to bradykinin, a B2 agonist, without itself producing any direct constrictor responses. These findings suggest a complex and probably indirect action of kinins on asthmatic airways probably involving more than one receptor subtype.

The general criteria used to characterise receptors include comparison of order of potency of agonists, the use of desensitization experiments and evaluation of the efficacy of potent and specific antagonists. In asthmatic airways, *in vivo* structure activity studies suggested that the kinin receptor subtype responsible for the bronchoconstrictor response is of the B2 variety (9). Since no specific and competitive B2 receptor blockers are as yet available for *in vivo* studies in humans to block the bronchoconstrictor action of bradykinin and kallidin in asthma, the only criterion that we could use in order to extend our previous findings and to find out whether bradykinin and kallidin act on the same or on different receptors, was to establish whether cross-tachyphylaxis could be shown between the two.

The present study confirms that bradykinin and kallidin provoked concentration-related bronchoconstriction when administered by inhalation to asthmatic subjects, whilst removal of the N-terminal arginine to form [desArg^9]-bradykinin resulted in a total loss of this activity. On the basis of what is known about the receptor subclasses for kinins, a rank potency order of bradykinin > kallidin >> [desArg^9]-bradykinin is consistent with the view that the constrictor response is mediated through B2 receptors. A similar potency order has also been obtained for cat ileum *in vitro* (10), canine tracheal preparations (11) and nasal airways in atopic rhinitics and in normal volunteers (17). Two consecutive concentration-response challenges with bradykinin and kallidin resulted in a repeatable loss of responsiveness to these peptides without influencing the underlying level of "non-specific" bronchial reactivity when measured by histamine provocation. The mechanism of tachyphylaxis to inhaled kinins in asthma is poorly understood. Although secondary release of protective prostanoids with functional effects such as PGE_2 and PGI_2 could explain the phenomenon, we (16) and others (18) have failed to show that loss of kinin responsiveness was sensitive to blockade by prior treatment with cyclooxygenase inhibitors. In addition if functional antagonism mediated by prostanoids operated in the case of kinin tachyphylaxis, then a parallel loss of histamine responsiveness might be anticipated, but was not observed. Taken together these data suggest that refractoriness to these peptides is a receptor-specific mechanism and may represent a down-regulation of B2-receptor function or impaired neuropeptide release from sensory nerves.

Sequential bronchial provocation with bradykinin produced a significant reduction in the bronchoconstrictor response to kallidin and vice versa. These effects were obtained without influencing the underlying level of "non-specific" bronchial responsiveness when measured by histamine provocation and further supports a receptor-specific mechanism. Cross-tachyphylaxis between bradykinin and kallidin could be interpreted as an interreaction between the mechanism(s) responsible for the bronchoconstrictor activity of the two kinins in asthma is occurring at the same receptor level. Altough this is insufficient to allow us to draw definitive conclusions about the identity of the receptors stimulated by bradykinin and kallidin, our data offers some support to the view that both bradykinin and kallidin are acting via the same B2-receptor present at airway level.

When bradykinin concentration-response studies followed repeated inhalations with the B1 receptor agonist, $[\text{desArg}^9]$-bradykinin, the dose-response curves were shifted to the right in 5 of the 6 subjects studied. Taking into account the good repeatability of the bradykinin challenge, it is somewhat surprising that repeated bronchial challenges with nebulised $[\text{desArg}^9]$-bradykinin, an agonist of the B1 receptor subtype, reduced the bronchoconstrictor response to bradykinin without affecting the underlying level of "non-specific" bronchial responsiveness . We might speculate that, although not causing any overt airway response, exposure with $[\text{desArg}^9]$-bradykinin may be still capable of causing subsequent release of inhibitory prostaglandins such as PGE_2 and PGI_2 from the epithelium and other airway cells (1,19) which may be acting as functional antagonists. However, if functional antagonism accounted for loss of airways response to $[\text{desArg}^9]$-bradykinin, then a parallel reduction in histamine responsiveness might be anticipated. An alternative explanation is that $[\text{desArg}^9]$-bradykinin might have acted as a partial antagonist at the B2-receptor level. In a variety of pharmacological models, it has been shown that naturally occurring kinins are relatively non-selective in acting on both B2 and B1 receptors (20). It has also been demonstrated that B1 receptors can be induced during pathological states (21). Therefore, although symptoms induced by bronchial challenge with kinins are not due to stimulation of B1 receptors (9), we cannot rule the possibility that in the inflamed airways of asthmatic subjects repeated exposure with [desArg^9]-bradykinin might induce B1 receptors proliferation which might in turn provoke disruption of the receptor mediated response to subsequent inhalations with bradykinin. An additional possibility that can not be excluded is that different subsets of B2 receptors are expressed in inflamed asthmatic airways which, altough fully responsive to the classical B2 receptor agonists, may also interreact with [desArg^9]-bradykinin. Recent findings (22,23) indicate that multiple B2 kinin receptors may exist.

On the basis of this limited structure-activity study, we have confirmed the existence of a B2 receptor type in the asthmatic airways by showing a different order of potency of kinin analogues that can be accounted for by B1 and B2 receptor stimulation. In addition, cross-tachyphylactic studies with bradykinin and kallidin provide further support to the view that a B2 receptor is involved in this bronchoconstrictor response. However by showing no direct agonist but a potent antagonist property of the B1 agonist [desArg^9]-bradykinin against the subsequent response to inhaled bradykinin, it has been introduced a confounding factor in the interpretation of the results. It is possible that desArg derivatives may have effects on the neuroeffector mechanism of bronchoconstriction which are more complex than initially proposed and might involve different receptor subtypes. Characterization of kinin receptors by the simple criteria of the order of potency of agonists and of the cross-tachyphylactic response to B2 agonists provides just limited information and only with the availability of potent and selective antagonist of kinin receptors will it become possible to help to identify kinin receptors subtypes.

REFERENCES

1. Regoli D, Barabe J. Pharmacology of bradykinin and related peptides. Pharmacol. Rev. 1980;31:1-46.

2. Proud D, Togias A, Naclerio RM, Crush SA, Norman PS, Lichtenstein LM. Kinins are generated *in vivo* following nasal airway challenge of allergic individuals with allergen. J. Clin. Invest. 1983; 72:1678-85.

3. Naclerio RM, Proud D, Togias AG, et al. Inflammatory mediators in late antigen-induced rhinitis. N. Engl. J. Med. 1985;313:65-70.

4. Christiansen SC, Proud D, Cochrane CG. Detection of tissue kallikreinin the bronchoalveolar lavage fluid of asthmatic subjects. J. Clin. Invest.1987;75:188-97.

5. Christiansen SC, Proud D, Sarnoff RB, Cochrane CG, Zuraw BL. Human bronchial kallikrein in bronchoalveolar lavage fluid following local allergen challenge. J. Allergy Clin. Immunol. 1989;83:285 (abstract).

6. Liu MC, Hubbard WC, Proud D, et al. Immediate and late inflammatory responses to ragweed antigen challenge of the peripheral airways in allergic asthmatics. Cellular, mediator and permeability changes. Am. Rev. Respir. Dis. 1991;144:51-58.

7. Saria A, Lundberg JM, Skofitsch G, Lembeck F. Vascular protein leakage in various tissues induced by substance P, capsaicin, bradykinin, serotonin, histamine and by antigen challenge. Naunyn Schmiedebergs Arch. Pharmacol. 1983;324:212-18.

8. Griesbacher T, Lembeck F. Effect of bradykinin antagonists on bradykinin-induced plasma extravasation, venoconstriction, prostaglandin E_2 release, nociceptor stimulation and contraction of the iris sphincter muscle in the rabbit. Br. J. Pharmacol. 1987;92:333-340.

9. Polosa R, Holgate ST. Comparative airway responses to inhaled bradykinin, kallidin and [desArg[9]]-bradykinin in normal and asthmatic subjects. Am. Rev. Respir. Dis. 1990;142:1367-71.

10. Drouin J-N, St-Pierre SA, Regoli D. Receptors for bradykinin and kallidin. Can. J. Physiol. Pharmacol. 1979;57:375-9.

11. Rangachari PK, McWade D, Donoff B. Luminal receptors for bradykinin on the canine tracheal epithelium: functional subtyping. Regul. Pept. 1988;21:237-44.

12. Lewis RA. Therapeutic aerosols. In: Bonsignore GC&C ed. Drugs and the lung. London: Plenum, 1984: 63-86.

13. Chai H, Farr RS, Froehlich LA, et al. Standardization of bronchial inhalation challenge procedures. J. Allergy Clin. Immunol. 1975;56:323-27.

14. Altman DG, Bland JH. Statistical methods for assessing agreement between two methods of clinical measurement. Lancet 1986;8(2):307-10.

15. Polosa R, Phillips GD, Lai CKW, Holgate ST. Contribution of histamine and prostanoids to bronchoconstriction provoked by inhaled bradykinin in atopic asthma. Allergy 1990;45:174-182.

16. Polosa R, Lai CKW, Robinson C, Holgate ST. The influence of cyclooxygenase inhibition on the loss of bronchoconstrictor response to repeated bradykinin challenge in asthma. Eur. Respir. J. 1990;3:914-21.

17. Rajakulasingam K, Polosa R, Holgate ST, Howarth PH. Comparative nasal effects of bradykinin, kallidin and [desArg^9]-bradykinin in atopic rhinitic and normal volunteers. J. Physiol. 1991;437:577-87.

18. Fuller RW, Dixon CMS, Cuss FMC, Barnes PJ. Bradykinin-induced bronchoconstriction in humans. Mode of action. Am. Rev. Respir. Dis. 1987;135:176-80.

19. Cahill M, Fishman JB, Polgar P. Effect of [desArg^9]-bradykinin and other bradykinin fragments on the synthesis of prostacyclin and the binding of bradykinin by vascular cells in culture. Agents Actions 1988;24:224-31.

20. Rhaleb N-E, Drapeau G, Dion S, Jukic D, Rouissi N-E, Regoli D. Structure-activity studies on bradykinin and related peptides: agonists. Br. J. Pharmacol. 1990;99:445-48.

21. Marceau F, Lussier A, St-Pierre S. Selective induction of cardiovascular responses to [desArg^9]-bradykinin by bacterial endotoxin. Pharmacology 1984;29:70-74.

22. Farmer SG, Burch RM, Meeker SN, Wilkins DE. Evidence for a pulmonary bradykinin B_3 receptor. Mol. Pharmacol. 1989;36:1-8.

23. Farmer SG, Ensor JE, Burch RM. Evidence that cultured airway smooth muscle cells contain bradykinin B_2 and B_3 receptors. Am. J. Respir. Cell Mol. Biol. 1991;4:273-77.

AAS 38/III
Recent Progress on Kinins
© 1992 Birkhäuser Verlag Basel

KALLIKREIN-LIKE ACTIVITY IN THE BRONCHOALVEOLAR LAVAGE (BAL) FLUID OF SENSITIZED, CHALLENGED GUINEA-PIGS

R.L. Featherstone, J.E. Parry & M.K. Church

Clinical Pharmacology, Southampton General Hospital, England. SO9 4XY

SUMMARY: Sensitization and challenge of guinea-pig airways with aerosols of ovalbumin solution administered through an endotracheal tube leads to a significant increase in kallikrein-like enzyme activity in bronchoalveolar lavage (BAL) fluid. This increase is present up to six hours after challenge and may be related to changes in airways resistance that persist for similar periods.

INTRODUCTION

Tissue kallikrein is found in increased amounts in BAL fluid obtained from asthmatic patients (1) and, it is hypothesised, contributes to the allergic inflammatory response by forming kinins from kininogen precursors.

To develop an animal model for studying the kallikrein/kinin system in allergic airways disease we have investigated the appearance of kallikrein activity in BAL fluid of guinea-pigs sensitized and challenged with ovalbumin by different routes. For the method producing the greatest increase in kallikrein levels, a time course of the response was determined and also changes in pulmonary resistance measured.

METHODS

Animals were sensitized on two occasions (day 0 and day 7) by three minutes exposure to an aerosol of 1% ovalbumin solution (w/v) generated by a nebuliser. Exposure to the aerosol took place either whilst under Metofane anaesthesia with an endotracheal tube

(ETT) inserted or whilst conscious and freely breathing. An alternative sensitization procedure was intraperitoneal (i.p.) injection of ovalbumin, 5mg on day 0 and 10 mg on day 2.

Challenge of sensitized animals took place on day 14 by exposure for 5 minutes to aerosols of either 2% ovalbumin or saline. Exposure was either via ETT or whilst conscious and freely breathing. All animals received mepyramine (14mg/kg i.p.) 30 minutes prior to challenge.

Fifteen minutes after challenge animals were killed by anaesthetic overdose. Following this they were exsanguinated and a cannula placed into the trachea. Bronchoalveolar lavage was then performed with 3 x 5ml of normal saline. The BAL fluid collected from the challenged animals was first spun at 100 x g for 10 minutes. The supernatants were then stored at -20°C until assay. Immediately prior to assay the BAL supernatants were concentrated 4-5 fold in an Amicon ultrafiltration unit with 10kDa cut off. For assay, 90µl of supernatant concentrate were added to 100µl of buffer (0.2M TRIS-HCl/0.2mM EDTA) and the reaction started by addition of 10µl of the artificial substrate S2266 (0.1mM). Change in optical density at 410nm over 30 minutes was measured in a microtiter plate reader and compared to that produced by human urinary kallikrein (hUK) standards. Results are expressed as ng hUK/ml BAL.

Preliminary experiments having revealed that sensitization and challenge via ETT gave the best response in terms of kallikrein released, the time course of this reaction was studied. Animals sensitized and challenged by this method were killed at various time points up to six hours after challenge for measurement of kallikrein in BAL. Another group of animals were assessed for changes in pulmonary resistance over this time period using a whole-body plethysmographic technique described previously (2). Results were analysed using Student's 't' test.

RESULTS

Only animals sensitized by the endotracheal route show a significant increase in levels of kallikrein-like activity 15 minutes after ovalbumin challenge compared with naive control animals (Figure 1), although at this time point the increase is not significant compared to that in sensitized animals challenged with saline through ETT.

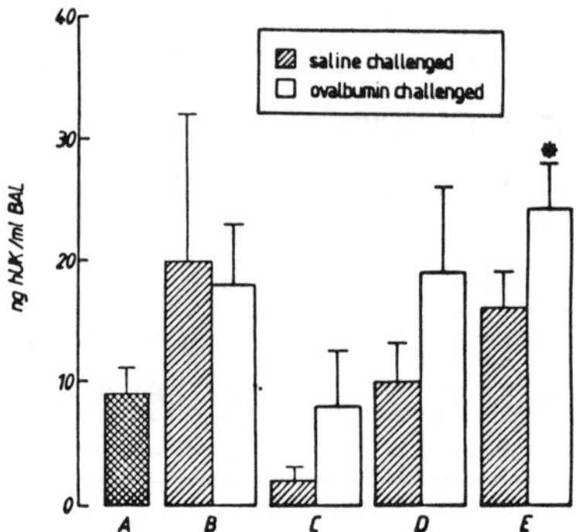

Figure 1. Levels of kallikrein-like activity in BAL fluid from: A, naive animals; B, aerosol sensitized and challenged (free breathing) animals; C, I.p. sensitized, aerosol challenged (free breathing) animals; D, I.p. sensitized, aerosol challenged (via ETT) animals; E, aerosol sensitized and challenged (via ETT) animals. * = significantly greater than naive, unchallenged controls, $p < 0.05$. Each point is the mean of 5-10 determinations ± SEM.

By 1 hour after challenge levels of kallikrein activity in BAL from animals sensitized and challenged with ovalbumin via ETT are significantly raised (53.2 ± 11.9 ng hUK/ml BAL, n = 5, p<0.05) compared to saline challenged controls (Figure 2) and are still so at 6 hours after challenge (45.0 ± 13.0, n = 8, p< 0.05).

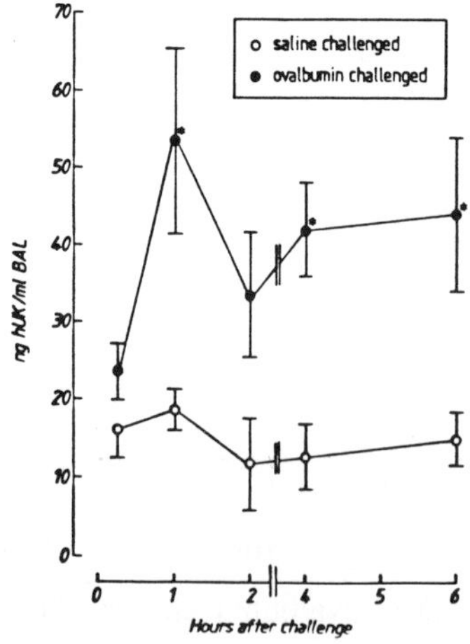

Figure 2. Levels of kallikrein-like activity in BAL fluid at different time points after challenge from ETT sensitized animals challenged with ovalbumin or saline. * = significantly greater than saline challenged control, $p < 0.05$. Each point is the mean of 5-10 determinations ± SEM.

Animals sensitized by the endotracheal route respond to challenge with an initial increase in airways resistance of between 400 and 500%, this rapidly decreases to between 100 and 175% by 45 minutes after challenge and is maintained at this level for up to 6 hours (Figure 3).

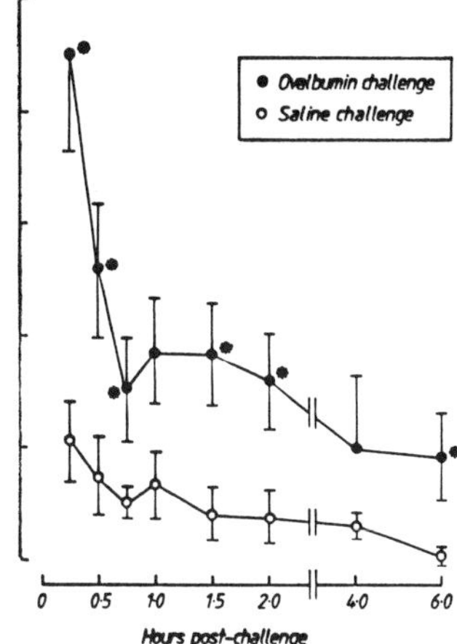

Figure 3. Changes in airways resistance with time after challenge in ovalbumin sensitized guinea-pigs challenged with ovalbumin or saline via ETT. * = significantly greater than saline control, p < 0.05. Each point is the mean of 5-10 determinations ± SEM.

DISCUSSION

The data presented here show a significant increase in kallikrein-like activity in BAL fluid from guinea-pig airways following sensitization and challenge with ovalbumin only when using the endotracheal route, which delivers antigen to the lower airways with maximum efficiency.

Animals sensitized endotracheally respond to antigen challenge with an initial, large increase in airways resistance which rapidly fades to a more maintained level lasting at least 6 hours This increase is parallel to that found in kallikrein-like activity in BAL fluid. It is tempting to speculate that the kallikrein-like activity seen contributes to the maintained rise in airways resistance, probably by the generation of kinins. Further studies are planned using protease inhibitors, kinin antagonists and radioimmunoassays for kinins to validate this hypothesis.

ACKNOWLEDGEMENTS: This work was supported by a grant from the Ferring Peptide Research Partnership.

REFERENCES

1. Christiansen SC, Proud D, Cochrane CG. Detection of tissue kallikrein in the bronchoalveolar lavage fluid of asthmatic subjects. J. Clin. Invest. 1987; 79: 188-197.

2. Varley JG, Heath JR, Bacon R, Holgate ST, Church MK. Reproducibility of a new automated guinea-pig body plethysmograph. Am. Rev. Respir. Dis., 1991; 143(suppl): A353.

INCREASE IN THE KININ LEVELS IN THE BRONCHIAL WASHINGS AFTER INTRAVENOUS INJECTION OF LEUKO-TRIENE C_4 IN GUINEA PIGS

M. Majima, H.Y. Jin, M. Katori and *N. Sunahara

Department of Pharmacology, Kitasato University School of Medicine, Sagamihara, Kanagawa 228, Japan and *Dainippon Pharmaceutical Co., Enoki, Osaka 661, Japan

SUMMARY: In guinea pig plasma, bradykinin (BK) was degraded mainly to des-Arg^1-BK by an aminopeptidase-like enzyme, which was inhibited by 2-mercaptoethanol. Besides this degradation, BK was also hydrolyzed by kininase I and kininase II from C-terminal end to des-Arg^9-BK, des-Phe^8-Arg^9-BK and Arg-Pro-Pro-Gly-Phe ([1-5] BK). The formation of des-9-BK was strongly blocked by DL-2-mercaptomethyl-3-guanidinoethylthiopropanoic acid (MGPA) and that of des-8,9-BK and [1-5] BK was inhibited by captopril. When guinea pigs were pretreated with a cocktail of 2-mercaptoethanol, MGPA and captopril, intravenous administration of leukotriene (LT) C4 (10 nmol/kg) caused an increase in the levels of free kinin in the bronchial washings of guinea pigs. This increase was accompanied with the increase in glandular-kallikrein activity, which could be inhibited by aprotinin. As BK is reported to induce both bronchoconstriction and bronchial secretion, the increased free BK induced by LTC4 might enhance the effect of LTC4.

INTRODUCTION

Peptide leukotrienes (P-LTs), released from lungs of sensitized guinea pig, were reported to be potent bronchoconstrictors (1, 2). In addition to bronchoconstriction, LTs have been reported to increase bronchial secretion (3, 4). The increased levels of glandular-kallikrein, a kinin-generating enzyme, was reported to increase in nasal or bronchial lavage fluid after provocation with allergen of the nasal (5) or bronchial mucosa (6) of atopic subjects. We have previously reported that intravenous injection of LTC4 increased the activity of glandular-kallikrein in the bronchial washings of guinea pigs via the formation of thromboxane and the release of acetylcholine (7). In the present experiment, we describe that intravenous injection of LTC4 to guinea pigs increased the levels of endogenous free kinin in bronchial washings, which were accompanied with increased secretion of glandular-kallikrein.

MATERIALS AND METHODS

Male Hartley guinea pings (specific pathogen free, weighing 450-550g) were anaesthetized with sodium pentbarbital (50 mg/kg, s.c.). The right jugular vein was cannulated with a polyethylene cannula and LTC4 (3-30 nmol/ml saline /kg) was administered as a single bolus injection through the jugular vein. Twenty minutes later, each animal was killed by exsanguination and the lung and trachea with cannula were isolated. Half of the peripheral portion of the lung was removed by cutting with small scissors. Physiological saline (2.4 ml/animal), injected into the trachea, was able to flow out through the cut surface of the lung.

The activity of glandular-kallikrein in the bronchial washings was determined by kinin releasing activity from low molecular weight kininogen: Three hundred µl of bronchial washing were incubated with 200 µl of the solution of low molecular weight kininogen (0.5 mg/ml in 0.1M Tris-HCl buffer, pH 8.0) at 37°C for 60 min. After addition of 100 µl of a 20% solution of trichloroacetic acid to the reaction mixture, the amount of bradykinin released was measured by a bradykinin enzyme immunoassay kit (Markit-A, Dainippon Pharmaceutical Co., Osaka, Japan) after neutralization with buffer B in the kits .

Degradation pathway of bradykinin by guinea pig plasma and the inhibition of kininases were examined *in vitro* in the following way: Citrated plasma from guinea pigs (10 µl) was incubated in the presence or absence of a solution of kininase inhibitors (340 µl) with 40 nmol of bradykinin (in 100 µl of physilogical saline) and bradykinin degradation products were analyzed by reversed-phase HPLC, according to the method reported previously (8). The kininase inhibitors used were captopril for kininase II, DL-2-mercaptomethyl-3-

guanidinoethylthio-propanoic acid (MGPA) for kininase I and 2-mercapto-ethanol for aminopeptidase inhibitor.

To determine the levels of free kinin in bronchial washings *in vivo*, a cocktail of captopril (10 mg/kg), MGPA (3 mg/kg) and 2-mercaptoethanol (30 mg/kg) was administered intravenously 10 min before LTC$_4$ stimulation (3 nmol/kg, i.v.). After sacrifice of animals by exsanguination, the bronchial washings from isolated lungs were directly collected into the plastic tubes containing absolute ethanol. After heating at 70°C for 10 min, ethanol solution was centrifuged (1500g, 4°C for 15 min). The supernatant was dried under reduced pressure, and the residue was dissolved with acidified water (pH 3) and applied to Sep Pak C$_{18}$ column. Kinin fractions eluted with acetonitril was dried and the residue was applied reversed-phase HPLC (9). Bradykinin and kallidin fractions were pooled and evaporated. The residue of HPLC elute was dissolved with assay buffer A of the bradykinin enzyme immunoassay kits and the amounts of bradykinin were determined.

Student's t-test was used to evaluate the significance of differences. When variances were heterogeneous, statistical analysis was performed by the Aspine-Welch method or by the Wilcoxon's rank sum test. A value of p of less than 0.05 was regarded as significant.

RESULTS

Effects of 2-mercaptoethanol, DL-2-mercaptomethyl-3-guanidinoethylthio-propanoic acid (MGPA) and captopril on bradykinin (BK) degradation by guinea pig plasma

When bradykinin was incubated *in vitro* with guinea pig plasma, des-9-BK, des-8,9-BK, [1-5] BK and des-1-BK was generated (Figure 1). Among these, des-1-BK was a major degradation product. The generation of des-1-BK was completely inhibited by 2-mercaptoethanol (100 µM) without effects on the generation of des-9-BK, des-8,9-BK and [1-5] BK, whereas MGPA (10 µM) and captopril (30 µM) did not show the inhibitory effects. The decrease in the residual levels of BK by degradation was inhibited by 2-mercaptoethanol by 46%. Although the inhibitory effects of MGPA and captopril on the BK

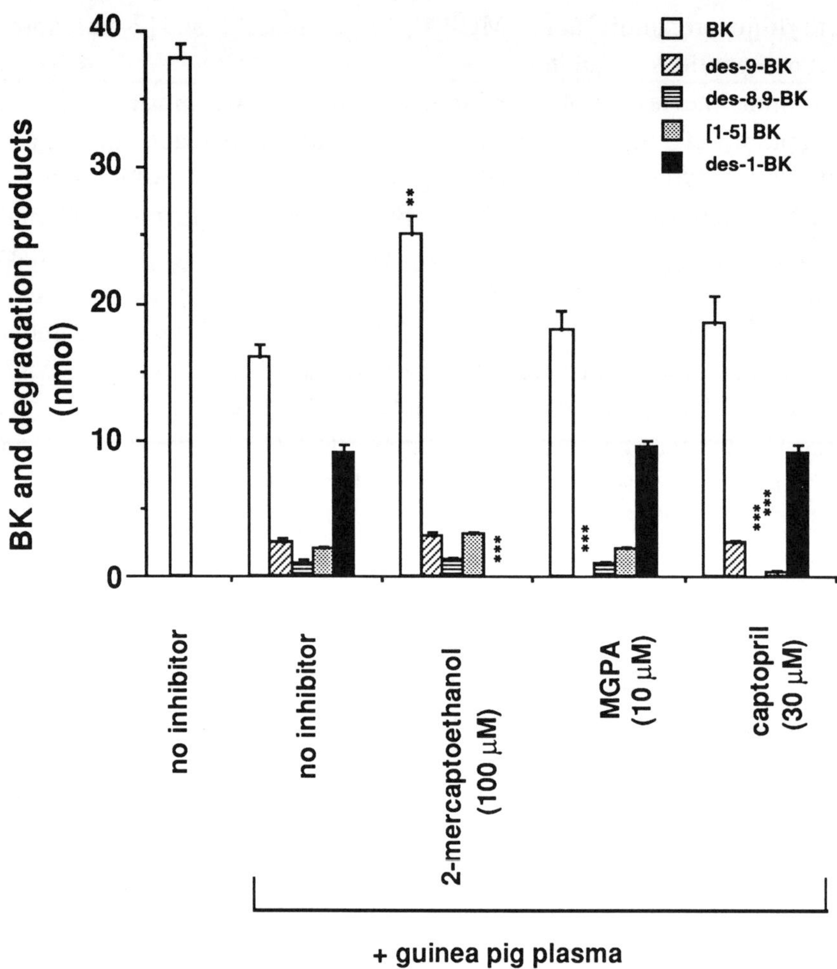

Figure 1. Effects of 2-mercaptoethanol, MGPA and captopril on the bradykinin (BK) degradation by guinea pig plasma.

Each point represents the mean value (± s.e. mean) from four experiments and compared with the value for no inhibitor in guinea pig plasma. **p<0.01, ***p<0.001.

degradation was not significant, the formation of des-9-BK was completely inhibited by MGPA without changes in the levels of other metabolites. Captopril inhibited the formation of both des-8,9-BK and [1-5] BK from BK strongly, but the levels of des-9-BK was not affected.

Increase in the free kinin levels in the bronchial washings after leukotriene C4

No kinin was able to be detected in the bronchial washings without treatment with combination of the three kininase inhibitors. The pretreatment of animals with a cocktail of kininase inhibitors resulted in the detection of free kinin in the bronchial washings (Figure 2). In saline-injected groups, the levels of free kinin was 9.2 ± 0.8 pmol/animal. LTC_4 (3 nmol/kg, i.v.) caused the significant increase in the levels of free kinin in the washings to 34.5 ± 6.0 pmol/animal. This increased levels of free kinin was accompanied with the increase in glandular-kallikrein activity. The glandular-kallikrein activity in bronchial washings after injection of saline was 22 ± 6 pmol/min/ml washing (n=6), which was increased to 8.4 fold (185 ± 16 pmol/min/ml washing, n=4), after injection of LTC_4 (3 nmol/kg, i.v.). The kinin-generating activities in the bronchial washings were attributable to kallikrein, since the addition of

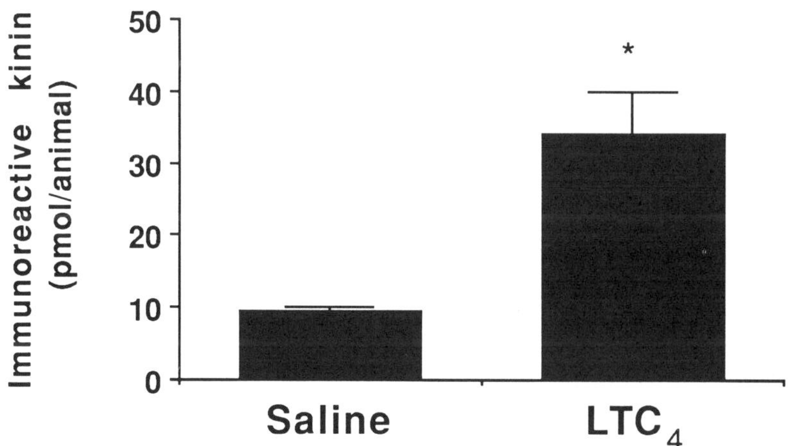

Figure 2. Immunoreactive kinin levels in bronchial washings.
Each point represents the mean value ($\pm$ s.e. mean) from four experiments and compared with the value for saline control. *p<0.05.

aprotinin (5000KIU/ml) to reaction mixtures caused marked inhibition of the kinin-generating activity to 18 ± 2 pmol/min/ml washing (n=4).

DISCUSSION

Asthmatic patients are affected by not only bronchoconstriction, but also by increased bronchial secretion. Besides bronchoconstriction, LTC4 increase the secretion of mucus glands in the air way (3). In our previous study, intravenous injection of LTC4 was found to induce an increase in the level of glandular-kallikrein in bronchial washings in guinea pigs. This increase was blocked by pretreatment of animal with a thromboxane synthetase inhibitor and atropine, suggesting that the increased levels of glandular-kallikrein induced by LTC4 were mediated by thromboxane formation and acetylcholine release (7). As the increase in glandular-kallikrein in bronchial washings was inhibited by pretreatment with a bradykinin antagonist (7), it was plausible that kinin generated accelerated the secretion of glandular-kallikrein *in vivo*. Degradation of bradykinin in human and rat plasma was conducted by kininase I and kininase II and degradation products were des-9-BK, des-8,9-BK and [1-5] BK (8). In contrast, in guinea pigs, bradykinin was mainly degraded by aminopeptidase-like enzymes. Although kininase I and kininase II activities were identified by the inhibitory effects of MGPA and captopril on the the formations of des-9-BK, des-8,9-BK and [1-5] BK, the contributions of these kininases were less than that of aminopeptidase-like enzymes, as seen in Figure 1. Bradykinin was hardly able to be detected in the bronchial washings in the absence of kininase inhibitors, but when the animals were pretreated with a cocktail of 2-mercaptoethanol, MGPA and captopril, the free kinin levels in the air way were increased after the LTC4 injection, acompanied with the increased secretion of glandular-kallikrein. As bradykinin administration could introduce bronchial secretion and bronchoconstruction (2), the increased formation of kinin in the airway might be involved in the pathogenesis of some kinds of airway disease, such as bronchial asthma, airway infection, stimulation by cold and dry breathing air, via the secretion of glandular-kallikrein. Our present observations also support that kinin formation stimulated by LTC4 results in a positive feedback loop that controls the release of the glandular-kallikrein into bronchial secretions.

ACKNOWLEDGMENTS

This work was supported in part by a Grant-in-Aid for the Encouragement of Young Scientists (no. 63772021) from the Ministry of Education, Science and Culture of Japan. The authors wish to thank Mr. Hiroshi Ishikawa for preparing a fine experimental system, and Ms. Harue Mihara, Mr. Osamu Yoshida, Ms. Michiko Takahara and Ms. Maki Saito for their skillful technical assistance.

REFERENCES

1. PIPER, P. J., SAMHOUN, N. M. The mechanism of action of leukotriene C_4 and D_4 in guinea-pig isolated lung and parenchymal strips of guinea pig, rabbit and rat. Prostaglandins, 1981; 21:793-803.
2. UENO, A., TANAKA, K., HIROSE, R., SHISHIDO, M., KATORI, M. Possible involvement of thromboxane in hypertensive and bronchocontrictive effect of leukotriene C_4. Adv. Prostaglandins Thromboxane Leukotriene Res. 1983; 12:139-144.
3. MAROM, Z., SHELHAMER, H., BACH, M. K., MORTON, D. R., KALINER, M. Slow-reacting substances, leukotriene C_4 and D_4, increase the release of mucus from human airways in vitro. Am. Rev. Respir. Dis. 1982; 126:449-451.
4. COLES, S.J., NEILL, K.H., REID, L.M., AUSTEN, K.F., NII, Y., COREY, E.J., LEWIS, R.A. Effects of leukotriene C_4 and D_4 on glycoprotein and lysozyme secretion by human bronchial mucosa. Prostaglandins, 1983; 25:155-170.
5. PROUD, D., TOGIAS, A., NACLERIO, R. M., CRUSH, S. A., NORMAN, P. S., LICHTENSTEIN, L. M. Kinins are generated in vivo following nasal airway challenge of allergic individuals with allergen. J. Clin. Invest. 1983; 72:1678-1685.
6. CHRISTIANSEN, S. C., PROUD, D., COCHRANE, C. G. Detection of tissue kallikrein in the bronchoalveolar lavage fluid of asthmatic subjects. J. Clin. Invest. 1987; 79:188-197.
7. JIN, H. Y., KATORI, M., MAJIMA, M., SUNAHARA, N. Increased secretion of glandular-kallikrein in the bronchial washings induced by intravenous injection of leukotriene C4 in guinea pigs. British J. Pharmacol. 1992; 105: 632-638.

8. MAJIMA, M., UENO, A., SUNAHARA, N., KATORI, M. Measurement of des-Phe^8-Arg^9-bradykinin by enzyme-immunoassay a useful parameter of plasma kinin release. 1988; In Kinin V, Part B, pp.331-335 Adv. Exp. Med. Biol., New York, Plenum Press.
9. MAJIMA, M., KATORI, M., HANAZUKA, M., MIZOGAMI, S., NAKANO, T., NAKAO, Y., MIKAMI, R., URYU, H., OKAMURA, R., MOHSIN, S.S.J., OH-ISHI, S. Suppression of rat deoxycorticosterone-salt hypertension by kallikrein-kinin system. Hypertension, 1991; 17:806-813.

BRADYKININ AND OTHER INFLAMMATORY MEDIATORS IN BAL-FLUID FROM
PATIENTS WITH ACTIVE PULMONARY INFLAMMATION

C. R. Baumgarten, B. Lehmkuhl, **R. Henning, T. Brunnee,
*P. Dorow, **W. Schilling, and G. Kunkel

Dept. of Clinical Immunology and Asthma Poliklinik, Free University, Berlin,
*DRK-Hospital, Berlin-Wedding; **FLT Berlin-Buch, Germany

SUMMARY: We evaluated the levels of bradykinin, albumin, TAME-esterase activity, histamine, PGD2 and LTC4 in bronchoalveolar lavage fluid from asthmatics and from patients with pneumonia, sarcoidosis, fibrosis, and chronic bronchitis. Compared with the results of healthy volunteers and atopic asymptomatic asthmatics the bradykinin levels and TAME-esterase activity were significantly elevated. In all other groups, histamine was additionally elevated in asymptomatic asthmatics, whereas albumin was elevated in symptomatic asthmatics and fibrosis patients, and decreased in chronic bronchitis and pneumonia patients. Following local intrabronchial allergen challenge of mild grass pollen asthmatics out of season bradykinin levels increased significantly, correlated with albumin, histamine and TAME-esterase activity. In contrast to the increased mediator concentrations in the early phase reaction there was no change of BAL cells in asthmatics compared to baseline and healthy volunteers . The presence of bradykinin in the bronchoalveolar space of patients with active pulmonary inflammations and bradykinin generation in asthmatics as a result of intrabronchial allergen challenge provides strong evidence that kinins are involved in inflammatory disorders of the lower airways.

INTRODUCTION

Bradykinin (Bk) may have a role as a mediator in allergic and other inflammatory respiratory diseases in man. The known pharmacological properties of kinins suggest that these peptides may contribute to the inflammatory response in pulmonary inflammation. Support for a potential role in asthma is provided by the observations that provocation of the lower airways of asthmatics with Bk results in bronchoconstriction (1), and Bk administered to the nasal mucosa induces the symptoms of rhinitis (2). The presence of kinins in the upper airways and the demonstrated generation during intranasal allergen challenge of patients with allergic rhinitis (3, 4) also supports a role in the lower airways in analogy.

The present study was undertaken to characterize the mediators in bronchoalveolar lavage fluid of patients with different pulmonary inflammation and the mediator and cellular changes occurring immediately following bronchoscopic segmental challenge of the peripheral airways with grass antigen in allergic, mildly asthmatic subjects.

METHODS

The study protocols were approved by the local ethic committee and all subjects gave written consent. 67 patients with active pulmonary inflammation (pneumonia, sarcoidosis, fibrosis, chronic bronchitis, symptomatic and asymptomatic asthma) were submitted to a bronchoalveolar lavage. Lavage samples were obtained using the technic of segmental airway lavage. Intramuscular atropine and nebulised 4 % lidocaine were given for general preparation. The fiberoptic bronchoscop was inserted into a subsegment bronchus of the right middle lobe. Lavage was performed by injecting five 20-ml aliquots of normal saline, prewarmed to 37 C, with immediate aspiration after instillation of each aliquot. In each individual, mediators were present in similar concentrations in each of the five 20-ml BAL aliquots. Therefore, these values were averaged to obtain the values used in all subsequent calculations. Antigen challenge was performed in 8 asymptomatic asthmatics with mild seasonal asthma. The allergen dose for the intrabrochial provocation was choosen to be 10% of the individual threshold dose inducing a 20% drop of FEV1 and a 100% rise of Raw following bronchial provocation. Antigen challenge was performed by instilling 5 ml grass antigen diluted in normal saline and performing BAL after 6-8 min. A total of 8 asthmatic subjects were challenged with 5 ml of 100 to 2500 BU of grasspollen extract. The nonallergic controls obtained 10000 BU of grasspollen extract. BAL fluid was collected before, immediately after local response whithin 6-8 min following local allergen challenge. All asthmatic patients had a mild seasonal grasspollen asthma; diagnosis was confirmed by history, skin prick test and RAST. The history of seasonal asthma lasted from 2 to 8 years. They had a functional evidence of reversible obstructive airway disease by acetylcholin and specific grasspollen allergen provocation test and normal pulmonary function before bronchoscopy without any medication. No drugs were allowed during the study.

MEASUREMENT OF MEDIATORS:
Kinin analysis was done in a 0.8 ml sample of lavage fluid in which an inhibitor cocktail was added immediately following recovery containing lysozyme (0,1 %), phenanthrolene (5 mM), benzamidine (5 mM) and EDTA (200 mM) at pH 7.4. The minimum detectable level of the specific radio immunoassay is 20 pg/ml bradykinin (3). The crossreactivity is < 2% to highly purified human kininogen and < 1% to des-arg9-bradykinin.

N- -tosyl-L-arginine methyl ester (TAME) esterase activity was determined by the method of Imanari et al. (5).

Albumin was detected by double-antibody RIA using sheep anti-albumine for substrate binding and goat anti-sheep IgG for detection. The minimum detectable dosis was less than .1ng/ml, no crossreaction with bovine albumin to 10 mg /ml was seen (4).

For histamine analysis 20 % of 10 mg/ml $HClO_4$ was added for protein removal at the end of trial prior to storing at -70 degrees Celsius. After thawing and centrifugation the supernatant was analysed with an automatic fluorimeter with a sensitivity of 1.8 pmol/ml using the Siraganian method (6). PGD2 and LTC4 were assayed with a commercial RIA assay (Amersham, Braunschweig, Germany) using dextran-coated charcoal as a separation technique. Lavages were extracted for high performance liquid chromatography (HPLC) using C18 Sep-Pak cartridges (Millipore, Bedford, MA), which were activated with methanol and washed with water before use. Leukotrienes were eluted with methanol as previously described (7).

Statistical analysis: For independent variables (allergics versus controls) the two-tailed U-test (Wilcoxon, Mann-Whitney), for dependent variables (pre/post-challenge), the two-tailed matched pairs signed rank test (Wilcoxon) and for correlation between different parameters in the same collective, the Spearman rank correlation coefficient were applied. Prior to calculation baseline values were subtracted.

RESULTS

Comparing baseline concentrations of kinins between asymptomatic asthmatics, patients with active pulmonary inflammation and healthy subjects, no significant differences were seen between controls and asymptomatic asthmatics, whereas active pulmonary inflammation was associated with elevated amounts of kinins (Fig. 1), especially patients with fibrosis and chronic bronchitis demonstrated high amounts of kinins. Concomitant with the elevated kinin levels high TAME-esterase activity was detected (Fig. 2). There was a significant correlation between the levels of kinins and TAME-esterase activity in each individual (p< 0.001). Histamine was significantly elevated in symptomatic asthmatics, in the other investigated patients with pulmonary inflammation and in addition also in asymptomatic asthmatics (Fig. 3). In contrast to the kinin findings elevated amounts of albumin were only found in symptomatic asthmatics and fibrosis patients with borderline significance. The other groups were found to be similar to controls with the exception of patients with chronic bronchitis and pneumonia. In both groups significant lower levels were found (Fig. 4).

Following local intrabronchial allergen challenge, edema of the bronchial mucosa and a reduction of bronchial diameter were observed through the bronchoscope within 2 to 3 min in all allergic subjects. The control levels and acute changes in the mediators and markers of

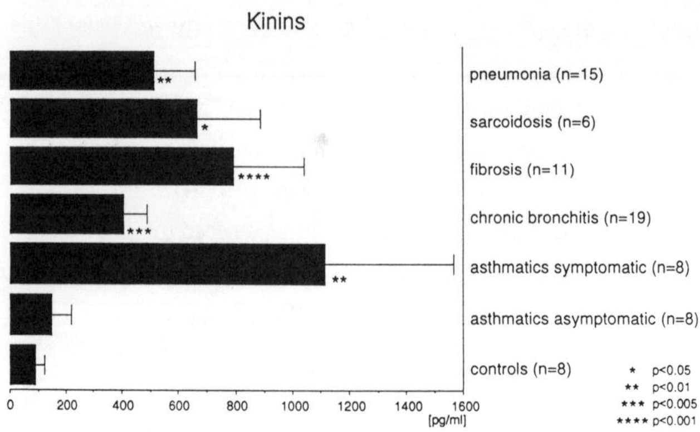

Figure 1

Baseline concentrations of immunoreactive kinins in BAL-fluids of healthy nonatopic subjects and patients with different active pulmonary inflammation (mean $\pm$ SEM).

p   patients vs. controls

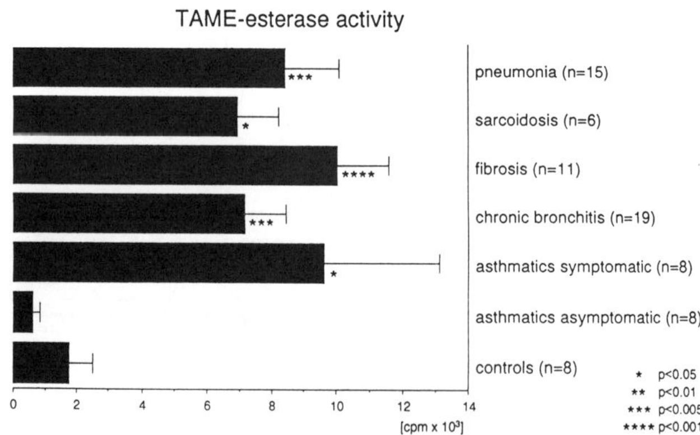

Figure 2

Baseline concentrations of TAME-esterase activity in BAL-fluids of healthy nonatopic subjects and patients with different active pulmonary inflammation (mean $\pm$ SEM).

p   patients vs. controls

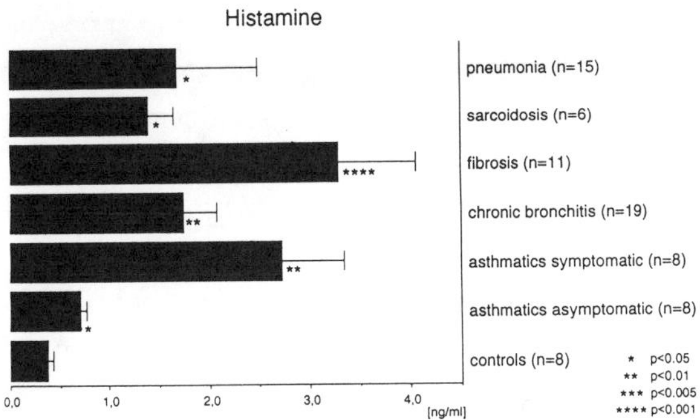

Figure 3

Baseline concentrations of histamine in BAL-fluids of healthy nonatopic subjects and patients with different active pulmonary inflammation (mean ± SEM). Compared to controls the histamine levels were significantly higher in all other investigated groups. p patients vs. controls

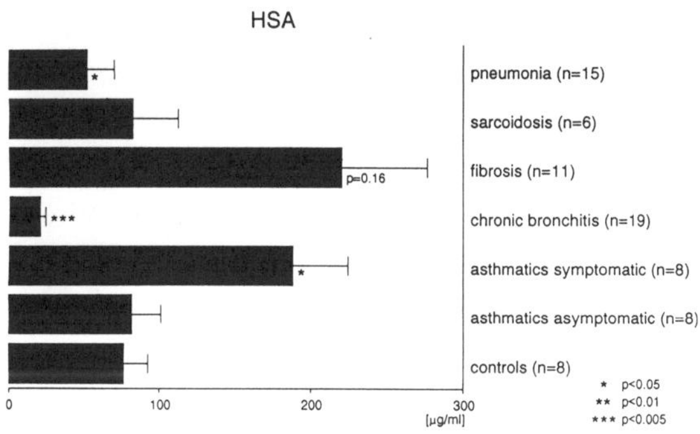

Figure 4

Baseline concentrations of immunoreactive human serum albumin (HSA) in BAL-fluids of healthy nonatopic subjects and patients with different active pulmonary inflammation (mean ± SEM). In contrast to the higher kinin levels the albumin concentrations were lower in patients with chronic bronchitis and pneumonia compared to controls. p patients vs. controls

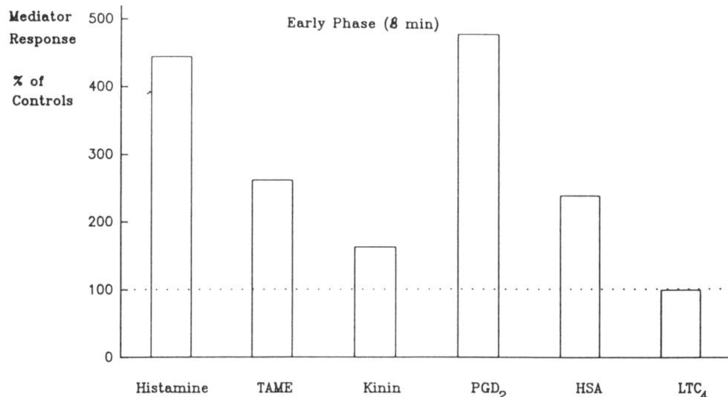

Figure 5

Mediator concentrations in BAL-fluids of asymptomatic allergic patients with mild grasspollen asthma immediately following local intrabronchial antigen challenge in percentage of the responses of healthy nonatopic subjects.

airway permeability in normal and asthmatic subjects after antigen challenge are shown in fig. 5. Despite the 10-fold higher dose of antigen used to challenge normal versus asthmatic subjects, no significant changes in mediators and albumin occured. Although the mediator response of asthmatic subjects were quite variable, mean levels of kinins, albumin and TAME-esterase activity increased significant. There was no change in the levels of LTC4 but a marginal increase in histamine and PGD2 immediately after antigen challenge.

The fluid recoveries from BAL pre- or postantigen challenge were the same. Differential cell counts were determined, and the absolute number of each cell type was calculated. The composition of the cells recovered was not altered by antigen challenge in either the normal or asthmatic group.

DISCUSSION

The symptoms of pulmonary diseases including asthma involve chronic inflammation of the lower airways with a bronchial inflammation as a major characteristic of the asthmatic disease. Recent studies using BAL to sample mediators following antigen challenge of allergic subjects with and without asthma have demonstrated the release of multiple mediators including histamine, PGD2 and tryptase, suggesting a central role for mast cells. Our examination of inflammatory mediators in different kinds of pulmonary diseases has confirmed these observations in asthmatics and extended these to include patients with other pulmonary diseases. Not surprisingly to the demonstrated high kinin levels in patients with ongoing pulmonary inflammation was the concomitant elevated TAME-esterase activity, since the ability to cleave this synthetic substrate would be expected in kinin-liberating enzymes (8). It was unexpected, however, that there was no evidence of increased airway permeability, assessed by albumin, parallel to the kinin concentration. There is no explanation for lower levels of albumin in patients with chronic bronchitis and pneumonia.

Exposure to antigen in sensitive individuals demonstrates the immediate release of histamine and PGD2 as a result of mast cell degranulation. This release is independent of cellular changes which were seen in chronic antigenic exposure and the late phase reaction(9). It was also expected, that there was evidence of increased airway permeability following antigen challenge, assessed by albumin, as has been found in the upper airways (4).

Although there were high concentrations in kinins, histamine and TAME-esterase activity found in fibrosis patients the albumin concentration, as a marker of increased airway permeability , was unexpected low. In addition, since fibrosis is thought to be a final stage of inflammation, the high concentrations of inflammatory mediators disprove this. The high bradykinin concentrations in BAL fluid of fibrosis patients may explain the increased generation of collagen (10), resulting in the characteristic extensive interstitial proliferation of fibrous tissue.

In Summary, this study demonstrates elevated kinin levels in BAL fluid of patients with different kind of inflammatory diseases including asthma. The demonstrated kinin generation and its interrelationship with additional preformed and newly generated mediators strongly suggests that the kinins are involved in inflammatory disorders of the lower airways in analogy with the upper airways.

REFERENCES

1. Fuller RW, Dixon CMS, Cuss FMC, Barnes PJ. Bradykinin-induced bronchoconstriction in humans: mode of action. Am Rev Respir Dis 1987; 135:176-80.

2. Brunnee T, Nigam S, Kunkel G, Baumgarten CR. Nasal challenge studies with bradykinin: influence upon mediator generation. Clin and Experim Allergy 1991; Vol 21: 425-431.

3. Proud D, Togias AG, Naclerio RM, Crush SA, Norman PS, Lichtenstein LM. Kinins are generated in vivo following nasal airway challenge of allergic individuals with allergen. J Clin Invest 1983; 72:1678-85.

4. Baumgarten CR, Togias AG, Naclerio RM, Lichtenstein LM, Norman PS, Proud D. Influx of kininogens into nasal secretions after antigen challenge of allergic individuals. J Clin Invest 1985; 76:191-7.

5. Imanari T, Tokio K, Yoshida H, Yates K, Pierce JV, Pisano JJ. Radiochemical assays for human urinary, salivary and plasma kallikreins. In: Pisano JJ, Austen KF, eds. Chemistry and biology of the kallikrein-kinin system in health and disease. Washington, DC: Department of Health, Education and Disease (NIH publication no. 76-791), 1976: 205-13.

6. Siranganian R. An automated continuous flow system for the extraction and fluorimetric analysis of histamine. Anal Biochem 1974 57:283-94.

7. Peters SP, MacGlashan DW Jr, Schulman ES, et al. Arachidonic acid metabolism in purified human lung mast cells. J Immunol 1984; 132:1972-9.

8. Baumgarten CR, Nichols RC, Naclerio RM, Proud D. Plasma kallikrein during experimentally induced allergic rhinitis: role in kinin formation and contribution to TAME-esterase activity in nasal secretions. J Immunol 1986; 137:977-82.

9. Liu MC, Hubbard WC, Proud D, Stealey BA, Galli SJ, Kagey-Sobotka A, Bleecker ER, Lichtenstein LM. Immediate and late inflammatory responses to ragweed antigen challenge of the peripheral airways in allergic asthmatics. Am Rev Respir Dis 1991; 144:51-58.

10. Regoli, D. Kinins. Br Med Bull 1987; 43:270-84.

AAS 38/III
Recent Progress on Kinins
© 1992 Birkhäuser Verlag Basel

COUGH INDUCED BY ACE-INHIBITORS. A KININ RELATED PHENOMENON?

A. Overlack, B. Müller, L. Schmidt, M.L. Scheid, M. Müller, K.O. Stumpe

Med. Univ. Poliklinik, Wilhelmstr. 35, 5300 Bonn 1, Germany

SUMMARY: Cough induced by ACE-inhibitors may be related to bronchial hyperreactivity and/or to an accumulation of kinins. In a placebo-controlled, double-blind randomized study in asthmatic and hypertensive patients lung function and bronchial reactivity to histamine and bradykinin remained unaltered although in hypertensive patients with cough, reactivity to histamine tended to be more pronounced and bronchial hyperreactivity to be more frequent than in those without cough. The findings do not support a major role of kinins in ACE inhibitor-induced cough.

INTRODUCTION

Why ACE-inhibitors cause cough is not completely understood. ACE inhibitors increase the cough response to inhaled capsaicin, but this response occurs in patients with cough as well as in normal volunteers (1,2,3) and the evidence that the cough reflex is altered is conflicting (3,4). Preexisting or during ACE-inhibition evolving bronchial hyperreactivity may contribute to the development of cough (4,5,6). Cough following ACE-inhibition may also be a kinin-related phenomenon. ACE is identical to kininase II, which is partly responsible for the degradation of kinins (7). Therefore, ACE-inhibition may lead to an accumulation of kinins. Kinins participate in inflammatory processes (8) and inhalation of bradykinin is followed by bronchoconstriction in asthmatic patients (9).

PATIENTS AND METHODS

STUDY 1: Eight patients with bronchial asthma were included in the study. They had a history of bronchial asthma for 3 to 20 years. Their mean age was 25 ± 2 years. In a previous clinical workup they had been found to have perennial allergic asthma with a marked bronchial hyperreactivity when challenged with histamine. Their disease was clinically mild and stable and well controlled by inhaled salbutamol or fenoterol alone. The betasympathomimetic drug was withheld for at least 8 hours prior to the inhalation challenges. For inclusion in the study, FEV_1 had to be $\geq 80\%$ of predicted.

The study was conducted in a randomised, double-blind, cross-over fashion. It consisted of two parts: acute and short-term. First, the patients received a single dose of 25 mg captopril or placebo and histamine challenge was performed after 60 min. Two days later, bradykinin challenge was performed 60 min after the same drug (captopril or placebo). One week later, the other drug was given and the patient was challenged with histamine and bradykinin in the same way. The second part of the study consisted of a short-term treatment with captopril (2 x 25 mg/d) and matching placebo for two weeks each. Histamine and bradykinin challenges were performed on day 12 and day 14, respectively, 60 min after the last administration of captopril or placebo. To minimize circadian influences on lung function and bronchial reactivity, the patients were studied at the same time in the morning between 8 and 10 a.m.. Captopril and matching placebo were generously supplied by Squibb van Heyden GmbH (Munich, Germany). All patients gave informed consent to take part in the study, which was approved by the Board Ethics Committee.

Specific airway resistance (SR_{Aw}) was measured by a computer-assisted body plethysmograph (Masterlab II, Jaeger GmbH, Würzburg, Germany) and each value reported is the mean of 3 sequential recordings. In addition to the measurements during inhalation challenge, SR_{Aw} was determined before and 60 min after administration of captopril and placebo. Bronchial challenges with histamine and bradykinin were performed according to the method of Chai et al (10). A DeVilbiss nebulizer (No. 646) controlled by a dosimeter was used. Increasing concentrations (0.05, 0.1, 0.2, 0.4, 0.75, 1.5, 3, 6, 12 and 24 mg/ml) of histamine diphosphate and bradykinin acetate (Sigma Chemie, Deisenhofen,

Germany) were freshly prepared with 0.9% saline as diluent. The subjects were instructed to take five breaths starting from functional residual capacity and to inhale to total lung capacity at a constant inspiratory flow of about 2 L/sec. After measuring baseline values and the airway response to the diluent, lung function was determined 3 minutes after the inhalation of each concentration of histamine or bradykinin. The challenges were stopped when SR_{AW} had increased by at least 100%. Dose-response curves were constructed by plotting SR_{AW} against the logarithms of histamine and bradykinin concentrations. The provocation concentration (PC) was defined as the concentration necessary to increase SR_{AW} by 100% compared to saline ($PC_{100}SR_{AW}$). When SRAW had not increased by 100% after inhaling the maximum concentration (24 mg/ml), $PC_{100}SR_{AW}$ was arbitrarily chosen as equal to 24 mg/ml. According to the criteria of Magnussen (11) the level of bronchial hyperreactivity was divided into marked ($PC_{100}SR_{AW}$ below 0.5 mg histamine/ml), moderate ($PC_{100}SR_{AW}$ 0.5 to 2 ml histamine/ml) or mild ($PC_{100}SR_{AW}$ 2 to 8 mg histamine/ml). Consequently, a normal response was defined as a $PC_{100}SR_{AW}$ of above 8 mg histamine/ml.

STUDY 2: Twelve hypertensive patients were included in the study. They had been diagnosed as having mild to moderate essential hypertension. Six of these patients had been treated with captopril for more than one year without complaining of any side effects, especially of cough. In the other six patients, previous therapy with captopril had to be withdrawn because of cough developing within 2 weeks after starting therapy. In all patients, antihypertensive medication was withheld for at least 4 weeks prior to the study. None of them had a history of atopy or pulmonary disease and all had normal lung function. The hypertensive patients received captopril and placebo for 2 weeks each according to the short-term part of study 1. Procedure and measurements were identical to those performed in the athmatic patients.

Results in the text are expressed as the mean $\pm$ SD. Statistical analysis was performed by Wilcoxon matched pairs signed rank test, the U-test of Wilcoxon, Mann and Whitney and by Spearman's rank correlation. Statistical significance was considered for $p < 0.05$.

RESULTS

<u>STUDY 1</u>: In the asthmatic subjects baseline lung function was not changed significantly by captopril. Before single acute administration of captopril SR_{AW} was 9.93 $\pm$ 3.76 and after 60 min 8.5 $\pm$ 3.61 cmH_2O x sec, respectively. At the end of the two weeks treatment period, SR_{AW} was 8.97 $\pm$ 3.66 cmH_2O x sec with administration of captopril and 8.56 $\pm$ 2.03 cmH_2O x sec with placebo. All asthmatic subjects showed bronchoconstriction with low doses of histamine (Fig. 1,2), exhibiting marked bronchial hyperreactivity. Reactivity to histamine was not changed by the ACE−inhibitor captopril when given in a single dose or short−term for two weeks. The individual responses to inhaled histamine appeared to be very consistent throughout the study. All patients showed bronchial reactivity to bradykinin (Fig. 1,2). During placebo but not during captopril, $PC_{100}SR_{AW}$ was significantly higher (p=0.031) for bradykinin as compared to histamine. There was a tendency to an increase in reactivity to bradykinin when captopril was given acutely or short−term. However, because of the high variation in the response to bradykinin the difference to placebo did not reach statistical significance. During captopril treatment none of the asthmatic subjects complained of side effects like dry cough, worsening of asthma or symptoms suggesting hypotension.

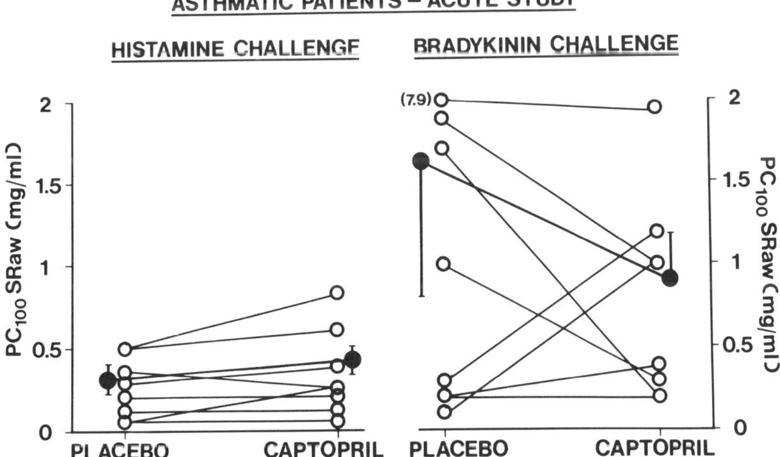

Fig. 1: Provocation concentrations of histamine and bradykinin which caused a 100% rise in specific airways rsistance ($PC_{100}SRaw$) in asthmatic subjects after acute administration of placebo and 25 mg captopril.

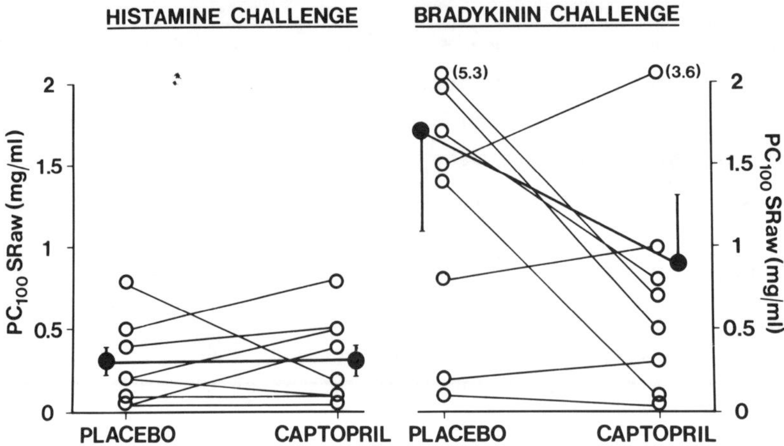

Fig. 2: Provocation concentrations of histamine and bradykinin which caused a 100% rise in specific airways resistance (PC$_{100}$SRaw) in asthmatic subjects after administration of placebo and captopril for 2 weeks.

<u>STUDY 2</u>: The two groups of hypertensive patients were similar with regard to age, sex distribution, smoking habits, lung function and blood pressure. In all patients who had exhibited cough during previous therapy with captopril, dry cough reappeared within the study period when captopril was given and disappeared promptly after stopping captopril. In both groups of hypertensive patients, lung function did not change significantly. At the end of the treatment periods, SR$_{AW}$ was 7.6 $\pm$ 3.29 and 8.38 $\pm$ 3.85 cmH$_2$O x sec in the patients with cough and 7.21 $\pm$ 2.08 and 7.52 $\pm$ 2.01 cmH$_2$O x sec in the patients without cough during placebo and captopril, respectively. In both hypertensive groups, treatment with captopril did not influence the degree of bronchial reactivity to histamine and bradykinin significantly (Fig. 3,4). During placebo three of the patients with cough showed moderate and two mild hyperreactivity to histamine, in one patient a normal response was observed. During captopril treatment, all of them exhibited bronchial hyperreactivity to histamine. In three patients hyperreactivity was moderate and in three of them mild. During placebo, three of the patients without cough showed a normal response to histamine, one exhibitad a moderate and two a mild degree of hyperreactivitiy. Except for one patient, who developed mild hyperreactivity, this did not change during captopril. The mean value of

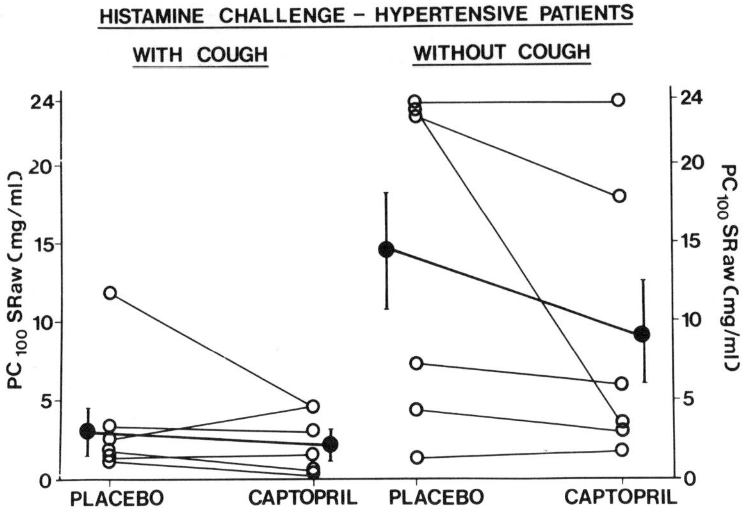

Fig. 3: Provocation concentration of histamine which caused a 100% rise in specific airways resistance (PC$_{100}$SRaw) after administration of placebo and captopril for 2 weeks in hypertensive patients with and without ACE inhibitor-induced cough.

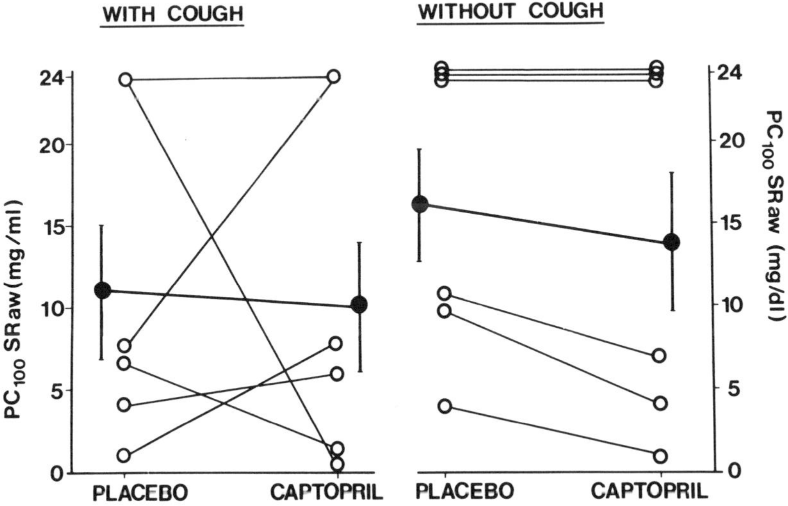

Fig. 4: Provocation concentration of bradykinin which caused a 100% rise in specific airways resistance (PC$_{100}$SRaw) after administration of placebo and captopril for 2 weeks in hypertensive patients with and without ACE inhibitor-induced cough.

PC$_{100}$SR$_{AW}$ was lower in the patients with cough, but because of the high interindividual variability in the reaction to histamine this difference did not reach statistical significance (p=0.091). No marked differences in the reactivity to bradykinin could be observed between patients with and without cough. In both groups of patients, PC$_{100}$SR$_{AW}$ of histamine was not significantly different from PC$_{100}$SR$_{AW}$ of bradykinin.

Inhaled bradykinin caused some cough and the feeling of a sore throat in all subjects. When the patients of study 1 and study 2 were compared, PC$_{100}$SR$_{AW}$ for histamine and bradykinin was significantly (p < 0.001) higher in the hypertensive patients. There was a significant positive correlation between reactivity to histamine and bradykinin when the patients of both studies were taken together (during placebo: r=0.82; during captopril: r=0.64).

DISCUSSION

In recent years, cough has been recognized as one of the most common side effects of ACE-inhibitors. It appears to be caused by all ACE-inhibitors (12). The reported frequency of cough thought to be related to ACE-inhibition usually ranges somewhere between 1 and 3% (13,14,15,16), although smaller studies of patients attending hypertension clinics suggest a higher incidence (17,18). Why ACE-inhibitors cause cough is not completely understood.

Patients who cough may have bronchial hyperreactivity to histamine or metacholine (4,6) which may increase even further during ACE-inhibition (4). Hinojosa et al (5) recently demonstrated in two hypertensive sisters that de novo bronchial hyperreactivity with accompanying cough may be induced by ACE-inhibition. In contrast, in the study reported by Boulet et al (19) the development of cough was not associated with bronchial hyperresponsiveness. In our study, 5 of 6 hypertensive patients with cough during ACE-inhibition exhibited mild to moderate hyperreactivity to histamine during placebo and all of them during captopril treatment. However, in contrast to the results of Bucknall et al (4) the degree of hyperreactivity was not significantly increased during ACE-inhibition. In the group of hypertensive patients without cough, fewer showed hyperreactivity to histamine and the concentration of histamine causing a 100% increase in specific airways resistance was higher. Although these differences were not statistically significant these results may raise the possibility that bronchial hyperreactivity may be of importance in some of the patients developing cough during ACE-inhibition. However, in our asthmatic patients, who all had marked bronchial hyperreactivity, neither cough nor airflow obstruction developed during treatment with captopril. Therefore, the role of bronchial hyperreactivity in the development of cough during ACE-inhibition remains doubtful. The present results support the findings of Dixon et al (20) and Sala et al (21), who also did not observe changes in bronchial reactivity during acute or short-term treatment with ACE-inhibitors in asthmatic patients.

The duration of our study and of the study of Sala et al (21) could have been too short for the appearance of cough. Cough mostly develops early in the treatment but may not become apparent for several weeks or even months after starting ACE-inhibition (22). Because of the small numbers of reported patients it is not known whether asthmatic patients are more prone to develop cough than non-asthmatic subjects. However, there are only few reports of a deterioration of lung function or symptoms due to ACE-inhibitors in pre-existing asthma (13,17,23). In our study, baseline lung function remained unaltered in the asthmatic patients when treated with captopril. These results are in agreement with the few studies looking specifically at the effect of ACE-inhibitors in patients with bronchial asthma or chronic bronchitis (21,24,25,26). These studies demonstrated either no change or even a slight improvement in lung function. These observations suggest that treatment with an ACE-inhibitor is safe in most patients with bronchial asthma.

Kinins could participate in the development of cough following ACE-inhibition. They may accumulate during treatment with an ACE-inhibitor, since ACE is identical to Kininase II which breaks down bradykinin and other peptides participating in inflammation (7,27). The wheal and flare response to intradermal bradykinin is enhanced by enalapril (28). Inhalation of bradykinin causes cough and throat irritation in normal subjects and asthmatic patients (9) and bronchoconstriction in patients with bronchial hyperreactivity (9,20). Ujiie et al (29) demonstrated an increased cough reflex to inhaled bradykinin in patients with ACE-inhibitor associated cough and suggested that impaired metabolism of bradykinin by ACE inhibitors relates to the manifestation of cough in hypertensive patients receiving ACE inhibitors. In the present study, bradykinin challenge was followed by bronchoconstriction in all asthmatic patients. However, reactivity to bradykinin was not significantly increased after acute or short-term administration of captopril. Also, reactivity to bradykinin in the hypertensive patients appeared to be unrelated to cough. It has been demonstrated previously that captopril does not enhance the effect of bradykinin on human bronchial smooth muscle in vitro (9) and that single administration of ramipril does not increase reactivity to inhaled bradykinin in subjects with mild bronchial asthma (20). In our study, even two weeks of treatment with an ACE-inhibitor, captopril, did not influence the effect of bradykinin in patients with bronchial asthma or hypertension. From these results a major importance of kinins in the development of cough related to ACE-inhibition appears to be unlikely.

CONCLUSION

In the present study reactivity to histamine and bradykinin was not significantly altered by captopril in asthmatic patients and in hypertensive subjects with or without ACE-inhibitor related cough. Reactivity to histamine tended to be more pronounced in hypertensive patients with cough than in those without cough. Although it can not be excluded that bronchial hyperreactivity may be important for the development of ACE inhibitor-related cough in some patients its role remains doubtful especially because in the asthmatic patients, who all had clear bronchial hyperreactivity, cough or bronchoconstriction could not be observed. On the other hand, these results demonstrate that treatment with ACE-inhibitors is safe in the majority of asthmatic patients. Our findings do not support a major role for kinins in ACE inhibitor-induced cough.

REFERENCES

1. Fuller RW, Choudry NB. Increased cough reflex associated with angiotensin converting enzyme inhibitor cough. Br Med J 1987; 295:1025–1026

2. McEwan JR, Choudry N, Street R, Fuller RW. Change in cough reflex after treatment with enalapril and ramipril. Br Med J 1989; 299:13–16

3. Morice AH, Lowry R, Brown MJ, Higenbottam T. Angiotensin–converting enzyme and the cough reflex. Lancet 1987; ii:1116–1118

4. Bucknall CE, Neilly JB, Carter R, Stevenson RD, Semple PF. Bronchial hyperreactivity in patients who cough after receiving angiotensin converting enzyme inhibitors. Br Med J 1988; 296:86–88

5. Hinojosa M, Quirce S, Puyana J, Codina J, Rull SG. Bronchial hyperreactivity and cough induced by angiotensin–converting enzyme–inhibitor therapy. J Allergy Clin Immunol 1990; 85:818–819

6. Kaufman J, Casanova JE, Riendl P, Schlueter DP. Bronchial hyperreactivity and cough due to angiotensin–converting enzyme inhibitors. Chest 1989; 95:544–548

7. Erdös EG. The angiotensin I converting enzyme. Fed Proc 1977; 36:1760–1765

8. Proud D, Kaplan AP. Kinin formation: Mechanisms and role in inflammatory disorders. Ann Rev Immunol 1988; 6:49–83

9. Fuller RW, Dixon CMS, Cuss FMC, Barnes PJ. Bradykinin–induced bronchoconstriction in humans. Mode of action. Am Rev Respir Dis 1987; 135:176–180

10. Chai H, Farr RS, Froehlich LA, Mathison DA, McLean JA, Rosenthal RR, Sheffer AL, Spector SL, Townley RG. Standardization of bronchial inhalation challenge procedures. J Allergy Clin Immunol 1975; 56:323–327

11. Magnussen H. Überempfindlichkeit der Atemwege. Messung, Vorkommen, klinische Bedeutung. Dtsch med Wschr 1990; 115:1604–1610

12. Berkin KE, Ball SG. Cough and angiotensin converting enzyme inhibition. Br Med J 1988; 296:1279–1280

13. Coulter DM, Edwards IR. Cough associated with captopril and enalapril. Br Med J 1987; 294:1521–1523

14. Rush JE, Merrill DD. The safety and tolerability of lisinopril in clinical trials. J Cardiovasc Pharmacol 1987; 9 (suppl 3):99–107

15. Santoni JP, Richard C, Pouyollon F, Castaings C, Brown C. Tolerance et securite d'emploi du perindopril. Arch Malad Coeur 1989; 82:87–92

16. Todd PA, Benfield P. Ramipril. A review of its pharmacological properties and therapeutic efficacy in cardiovascular disorders. Drugs 1990; 39:110–135

17. Hood S, Nicholls MG, Gilchrist NL. Cough with angiotensin converting-enzyme inhibitors. N Z Med J 1987; 100:161–163

18. Town GI, Hallwright GP, Maling TJB, O'Donnell TV. Angiotensin converting enzyme inhibitors and cough. N Z Med J 1987; 100:161–163

19. Boulet LP, Milot J, Lampron N, Lacourciere Y. Pulmonary function and airway responsiveness during long-term therapy with captopril. J Am Med Ass 1989; 261:413–416

20. Dixon CMS, Fuller RW, Barnes PJ. The effect of an angiotensin converting enzyme inhibitor, ramipril, on bronchial responses to inhaled histamine and bradykinin in asthmatic subjects. Br J Clin Pharmacol 1987; 23:91–93

21. Sala H, Abad J, Juanmiquel Ll, Plans C, Ruiz J, Roig J, Morera J. Captopril and bronchial reactivity. Postgrad Med J 1986; 62 (suppl 1):76–77

22. Inman WHW, Rawson NSB, Wilton LV, Pearce GL, Speirs CJ. Postmarketing surveillance of enalapril. I: Results of prescription–event monitoring. Br Med J 1988; 297:826–829

23. Semple PF, Herd GW. Cough and wheeze caused by inhibitors of angiotensin–converting enzyme N Engl J Med 1986; 314:61

24. Bertoli L, Fusco M, Lo Cicero S, Micallef E, Busnardo I. Influence of ACE inhibition on pulmonary haemodynamics and function in patients in whom beta–blockers are contra–indicated. Postgrad Med J 1986; 62 (suppl):47–51

25. Peacock AJ, Matthews AW. The effect of captopril on pulmonary haemodynamics and lung function in patients with chronic airflow obstruction. Thorax 1986; 41:225

26. Riska H, Stenius–Aarniala B, Sovijarvi ARA. Comparison of the efficacy of an ACE–inhibitor and a calcium channel blocker in hypertensive patients. A preliminary report. Postgrad Med J 1986; 62 (suppl 1):52–53

27. Casceiri MA, Bull HG, Mumford RA Patchett AA, Thornberry NP, Liang T. Carboxyl–terminal tripeptidyl hydrolysis of substance P by purified rabbit lung angiotensin converting enzyme and the potentiation of substance P in vivo by captopril and MK422. Mol Pharmacol 1984; 25:287–293

28. Fuller RW, Warren JB, McCusker M, Dollery CT. Effect of enalapril on the skin response to bradykinin in man. Br J Clin Pharmacol 1987; 23:88–90

29. Ujiie Y, Sekizawa K, Katsumata U, Sasaki H, Takishima T. Bradykinin-induced cough reflex markedly increases in patients with cough associated with captopril or enalapril. Am Rev Respir Dis 1990; 141:A923

THE KALLIKREIN-KININ SYSTEM OF SWEAT IN NORMAL AND CYSTIC FIBROSIS SUBJECTS

A. Zucollo, J. Martiarena, C. Luna, O. Pivetta, A. Villagra and O. Catanzaro

Cátedra de Fisiología y Programa de Péptidos, Facultad de Farmacia y Bioquímica, UBA y CONICET, Dept. de Genética Humana, Buenos Aires, Argentina

SUMMARY: The mechanisms of kallikrein secretion was studied in 25 normal subjects (11 male, 14 female) and 5 subjects with cystic fibrosis (identified by the usual criteria). The results obtained show a decreasing of prekallikrein and total kallikrein in cystic fibrosis stimulated by IPR or pilocarpine compare to controls. On the other hand sweat from males shows more prekallikrein and total kallikrein than females. According with this results we can concluded that the basic defect in sweat cystic fibrosis gland could be the possible involvment of the kallikrein-kinin system on the high osmolality of the sweat secretion.

INTRODUCTION

The sweat gland had been largely used for the identification of the basic defect in Cystic Fibrosis. Features of the disease included abnormal electrolyte composition in sweat, saliva and pancreatic fluid; elevated sweat Na^+ Cl^- is the principal factor of the disease. Kinin-forming enzyme in normal sweat was reported in 1958 (1). According to Scharter (2) kininogenase enzyme in sweat was detected as a proteases like Kallikrein. The enzyme was partially purified by Hibino et al. 3). The high concentration of Na^+ and Cl^- in homozygotes sweat (4), results from impermeability to Cl^- in the reabsorptive duct (5,6). Cultures derived from whole Cystic Fibrosis glands demonstrated electrogenic sodium absorption, but they showed a reduced sensitivity to amiloride (7), an effect considered to result from the reduced apical Cl^- conductance (8,9). In the present paper we shows the mechanisms of kallikrein secretion of normal and Cystic Fibrosis sweat glands.

MATERIALS AND METHODS

Sweat was collected from 11 male, 14 female normal subjects and 5 subjects with Cystic Fibrosis (homozygotes identified by the usual criteria). Subjects were registered at the Dep. of Human Genetic, Hospital Rivadavia. Sweating was induced on the volar surface of the forearm by intradermal inyection of 8×10^{-5} M Isoproterenol (IPR), or by iontophoretic stimulation of Pilocarpine 2% with a current of 200 mA passed for 5 min. Samples drops were collected in a special preweight pads. Secreted glands number of secreting sweat glands were counted according to Behm et al., (10). Pads were eluted, centrifuged, dialized, and liofilized. All the samples were recovery in 200 ul distilled water. Kininogenase activity was measured using LMWK from dog plasma according to Nasjletti et al. (11). Kallikrein was determined with amidolytic assay H-D-Val-Leu-Arg-pNA (S-2266, Kabi) substrate.

RESULTS

Figure 1 shows the prekallikrein and active kallikrein stimulated by IPR and Pilocarpine. As a matter of fact the concentration of active kallikrein, measured as μmpNA/mg P/h was significantly increased with IPR stimulation than pilocarpine (+100%). However the same stimulation did not alter the prekallikrein activity. The relative concentration of kallikrein (active, prekallikrein in total kallikrein) in controls and Cystic Fibrosis measured in μmpNA/gland or mg protein shows a significants difference for males compared to females and males Cystic Fibrosis stimulated with pilocarpine (Fig. 2).

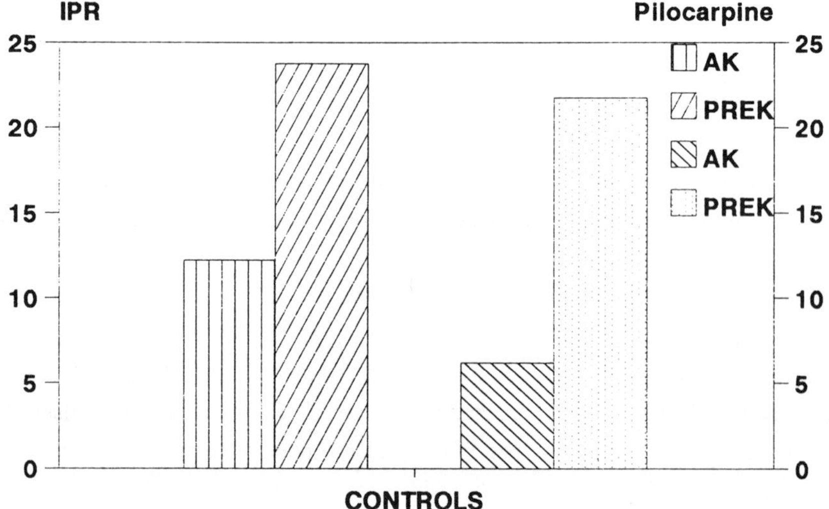

Fig. 1. Means sweat kallikrein activity response to β-adrenergic and cholinergic drugs in controls. PREK and AK = μmpNA/mgP/h

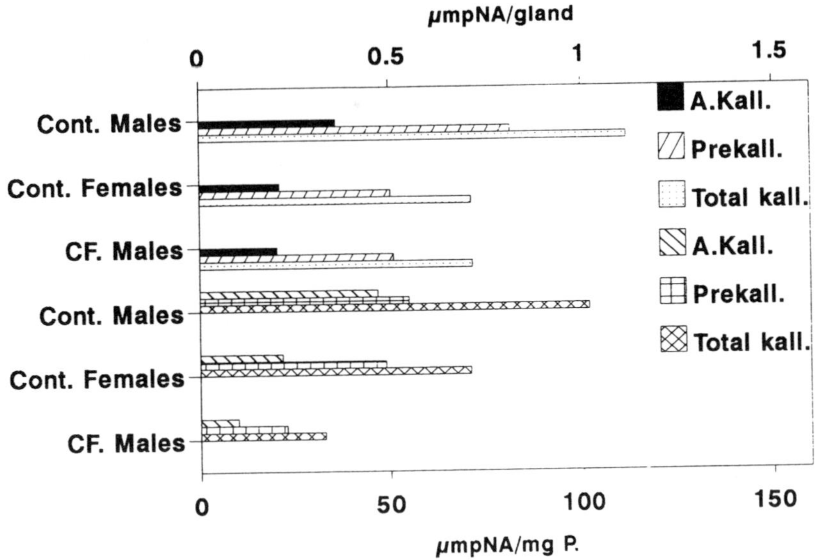

Fig. 2. Relative concentration and active sweat secretion in controls and CF patients

When cholinergic and adrenergic kallikrein stimulation was compared between controls and CF subjects a significant increasing was observed in controls pK and active Kallikrein compared to CF, but β-adrenergic stimulation do not sweat at all in CF (p<0.001) (Fig. 3). Kininogenase activity also showed differences between controls and CF subjects in any form of expression. Table 1 summary the sweat secretion in a defined area; the volume of sweat in mg showed significantly differences between adrenergic and cholinergic stimulation in controls and CF.

Table 1. Kininogenase activity in sweat of normal and CF childrens

	NgBK/min/mgP	PgBK/gl/min
Normals (14)	9.62 ± 2.92	77.6 ± 23.62
CF (7)	2.79 ± 0.47	16.7 ± 2.85
	$p \leq 0.05$	$p \leq 0.05$

On the other hand protein concentration did not differ between cholinergic and β-adrenergic stimulation in control and CF. The response of kallikrein secretion to β-adrenergic and cholinergic in controls show an increasing of kallikrein activity of 156 % with β-adrenergic vs pilocarpine. CF sweat kallikrein did show only secretion with pilocarpine (53 % respect to control Pilocarpine).

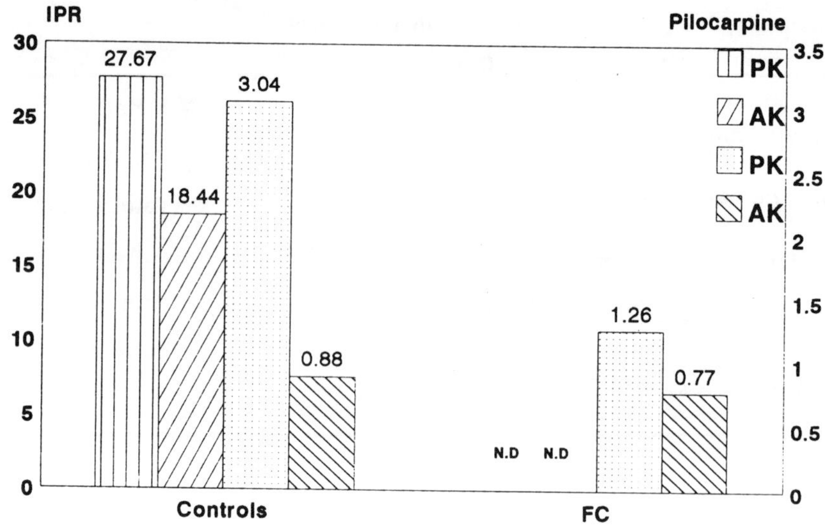

Fig. 3. Means sweat kallikrein activity response to β-adrenergic and cholinergic drugs. PK and AK units are μmolpNA/100mg sweat/h.

Table 2. Means of active sweat glands in 12 mm^2 area

	Drugs	Sweat mg	Protein μg	Act. Kall. μmpNA
CONTROLS	IP	17.71 ± 6.21	182 ± 9.31	2.92 ± 0.32
	PIL	190.04 ± 24.10	170 ± 24.10	1.14 ± 0.41
CF	IPR	--	--	--
	PIL	89.23 ± 9.31	133 ± 14.20	0.61 ± 0.11

DISCUSSION

Our results confirm that the kallikrein kinin system is involved in the physiology of the sweat glands. They have then played a key role in research aimed at identifying the basic defect in CF.

According to Quinton et al. (12) the secretory process of sweat gland was thought to be normal in CF (13). Primary sweat of CF homozygotes has a normal ion content and secretory rate than controls (10). However, while sweat rates in response to cholinergic stimulation are normal, CF homozygotes do not sweat at all in response to β-adrenergic agonist (14).

A series of observations have characterize to a greater extent the abnormalities in epithelial ion transporting events in patients with Cystic Fibrosis. Quinton (15,16) has suggested that there is reduced chloride absorption in the sweat glands ducts of CF patients, and this relative chloride impermeability may form the basis for the observed electrolyte abnormality in sweat ducts, as well as in airway. In CF epithelial tissues (17) Cl^- secretion is not induced by B-agonist, prostaglandins or Calcium Ionophore A-23187.

Recently Poblete et al. (18) observed the cellular localization of tissue kallikrein and its substrate Kininogen in sweat glands secretory units and along the luminal microvilli of the ducts cells.

Moreover, kallidin increases Na^+ absorption across cultured sweat epithelium, and iontophoretic application of bradykinin produces an abrupt fall of sweat osmolality (19).

This statements are relevant, because in the past five years, tissues kallikrein and their products kinins are reasonable In Vivo candidates for participation in the regulation of transephitelial Na^+ and chloride transport. The results observed by us could explain the possible mechanisms of chloride impermeability in the sweat gland duct. If kinins are formed in the duct lumen, permeation to the serosal side of the luminal cells could modulate the excretion of water by the gland.

On the other hand, if Kinins produce a fall of sweat osmolality, decreased formation of the enzyme kallikrein could produce a modification of the ionic absorption. Sweat gland receive both cholinergic and adrenergic innervation (20,21), while vasoactive intestinal peptide (VIP) immunoreactive nerve endings appear to innervate both the coil and duct (22,23).

CONCLUSION

A reasonable interpretation of the present data is that:

1) The mechanisms of kallikrein kinin secretion are under the control of β-adrenergic more than cholinergic.

2) The high concentration of Na^- and Cl^- in CF sweat (the most reliable indicator of the disease) could be due to a defective kallikrein kinin system.

ACKNOWLEDGEMENTS

This work was support by a Grant of Consejo Nacional de Investigaciones Científicas y Técnicas, Argentina.

REFERENCES

1. Fox R.H., Hilton S.M. J.Physiol. 1958; 142:219-32.

2. Schachter M. Pharmacol.Rev. 1980; 31:1-17.

3. Hibino T, Takemura T, Sato K., J.Invest.Dermatol. 1988; 90:569.

4. di Sant' Agnese P.A., Darling R.C., Perera G.A., Shea E. Am.J.Med. 1953; 15:777-784.

5. Quinton P.M. Nature (Lond) 1983; 301:421-422.

6. Bijman J., Quinton P.M. Am.J.Physiol.1984; 247:C3-C9.

7. Pedersen P.S. International Research Communications System Medical Science. 1984; 12:752-753.

8. Collie G., Buchwald M., Harper P., Riordan J.R. In Vivo Cellular and Developmental Biology 1985; 21:597-602.

9. Lee C.M., Carpenter F., Coaker T., Kealey T. Journal of Cell Science. 1986; 83:103-118.

10. Behm J.K., Hagiwara G., Lewiston N.J., Quinton P.M., Wine J.J. Pediatric Research 1987; 22:271-276.

11. Nasjletti A., Colina-Chourio J., Mc.Giff J.C. Circ.Res. 1975; 37:59-65.

12. Quinton P.M. In: Quinton P.M., Martinez J.R., Hopfer V. (eds). Fluid and Electrolyte Abnormalities In Exocrine Glands In Cystic Fibrosis. San Francisco Press, San Francisco. 1982; pp 53-76.

13. Quinton, P.M. Am.J.Physiol. 1986; 251:C649-C652.

14. Shultz I.J. J.Clin.Invest. 1969; 48:1470-1477.

15. Quinton P.M. In: Taussig L.M., ed. Cystic Fibrosis. New York, NY. Thieme-Stratton, 1984; 303-47.

16. Quinton P.M., Bijman J., N.Engl.J.Med. 1983; 308:185-9.

17. Cuthbert A.W., Egleme C., Greenwood H., Hickman M.E., Kirkland S.C., MacVinish L.J., Br.J.Pharmacol. 1987; 91:503-515.

18. Poblete M.T., Reynolds N.J., Figueroa C.D., Burton J.L., Müller-Esterl W., Bhoola K.D. British Journal of Dermatology, 1991; 124:236-241.

19. Gordon C., Schwartz V. J.Physiol. 1971; 213:68-69.

20. Sato K. Rev.Physiol.Biochem.Pharmacol. 977; 79:51-131.

21. Reddy M.M., Quinton P.M. Biophys.J. 1986; 49:157a.

22. Uno H. J.Invest.Dermatol. 1977; 69:112-130.

23. Heinz-Erian P., Dey R.D., Flux M., Said S.I. Science 1985; 229:1407-1408.